Cranfi

at Cranfield
sity
L

A Specialist Periodical Report

Dielectric and Related Molecular Processes

Volume 1

A Review of Selected Developments
in the period 1966–1971

Senior Reporter
Mansel Davies, *Edward Davies Chemical Laboratories, The University College of Wales, Aberystwyth*

Reporters
J. B. Hasted, *Birkbeck College, University of London*
S. Kielich, *University of Poznan, Poland*
B. K. P. Scaife, *Trinity College, Dublin*
G. Schwarz, *University of Basel, Switzerland*
A. Suggett, *Unilever Research Ltd., Colworth, Bedford*
H. Sutter, *University of Cambridge*
G. Wyllie, *University of Glasgow*

The Chemical Society
Burlington House, London, W1V 0BN

ISBN: 0 85186 505 4

Library of Congress Catalog Card No. 72-83457

Organic formulae composed by Wright's Symbolset method

PRINTED IN GREAT BRITAIN BY JOHN WRIGHT AND SONS LTD., AT THE STONEBRIDGE PRESS, BRISTOL

Preface

It could well be that prefaces are the least read of the greatly unread productions of modern publishing but the following paragraphs are written to try to help the reader decide what in this particular volume has any claim to his attention. This volume is the first of a new biennial series on dielectric and related studies, so some comments are offered on the area embraced by the title and some of the prospects in this area of molecular science are described. Then the specific characters of the contributions in the volume are briefly indicated, and this could help the reader find those of greatest immediate interest to him. In any case the preface should be helpful to a reviewer.

The term 'dielectric properties' in its broad usage refers to those responses other than the simple ohmic currents which develop in a medium on the application of an electric potential. Thus an ideal dielectric medium shows no direct current when a static electric potential is applied, but as only gases approach this condition it is usual to speak of the dielectric properties of what may also be a semi-conductor.

No material medium is known which can be adequately described as having a 'dielectric constant', *i.e.* polarization which is strictly proportional to the electric field, and independent of its frequency, the temperature, and pressure. Both for general physical and molecular interests the study of dielectric properties becomes the investigation of the variation of the electric permittivity with the several variables to which it responds: thus the study becomes field-dependent dielectric spectroscopy, with the thermodynamic variables of state (concentration, temperature, pressure, *etc.*) as other necessary parameters. Such attempts to define a specific area in the scientific field are certain to fail in practical use by being too narrow or too broad. Dielectric spectroscopy necessarily has a very broad frequency range in the electromagnetic spectrum and an arbitrary and appropriately variable limit can be set by confining it to the range of non-quantal interactions of electric fields with material systems. This serves to separate it from most aspects of optical spectroscopy, although a reference to the Maxwell relation, (electric permittivity) = (refractive index)^2, suffices to show this is an artificial border. Also, for present purposes, the magnetic field-dependent aspects should be excluded as they embrace the special topic of n.m.r. spectroscopy. With these exclusions, dielectric spectroscopy may be taken to extend up to frequencies of *ca.* 10^{13} Hz, *i.e. ca.* 300 cm^{-1}.

The range of subject matter and the variety of techniques in this broad area have meant that in the past dielectric studies have tended to develop in almost independent patches and even now—except in some theoretical aspects—a deliberate attempt has to be made to bring the diverse but cognate features together. The emergence of 'materials science' as a recognized discipline embracing aspects of physics, chemistry, and engineering science has hastened the merging of dielectric studies, and von Hippel must be credited with a major pioneering role in this development. In the U.K. the flourishing of a Dielectrics Discussion Group with members from three major scientific disciplines is a significant feature, the more so because as a dielectric study group they do not find it appropriate to be formally associated with any of the established societies to which they otherwise belong. However, this series of volumes, which it is intended will appear biennially, will be concerned primarily with molecular aspects of dielectric studies. This net is already a large one and the Senior Reporter will be happy to consider suggestions from colleagues as to the selection of topics for later volumes.

It cannot be said that this is the first attempt to provide a regular series of reports on progress in dielectrics but, insofar as they continue, some of the earlier ventures are now very irregular in their appearance. At the same time dielectric studies are acquiring increased momentum and range. As symptoms of these features one can mention the arrival of dielectric spectrometers which scan very many decades of frequency to provide an almost immediate spectrum of both the permittivity and absorption changes as functions of frequency. The Reddish–Hyde dielectric spectrometer is certain to initiate a revolution in the measurement of time-dependent properties, as the variation in any bulk property, *e.g.* in chemical composition, density, viscosity, *etc.*, which can be converted into a capacitance change can now be automatically scanned and recorded over times from 10^{-6} s to 10^{3} s. In view of the dominant role which reaction kinetics has had for some decades in British physical chemistry, it is safe to predict that the 1980's will see very many of our laboratories equipped with dielectric spectrometers and the area will receive that accolade of well-established topics, a special committee and allocation of funds by the S.R.C. On the interpretative side, recent years have also seen the replacement of the simple dipole relaxation model of Debye (1913), which provided the pattern of discussion for more than half a century, by the more general representation of the Kubo correlation function. Formally this provides a means of unifying the total time pattern of the changes in a dielectric, be they of single or multiple origins, and independently of the sequential or simultaneous occurrence of the component (model) processes. Finally, and not least significantly, the more than thirty years of pioneering work of Piekara in the study of high electric field effects is now proliferating in a variety of directions, and another prediction is easy: before the end of the century Piekara's work will undergo the same expansion as we have seen in Aston's lone exploitation of the mass-spectrometer.

Coming to the present volume, one aim has been to provide a basis on which the student and researcher in molecular science can build a sound appreciation of the present and future developments. Accordingly, the chapters do not presume too much previous knowledge of their subjects. Professor Scaife is concerned, *inter alia,* to make clear what is the character of those aspects of the macroscopic dielectric behaviour which can be precisely delineated in the theoretical representations which rest on Maxwell's analysis, and he relates these to some of the general microscopic features. The time-dependent aspects of these features are the particular concern of Chapter 2 in which Dr. Wyllie gives an exposition of the essentials of molecular correlation functions. As dielectric relaxation methods provided one of the clearest models of relaxation studies, there is reason to suggest that dipole reorientation provides one of the clearest examples of the correlational treatment. If only for this reason, Dr. Wyllie's chapter could well provide valuable insights for many whose primary interest is not in dielectrics.

The gas phase is invariably regarded as the simplest for a quantitative representation of molecular behaviour, and undoubtedly the dilute gas is one where the Debye treatment in terms of the dipole moment receives its best justification. Despite the century-old insights of van der Waals on the gas deviations which so strikingly help in appreciating the appearance and behaviour of the liquid state, the parallel links in our understanding of the dielectric behaviour of liquids have still to be forged. In Chapter 3 Dr. Sutter summarizes the present analysis of the permittivities of gases at higher pressures. Their study serves to reveal some of the essential detail in two- and three-molecule encounters. Furthermore, the difficulties, both experimental and theoretical, in the dielectric appraisal are explicitly illustrated, and much if not all of the currently valid molecular data from this source are summarized in the account.

It is a trite observation that our understanding of water is in many respects even less adequate than that of other liquids. Recent monographs have shown that the past decade has seen significant advances, but a satisfactory treatment of the dielectric spectrum is still elusive. Professor Hasted, who has long-developed interests in this problem, contributes his own assessment of the present situation in Chapter 4. He includes an account of some computer-model calculations as well as a summary of recent far-i.r. observations which are of immediate relevance to the quantitative appreciation of the dielectric spectrum of water.

The newly established methods in time-domain dielectric studies are reviewed by Dr. Suggett in Chapter 5. In general principle these are extensions of the step- or ramp-voltage methods which in various forms have been in occasional use over some thirty years, but the differences in the speed of operation are such as to make them qualitatively different. For the high-frequency region the first system was described by Fellner-Feldegg in 1969, whilst a spectrometer operating from 10^{-4} Hz to 10^{6} Hz was only described in 1970. Accordingly, we are still in the very early stages of these newer methods and many extensions and improvements will certainly follow.

Of special systems, those in the biomolecular field offer a special challenge and also special opportunities. The challenge is many headed: the difficulties of handling small amounts of complex material in conducting (usually aqueous) media are considerable. The largest molecules can acquire some properties of macroscopic particles so that interfacial effects can add to the complications. Furthermore, various intramolecular configurations may not only be present but may be in dynamic equilibrium. Changes in the latter may be equivalent to chemical processes and can often be of the greatest importance in the biological function of these molecules. Professor Schwarz in Chapter 6 has given a systematic presentation of many of these aspects. He deals firstly with conditions under which a meaningful electric dipole moment can be evaluated for a solute species. Accordingly this chapter might well be the first choice for many chemists and biophysicists. In it Professor Schwarz also explains the possible measurement of chemical rate processes in the form of a directly observable dielectric absorption. For the biomolecular scientist the possibilities here must be quite exciting and Professor Schwarz refers to some instances where the possibilities have been realized.

Finally there is Chapter 7 by Professor Kielich which, in fact, might form a small monograph of its own. The introductory survey of this chapter serves to orient the student in the general field of dielectric studies and it provides a selected bibliography for the whole subject which many readers, it is hoped, will find of lasting value. The special aim of the chapter is to outline systematically the present possibilities in high electric field studies both in terms of the macroscopic factors and at the molecular level. This whole new area is ripe for systematic exploration and Professor Kielich has made a major contribution to the prospects in future work by presenting the general theory and much specific detail for several aspects which can already be envisaged. No such survey of high electric field effects has previously been available and many colleagues will be grateful to Professor Kielich for the major effort he has given to this thorough presentation. The chapter includes a number of features which have not been generally expounded: *e.g.* the significance of fluctuating local fields present in condensed vapours and liquids. Few will read the summary contents of Chapter 7 without finding some worth-while items with which they are not familiar.

Such, briefly, is the content of the chapters which follow. Apart from thoroughly recommending them to others, it only remains for this first reader to offer his sincere thanks—not only for the very ready collaboration of the authors, but also for their valuable contributions to the not excessive corpus of significant literature on dielectric studies. The interest and help of the Chemical Society editors are also much appreciated.

MANSEL DAVIES

Aberystwyth, May 1972

Contents

Chapter 7 General Molecular Theory and Electric Field Effects in Isotropic Dielectrics
By S. Kielich

1

The Theory of the Macroscopic Properties of Isotropic Dielectrics

BY B. K. P. SCAIFE

1 Introduction

A chemist is anxious to obtain as much information as possible about the molecular conformation, interaction, and dynamics of the substances which he studies. Because all materials are built up from combinations of electric charges, protons, and electrons, it is not surprising that certain molecular properties and motions give rise to macroscopic properties which determine the reaction of a particular material to the imposition of an electromagnetic field.

To relate the observed macroscopic phenomena to various molecular processes is, in general, a task of great difficulty. Such an undertaking requires a fully developed macroscopic theory to enable proper design and assessment of experimental procedures to be made. Furthermore, without an adequate theory of molecular processes and without a precise relationship between macroscopic and microscopic parameters, it will not be possible to obtain accurate information about events at a molecular level from experimental data.

The primary concern of this chapter is to provide an introduction to the theory of the macroscopic properties of dielectric bodies. Attention is restricted to isotropic materials and it is assumed that the external electric field is always sufficiently weak to allow any non-linear effects which might arise to be ignored.

The subject matter is such that it lends itself to mathematical description; nevertheless, the emphasis here will be on the physical meaning of the mathematical equations and we shall not be particularly concerned with matters of mathematical technique.

This introductory survey commences with a derivation of the macroscopic equation

$$\boldsymbol{D} = \varepsilon_0 \boldsymbol{E} + \boldsymbol{P}$$

based on Maxwell's equations for charges in free space. The concept of the relative permittivity for static fields, ε_s, is introduced in Section 3. In the following section it is shown how this concept may be generalized for the case of harmonically varying applied fields.

A frequency ($= \omega/2\pi$) dependent complex susceptibility,

$$\chi(\omega) = \chi'(\omega) - i\chi''(\omega) = [\varepsilon(\omega) - 1]\,\varepsilon_0,$$

is defined and the physical significance of $\chi'(\omega)$ and $\chi''(\omega)$ is discussed. It is also shown how $\chi(\omega)$ is related to the temporal behaviour of the dielectric polarization following the sudden application, or removal, of an electric field. Various forms of the Kramers–Kronig dispersion relations are introduced for $\chi'(\omega)$ and $\chi''(\omega)$ and for a number of functions of $\chi(\omega)$. The section closes with the definition of the frequency-dependent complex refractive index $\bar{n}(\omega) = n(\omega) - i\kappa(\omega)$ and a discussion of its relation to $\varepsilon(\omega)$.

The final section discusses, in outline, various macroscopic manifestations of the underlying dynamic nature of a dielectric material. It is shown how certain macroscopically observable quantities are of direct significance on a microscopic scale.

2 Basic Considerations

The electric and magnetic fields set up in free space by a system of charges is described by Maxwell's equations, which read, in SI units, as follows:

$$\operatorname{curl} \tilde{\boldsymbol{E}}(\boldsymbol{r}, t) = -\,\partial_t \tilde{\boldsymbol{B}}(\boldsymbol{r}, t) \tag{1}$$

$$\operatorname{curl} \tilde{\boldsymbol{H}}(\boldsymbol{r}, t) = \partial_t \tilde{\boldsymbol{D}}(\boldsymbol{r}, t) + \tilde{\boldsymbol{J}}(\boldsymbol{r}, t) \tag{2}$$

$$\operatorname{div} \tilde{\boldsymbol{D}}(\boldsymbol{r}, t) = \tilde{\rho}(\boldsymbol{r}, t) \tag{3}$$

$$\operatorname{div} \tilde{\boldsymbol{B}}(\boldsymbol{r}, t) = 0 \tag{4}$$

$$\tilde{\boldsymbol{B}}(\boldsymbol{r}, t) = \mu_0 \tilde{\boldsymbol{H}}(\boldsymbol{r}, t) \tag{5}$$

$$\tilde{\boldsymbol{D}}(\boldsymbol{r}, t) = \varepsilon_0 \tilde{\boldsymbol{E}}(\boldsymbol{r}, t) \tag{6}$$

The absolute permittivity and absolute permeability of free space are denoted by ε_0 and μ_0 respectively. The sources of the electromagnetic-field vectors $\tilde{\boldsymbol{D}}$, $\tilde{\boldsymbol{E}}$, $\tilde{\boldsymbol{B}}$, and $\tilde{\boldsymbol{H}}$ are the charge density, $\tilde{\rho}(\boldsymbol{r}, t)$, and the current density, due to charges in motion, $\tilde{\boldsymbol{J}}(\boldsymbol{r}, t)$. The radius vector of a point with co-ordinates x, y, and z is denoted by $\boldsymbol{r} = \boldsymbol{i}x + \boldsymbol{j}y + \boldsymbol{k}z$, and the time is denoted by t. The use of the tilde, $\tilde{}$, on the field variables in equations (1)—(6) is to emphasize that these equations describe the electromagnetic field in *free space*. Quantities without the tilde will have meanings to be defined below.

In view of equations (5) and (6) it would appear an unwarranted extravagance to use four variables when two would suffice. As we shall see presently, the distinction between $\tilde{\boldsymbol{D}}$ and $\tilde{\boldsymbol{E}}$ in free space, which stems from the SI system of units, will prove of considerable benefit when we come to setting up the equations for the potential gradient in polarizable matter.

For a great many applications in dielectrics, Maxwell's equations may be greatly simplified by neglecting the magnetic field on the ground that the motion of the charges is not sufficiently rapid to give rise to appreciable radiation. With this approximation equations (1) and (2) are uncoupled and we may use the following equations to describe the spatial variation of $\tilde{\boldsymbol{D}}$ and $\tilde{\boldsymbol{E}}$:

$$\text{curl}\,\tilde{\boldsymbol{E}}(\boldsymbol{r}, t) = 0 \tag{7}$$

$$\text{div}\,\tilde{\boldsymbol{D}}(\boldsymbol{r}, t) = \tilde{\rho}(\boldsymbol{r}, t) \tag{8}$$

$$\tilde{\boldsymbol{D}}(\boldsymbol{r}, t) = \varepsilon_0\,\tilde{\boldsymbol{E}}(\boldsymbol{r}, t) \tag{9}$$

It is convenient and fruitful to regard $\tilde{\boldsymbol{D}}$ as the electric flux density and $\tilde{\boldsymbol{E}}$ as the force that would be exerted on a unit point charge.

The implications of equations (7) and (8) are[1] (*i*) that $\tilde{\boldsymbol{E}}(\boldsymbol{r}, t)$ may be described in terms of a scalar potential $\tilde{\phi}(\boldsymbol{r}, t)$, such that

$$\tilde{\boldsymbol{E}}(\boldsymbol{r}, t) = -\,\text{grad}\,\tilde{\phi} = -\,\nabla_{\mathbf{r}}\tilde{\phi} = -\,[\boldsymbol{i}\partial_x\tilde{\phi} + \boldsymbol{j}\partial_y\tilde{\phi} + \boldsymbol{k}\partial_z\tilde{\phi}] \tag{10}$$

and (*ii*) that the electric flux lines, whose density is $\tilde{\boldsymbol{D}}$, begin on positive charges and end on negative charges. Notice that, for a system of charges in free space, at each and every point the electric flux density, $\tilde{\boldsymbol{D}}$, is related to the electric field intensity, $\tilde{\boldsymbol{E}}$, by the simple equation (9).

The charge system that we are concerned with is one in which the charges are the sub-atomic charged particles, electrons and protons, which make up dielectric materials. It is clear that the quantities $\tilde{\boldsymbol{D}}$, $\tilde{\boldsymbol{E}}$, and $\tilde{\rho}$ must now be regarded as inaccessible to simple macroscopic measurements. Indeed we shall regard $\tilde{\boldsymbol{D}}$, $\tilde{\boldsymbol{E}}$, and $\tilde{\rho}$ as variables at a microscopic level and our immediate task is to find a means of relating them to corresponding macroscopic quantities.

We must realize at once that for most purposes a 'small' distance – 'small' in a macroscopic sense – when expressed in terms of, say, the radius of a hydrogen atom appears to be very large on a microscopic or atomic scale. Thus when we speak of the macroscopic charge density 'at a point' we mean: the ratio of electric charge contained in a small (on a macroscopic scale) volume around the point in question divided by the volume. Expressed mathematically this last statement reads:

$$\frac{1}{V}\iiint \tilde{\rho}(\mathbf{r}', t)\,\mathrm{d}V = \rho(\mathbf{r}, t) \tag{11}$$

in which the *microscopic* charge density, $\tilde{\rho}$, is distinguished from the macroscopic charge density, ρ, by the tilde. In the rest of this chapter only microscopic quantities will carry the tilde. It is to be understood that the volume V, surrounding the point $\boldsymbol{r}$ and which appears in equation (11), is to be regarded as 'macroscopically small but microscopically large'. This method of deriving macroscopic variables from microscopic variables was introduced by Lorentz.[2]

[1] W. K. H. Panofsky and M. Phillips, 'Classical Electricity and Magnetism', Addison-Wesley, Cambridge, Mass., 1955.

[2] H. A. Lorentz, 'The Theory of Electrons', Teubner, Leipzig, 1909.

Equations (8) and (11) lead to the equation

$$\operatorname{div} \boldsymbol{D}(\boldsymbol{r}, t) = \rho(\boldsymbol{r}, t) \tag{12}$$

which is a relation between macroscopic quantities.

An alternative definition of $\boldsymbol{D}$, the macroscopic flux density, is embodied in the equation (see Figure 1)

$$D_n = \frac{1}{Sh} \iiint \tilde{\boldsymbol{D}} \cdot \boldsymbol{n} \, \mathrm{d}V \tag{13}$$

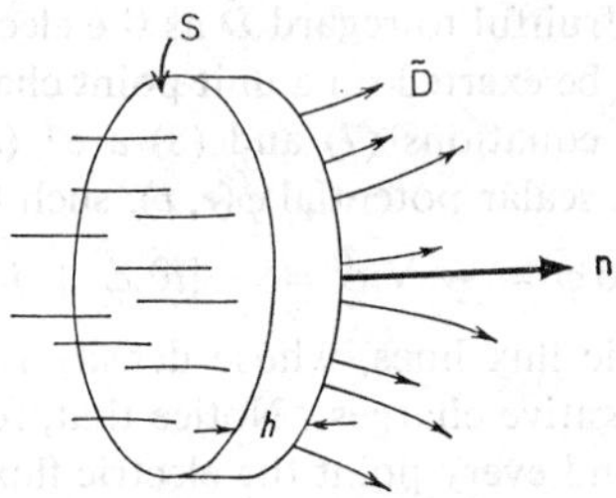

Figure 1 *The macroscopic flux density,* $\boldsymbol{D}$*, is obtained by averaging the microscopic flux density,* $\tilde{\mathbf{D}}$*, over a macroscopically small volume Sh*

in which the volume, Sh, over which the integration is taken is that of a thin disc of area S and thickness h. Both S and h are macroscopically small. D_n is the component of the macroscopic flux density normal to the disc and $\mathbf{n}$ is the unit normal pointing outwards from one of the faces of the disc.

However, it must not be concluded that every average of the kind used in equations (11) and (13) is related to a meaningful macroscopic quantity. A case in point is $\tilde{\boldsymbol{E}}$: whereas it is perfectly possible to consider its volume average, as we did for $\tilde{\rho}$, the result is of no relevance on a macroscopic level. In contrast to the Lorentzian approach, we choose to relate $\tilde{\boldsymbol{E}}$, the microscopic negative potential gradient, to the macroscopic negative potential gradient which we shall denote by $\boldsymbol{E}$. Thus we are led to the following expression for $\boldsymbol{E}$:

$$\int_{\boldsymbol{r}-\frac{1}{2}\boldsymbol{L}}^{\boldsymbol{r}+\frac{1}{2}\boldsymbol{L}} \tilde{\boldsymbol{E}}(\boldsymbol{r}', t) \cdot \mathrm{d}\boldsymbol{s} = \boldsymbol{L} \cdot \boldsymbol{E}(\boldsymbol{r}, t) \tag{14}$$

in which $\boldsymbol{L}$ is a distance at once macroscopically small and microscopically large. In view of equations (13) and (14) it is apparent that any relationship between the macroscopic quantities $\boldsymbol{D}$ and $\boldsymbol{E}$ must involve the charge configuration and dynamics of the system.

For simplicity we shall restrict our attention to those dielectric materials which are good insulators, that is materials for which transport of charge (over macroscopic distances) is practically impossible. In all such materials the constituent charges are strongly localized. This strong localization does not prevent the possibility of relative displacement between the constituent charges of opposite sign. Such displacements might arise from the imposition of an external potential gradient, or from the placing of charged particles in the midst of the material, or even from thermal agitation. Whatever their cause, these displacements manifest themselves as electrically polarized particles. It is important to note that electrical neutrality is preserved which means that, on a macroscopic scale, there is never an excess of charge of one sign.

The charge configuration on an atomic level can be described mathematically in terms of, so-called, multipole moments.[3] As far as macroscopic effects are concerned, only the dipole moment is usually of importance. The dipole moment, $\boldsymbol{m}$, of a group of charges e_i ($i = 1, \ldots, N$) at points $\boldsymbol{r}_i$ satisfies the equation

$$\boldsymbol{m} = \sum_{i=1}^{N} e_i \boldsymbol{r}_i$$

provided the total charge, $\sum_{i=1}^{N} e_i$, is zero. In addition, and of supreme importance, is the fact that the dipole moment $\boldsymbol{m}$ is independent of the co-ordinate system used. This means that no matter how we set about calculating the dipole moment of a group of charges we shall always obtain the same result, provided only that the total charge vanishes.[4] As a result it is possible to describe the polarization of a material in two entirely equivalent ways, the one using macroscopic variables the other using microscopic variables.

Let us introduce a new vector quantity, denoted by $\boldsymbol{P}(\boldsymbol{r}, t)$, called the polarization or the dipole moment per unit volume. The total dipole moment of a (macroscopically) small volume, V, is equal to the sum of the molecular dipole moments, $\sum_j \boldsymbol{m}_j$, contained therein. Thus,

$$\boldsymbol{P}(\boldsymbol{r}, t) = \frac{1}{V}\sum_{i} \boldsymbol{m}_j \tag{15}$$

With the aid of the polarization, $\boldsymbol{P}$, we may describe the electrical state of a dielectric body. It is possible to calculate the electric potential, ϕ, in terms of $\boldsymbol{P}$ at points both inside and outside the body. We do this by adding (see Figure 2) the elementary contribution to ϕ from each polarized element of volume, $\mathrm{d}V$, of the body. Treating each element of volume as a dipole of moment $\boldsymbol{P}\,\mathrm{d}V$ and of negligible spatial extension, we may write[5]

$$\phi(\boldsymbol{r}) = \frac{1}{4\pi\varepsilon_0}\iiint -\frac{(\boldsymbol{r}' - \boldsymbol{r})\cdot\boldsymbol{P}}{|(\boldsymbol{r}' - \boldsymbol{r})|^3}\mathrm{d}V = \frac{1}{4\pi\varepsilon_0}\iiint\left[\boldsymbol{P}\cdot\nabla_{\boldsymbol{r}'}\left(\frac{1}{|\boldsymbol{r}' - \boldsymbol{r}|}\right)\right]\mathrm{d}V$$

Poisson,[6] in his celebrated memoir on the theory of magnetism, showed that, following integration by parts, this last equation could be written:

$$\phi(\boldsymbol{r}) = \frac{1}{4\pi\varepsilon_0}\iint\frac{\boldsymbol{n}\cdot\boldsymbol{P}}{|\boldsymbol{r}' - \boldsymbol{r}|}\mathrm{d}S + \frac{1}{4\pi\varepsilon_0}\iiint\frac{[-\operatorname{div}\boldsymbol{P}]}{|\boldsymbol{r}' - \boldsymbol{r}|}\mathrm{d}V \tag{16}$$

[3] J. A. Stratton, 'Electromagnetic Theory', McGraw-Hill, New York, 1941.
[4] H. Fröhlich, 'Theory of Dielectrics', Clarendon Press, Oxford, 1958.
[5] B. K. P. Scaife, 'Problems in Physical Electronics', ed. A. K. Jonscher and R. Ferrari, Pion, London, 1972, Chap. 7.
[6] S. D. Poisson, *Mémoires de l'Académie*, 1826, **5**, 247.

The interpretation of this result is straightforward. It tells us that in calculating $\phi(\boldsymbol{r})$ we may do so by assuming a surface charge density, $\boldsymbol{n}\cdot\boldsymbol{P}$, at the surfaces of the dielectric body and a volume charge density, of amount

$$\rho_P(\boldsymbol{r}') = -\operatorname{div}\boldsymbol{P} = -\nabla_{r'}\cdot\boldsymbol{P} = -(\partial_{x'}P_x + \partial_{y'}P_y + \partial_{z'}P_z) \quad (17)$$

The occurrence of the surface charge density is easy to understand since the surface will contain the ends of the polarized particles. The volume charge distribution is a much more subtle concept and it will be profitable to devote some attention to it.

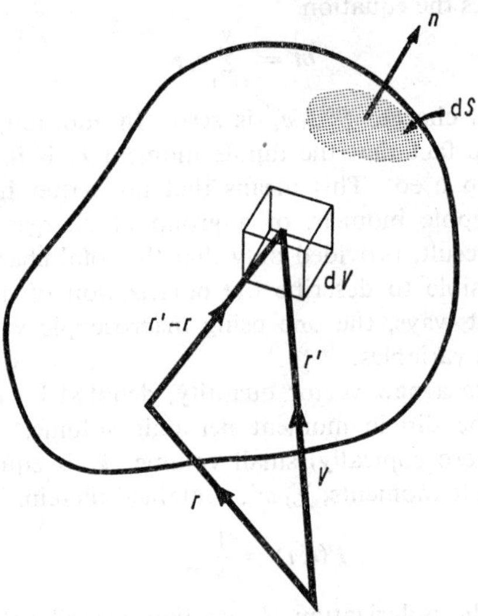

Figure 2

In discussing this charge density, $-\operatorname{div}\boldsymbol{P}$, it is important to realize that equation (16) does not imply that isolated charges, of one sign or another, actually exist in the material. Remember that we set out with the understanding that, on a macroscopic level, electrical neutrality was preserved. The solution to the apparent paradox of electric charge appearing as a result of combining electrically neutral polarized particles can be obtained by considering the very simple system exhibited in Figure 3 in which are shown spatially compact but widely separated electric dipoles of varying magnitudes.

Suppose that the inter-dipole distance a is a macroscopically small quantity. The linear density of dipole moment at $x = na$ is simply $(m_n/a) = p(x)$, say. The dipole moment m_n merely specifies the product of charge and separation. It is possible to imagine a dipole of moment m_n made up of two equal and opposite charges (m_n/a) separated by a distance a. Applying the same reasoning to each dipole we find the net charge, midway between each dipole pair, to be $(m_{n-1} - m_n)/a = p(x - a) - p(x) = q(x)$, say. Because the distance a is small we may write, by Taylor's theorem,

$$q(x) = p(x) - a\,\partial_x p(x) - p(x)$$

and hence the linear charge density, $\rho(x)$, is given by

$$\rho(x) = -\,\partial_x p(x) \tag{18}$$

Thus, in this one-dimensional example at least, we have shown how the potential set up by a system of dipoles can be produced by a particular distribution of isolated charges. Equation (18) is the one-dimensional

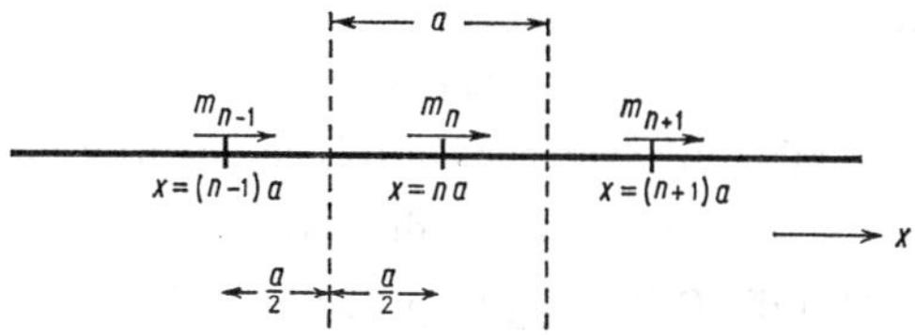

Figure 3

version of equation (17), $\rho_P = -\operatorname{div} \boldsymbol{P}$. It is well to remember that a mere spatial non-uniformity is not sufficient to give rise to a non-vanishing divergence of $\boldsymbol{P}$. For example, the polarization surrounding a point charge, embedded in a dielectric medium, is divergence-free. Examples of the fact that div $\boldsymbol{P}$ does not always vanish will be found in Fröhlich's monograph.[4, 5]

The fact that the potential distribution, in and around a polarized body, can be described in terms of a surface and a volume distribution of charge does not in any way contradict the fact that electrical neutrality, on a macroscopic scale, can be maintained at every point of the body.

Once the potential, $\phi(\boldsymbol{r}, t)$ has been obtained we may calculate the value of the negative potential gradient, $\boldsymbol{E}(\boldsymbol{r}, t)$, by means of equation (10).†

We now turn to the problem of relating $\boldsymbol{D}$, $\boldsymbol{E}$, and $\boldsymbol{P}$. Envisage a dielectric body of arbitrary shape which has a distribution of polarization, $\boldsymbol{P}$, partly produced by a distribution of charges of density ρ and partly by particular charge configurations at a microscopic level resulting from thermal fluctuations. In any very small region of the body the polarization $\boldsymbol{P}$ will be sensibly spatially uniform. If we imagine a small surface drawn in the body perpendicular to $\boldsymbol{P}$ it is clear that an electric flux density of amount $\boldsymbol{P}$ must pass through this surface. This may be seen by imagining the polarization to be frozen and the small surface turning into a thin disc-shaped fissure. $\boldsymbol{P}$ is only one contribution to the total flux density, $\boldsymbol{D}$, at a

† Notice that we nowhere speak of $\boldsymbol{E}(\boldsymbol{r}, t)$ as 'an electric field'. The reason is simple: as soon as the word 'field' is mentioned the tendency is to think of measuring this field by placing a charge in the field and measuring the force exerted on it. Such a procedure is eminently reasonable for charges and fields in *free space*, but when a charge is embedded in a dielectric medium the notion of the force exerted on it is certainly a microscopic entity and we are thus faced with the problem of relating this force, observed on a microscopic scale, with some macroscopic observable.

particular point. The Poisson analysis, outlined earlier, shows that all polarization effects can be described by surface and volume distributions of charge. Once the magnitude and position of these equivalent charges have been determined the system is treated as a system of charges in free space without any dielectric material present. In the vicinity of the point of interest a negative potential gradient, $\boldsymbol{E}$, is set up which is due to the imposed charges and the equivalent charges. This total negative potential gradient exists in free space and therefore we may associate with it a flux density $\varepsilon_0 \boldsymbol{E}$.

The total electric flux density $\boldsymbol{D}$ is the sum of the two contributions $\boldsymbol{P}$ and $\varepsilon_0 \boldsymbol{E}$, hence

$$\boldsymbol{D} = \varepsilon_0 \boldsymbol{E} + \boldsymbol{P} \tag{19}$$

It is instructive to take the divergence of this equation; we find

$$\mathrm{div}(\boldsymbol{D} - \boldsymbol{P}) = \mathrm{div}\, \varepsilon_0 \boldsymbol{E}$$

From equation (12) div $\boldsymbol{D}$ is equal to ρ, the true, free, or external charge density. In view of equation (17), div $\boldsymbol{P}$ may be replaced by $-\rho_{\mathrm{P}}$ the polarization or bound charge density.

Consequently

$$\mathrm{div}\, \varepsilon_0 \boldsymbol{E} = \rho + \rho_{\mathrm{P}} \tag{20}$$

which means that the flux density, $\varepsilon_0 \boldsymbol{E}$, in the Poisson equivalent system, is determined by both the true and the bound charge densities.

3 The Relative Permittivity for Static Fields

For most purposes, a dielectric body can be considered as a purely passive system and therefore any polarization and consequent potential gradients that exist must have been produced by the imposition of some charge distribution. It must, therefore, be possible to express $\boldsymbol{D}$ and $\boldsymbol{P}$ in terms of $\boldsymbol{E}$. It is usual to think of $\boldsymbol{E}$, the negative potential gradient, as the external agency which produces $\boldsymbol{D}$ and $\boldsymbol{P}$. That this procedure is reasonable will be seen by considering the polarization of a dielectric slab placed between two parallel conducting plates which are maintained at a constant difference in potential.

In Figure 4(*a*) is shown a pair of parallel conducting plates, of area A and separation h, connected to an ideal voltage generator of constant voltage V. Immediately following the connection of the plates to the generator a surge of current flows to produce a charge $+ Q_0 = + \sigma_0 A$ on the inner face of the upper plate and a charge $- Q_0 = - \sigma_0 A$ on the inner face of the lower plate. These surface charge densities give rise to an electric flux density $D_0 = \sigma_0$ which is associated with a negative potential gradient $E = (D_0/\varepsilon_0)$ which is simply (V/h) because D_0 and E are spatially uniform. The ratio (Q_0/V) is called the capacitance, C_0, of the system and is entirely determined by the geometry of the plates. In fact $C_0 = \varepsilon_0 A/h$.

When a dielectric slab is inserted between the plates a further flow of charge to the upper plate takes place bringing the total charge from $+ Q_0$ to $+ Q$. The extra charge $Q - Q_0$ is a result of the polarization of the dielectric. The situation is depicted in Figure 4(*b*) in which the gap between

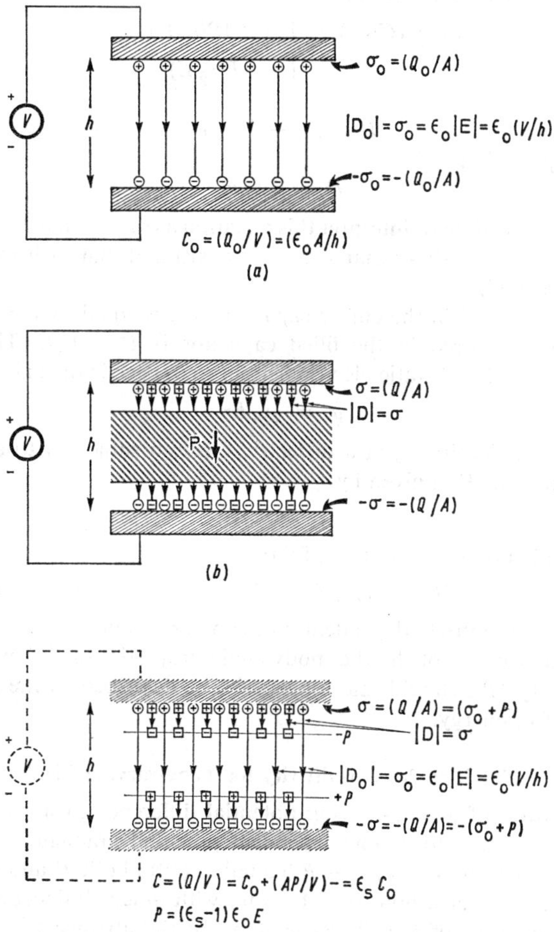

Figure 4 (*a*) *An empty parallel plate capacitor connected to a source of constant voltage, V;* (*b*) *the same capacitor after the insertion of a slab of dielectric;* (*c*) *the Poisson equivalent charge distribution for the case shown in* (*b*)

the dielectric slab and the conducting plates has been greatly exaggerated for clarity. Figure 4(*c*) shows the Poisson system equivalent to the situation shown in Figure 4(*b*). The new capacitance $C = Q/V$ is greater than the empty capacitance C_0 by an amount AP/V. In most cases it is found that the ratio depends only upon the material from which the slab is made. Throughout this chapter we shall assume that there is no dependence of (C/C_0) on the magnitude of the negative potential gradient, E, imposed on it. The relative permittivity for static fields, denoted by ε_s, is defined as the ratio (C/C_0). Consequently

$$\varepsilon_s = (C/C_0) = 1 + (AP/\varepsilon_0 V)\, C_0$$
$$= 1 + (P/\varepsilon_0 E)$$

and hence

$$P = (\varepsilon_s - 1)\,\varepsilon_0 E \qquad (21)$$

and this implies that

$$D = \varepsilon_s\,\varepsilon_0 E \qquad (22)$$

It can be misleading to interpret this equation as saying that E 'causes' D. The equation merely states that D and E coexist and that their magnitudes are simply related.

The energy stored in the empty capacitor, U_0, is equal to $V\int i\,dt = VQ_0$ and the energy stored in the filled capacitor is $U = VQ$. The energy density, u_0, stored in a static electric field in free space is thus given by

$$u_0 = \tfrac{1}{2}\varepsilon_0 E^2 \qquad (23)$$

and the energy density, u, in a dielectric medium subjected to a negative potential gradient E, is given by

$$u = \tfrac{1}{2}\varepsilon_s\,\varepsilon_0 E^2 \qquad (24)$$

which is made up of two parts, as follows,

$$u = \tfrac{1}{2}(\varepsilon_s - 1)\,\varepsilon_0 E^2 + \tfrac{1}{2}\varepsilon_0 E^2 = \tfrac{1}{2}PE + u_0$$

The term $\tfrac{1}{2}PE$ is entirely dependent on the polarization.

If we take account of the thermodynamic properties of the material the energy density $\tfrac{1}{2}PE$ should, more properly, be described as the density of Helmholtz free energy.

4 The Relative Permittivity for Time-varying Fields

The application of a time-varying potential difference to a parallel-plate capacitor must lead to a time-dependent potential gradient, $-E(t)$. In view of the relation $D = \varepsilon_0 E + P$ [equation (19)] both D and P must be functions of time. Just how D and P vary with time will depend upon the dynamic properties of the dielectric slab. In the absence of a dielectric ($P = 0$), of course, $D_0(t) = \varepsilon_0 E(t)$, and hence the ratio

$$Q_0(t)/V(t) = AD_0(t)/hE(t)$$

is equal to $\varepsilon_0 A/h$ which is independent of time. However, when a dielectric is present the ratio of charge to potential difference, $Q(t)/V(t)$, in general will be a complicated function of time. Such a state of affairs may be expressed mathematically by writing

$$Q(t) = \hat{f}\,V(t)$$

in which $\hat{f}$ is an operator which will convert the time-varying voltage, $V(t)$, into the time-varying charge, $Q(t)$. An alternative way, which is independent of the particular capacitor geometry, of expressing this idea is as follows:

$$D(t) = \varepsilon_0 \hat{\varepsilon} E(t) \tag{25}$$

in which $\hat{\varepsilon}$ is the (dimensionless) permittivity operator which converts $\varepsilon_0 E(t)$ into $D(t)$. Notice that we have tacitly assumed in equation (25) that $D(t)$ depends only upon the first power of $E(t)$. Therefore the ratio $D(t)/E(t)$ is independent of $E(t)$. A system in which the ratio (response/stimulus) is a function of time only and is independent of the magnitude of the stimulus is said to be a linear system. [The fact that the graph of $D(t)$ against $E(t)$ is, in general, not a straight line has no bearing on the linearity of the system.]

No matter what the actual behaviour of the dielectric material is, there will always be some part of $D(t)$ which follows the variation of $E(t)$ precisely. The amount of this contribution to the total flux density is $\varepsilon_0 E(t)$. Therefore it is possible and useful to introduce a susceptibility operator, $\hat{\chi}$, which is defined by the equation

$$\hat{\chi} = \varepsilon_0(\hat{\varepsilon} - 1) \tag{26}$$

and hence, from equations (19), (25), and (26),

$$P(t) = \hat{\chi} E(t) \tag{27}$$

For $E(t)$ equal to the unit step-function $\theta(t)$,

$$[\theta(t) = 0,\ t < 0;\ \theta(t) = 1,\ t > 0],$$

we shall denote the response of the dielectric by $a(t)$. Thus,

$$a(t) = \hat{\chi}\theta(t) \tag{28}$$

which implies that $a(t)$ must vanish for $t < 0$. As we have shown elsewhere,[7, 8] in some detail, the value of $P(t)$ caused by a potential gradient, $-E(t)$, which vanishes for $t < 0$ but which otherwise has an entirely arbitrary time dependence, may be calculated by means of the equation

$$P(t) = \int_0^t E(t - t') \frac{\mathrm{d}a(t')}{\mathrm{d}t'} \mathrm{d}t' \tag{29}$$

This equation is derived on the assumption that the dielectric responds to E in a linear fashion. Linearity means that the total response to several stimuli is the sum of the responses to each stimulus acting on its own.

It should be noted that equation (29) involves $\mathrm{d}a/\mathrm{d}t$, the derivative of the step-function response. It is usually referred to as the pulse or delta-function response because

$$\frac{\mathrm{d}\theta}{\mathrm{d}t} = \delta(t)$$

[7] B. K. P. Scaife, *Progr. Dielectrics*, 1963, **5**, 143.

[8] B. K. P. Scaife, 'Complex Permittivity', English Universities Press, London, 1971; *cf.* N. E. Hill, *J. Phys.* (*C*), 1972, **5**, 415.

where $\delta(t)$ is the Dirac function,† and hence

$$c(t) \equiv \frac{\mathrm{d}a}{\mathrm{d}t} = \frac{\mathrm{d}}{\mathrm{d}t}\hat{\chi}\theta(t) = \hat{\chi}\frac{\mathrm{d}}{\mathrm{d}t}\theta(t) = \hat{\chi}\delta(t) \qquad (30)$$

From an experimental point of view it is much easier to measure $a(t)$ rather than $c(t)$. To measure $c(t)$ it is necessary to produce a pulse which approximates closely to a delta function, *i.e.* a pulse which has an exceedingly large value for a vanishingly small period. Measurement of $a(t)$ is extensively used as a means of investigating the low-frequency behaviour of dielectrics.

For harmonically varying potential gradients of the form

$$E(t) = E_m \cos \omega t \, \theta(t),$$

equation (29) simplifies to

$$P(t) = E_m[\chi'(\omega) \cos \omega t + \chi''(\omega) \sin \omega t] \qquad (31)$$

for values of t well in excess of the longest relaxation time of the material. The quantities $\chi'(\omega)$ and $\chi''(\omega)$ are, respectively, the cosine and sine transforms of the pulse response $c(t)$. Thus

$$\chi'(\omega) = \int_0^\infty c(t) \cos \omega t \, \mathrm{d}t \qquad (32)$$

and

$$\chi''(\omega) = \int_0^\infty c(t) \sin \omega t \, \mathrm{d}t$$

The term $E_m \chi'_m(\omega) \cos \omega t$ in equation (31) is that part of the polarization which is in phase with $E(t)$ and therefore is associated with storage of energy. Energy dissipation is associated with the second term, $E_m \chi''(\omega) \sin \omega t$, which differs in phase from $E(t)$ by precisely $\pi/2$, that is, the second term is in quadrature with the first, in-phase, term.

For a parallel-plate capacitor the rate at which energy is supplied is

$$V\frac{\mathrm{d}Q}{\mathrm{d}t} = AhE(t)\frac{\mathrm{d}D}{\mathrm{d}t} \qquad (33)$$

Only a portion of this power is absorbed by the dielectric proper, indeed an amount $\frac{1}{2}\varepsilon_0 Ah(\mathrm{d}E^2/\mathrm{d}t)$ is necessary to establish the potential gradient – $E(t)$. Thus the power, W_a, absorbed by each unit volume of the dielectric is given by

$$W_a = E(t)\frac{\mathrm{d}[D(t) - \varepsilon_0 E(t)]}{\mathrm{d}t} = E(t)\frac{\mathrm{d}P(t)}{\mathrm{d}t} \qquad (34)$$

From this equation and equation (31) we find that

$$W_a = E_m^2[-\,\omega\chi'(\omega) \sin \omega t \cos \omega t + \omega\chi''(\omega) \cos^2 \omega t] \qquad (35)$$

† The Dirac delta function may be defined by the following statements: $\int_{-\infty}^{\infty} \delta(t)\,\mathrm{d}t = 1$ and $\int_{-\infty}^{\infty} f(t')\,\delta(t - t')\,\mathrm{d}t' = f(t)$. Very crudely one can say $\delta(t) \to \infty$, $t = 0$; $\delta(t) = 0$, $t \neq 0$.

The first term oscillates in sign at *twice* the frequency of $E(t)$, and has a zero time average. The second term varies rhythmically, it is always positive, and its time average is $\frac{1}{2}\omega\chi''(\omega) E_m^2$. This is the mean power absorbed per unit volume of the dielectric material and converted into heat. The reversible part of the polarization process is associated with the pulsating power flow described by the first term.

From equation (32) it will be seen that χ' and χ'' must satisfy the following equations

$$\chi'(\omega) = \chi'(-\omega)$$
$$\chi''(\omega) = -\chi''(-\omega) \tag{36}$$

hence the complex susceptibility,† $\chi(\omega) \equiv \chi'(\omega) - i\chi''(\omega)$, $(i = \sqrt{-1})$ must satisfy the relation

$$\chi(\omega) = \chi^*(-\omega) \tag{37}$$

where the asterisk indicates the conjugate of a complex quantity. Equation (36) has a rather deep significance because it may be demonstrated that these conditions are a direct result of the, so-called, causal nature of physical systems by which the response to a stimulus never precedes the stimulus.

By making use of the symbol R_e to denote 'real part of' we may write equation (31) in the form

$$P(t) = E_m R_e[\chi(\omega) \exp(+i\omega t)] \tag{38}$$

and equation (32) becomes

$$\chi(\omega) = \int_0^\infty c(t) \exp(-i\omega t)\, dt \tag{39}$$

The decay function, $b(t)$, which is defined by the equation

$$b(t) = [a(\infty) - a(t)]\, \theta(t)$$

when inserted into equation (39) yields

$$\chi(\omega) = -\int_0^\infty \frac{db}{dt} \exp(-i\omega t)\, dt \tag{40}$$

and, following integration by parts and substitution of χ_s for $\chi'(0)$, we obtain

$$\chi(\omega) = \chi_s - i\omega \int_0^\infty b(t) \exp(-i\omega t)\, dt. \tag{41}$$

One may deduce from equation (32) the well-known Kramers–Kronig dispersion relations which may be written in the following form:

$$\chi'(\omega) = \frac{2}{\pi}\int_0^\infty \frac{[\chi''(\mu)\,\mu - \chi''(\omega)\,\omega]}{(\mu^2 - \omega^2)}\, d\mu \tag{42}$$

$$\chi''(\omega) = \frac{2}{\pi}\int_0^\infty \frac{[\chi'(\mu) - \chi'(\omega)]\,\omega}{(\omega^2 - \mu^2)}\, d\mu \tag{43}$$

† Because we adopt the convention of treating counter-clockwise rotations in the Argand diagram as positive, the imaginary co-ordinate of $\chi(\omega)$ must carry a negative sign. This convention means that we write $\cos \omega t$ = Real part of $(\exp + i\omega t)$.

In these equations, μ denotes an arbitrary value of the pulsatance, whereas ω denotes a particular value. By writing the dispersion relations in this way we avoid having to introduce the notion of the Cauchy principal value. It will be noticed that there are no discontinuities in the integrands, when $\mu = \omega$.

The Fourier transforms in equation (32) may each be inverted with the result that

$$c(t) = \frac{2}{\pi}\int_0^\infty \chi'(\omega)\cos\omega t\,\mathrm{d}\omega \quad (t > 0)$$

$$c(t) = \frac{2}{\pi}\int_0^\infty \chi''(\omega)\sin\omega t\,\mathrm{d}\omega \quad (t > 0) \tag{44}$$

These equations mean that we can construct the pulse-response, $c(t)$, for all values of t, provided we have a sufficiently detailed knowledge of *either* $\chi'(\omega)$ *or* $\chi''(\omega)$ over the *whole* range of ω. It is possible to cast these two equations in a more useful form as follows: we may integrate each of the equations with respect to time to obtain

$$b(t) = \chi_s - \frac{2}{\pi}\int_0^\infty \frac{\chi'(\omega)}{\omega}\sin\omega t\,\mathrm{d}\omega$$

$$b(t) = \frac{2}{\pi}\int_0^\infty \frac{\chi''(\omega)}{\omega}\cos\omega t\,\mathrm{d}\omega \tag{45}$$

where we have used the facts that

$$\int_0^t c(t)\,\mathrm{d}t = -\int_0^t \frac{\mathrm{d}b}{\mathrm{d}t}\,\mathrm{d}t = b(0) - b(t),$$

and that $b(0) = a(\infty) = \chi_s$.

For some purposes it is desirable to have a relation between the current entering a capacitor and the dielectric susceptibility. To this end let us refer back to equation (31): if we differentiate this equation with respect to time we find that the current, $i_P(t)$, entering the capacitor, due to the polarization of the dielectric by the applied potential $V_m \cos\omega t$ ($V_m = hE_m$), is given by

$$i_P(t) = A\frac{\mathrm{d}P}{\mathrm{d}t} = (A/h)V_m[\omega\chi''\cos\omega t - \omega\chi'\sin\omega t] \tag{46}$$

It will be seen that the part of the current *in-phase* with the applied voltage, $(\omega\chi'' A/h)V_m\cos\omega t$ is associated with χ'', the dissipative term of the complex susceptibility χ. The wattless, or dissipation-free, component of the current has a phase which is $(\pi/2)$ ahead of the applied voltage. The factor $(\omega\chi' A/h)$ is called the capacitative reactance of the capacitor.

Figure 5 illustrates, in terms of rotating vectors in the complex plane, the phase relationship between $E(t)$, $P(t)$, and $\mathrm{d}P/\mathrm{d}t$. Equation (46) may be written

$$i_P(t) = \mathrm{R_e}\, Y_P(\omega) V_m \exp(+\,\mathrm{i}\omega t)$$

frequencies. Often $\chi'(\Omega)$ is abbreviated by χ_∞. Such a procedure is to be strongly deprecated because of its vagueness.

5 The Microscopic Significance of $\varepsilon(\omega)$

We shall conclude the brief survey of the theory of the macroscopic behaviour of dielectrics by drawing attention to the microscopic significance of some of the properties of $\varepsilon(\omega)$ or of functions of $\varepsilon(\omega)$.

The generalization of the work, by Kirkwood[10] and by Fröhlich,[4] on the relation between the mean-square fluctuation in dipole moment, $\langle \mathbf{M}^2 \rangle$, of a dielectric sphere and its zero-frequency polarizability, $\alpha'_{sph}(0)$, namely:

$$\langle M^2 \rangle = 3kT\alpha'_{sph}(0)$$

leads to the following equation

$$\langle M_\gamma^2 \rangle = kT\alpha'_\gamma(0) \qquad (56)$$

between the γ (= x, y, or z) component of dipole moment M and the polarizability of the body in the γ-direction. Boltzmann's constant and the absolute temperature are denoted by k and T respectively.

Equation (56) may be understood more easily by considering a thin rectangular slab of dielectric of volume Ah sandwiched between a pair of conducting plates. The polarizability of the slab, in a direction perpendicular to it, is $\alpha_\perp$ defined in equation (48). A dipole moment in this direction implies a polarization of amount $(M_\perp/Ah)$. By the usual Poisson-type analysis we find that there is a negative potential gradient $E = P/\varepsilon_0$ in a direction opposite to $P = (M_\perp/Ah)$. The difference in potential between the plates will therefore amount to $v = (M_\perp/\varepsilon_0 A)$. The mean-square value of this voltage is given by

$$\langle v^2 \rangle = \langle (M_\perp/\varepsilon_0 A)^2 \rangle$$

which by means of equations (48) and (56) reduces to

$$\begin{aligned} \langle v^2 \rangle &= kT(\varepsilon_s - 1)(h/\varepsilon_s \varepsilon_0 A) \\ &= (kT/C_0)[(\varepsilon_s - 1)/\varepsilon_s] \qquad (57) \end{aligned}$$

This equation gives the mean-square value of the voltage appearing across the terminals of a capacitor filled with a dielectric of zero-frequency relative permittivity, ε_s, in terms of C_0, the capacitance without a dielectric, and the absolute temperature, T. This noise voltage is caused by the thermally induced dipole-moment fluctuations which are themselves inextricably bound up with the dissipative processes. That this is so is indicated by the fact that equation (42) applied to $\alpha_\perp(\omega)$ leads to the relation

$$(\varepsilon_s - 1)/\varepsilon_s = \frac{2}{\pi}\int_0^\infty \frac{\varepsilon''(\mu)}{[\varepsilon'^2 + \varepsilon''^2]}\frac{d\mu}{\mu}$$

and since an ensemble average, $\langle v^2 \rangle$, and a time average, $\overline{v^2}$, are one and the same, we may say that

$$\overline{v^2} = \left(\frac{kT}{C_0}\right)\left(\frac{2}{\pi}\right)\int_0^\infty \frac{\varepsilon''(\mu)}{[\varepsilon'^2 + \varepsilon''^2]}\frac{d\mu}{\mu}$$

[10] J. G. Kirkwood, *J. Chem. Phys.*, 1939, **7**, 911.

It is possible to measure the intensity of each frequency component of the randomly varying $v(t)$. The mean-square voltage can be equated to the integral of the spectral density, $\Phi(\omega)$, as follows

$$\overline{v^2} = \frac{1}{\pi}\int_0^\infty \Phi(\omega)\, d\omega$$

Further analysis shows that the following relation is true:

$$\Phi(\omega) = \left(\frac{2kT}{\omega C_0}\right)\frac{\varepsilon''(\omega)}{[\varepsilon'^2 + \varepsilon''^2]} \tag{58}$$

This is but one form of the celebrated fluctuation–dissipation theorem which in this case, provides a link between the *equilibrium* fluctuations in capacitor voltage and $\varepsilon''(\omega)$ which is a measure of the rate at which electrical energy is dissipated into heat.

Rather than deal with the frequency spectrum of the fluctuating voltage we may study the autocorrelation function $\rho_v(\tau)$ defined as the mean value of the product of v at some instant with v at some time τ units of time earlier or later. Thus

$$\rho_v(\tau) = \overline{v(t)\, v(t + \tau)}$$

and it may be shown that

$$\rho_v(\tau) = \frac{1}{\pi}\int_0^\infty \Phi(\omega) \cos \omega\tau\, d\omega$$

This equation with equation (58) leads us to

$$\rho_v(\tau) = \left(\frac{kT}{C_0}\right)\left(\frac{2}{\pi}\right)\int_0^\infty \frac{\alpha''_\perp(\omega)}{\omega} \cos \omega\tau\, d\omega$$

which, in the light of equation (45), indicates that

$$\rho_v(\tau) = \left(\frac{kT}{C_0}\right) \quad \text{(decay function of polarization perpendicular to a thin slab)}$$

The significance of this result lies in the fact that $\rho_v(\tau)$ arises spontaneously from the dynamic activity of the material, whereas the decay function describes the behaviour of the system when it reacts to the sudden removal of an *externally* applied electric field.

2
Dielectric Relaxation and Molecular Correlation

BY G. WYLLIE

1 Introduction

In the past decade a great deal of work has been done pursuing the dielectric properties of materials through the sub-millimetre waveband and into the traditional domain of spectroscopy. The rotary movements of molecules are studied through i.r. and Raman spectroscopy as well as by direct dielectric and magnetic-resonance measurements.[1–3]

The unification of the frequency domain up to 10^{13} Hz has been primarily the result of advances in experimental technique, notably the development of laser sources and of interferometric spectroscopy[4] in the far-i.r. As a result it has become necessary, for the theoretical interpretation of the measurements, to take account in detail of the ways in which the random motions of molecules differ from simple diffusion of position or of orientation.[5]

The need for an account of inertial effects in relaxation at high frequencies has been known for a considerable time.[6–8] In dilute gases a description in terms of free rotation perturbed by collision is substantially developed to account for the shapes of i.r. and Raman lines.[9, 10] In crystals at sufficiently low temperature, description in terms of bands of regular oscillations perturbed by anharmonicity provides a regular sequence of approximations.[11, 12] In dense gases and in liquids[13–16] it is necessary to face simultaneously a complex pattern of molecular motions and a difficult question of the delayed electrical interaction which gives rise to the 'internal

1 R. G. Gordon, *Adv. Magn. Resonance*, 1968, **3**, 1.
2 R. P. Young and R. N. Jones, *Chem. Rev.*, 1971, **71**, 219.
3 M. Davies, *Ann. Reports (A)*, 1970, **67**, 65.
4 H. A. Gebbie and R. Q. Twiss, *Reports Prog. Phys.*, 1966, **29**, 729.
5 Papers in *Adv. Chem. Phys.*, 1969, **15**. The whole volume is relevant.
6 Y. Rocard, *J. Phys. Radium*, 1933, **4**, 247.
7 J. G. Powles, *Trans. Faraday Soc.*, 1948, **44**, 802.
8 R. A. Sack, *Proc. Phys. Soc.*, 1957, **70B**, 402, 414.
9 A. Ben-Reuven, *Adv. Atomic Mol. Phys.*, 1969, **5**, 201.
10 G. Birnbaum, *Adv. Chem. Phys.*, 1967, **12**, 487; F. J. Bartoli and T. A. Litovitz, *J. Chem. Phys.*, 1972, **56**, 404, 413.
11 D. H. Martin, *Adv. Phys.*, 1965, **14**, 39.
12 R. A. Cowley, *Adv. Phys.*, 1963, **12**, 421.
13 R. G. Gordon, *J. Chem. Phys.*, 1963, **39**, 2788; 1964, **41**, 1819; 1965, **43**, 1307.
14 H. Shimizu, *J. Chem. Phys.*, 1965, **43**, 2453.
15 S. Bratož, J. Rios, and Y. Guissani, *J. Chem. Phys.*, 1970, **52**, 439.
16 G. Ewing, *Accounts Chem. Res.*, 1969, **2**, 168.

field'. The latter problem has now received an acceptable analysis[17] for the simplest form of the particular model of a dielectric which was introduced by Onsager.

It is always legitimate to describe an irregular motion in terms of a spectrum of frequencies, without implying that any particular object is executing a regular oscillation at any of the frequencies of the spectrum. It is equally legitimate to use the 'auto-correlation function' of the motion, which is the expectation value of a time-delayed product of values of the variable under consideration. This function of the time delay is a Fourier transform of the spectral density function, so it carries information about the higher frequencies of the spectrum primarily in its behaviour for small values of the time – of the order of the corresponding periods. The principal object of our attention is the auto-correlation $\langle \boldsymbol{u}(0)\cdot\boldsymbol{u}(t)\rangle$, where angle brackets signify the appropriate (ensemble) average, of a unit vector $\boldsymbol{u}$ fixed in a moving molecule. Where $\boldsymbol{u}$ is the direction of a permanent dipole, this function determines the complex frequency-dependent electric permittivity. Where $\boldsymbol{u}$ is the direction of the oscillating dipole of an i.r. transition, the correlation function determines, and may be derived from, the shape of the corresponding i.r. line.

The mechanical behaviour of a molecular system is most simply described within periods too short for perturbations to have much effect, so the behaviour of the correlation function over very short times is generally simple. Various forms of the dipole correlation function appear in the literature, some based on self-consistent simple models, others as convenient semi-empirical expressions. It is difficult to appreciate them without working out at least some simple cases, and a good part of this chapter is devoted to such an exposition. Little attention will be given to the general theory, which is fully described in numerous reviews,[18] but some vital results are summarized in an Appendix.

This is intended as an expository article rather than an extended review, and little effort has been made to obtain historical balance through the references. Most of the work cited occurs in the English language literature, whereas a balanced review of the development of the field would emphasize the important contributions of French, other European, and Soviet workers.

2 Dielectric Response

Experimental observations give the dielectric response of bulk material. Individual constituents of the sample may have different time-average behaviour, so that the bulk response need not coincide with the response of any single molecule. Moreover, in pure, strongly polar, materials the relation of the macroscopic response to molecular behaviour depends

[17] Tsu-Wei Nee and R. Zwanzig, *J. Chem. Phys.*, 1970, **52**, 6353.
[18] R. Zwanzig, *Ann. Rev. Phys. Chem.*, 1965, **16**, 67.

significantly on the mutual correlation of the motion of molecules through the long-range electrostatic interactions. In electrically dilute materials, however, the discrepancy between macroscopic and molecular behaviour is often unimportant.

This article is concerned with dielectric relaxation in the linear range, where, for example, the application and removal of a test field produce effects differing only in sign, with an otherwise identical variation in time. It is generally a condition of such linear response that the orientation energy $\mu \boldsymbol{E}$ of a dipole μ in a field $\boldsymbol{E}$ should be much less than the characteristic thermal energy kT, so that the applied field causes a small perturbation of the thermal motions already present in the system. In such a case the relaxation of the polarization, following removal of the test field, has a course identical with the average relaxation of a spontaneous fluctuation occurring in the entire absence of an external field.

In principle, complete information on the dielectric behaviour may be obtained by studying the noise output of a condenser.* The advantage of testing with a substantial applied field is simply to produce signals well above the noise from other possible, uncontrolled, sources in the system.

The analytical description of a linear response may take several forms. If we consider the applied field resolved into harmonic components of typical form $\boldsymbol{E}(\omega) \exp(\mathrm{j}\omega t)$, the corresponding displacement component $\boldsymbol{D}(\omega) = \varepsilon(\omega)\, \boldsymbol{E}(\omega) \exp(\mathrm{j}\omega t)$ where ε is in general complex and equal to $\varepsilon' - \mathrm{j}\varepsilon''$ where ε' and ε'' are real. The choice of sign of the imaginary exponent is arbitrary. A frequent, but unfortunately not universal, rule of translation is $\mathrm{j} = -\mathrm{i}$. The signs in the exponent and in ε must be opposite to obtain a positive sign for the power loss for a positive value of ε''.

For an arbitrary varying field, then, for which

$$\boldsymbol{E}(t) = \int_{-\infty}^{\infty} \boldsymbol{E}(\omega)\, \mathrm{e}^{\mathrm{j}\omega t}\, \mathrm{d}\omega$$

the displacement

$$\boldsymbol{D}(t) = \int_{-\infty}^{\infty} \varepsilon(\omega)\, \boldsymbol{E}(\omega)\, \mathrm{e}^{\mathrm{j}\omega t}\, \mathrm{d}\omega$$

The response is then described in terms of a knowledge of ε for all ω. In fact, since we require a real $\boldsymbol{E}(t)$ to produce a real $\boldsymbol{D}(t)$, $\varepsilon'(-\omega) = \varepsilon'(\omega)$, $\varepsilon''(-\omega) = -\varepsilon''(\omega)$ and we need know ε only for zero and positive ω.

Heaviside's description of the response arises from the idea of approximating $\boldsymbol{E}(t)$ by a staircase function. Then if $\theta(t)$ is the function which is zero for $t < 0$ and 1 for $t > 0$, we see that

$$\boldsymbol{E}(t) = \int_{-\infty}^{\infty} \frac{\mathrm{d}\boldsymbol{E}}{\mathrm{d}\tau}\, \theta(t - \tau)\, \mathrm{d}\tau$$

If now $\boldsymbol{D}_0(t) = \phi_D(t)$ is the response to $\boldsymbol{E}(t) = \theta(t)$, we have in general

$$\boldsymbol{D}(t) = \int_{-\infty}^{\infty} \frac{\mathrm{d}\boldsymbol{E}}{\mathrm{d}\tau}\, \phi_D(t - \tau)\, \mathrm{d}\tau$$

*See Chapter 1, equation (57) *et seq.*

where the upper limit may be taken at t since $\phi_D(t)$ must vanish for $t < 0$, as effects do not precede their causes.

Since $\boldsymbol{D}$ contains $\boldsymbol{E}$ itself, ϕ_D by definition contains some multiple of $\theta(t)$. Moreover, some component of the polarization – that associated with the deformation of bound-electron orbits – follows the applied field immeasurably fast. Separating the fast component we write

$$\begin{aligned} \boldsymbol{D}(t) &= \varepsilon_\infty \boldsymbol{E}(t) + \int_{-\infty}^{\infty} \frac{\mathrm{d}\boldsymbol{E}}{\mathrm{d}\tau} \phi_1(t-\tau)\,\mathrm{d}\tau \\ &= \varepsilon_\infty \boldsymbol{E}(t) + \int_{-\infty}^{t} \boldsymbol{E}(\tau)\dot{\phi}_1(t-\tau)\,\mathrm{d}\tau \end{aligned}$$

on integrating by parts and taking $\boldsymbol{E}(-\infty) = 0$, $\phi_1(0) = 0$.

Again, the response to a steady field settles down in the end to a steady value so that $\phi_D(\infty) = \varepsilon_0$, say, and $\phi_1(\infty) = \varepsilon_0 - \varepsilon_\infty$. After a prolonged application of unit field, then, the course of $\boldsymbol{D}$ on switching off the field at $t = 0$ will be

$$\begin{aligned} \boldsymbol{D}(t) &= (\varepsilon_0 - \varepsilon_\infty)\,1 - \phi_1(t) \\ &\equiv (\varepsilon_0 - \varepsilon_\infty)\,\phi(t) \end{aligned}$$

The function $\phi(t)$ thus defined goes from 1 to 0 as t increases and is called the polarization relaxation function. We require $\phi(t) = 0$ for $t < 0$ to retain the proper causal sequence of events, so in general

$$\boldsymbol{D}(t) = \varepsilon_\infty \boldsymbol{E}(t) - (\varepsilon_0 - \varepsilon_\infty)\int_{-\infty}^{t} \boldsymbol{E}(\tau)\dot{\phi}(t-\tau)\,\mathrm{d}\tau$$

Regarding $\boldsymbol{E}(t)$, alternatively, as a sequence of impulses we see that

$$-(\varepsilon_0 - \varepsilon_\infty)\,\dot{\phi}(t)$$

is the delayed polarization response to unit impulse of $\boldsymbol{E}$ at $t = 0$. Substitution of

$$\boldsymbol{E}(\tau) = \int_{-\infty}^{\infty} \mathrm{e}^{\mathrm{j}\omega\tau}\,\boldsymbol{E}(\omega)\,\mathrm{d}\omega$$

then gives

$$\begin{aligned} \frac{\varepsilon - \varepsilon_\infty}{\varepsilon_0 - \varepsilon_\infty} &= -\int_{-\infty}^{t} \mathrm{d}\tau\,\mathrm{e}^{\mathrm{j}\omega(\tau - t)}\,\dot{\phi}(t-\tau) \\ &= -\int_{0}^{\infty} \mathrm{e}^{-\mathrm{j}\omega\tau}\,\dot{\phi}(\tau)\,\mathrm{d}\tau \end{aligned}$$

This is the standard form of the equivalence between the electric permittivity and the relaxation function.

Relaxation of Thermal Fluctuations.—Following Onsager,[19] we now assert that thermal fluctuations of polarization in a thermal equilibrium system will, on average (over a suitable ensemble), relax with the same time course as any other polarization within the range of linear response. The essential

[19] L. Onsager, *Phys. Rev.*, 1931, **37**, 405; **38**, 2265.

reason for this is that the ensemble averaging builds in the physical condition that the entropy must remain maximal except so far as it is forced down by the specified polarization. Thus, given an initial fluctuant polarization $\boldsymbol{P}(0)$ (in particular with $\boldsymbol{E}$ remaining zero, as for a dielectric block in a short-circuited condenser) the average subsequent polarization

$$\langle \boldsymbol{P}(t) \rangle = \boldsymbol{P}(0)\, \phi(t)$$

Since the excursions of $\boldsymbol{P}$ vary in amount and sign, we multiply by $\boldsymbol{P}(0)$ and average again to obtain

$$\frac{\langle \boldsymbol{P}(0)\, \boldsymbol{P}(t) \rangle}{\langle \boldsymbol{P}(0)^2 \rangle} = \phi(t), \quad \text{for } t \geqslant 0$$

In other words, the relaxation function of polarization is the normalized auto-correlation function of the polarization in thermal equilibrium in zero applied field.

The fluctuation-dissipation theorem, of which this is one example, is referred to in the Appendix. Unlike $\phi(t)$, $\langle \boldsymbol{P}(0)\, \boldsymbol{P}(t) \rangle$ is finite for $t < 0$. In a non-rotating system in the absence of magnetic fields, the growth of thermal fluctuations, because of the perfect reversibility of the microscopic equations of motion, is the reverse of their decay. Thus

$$\langle \boldsymbol{P}(0)\, \boldsymbol{P}(t) \rangle = \langle \boldsymbol{P}(0)\, \boldsymbol{P}(-\, t) \rangle.$$

For a variable which contains an odd power of t, the sign reverses.

Relation to a Noise Current. The fluctuating polarization results from a movement of charge in the dielectric medium. In a short-circuited condenser, the polarization changes induce a corresponding current $I(t)$ in the external circuit. The mean square current $G(\nu)$ per unit frequency bandwidth at frequency ν can be related to a series of observations of $I(t)$ by the important Wiener–Khinchin theorem

$$G(\nu) = 4\int_0^\infty \langle I(0)\, I(\tau) \rangle \cos(2\pi\nu\tau)\, \mathrm{d}\tau$$

In thermal equilibrium $I(t)$ is a stationary process and

$$\langle I(0)\, I(\tau) \rangle = \langle I(\tau')\, I(\tau' + \tau) \rangle,$$

so may be derived from a single, sufficiently prolonged record of $I(t)$. As a rule $\langle I(0)\, I(t) \rangle$ diminishes at least as $\exp(-\, \alpha t)$ for large t, so the tail of the infinite integral can be reliably estimated from a finite experiment.

Electronic correlators work satisfactorily, but it is still cheaper, as a rule, to measure $G(\nu)$ and obtain $\langle I(0)\, I(t) \rangle$ by inverting the transform. This is equivalent to $\langle \dot{\boldsymbol{P}}(0)\, \dot{\boldsymbol{P}}(t) \rangle$. We have then, as shown in the Appendix,

$$\frac{\langle I(0)\, I(t) \rangle}{\langle I^2 \rangle} = \frac{\langle \dot{\boldsymbol{P}}(0)\, \dot{\boldsymbol{P}}(t) \rangle}{\langle \dot{\boldsymbol{P}}^2 \rangle} = \frac{\ddot{\phi}(t)}{\ddot{\phi}(0)}$$

In the following sections we shall consider the correlation functions which arise in the motion of various model systems. The correlation function of a well-defined model has at least the merit of being physically realizable. However, the simple systems at first discussed serve more or less to represent local conditions in a dielectric. The internal field continues to lurk in the background, and will not be considered till a later section of this chapter.

3 Relaxation in some Model Systems

We shall be concerned with thermal motion in classical mechanics, for the greater part. This implies a restriction of the frequencies considered at a given temperature, such that $\hbar\omega \ll kT$. One modification needed at higher frequencies is noted in the Appendix. We do not consider the dielectric response of cold many-body systems, which contains such fascinating results as Friedel's discovery [20] of the oscillating interaction potential of point charges in an electron gas. Although many dielectrics consist of distinct, electrically neutral molecules bearing a more or less rigid distribution of charge, we start by considering a more extended set of movements.

Diffusive Migration of a Charged Object, without Inertial Effects.—Free migration of charges yields an unbounded linear growth of polarization in a constant applied field, *i.e.* a conduction current. If charges migrate between impenetrable boundaries, the polarization is limited and the relaxation behaviour is controlled by the spacing of the boundaries and the detail of the migration mechanism. In the grossly macroscopic limit we have essentially a network of identifiable condensers and resistances and the appropriate description is that of the Maxwell–Wagner effect. We consider a smaller scale.

Two-site Model. Consider a particle bearing charge q, with two resting positions separated by a distance d, between which it makes very rapid jumps at random intervals with mean frequency ν. We suppose that in the absence of an applied field the probability, per unit time, of jumping out of whichever site is occupied is the same for both sites, so that on average each site is occupied for half of the time. The polarization with respect to the near position alternates between $qd/2$ and $-qd/2$, with time-average zero.

The polarization auto-correlation function is easily constructed if we assume that the duration of a jump is much less than the period of variation of any field we envisage. Over any particular interval t to $t + \tau$, the magnitude of $P(t)\,P(t + \tau)$ is $q^2 d^2/4$. It is positive if no jump, or an even number of jumps, occurred between t and $(t + \tau)$, and negative if an odd number of jumps occurred. If the jumps are independent events

[20] J. Friedel, *Phil. Mag.*, 1952, **43**, 153.

occurring quite at random, the number n occurring in time τ has the familiar Poisson distribution with the probability $p(n)$ of exactly n jumps given by

$$p(n) = (n!)^{-1} (\nu\tau)^n \exp(-\nu\tau)$$

We obtain for $\boldsymbol{P}(t)\,\boldsymbol{P}(t + \tau)$ the value $q^2 d^2/4$ with probability

$$p_{\text{even}} = \sum_{n \text{ even}} p_n$$

and the value $-q^2 d^2/4$ with probability

$$p_{\text{odd}} = \sum_{n \text{ odd}} p_n$$

Thus the expectation value is

$$\langle \boldsymbol{P}(t)\,\boldsymbol{P}(t + \tau)\rangle = \frac{q^2 d^2}{4} \mathrm{e}^{-\nu\tau} \sum_{n=0}^{\infty} \frac{(-\nu\tau)^n}{n!}$$

$$= \frac{q^2 d^2}{4} \mathrm{e}^{-2\nu\tau}$$

and the normalized relaxation function $\langle \boldsymbol{P}(t)\,\boldsymbol{P}(t + \tau)\rangle / \langle \boldsymbol{P}^2\rangle$ is simply $\gamma(\tau) = \exp(-2\nu\tau)$, giving a pure exponential decay with relaxation time $\tau_r = (2\nu)^{-1}$. If the correlation function γ for this model is taken as the correlation ϕ for a possible substance, we obtain

$$\frac{\varepsilon - \varepsilon_\infty}{\varepsilon_0 - \varepsilon_\infty} = -\int_0^\infty \mathrm{e}^{-\mathrm{j}\omega\tau}\, \dot{\gamma}(\tau)\, \mathrm{d}\tau$$

$$= \frac{1}{1 + \mathrm{j}\omega\tau_\mathrm{r}}$$

The value of $(\varepsilon_0 - \varepsilon_\infty)$ depends only on the relative probability of occupation of the sites in thermal equilibrium in a constant applied field. Provided that the difference in free energy between the sites, due to the field, is given only by the potential energy difference qEd, the ratio of the probabilities of occupation of the sites is

$$p_+/p_- = \exp(-qEd/kT)$$

where the subscript + refers to the site of higher energy. Then $p_+ = x^{-1}(x + x^{-1})^{-1}$ and $p_- = x(x + x^{-1})^{-1}$, where $x = \exp(qEd/2kT)$, and the expectation value of polarization is

$$\tfrac{1}{2}qd(p_- - p_+) = \tfrac{1}{2}qd \tanh(qEd/2kT)$$

$$\approx \boldsymbol{E}q^2 d^2/(4kT) \quad \text{for } qEd \ll kT$$

Alternative treatment. In some simple models the relaxation behaviour can be analysed directly. For example, suppose we have the above system polarized in field $\boldsymbol{E}$ and that at time $t = 0$ the field is removed. The probabilities p_+ and p_- will then drift towards equality. The probability that a particle should not have left its initial site at all over time t decreases as $\exp(-\nu t)$. Allowing for the two-way flow between the sites we have then, in the absence of the field,

$$\frac{\mathrm{d}p_+}{\mathrm{d}t} = \nu(p_- - p_+)$$

$$\frac{\mathrm{d}p_-}{\mathrm{d}t} = \nu(p_+ - p_-)$$

so

$$\frac{\mathrm{d}}{\mathrm{d}t}(p_- - p_+) = -2\nu(p_- - p_+)$$

and the expectation polarization, which is proportional to $p_- - p_+$, decays as $\langle \boldsymbol{P}(t)\rangle = \boldsymbol{P}(0)\exp(-2\nu t)$.

Asymmetric two-site model. If the two sites are not equivalent they may be unequally populated in equilibrium under no field. This will in general give rise to a temperature-dependent polarization in zero field, that is, to a variety of pyroelectricity. A set of pyroelectric elements can be arrayed in a material so that the behaviour cancels, whereas they all contribute to the polarization response to an electric field.

To discuss the relaxation, the asymmetry may be expressed by introducing unequal f_+, f_- for the probabilities per unit time of vacating occupied $+$ and $-$ sites respectively. Then the variation of the occupation probabilities is given by

$$\frac{\mathrm{d}p_+}{\mathrm{d}t} = -f_+p_+ + f_-p_-$$

$$\frac{\mathrm{d}p_-}{\mathrm{d}t} = f_+p_+ - f_-p_-$$

so

$$\frac{\mathrm{d}}{\mathrm{d}t}(f_+p_+ - f_-p_-) = -(f_+ + f_-)(f_+p_+ - f_-p_-)$$

and the variable $\pi = f_+p_+ - f_-p_-$ decays to zero with relaxation time $\tau_r = (f_+ + f_-)^{-1}$. Since $p_+ + p_-$ is constant (unity), any other linear combination of p_+ and p_- decays to a finite value with the same single relaxation time. In particular, the polarization so decays.

Relaxation in constant field. The relations of the last section do not depend on the source of asymmetry and will serve equally well to describe the relaxation even if the asymmetry is wholly due to the potential energy difference in a constant applied field. In that case the equilibrium value of p_+/p_- is $\exp(-qEd/kT)$, and the rate equations show that in the steady state, for which $\pi = 0$, p_+/p_- is f_-/f_+.

Moreover, in the region of linear response, τ_r must remain the same for relaxation in any constant field, so the sum $(f_+ + f_-)$ is constant and equal to 2ν. In other words f_+ and f_- change by equal and opposite amounts on the application of a field. This remains true even in the presence of other sources of asymmetry.

Reversible dipole. There is an interesting additional asymmetry in a system first discussed by Fröhlich.[21] We consider a linear molecule constrained by strong local forces due to its neighbours, which in the absence of an external field permit it to oscillate with frequency ν about two oppositely oriented equilibrium positions, making comparatively infrequent jumps from one equilibrium attitude to the other.

If the applied field acts along the equilibrium axis, there is a direct correspondence with the previous model. We have now two equal and opposite charges exchanging their positions at each reversal of the molecule, and the polarization alternates between qd and $-qd$, where the dipole moment μ of the molecule is equal to qd.

The new asymmetry enters[22] because the frequency is sensitive to the longitudinal field $\boldsymbol{E}$. The torque $\mu \boldsymbol{E} \sin \theta$ exerted for an angular deviation θ from the equilibrium line is a restoring torque in one orientation and a deviating torque in the other. The actual squared frequencies of oscillation about the two positions are $(\nu^2 \pm \mu E/4\pi^2 I)$, where I is the molecular moment of inertia. At high quantum numbers ($kT \gg h\nu$) the partition function for a two-dimensional classical oscillator is just $[kT/(h \times \text{frequency})]^2$. Taking into account that the mean axial dipole in the low-frequency orientation is reduced by the increased values of θ corresponding to the lower frequency, $\langle \boldsymbol{P} \rangle$ in field $\boldsymbol{E}$ is given by

$$\langle \boldsymbol{P} \rangle = \boldsymbol{E}\mu^2 \left[\frac{1}{kT} - \frac{1}{I\nu^2}\left(1 - \frac{1}{4\pi^2}\right) + \frac{1}{kT} \times (\text{higher powers of } kT/I\nu^2)\right]$$

The first term corresponds exactly to the previous model. The second is evaluated correctly only for small $kT/I\nu^2$, in which case it is itself relatively small.

The frequencies f_+ and f_- are in this case represented by

$$f_+ = (\nu^2 - \mu E/4\pi^2 I)^{\frac{1}{2}} \exp(-U/kT) \exp(\mu E/2kT)$$
$$f_- = (\nu^2 + \mu E/4\pi^2 I)^{\frac{1}{2}} \exp(-U/kT) \exp(-\mu E/2kT)$$

where any details we may add to the model jump mechanism are built into the structure of U. This quantity may be considered as the free energy of activation required for a jump from one orientation to the other at $\boldsymbol{E} = 0$. In a simple model, U may be taken independent of T. This would be improbable in a real material, where U will normally depend on both temperature and pressure.

Exponential decay. The pure exponential relaxation of the models just considered results from two simplifications. The more important is our assumption that the jump process is negligibly rapid. This is already inapplicable to the transverse relaxation of the Fröhlich model, which is obviously a resonant process dominated by the oscillation frequency ν. The other is the restriction to only two sites.

[21] H. Fröhlich, 'Theory of Dielectrics', Oxford University Press, Oxford, 1949.
[22] G. Wyllie, *Progr. Dielectrics*, 1960, **2**, 1.

Three Sites in a Row.– Two relaxation rates. Let the sites be equally spaced in line and denote them by 1, 2, 3 in succession, with occupation probabilities p_1, p_2, p_3. The evolution of the p for this model is described by

$$-\frac{\mathrm{d}p_1}{\mathrm{d}t} = f_1 p_1 - f_{21} p_2$$

$$-\frac{\mathrm{d}p_2}{\mathrm{d}t} = -f_1 p_1 + f_{21} p_2 + f_{23} p_2 - f_3 p_3$$

$$-\frac{\mathrm{d}p_3}{\mathrm{d}t} = f_3 p_3 - f_{23} p_2$$

The polarization $qd(p_3 - p_1)$ does not involve p_2, so we extract from the first and last equations

$$-\frac{\mathrm{d}}{\mathrm{d}t}(f_{23} p_1 - f_{21} p_3) = f_1 f_{23} p_1 - f_3 f_{21} p_3$$

so that if $f_1 = f_3$ the variable $(f_{23} p_1 - f_{21} p_3)$ decays with the single rate constant f_1. If moreover $f_{23} = f_{21}$, the polarization decays with that single rate coefficient.

If these conditions on the f are not satisfied, or the sites are unequally spaced, we find, in general, two relaxation times: the deviations of the p from their equilibrium values contain two terms of the form $\exp(-rt)$ where the values of r are solutions of

$$[r - \tfrac{1}{2}(f_1 + f_3 + f_{21} + f_{23})]^2 = \tfrac{1}{4}(f_1 - f_3 + f_{21} - f_{23})^2 + f_{21} f_{23}$$

The simplest special case is obtained when all f are equal, giving $r = f$ and $3f$. Then $(p_3 - p_1)$ decays at rate f and $(p_3 + p_1)$ at $3f$. Both decays appear in the polarization if the site separation d_{12} is not equal to d_{23}. More elaborate cases of this sort of multiple relaxation, on rings as well as rows of sites, have been analysed by Hoffman and Axilrod,[23–25] and other workers.[26, 27]

Continuous Diffusion. If there are many sites available to the charge carrier or many orientations accessible to the turning dipole, discussion of the motion among the individual sites becomes intolerably complex (though accessible to mechanized calculation). If the sites are very numerous and close together, the random walk of the charge carrier can with propriety be approximated as a continuous diffusion process. The discrete occupation probabilities are replaced by a probability density function $p(x)$ such that the probability of occupation of a region R is $\int_R p(x)\,\mathrm{d}x$, where x represents the relevant position co-ordinates.

23 J. D. Hoffman, *J. Chem. Phys.*, 1955, **23**, 1331.
24 J. D. Hoffman and B. M. Axilrod, *J. Res. Nat. Bur. Stand.*, 1955, **54**, 357.
25 B. M. Axilrod, *J. Res. Nat. Bur. Stand.*, 1956, **56**, 81.
26 L. Onsager and L. K. Runnels, *J. Chem. Phys.*, 1969, **50**, 1089.
27 G. Williams and M. Cook, *Trans. Faraday Soc.*, 1971, **67**, 990.

The evolution of $p(x)$ as a result of the random walk is then expressed by a diffusion equation whose simplest form is

$$\frac{\partial p}{\partial t} = D\frac{\partial^2 p}{\partial x^2}$$

The only characteristic of the carrier motion embodied in this equation is the mean square displacement per unit time, which gives D with a simple numerical factor dependent on the number of degrees of freedom summarized in x. All other detail of the motion is lost, and solutions of the diffusion equation represent the true evolution of $p(x)$ only for times long compared with the jump time.

The one-dimensional diffusion equation given above has a set of solutions of the form

$$p \propto \exp(jkx)\exp(-k^2 Dt)$$

Thus a density distribution harmonic in x decays, without change of shape, with a single relaxation time which varies inversely as the square of the wave number.

Relaxation by rectilinear diffusion. To find relaxation rather than conduction, the range of diffusion must be restricted, say along a straight track from $x = -d$ to $x = +d$. We take the binding sites to be uniform in character and distribution, so that the equilibrium density function $p_0(x)$ in zero applied field is constant and equal to $(2d)^{-1}$.

In the presence of a field $\boldsymbol{E}$ in the positive direction of x, the potential energy of the carrier at point x is $-\boldsymbol{E}qx$, and the equilibrium density function is

$$p_0(x, \boldsymbol{E}) = p_0 \exp(\boldsymbol{E}qx/kT)$$

where p_0 is fixed by the normalizing condition

$$\int_{-d}^{d} p_0(x, \boldsymbol{E})\,\mathrm{d}x = 1$$

This is the density distribution of the 'isothermal atmosphere', and there are two simple limiting cases: (*i*) If $q\boldsymbol{E}d \gg kT$, we have $d \gg \delta$ where $\delta = kT/q\boldsymbol{E}$ is the distance in which the potential energy changes by kT. Then the charge has 95% probability of being within 3δ of $x = d$ and the relaxation behaviour is very non-linear in $\boldsymbol{E}$. (*ii*) If $q\boldsymbol{E}d \ll kT$, the exponential is well enough approximated by

$$\begin{aligned} p_0(x, \boldsymbol{E}) &= (2d)^{-1}(1 + \boldsymbol{E}qx/kT) \\ &= p_0(x, 0) + \boldsymbol{E}qx/(2dkT) \end{aligned}$$

and the relaxation of this when $\boldsymbol{E}$ is switched off can be followed.

The density p is equivalent to the concentration c at x for a large number of non-interacting particles, and the boundary conditions on p at the impenetrable barriers at $x = \pm d$ are the conditions on c for zero flux – that the gradient should be zero. We must have $\partial p/\partial x = 0$ at $x = d$ and $-d$.

The initial distribution can be expanded in terms of $\sin kx$, where the boundary conditions force us to choose $\cos kd = 0$, that is, $kd = (2n + 1)\pi/2$. Then as each component relaxes, the distribution at time t is given by

$$p(t, x) = \frac{1}{2d} + \frac{4Eq}{\pi^2 kT} \sum_{n=0}^{\infty} \frac{(-1)^n}{(2n+1)^2} \sin \frac{(2n+1)\pi x}{2d} \times \exp\left[-\frac{(2n+1)^2 \pi^2}{4d^2} Dt\right]$$

The initial polarization in field E is given by

$$\int_{-d}^{d} qxp_0(x, E)\,\mathrm{d}x = Eq^2 d^2/3kT$$

The relaxing polarization is

$$\frac{4Eq^2}{\pi^2 kT} \cdot \frac{8d^2}{\pi^2} \sum_{n=0}^{\infty} (2n+1)^{-4} \exp\left[-\frac{(2n+1)^2 \pi^2}{4d^2} Dt\right]$$

Since the coefficients in this sum fall away so rapidly, one would hardly observe more than two or three relaxation times. The low-frequency behaviour is dominated by the first relaxation time $\tau_1 = 4d^2/\pi^2 D$.

Rectilinear movement of the charge carrier does not correspond directly to diffusive rotation of a rigid dipole. Such rotation in a fixed plane may be described by the diffusion of (equal and opposite) charges round a fixed circle. The relaxation in fact takes a simpler course in these conditions.

Diffusion on a circular track. Consider charges $+q$, $-q$ at opposite ends of a rotating diameter of a fixed circle of diameter d. The rotating dipole qd has potential energy $-qdE\cos\theta$ when inclined at angle θ to the direction of an electric field E (taken in the plane of the circle). The equilibrium density function $p_0(\theta, E)$ is then $(2\pi)^{-1}\exp(qdE\cos\theta/kT)$ where θ ranges from 0 to 2π. Provided $qEd \ll kT$ we have approximately

$$p_0(\theta, E) = \frac{1}{2\pi} + \frac{qdE}{2\pi kT}\cos\theta$$

The component dipole moment parallel to E is $qd\cos\theta$, so the expectation value of the polarization is

$$E\frac{q^2 d^2}{2\pi kT}\int_0^{2\pi} \cos^2\theta\,\mathrm{d}\theta$$

$$= E\frac{q^2 d^2}{2kT}$$

The diffusion equation remains simple in form

$$\frac{\partial p}{\partial t} = D\frac{\partial^2 p}{\partial \theta^2}$$

and a p proportional to $\cos\theta$ decays exponentially with rate constant D, so that

$$p(t, \theta) = \frac{1}{2\pi} + \frac{qdE}{2\pi kT}\cos\theta \exp(-Dt)$$

Evidently one obtains faster decays from the higher terms in the expansion of the exponential, but these contain higher powers of $\boldsymbol{E}$ and fall outside the bounds of linear response. The argument extends easily to a third dimension.

Diffusion on a spherical surface. Introducing the usual spherical polar co-ordinates we obtain the uniform probability density in solid angle $p_0(\theta, \phi, 0) = (4\pi)^{-1}$, when θ ranges from 0 to π and ϕ from 0 to 2π. In equilibrium with a field $\boldsymbol{E}$

$$p_0(\theta, \phi, \boldsymbol{E}) = p_0 \exp(qdE\cos\theta/kT)$$

$$\approx \frac{1}{4\pi} + \frac{qdE}{4\pi kT}\cos\theta \quad \text{for small } \boldsymbol{E}$$

Since p remains independent of ϕ, the diffusion equation has the form

$$\frac{\partial p}{\partial t} = \frac{D}{\sin\theta}\frac{\partial}{\partial\theta}\left(\sin\theta\frac{\partial p}{\partial\theta}\right)$$

and the component of p proportional to $\cos\theta$ decays exponentially with rate constant $2D$, while the initial polarization is $\boldsymbol{E}q^2 d^2/(3kT)$.

Pure rotary diffusion of rigid dipoles in two or three dimensions, then, gives exponential decay of polarization with a single relaxation time, provided the sites are uniformly distributed and D is constant. The description of the motion in terms of D alone breaks down, as we shall see, for very short times. A three-dimensional rigid body in any case executes a more complex motion. Even an internally uniform model of rectilinear charge-carrier diffusion automatically shows multiple relaxation. More realistic models must take account of the dynamics of molecular motion.

Random Motion of Objects with Inertia.—It has not so far been necessary to consider any explicit time-dependence of the forces, other than $\boldsymbol{E}$, acting on the moving charges. Further progress necessarily involves considering these forces and the inertia of the charge carriers.[28, 29] In view of the cardinal role of the momentum in dynamical discussions, it will occasionally be convenient to discuss the momentum auto-correlation function. The latter is indeed well defined even for free-Brownian motion, for which the position correlation function has no interest.

As in the last section, some aspects of relaxation behaviour are clear in the bounded rectilinear motion of a charged particle. We consider first

[28] S. Chandrasekhar, *Rev. Mod. Phys.*, 1943, **15**, 1.
[29] 'Noise and Stochastic Processes', ed. N. Wax, Dover, New York, 1954.

the idealized representation of Brownian motion given by the Langevin equation.

Linear Motion of a Brownian Particle. In the simplest case of Brownian motion, a massive particle is immersed in a medium of lighter particles whose rapid thermal motion produces a quickly fluctuating force on the massive Brownian particle. This force will be in part correlated with the motion of the Brownian particle itself. Langevin's simplifying hypothesis is that the correlated part of the force exerted by the medium is proportional and opposed to the velocity u of the particle. Langevin's equation of motion then has the form

$$\dot{u} + \beta u = A(t)$$

where $A(t)$ is a random function of time with zero expectation value and independent of the value of u. The over-simplification in this hypothesis is primarily the assumption that the correlated part of the reaction of the medium is independent of the previous values of u. It will be necessary to relieve that assumption later. Langevin's approximation is already sufficient to obtain a series of instructive results.

Direct integration of the equation of motion yields

$$u(t)\,e^{\beta t} = u(0) + \int_0^t A(t')\,e^{\beta t'}\,dt'$$

Averaging over many possible realizations of the random motion between 0 and t gives

$$\begin{aligned}\langle u(t)\,e^{\beta t}\rangle &= u(0) + \int_0^t \langle A(t')\rangle\,e^{\beta t'}\,dt' \\ &= u(0)\end{aligned}$$

so

$$\langle u(0)\,u(t)\rangle = \langle u(0)^2\rangle\,e^{-\beta t}$$

Since the average is to be taken over the unperturbed thermal distribution, $\langle u(0)^2\rangle = kT/m$ where m is the mass of the Brownian particle, so

$$\langle u(0)\,u(t)\rangle = (kT/m)\exp(-\beta t)$$

From the physical picture of Brownian motion, one would (rightly) expect another characteristic time, namely, the correlation time of the random acceleration $A(t)$. This may be formally defined by

$$\tau_A = \int_0^\infty \langle A(0)\,A(t)\rangle\,dt/\langle A^2\rangle$$

However, full analysis shows that Langevin's assumption that the damping force has no memory implies that $\langle A(0)\,A(t)\rangle = 0$ for $t > 0$. Thus if Langevin's equation appears valid within the limits of observation, τ_A is unobservably short. If, on the other hand, the particle of interest is not much more massive than the particles of the medium, we shall find $\beta\tau_A \sim 1$ and memory effects cannot be entirely neglected.

To begin with, we nevertheless consider a motion of Langevin type in a potential well. The discussion is completely elementary in the case of a harmonic oscillator potential without damping.

Undamped motion in an oscillator potential. If the equation of motion of an oscillating particle is $\ddot{x} = -\omega^2 x$ we have, for initial position $x(0)$ and initial velocity $u(0)$

$$x(t) = x(0) \cos \omega t + \omega^{-1} u(0) \sin \omega t$$

so

$$\langle x(0)\, x(t)\rangle = \langle x(0)^2\rangle \cos \omega t + \omega^{-1} \langle x(0)\, u(0)\rangle \sin \omega t$$

and we need only observe that $u(0)$ and $-u(0)$ are equally probable for any given $x(0)$ to obtain

$$\frac{\langle x(0)\, x(t)\rangle}{\langle x(0)^2\rangle} = \cos \omega t$$

It should be noted that only one property of the equilibrium distribution, $\langle x(0)\, u(0)\rangle = 0$, is needed for this result, which would be valid, for example, if the initial distribution were arbitrary in position and even, but otherwise arbitrary, in velocity.

Unless a system contains velocity-dependent forces, its Hamiltonian contains the velocity only through the kinetic energy, and for all configurations the thermodynamic-equilibrium density function is an even function of the velocity. Thus zero correlation of position and velocity in the initial distribution is a rather common property.

Velocity and acceleration, however, may be correlated – they certainly are in a Brownian motion – and the velocity self-correlation has therefore in many situations a non-zero initial slope. This is so even for a particle moving quite freely between reflecting walls.

Undamped thermal motion between reflecting boundaries. A full account of this can be constructed beginning with the evolution of a puff of gas released at an arbitrary point at an instant, say $t = 0$. Disregarding for the moment the reflecting boundaries, we know that a molecule with velocity u travels a distance ut in t. The distribution of u is Maxwellian, that is, Gaussian, so after time t the space distribution of the gas is Gaussian. The instantaneous form of this distribution is identical with one which might have arisen by random diffusion, but the mean square deviation $\langle u^2\rangle\, t^2 = t^2\, kT/m$ grows as t^2 instead of t.

The effect of the reflecting walls is fully described by folding the Gaussian back on itself at each wall in turn, or algebraically by taking the appropriate series of images. The space distribution for a given initial position is then given by a sum of Gaussian functions, with no sign of exponential relaxation.

The momentum self-correlation for this motion, with walls separated by L, has been calculated by Nossal,[30] who gives

$$\frac{\langle u(0)\,u(t)\rangle}{\langle u^2\rangle} = \frac{\langle p(0)\,p(t)\rangle}{mkT} = \frac{2}{\pi^2}\sum_{n=0}^{\infty}(n+\tfrac{1}{2})^{-2}\exp\left[-\frac{2\pi^2 kT}{mL^2}(n+\tfrac{1}{2})^2 t^2\right] - \frac{8t^2 kT}{mL^2}\sum_{n=0}^{\infty}\exp\left[-\frac{2\pi^2 kT}{mL^2}(n+\tfrac{1}{2})^2 t^2\right]$$

This expansion is ill behaved for small t, and it is in fact easy to show that for small positive values of t

$$\frac{\langle p(0)\,p(t)\rangle}{mkT} = 1 - 4\sqrt{\frac{2}{\pi}}\sqrt{\frac{kT}{m}}\frac{t}{L} - \ldots$$

The awkwardness of Nossal's expression at very small t then arises from fitting a finite slope at $t = 0$ with functions which are all flat at that point.

The straight line obtained from the two terms exhibited in the last equation is very close to Nossal's computed points down to the first zero of the correlation function. Beyond that zero the points lie equally close to the curve given by the lowest order terms ($n = 0$) of Nossal's expansion.

Recalling that in any steady state of the system

$$\langle\phi(-t)\,\phi(0)\rangle = \langle\phi(0)\,\phi(t)\rangle$$

simply by time displacement, the momentum correlation function for this problem must start

$$\frac{\langle p(0)\,p(t)\rangle}{mkT} = 1 - 4\sqrt{\frac{2}{\pi}}\sqrt{\frac{kT}{m}}\frac{|t|}{L} - \ldots$$

with a discontinuous first derivative at $t = 0$. The discontinuity arises in this case because a number of particles (those near to and approaching a boundary) proportional to δt undergo momentum reversal in any δt, however small. This results from the hypothesis of hard walls, which exert infinite forces. Softening of the walls in any realistic way flattens the correlation function at zero time. We see, however, that a discontinuity of the slope of the velocity self-correlation at zero time survives in Brownian motion of Langevin type, reinforcing one's suspicion of an artificiality in Langevin's assumptions, at least for very high frequency components of the random force. The position self-correlation function looks quite well-behaved under Langevin's assumptions.

Brownian motion in an oscillator potential. We consider a Brownian motion of Langevin type with a damping constant β independent of position, so that the residual random acceleration behaves as white noise, independent of both position and velocity. The Langevin equation is

$$\ddot{x} + \beta\dot{x} + \omega_0^2 x = A(t)$$

and the average position at time t of a particle which was at $x(0)$ with

[30] R. Nossal, *J. Math. Phys.*, 1965, **6**, 193.

velocity $u(0)$ at $t = 0$ is given by

$$x(t) = x(0)\,\mathrm{e}^{-\beta t/2}\cosh(\beta_1 t/2) + \beta_1^{-1}[\beta x(0) + 2u(0)]\,\mathrm{e}^{-\beta t/2}\sinh(\beta_1 t/2)$$

since the expectation of $\int_0^t A(t)\,\mathrm{d}t$ is zero. The value of β_1 is $(\beta^2 - 4\omega^2)^{\frac{1}{2}}$. In the case of underdamping, $\cosh(\beta_1 t/2)$ goes over to $\cos\omega_1 t$ and $\beta_1^{-1}\sinh(\beta_1 t/2)$ to $(2\omega_1)^{-1}\sin\omega_1 t$ with $\omega_1 = (\omega_0^2 - \beta^2/4)^{\frac{1}{2}}$: in critical damping $\cosh(\beta_1 t/2)$ goes to 1 and $\beta_1^{-1}\sinh(\beta_1 t/2)$ to $t/2$.

As before, $x(0)$ and $u(0)$ are uncorrelated and

$$\gamma(t) \equiv \frac{\langle x(0)\,x(t)\rangle}{\langle x^2\rangle} = \mathrm{e}^{-\beta t/2}[\cosh(\beta_1 t/2) + \beta\beta_1^{-1}\sinh(\beta_1 t/2)]$$

Expansion in powers of t verifies at once that the linear term vanishes, and $\gamma(t) = 1 - \frac{1}{2}\omega_0^2 t^2$ + higher powers of t for all three cases. Thus in this system the initial behaviour of the normalized correlation function depends neither on the damping nor the temperature, but only on the potential-energy function.

In underdamped motion, the popular functional form $\cos\omega_1 t\exp(-\alpha t)$ differs only in phase from the true correlation function, whose zeros are at $t = \omega_1^{-1}\tan^{-1}(-2\omega_1/\beta)$, so that the first zero occurs somewhat after the first quarter-cycle. The velocity auto-correlation function, which we shall need later, is obtained by differentiating γ twice and normalizing, giving

$$\frac{\langle u(0)\,u(t)\rangle}{\langle u^2\rangle} = \mathrm{e}^{-\beta t/2}[\cosh(\beta_1 t/2) - \beta\beta_1^{-1}\sinh(\beta_1 t/2)]$$

Brownian motion between reflecting walls. This problem has been discussed by Oppenheim and Mazur.[31] It is harder to visualize than the undamped motion since the Brownian particle may reverse its travel so as to return several times for reflection at one boundary before staggering off to the other. Nevertheless, if we consider a particular random walk executed by a particle starting at $t = 0$ at position $x(0)$ with velocity $u(0)$, we see that the sections of this walk between successive boundary reflections may be regarded as selected from one random walk executed from $x(0)$, $u(0)$ in the absence of boundaries, the same walk executed from displaced starting points, and the reverse walk, with initial velocity $-u(0)$, from another set of displaced starting points. If $p_0[x, t; x(0), u(0)]$ is the probability density in x for the particle at time t, with no reflecting boundaries, the probability density within boundaries at $x = 0$ and L is

$$p[x, t; x(0), u(0)] = \sum_{n=0}^{\infty}\{p_0[x, t; x(0) + 2nL, u(0)] + p_0[x, t; -x(0) + 2nL, -u(0)]\}$$

which is evidently periodic in x and $x(0)$ with period $2L$.

For a motion of the simple Langevin type, p_0 is well known to be the Gaussian $(b/\pi)^{\frac{1}{2}}\exp(-bl^2)$ where

$$l = x - x(0) - u(0)\,\beta^{-1}(1 - \mathrm{e}^{-\beta t})$$

[31] I. Oppenheim and P. Mazur, *Physica*, 1964, **30**, 1833.

and

$$b/(m\beta^2) = [2kT(2\beta t - 3 + 4e^{-\beta t} - e^{-2\beta t}]^{-1}$$

so the displacement $[x - x(0)]$ is symmetrically distributed around $u(0)\,\beta^{-1}[1 - \exp(-\beta t)]$ and, for given $u(0)$,

$$\langle[x - x(0)]^2\rangle = u(0)^2\,\beta^{-2}[1 - \exp(-\beta t)]^2 + \langle l^2\rangle$$

The final average over values of $u(0)$ gives

$$\langle[x - x(0)]^2\rangle = (2kT/m\beta^2)\,[\beta t - 1 + \exp(-\beta t)]$$

$$\equiv 2\Delta$$

The distribution is still normal, so the contracted density function

$$p_0[x, t; x(0)] = (4\pi\Delta)^{-\frac{1}{2}} \exp\{-[x - x(0)]^2/4\Delta\}$$

For large times, $\beta t \gg 1$, $\Delta \sim Dt$, where $D = kT/m\beta$ is the normal diffusivity of the particle, and p_0 for large t is the characteristic solution of the ordinary diffusion equation. For small times, $\beta t \ll 1$, we express Δ in powers of t, so that

$$\langle[x - x(0)]^2\rangle \approx (kT/m)\,t^2 = \langle[u(0)\,t]^2\rangle$$

since the perturbing forces have not had time to act.

It is clear that a sine-wave probability density of wavelength λ relaxes as $\exp(-4\pi^2\,\Delta/\lambda^2)$, so that the relaxation function for the confined Brownian particle, by analogy with ordinary diffusion, is

$$\gamma(t) = 96\pi^{-4} \sum_{n=0}^{\infty} (2n + 1)^{-4} \exp[-(2n + 1)^2\,\pi^2\,\Delta/L^2]$$

In view of the exponential in Δ, this is not a sum of exponential decays until we reach $\beta t \gg 1$, by which time the first term alone survives significantly.

The functional form $\exp(-\Delta/L^2)$ is attractive for a correlation function, giving a neat transition from the approximately Gaussian form, determined by the velocity distribution, for small times to exponential decay, determined by the damping, for long times. This form recurs in the consideration of rotary motion.

Brownian motion on circle and sphere. The equilibrium density on a circle or a sphere deviates from uniformity by an amount proportional to $\cos\theta$, where θ is the angle between the dipole and the field. For the circular motion, that is, a dipole confined to rotate in a plane, the cosine retains its form just as in the rectilinear motion discussed above, and relaxes in amplitude to give

$$\gamma(t) = \exp[(1 - \beta t - e^{-\beta t})\,kT/I\beta^2]$$

For small times, $\gamma \approx 1 - t^2\,kT/2I$, which corresponds to

$$\langle\cos\theta\rangle \approx 1 - \tfrac{1}{2}\langle\theta^2\rangle, \quad \text{with} \quad \langle\theta^2\rangle = \langle\dot\theta^2\rangle\,t^2$$

for small enough values of t. For large times, decay is exponential with $\tau_r = I\beta/kT$, the familiar Debye relaxation. A full discussion of this behaviour has been given by Sack.[8, 32]

The limiting form of this $\gamma(t)$ as β tends to 0 is $\exp(-t^2 kT/2I)$, which is obtained directly by averaging $\cos \omega t$ over all values of ω with weight proportional to $\exp(-I\omega^2/2kT)$.

The change of geometry from circle to sphere simply imports a factor of 2 in the decay rate for normal diffusion. We might expect that, at least for large enough β ($\beta^2 > kT/I$), we should have for Brownian motion on a sphere

$$\gamma(t) \approx \exp[2(1 - \beta t - e^{-\beta t})kT/I\beta^2]$$

but this does not have the correct limit as β goes to zero, except for small times.

Brownian Motion of a Rigid Body. Many molecules of interest are extended in three dimensions with three principal moments of inertia of comparable size. Gyroscopic forces complicate the discussion of Brownian motion for such bodies and a number of workers have developed convenient formal treatments without much physical novelty emerging. Steele[33] has given a clear treatment of this problem, but obtains tractable expressions only for spherical-top molecules, which have the same moment of inertia I about all axes through the molecular mass centre, so that the Euler equations of motion fall apart in independent variables.

A spherical-top molecule need not be geometrically spherical, so there may be three distinct principal axes for the viscous torque on the rotating molecule. Steele thus obtains three equations of motion

$$I\dot{\omega}_i = -\xi_i \omega_i + (\text{random torque})$$

of Langevin form, and through a diffusion equation, for not too small ξ, an auto-correlation function which is in turn of simple form only for $\xi_x = \xi_y$,

$$\gamma(t) = \cos^2\eta \exp(-2h_x) + \sin^2\eta \exp[-(h_x + h_z)]$$

where

$$h_i = (IkT/\xi_i^2)[\xi_i t/I + \exp(-\xi_i t/I) - 1]$$

and η is the angle between the z axis and the dipole axis in the molecule. We have thus a superposition of relaxations of the form found in the last section.

In the limit $\xi = 0$, on the other hand, we have continued free rotation of the spherical top, with fixed angular momentum. The dipole axis moves steadily on the surface of a fixed circular cone, and the limit of $\langle \boldsymbol{\mu}(0)\cdot\boldsymbol{\mu}(t)\rangle$ as $t \to \infty$ is $\mu^2/3$. Steele finds, for free rotation, the correlation

[32] R. A. Sack, *Physica*, 1956, **22**, 917.
[33] W. A. Steele, *J. Chem. Phys.*, 1963, **38**, 2404, 2411.

function

$$\gamma(t) = \frac{1}{3} + \frac{2}{3}\left(1 - \frac{kT}{I}t^2\right)\exp\left(-\frac{kT}{2I}t^2\right)$$

– whose second term corresponds rather closely to the leading pair of terms in Nossal's correlation function for free motion between reflecting walls. Steele's function has one minimum at $t \approx 1.7(I/kT)^{\frac{1}{2}}$. For small t, this γ is again approximated by $1 - t^2 kT/I$, in agreement with the contribution of two components of angular velocity to depolarization. The spin round the dipole axis makes no contribution.

The correlation function for free rotation of a rigid linear molecule goes to zero at large times, since the axis of rotation is always perpendicular to the dipole, and is given by

$$\gamma(t) = 1 - (kT/I)\, t \exp(-\, t^2 kT/2I) \int_0^t \exp(\tau^2 kT/2I)\, d\tau$$

For small t this is again just $(1 - t^2 kT/I)$. For larger t it approaches zero from below. Gordon has shown [34] that successive integration by parts yields the useful asymptotic form

$$\gamma(t) \sim \sum_{m=1}^{\min} 1, 3, \ldots (2m - 1)\,(t^2 kT/I)^{-m}$$

Here the summation is terminated at the minimum term, the series being clearly divergent for any given t.

For the free motion of a general rigid body with unequal non-zero principal moments of inertia, any particular state of motion, with constant energy and angular momentum, is describable by the rolling of a certain cone fixed in the body on another fixed in space. The cones do not generally have a circular right section, and the motion of the dipole axis has a large content of harmonics. Unless the molecule has a very special shape – such as a laminar symmetric top – averaging over all values of angular momentum will give a rather smooth correlation function.

In examining experimental results, the crude device of superposing not more than three sphere free-rotation functions, with values of I within the known range for the molecule, may give a useful trial function.

Scattered rotation of a linear molecule. Gordon [34] has determined the dipole self-correlation for linear molecules rotating freely between collisions of negligible duration which occur at random with mean period τ and either (*i*) randomize the direction of the rotation axis in the equatorial plane of the molecule without changing the kinetic energy, or (*ii*) randomize both the direction of the rotation axis and, preserving on average the appropriate thermal distribution law, the kinetic energy of rotation.

If τ is very short compared with the mean period of rotation, each case reduces to diffusion and the correlation function is that for a rotary

[34] R. G. Gordon, *J. Chem. Phys.*, 1966, **44**, 1830.

Brownian motion. If τ is at least of the order of the mean half-period of rotation, the correlation function acquires a region of negative values corresponding to that found for completely free rotation. In addition, the tendency of rapidly rotating molecules to resist scattering may be built in to model (*i*) by making τ increase with the angular speed ω as $\tau_0(1 + b\omega^4)$, with the result that the computed correlation function shows a small maximum following the negative region. Comparison with experimental results of Vodar and his co-workers,[35] for CO in Ar and for CO gas and liquid, gives very good agreement for reasonable values of τ_0 and b.

Scattered rotation of a spherical top. McClung[36] has extended Gordon's treatment to this type of motion, obtaining correlation functions of the spherical harmonics of the angular co-ordinates of an axis fixed in the rotating molecule. The massive expansions obtained are intractable except by computation. Their Fourier transforms are perhaps more transparent. McClung has concentrated on the behaviour near zero frequency, to obtain the long-period relaxation times, and shows how, in the transition from Debye (diffusive) motion to perturbed free rotation, the dipole relaxation time passes from inverse to direct dependence on the relaxation time for the angular momentum vector.

In a later paper McClung[37] presents experimental evidence on the correlation function for the transition dipole of an i.r. band in methane. In pure liquid methane, mechanism (*ii*) above gives good approximation to the experimental results, and (*i*) is definitely unsatisfactory: in solutions of methane in argon, xenon, and krypton, the reverse is the case.

A very useful mathematical paper by Fixman and Rider[38] follows a general discussion of the motion of a symmetric top with computed results for spherical top and linear molecules, using a scattering model which includes Gordon's (*i*) and (*ii*) situations as limiting cases. Their $\gamma(t)$ [$C_1(t)$ in their notation] extrapolates back from large t to a value less than unity for the spherical top – a useful property not shared by many well-defined mechanical models.

General Random Rotation in one Plane. Shimizu[39] has pointed out that an analysis by Kubo[40] of the shape of spectral lines can be applied very simply to this situation. If the angular velocity of the rotor at time t is $\omega(t)$, its angular displacement since $t = 0$ is $\theta = \int_0^t \omega(t')\,dt'$ and the normalized auto-correlation function of a vector rotating with it is

$$\langle \cos\theta(t)\rangle \equiv \left\langle \cos\left[\int_0^t \omega(t')\,dt'\right]\right\rangle \equiv \mathrm{Re}\left\langle \exp\left[i\int_0^t \omega(t')\,dt'\right]\right\rangle$$

[35] J. Vu, M. R. Atwood, and B. Vodar, *J. Chem. Phys.*, 1963, **38**, 2671.
[36] R. E. D. McClung, *J. Chem. Phys.*, 1969, **51**, 3842; 1971, **54**, 3248.
[37] R. E. D. McClung, *J. Chem. Phys.*, 1971, **55**, 3459.
[38] M. Fixman and K. Rider, *J. Chem. Phys.*, 1969, **51**, 2425.
[39] H. Shimizu, *J. Chem. Phys.*, 1968, **48**, 2494.
[40] R. Kubo, in 'Fluctuation, Relaxation and Resonance in Magnetic Systems', ed. D. ter Haar, Oliver and Boyd, Edinburgh, 1962, p. 23.

There are at least two averaging processes implied, one over $\omega(0)$ and one over different realizations of $\omega(t')$ between 0 and t. In the presence of position-dependent forces it may be necessary to average over $\theta(0)$ as well.

Following Kubo, we direct attention to the stochastic character of the process $\omega(t')$. Its equilibrium distribution, which is the distribution of $\omega(0)$, has density in ω proportional to $\exp(-\omega^2 I/2kT)$. We have so far considered the time dependence of ω in very simple ways.

The typical term in the expansion of $\langle \exp[i\int_0^t \omega(t')\,dt']\rangle$ has the form

$$\left\langle \mathrm{i}^n(n!)^{-1}\left[\int_0^t \omega(t')\,\mathrm{d}t'\right]^n\right\rangle = \mathrm{i}^n(n!)^{-1}\int_0^t \mathrm{d}t_1 \dots \int_0^t \mathrm{d}t_n \langle\omega(t_1)\dots\omega(t_n)\rangle$$

where the integrand is a multiple self-correlation of ω. These terms can be regrouped to show that

$$\left\langle \exp\left[\mathrm{i}\int_0^t \omega(t')\,\mathrm{d}t'\right]\right\rangle = \exp\left[\mathrm{i}\langle\omega\rangle_\mathrm{c} t - \int_0^t \mathrm{d}t_1\int_0^t \mathrm{d}t_2 \langle\omega(t_1)\,\omega(t_2)\rangle_\mathrm{c} \dots \right.$$
$$\left. + \,\mathrm{i}^n \int_0^t \mathrm{d}t_1 \dots \int_0^t \mathrm{d}t_n \langle\omega(t_1)\dots\omega(t_n)\rangle_\mathrm{c} \dots\right]$$

where $\langle \text{——} \rangle_\mathrm{c}$, called a cumulant average, denotes a combination of the multiple correlations of ω of the same structure as a cumulant or semi-variant of the same order, regarded as a combination of corresponding moments. Kubo emphasizes the important property that if the factors in any cumulant average fall into two or more uncorrelated groups then this quantity vanishes. The explicit form of the cumulant of general order is known. We require here only

$$\langle\omega\rangle_\mathrm{c} = \langle\omega\rangle \quad \text{and} \quad \langle\omega_1\,\omega_2\rangle_\mathrm{c} = \langle\omega_1\,\omega_2\rangle - \langle\omega_1\rangle\langle\omega_2\rangle.$$

If now $\langle\omega(0)\,\omega(t)\rangle$ has a well-defined lifetime so that it is exponentially small for $t > \tau_\mathrm{c}$, the factors in $\langle\omega_1 \dots \omega_n\rangle_\mathrm{c}$ must be uncorrelated unless the interval between the earliest and latest member is less than *ca.* $n\tau_\mathrm{c}$. In the multiple integral, then, provided $t \gg n\tau_\mathrm{c}$, only one co-ordinate retains independent freedom to move through the range from 0 to t and the integral is proportional to t. Provided we can justify neglecting large values of n ($> t/\tau_\mathrm{c}$) we must have

$$\left\langle \exp\left[\mathrm{i}\int_0^t \omega(t')\,\mathrm{d}t'\right]\right\rangle = \exp(at) \quad \text{for } t \gg \tau_\mathrm{c}$$

The constant a is in general complex, with negative real part. It will certainly be real if all the odd correlations of ω vanish, and in that case $\langle\cos\theta(t)\rangle \geqslant 0$ for all t.

Since the expectation value of ω^2 for a single rotational degree of freedom is kT/I, the real part of a is

$$-\,(kT/I)\,\tau_\mathrm{c}\left[1 + \sum_{n=1}^{\infty} O(\tau_\mathrm{c}^2\, kT/I)^n\right]$$

where τ_c is more precisely defined as $\int_0^\infty \langle \omega(0)\, \omega(t)\rangle\, dt/\langle \omega^2 \rangle$. In the limit $\tau_c^2\, kT/I \ll 1$, if a is in fact real, we have $\gamma(t) \approx \exp(-\, t\tau_c\, kT/I)$ for $t \gg \tau_c$. We have, of course, already seen a particular case of this with $\tau_c = \beta^{-1}$. It is interesting to see from another point of view how a system which is in fact rotating with a Gaussian distribution of angular velocities, with breadth of order $(kT/I)^{\frac{1}{2}}$, nevertheless shows a spectral response of relaxation type with characteristic rate $\tau_c\, kT/I$.

Gaussian process. The rotation rate $\omega(t)$ constitutes a Gaussian process if the multivariate density function for $\omega(t_1) \ldots \omega(t_n)$ has Gaussian form. If this is so, all cumulants beyond $\langle \omega_1\, \omega_2 \rangle_c$ vanish and the process ω is statistically determined by its mean (which is zero) and first self-correlation alone. In this case

$$\gamma(t) = \exp\left[-\int_0^t dt_1 \int_0^t dt_2 \langle \omega(t_1)\, \omega(t_2)\rangle\right]$$

$$= \exp\left[-\int_0^t (t - \tau) \langle \omega(0)\, \omega(\tau)\rangle\, d\tau\right]$$

For $t \ll \tau_c$ $\langle \omega(0)\, \omega(\tau)\rangle \approx \langle \omega^2 \rangle = kT/I$ so $\gamma(t) \approx \exp(-\, t^2\, kT/2I)$. For $t \gg \tau_c$ the upper limit of integration may be taken to be infinite and

$$\gamma(t) \sim \exp(-\, t\tau_c\, kT/I)$$

These results verify for the Gaussian process, at least, Kubo's provisional conjectures in the previous section. Setting

$$\psi(\tau) \equiv \langle \omega(0)\, \omega(\tau)\rangle / \langle \omega^2 \rangle = \exp(-\, \beta\tau)$$

regenerates the familiar Sack result, since this is indeed the velocity autocorrelation for a Langevin–Brownian motion. New forms of γ may now be had by using more interesting forms of ψ in Kubo's equation,

$$\gamma(t) = \exp\left[-\, (kT/I) \int_0^t (t - \tau)\, \psi(\tau)\, d\tau\right]$$

Libration in a Constraining Potential. In most crystals and many liquids the mean field of force due to the neighbours of a given molecule tends to hold it in a fixed orientation for longer or shorter times. Superposed on this time-smoothed force field is the rapid fluctuation due to the actual molecular movements. Even in a liquid, it may be worthwhile to approximate the force on a molecule as a random force of simple character superposed on a steady orienting field.

If the constraining field is strong, the molecule will move, in general, as a damped gyrostatic pendulum. If the constraint is highly anisotropic the motion may approach rotation round a fixed axis. Again, if the molecule is nearly a spherical top, its component rotations behave independently.

We consider, for simplicity, one angular co-ordinate obeying the Langevin equation

$$\ddot{\theta} + \beta\dot{\theta} + \omega_0^2\, \theta = A(t)$$

where $\dot{\theta}$ is the angular velocity ω, and ω_0 is the angular frequency of undamped oscillation. The self-correlation of ω is then

$$\frac{\langle\omega(0)\,\omega(t)\rangle}{\langle\omega^2\rangle} = e^{-\beta t/2}[\cosh(\beta_1 t/2) - \beta\beta_1^{-1}\sinh(\beta_1 t/2)]$$

in the overdamped case, and hence

$$\gamma(t) = \exp\left\{-\frac{kT}{I\omega_0^2}\left[1 - e^{-\beta t/2}\left(\cosh\frac{\beta_1 t}{2} + \frac{\beta}{\beta_1}\sinh\frac{\beta_1 t}{2}\right)\right]\right\}$$

and in the case of underdamping

$$\gamma(t) = \exp\left\{-\frac{kT}{I\omega_0^2}\left[1 - e^{-\beta t/2}\left(\cos\omega_1 t + \frac{\beta}{2\omega_1}\sin\omega_1 t\right)\right]\right\}$$

The expansion of the exponent in terms of t begins, as it should, with $-t^2 kT/2I$. Since the dipole stays within thermal range of its initial position, $\gamma(\infty) = \exp(-kT/I\omega_0^2)$, and this value is approached monotonically in the case of overdamping.

Clearly a possible potential for plane rotation is periodic in θ, so even in the unlikely situation that the whole circle contains only one potential minimum, this is separated from itself by a barrier of finite height which one would expect to be some moderate multiple of $I\omega_0^2$. If the motion is to be well described by the relations above, we require that $I\omega_0^2 \gg kT$, *i.e.* a stiff oscillation, and also that the structure of the set of neighbours which determines the size and direction of the orienting field should change with a relaxation rate less than β. The neighbour relaxation will, of course, appear in this model as a contributory part of β.

These correlation functions are double exponentials, and there is no single exponential relaxation so long as the motion is confined to one harmonic oscillator potential well, though, of course, the decay towards $\gamma(\infty)$ will not deviate much from one exponential for $t > \beta^{-1}$. As a rule each molecule will have at least two distinct possible positions of stable equilibrium. In the long run, thermally activated jumps between the potential wells centred on these equilibrium positions must give rise to exponential decay as in the initial two-site model of relaxation.

Simply to multiply correlations, such as have just been derived, by a suitable $\exp(-\alpha t)$ is obviously improper as it gives incorrect behaviour at small times. It may be empirically useful to allow such a correlation to run down nearly to its $\gamma(\infty)$ and then switch in the exponential decay with an arbitrary smoothing at the transition. One way to do this while preserving the initial slope is to multiply by $\exp[-\alpha(\beta t + e^{-\beta t} - 1)]$, but this spoils the initial curvature. In any case, the switching in of the exponential ought to arise naturally from the analysis of the problem.

Libration in a two-well potential. For simplicity we consider two equivalent equilibrium positions separated by π radians. The similar potential wells are taken to be harmonic to a height much greater than kT, and separated by equal potential barriers on either side.

The equilibrium distribution of the dipole will differ negligibly from the distribution over two strictly harmonic potential wells, and the dipole moment correlation function will be initially of the form derived in the last section. Let us call this $\gamma(t)$. As time goes on, the molecule has an increasing probability of escape over the potential barrier. If we regard the probability per unit time of this escape as strictly constant, the condition for which $\gamma(t)$ is the appropriate correlation function has now the diminishing probability $\exp(-ft)$.

If we have one condition with correlation function γ_1 decaying to another with correlation function γ_2, the resultant will, in general, be a convolution. For the present situation, however, γ_1 arises from oscillation in a small range of angles while γ_2 involves complete reversals. Shimizu [15, 39] has observed that in such a case

$$\gamma(t) = \gamma_1 e^{-ft} + \gamma_2(1 - e^{-ft})$$

may be a good approximation to the true correlation. Moreover, since the coefficient of γ_2 is ft for small times, errors in the expansion of γ_2 are shifted one order in time with respect to corresponding terms in γ_1. This observation enables us to use an artificial form for γ_2 without much falsifying γ.

For the present problem, we expect the form of γ_2 to be essentially $\exp(-2ft)$ for long times. An exact initial form of γ_2 is difficult to estimate, but it must begin with rotation through π in a time dependent mainly on the angular speed of the molecule at the crest of the potential barrier. The mean-square speed at this point will have the equipartition value kT/I, so we fit this with the artificial

$$\gamma_2(t) = \exp[-2f(I/kT)^{\frac{1}{2}}(\delta t + e^{-\delta t} - 1)]$$

with $\delta = (kT/I)^{\frac{1}{2}}$. This switches to the correct long-term behaviour in a time of order δ^{-1}, and is unphysical at the t^2 term in its expansion. This fault affects the t^3 term in the expansion of γ.

The consequent form of γ is

$$\gamma(t) = \gamma_1 e^{-ft} + d(e^{-2ft} - e^{-3ft}) \exp[-(2f/\delta) e^{-\delta t}]$$

where $d = \exp(2f/\delta)$. From our previous assumptions, d cannot be very large for a physically reasonable shape of the potential, so the long-term behaviour of γ is as $\gamma_1 \exp(-ft)$ unless $\Gamma_1 \equiv \gamma_1(\infty)$ is very small. But the approximations already used imply that $\Gamma_1 = \exp(-kT/I\omega_0^2)$ is not much less than unity. It is to be noted that $\gamma \sim \Gamma_1 \exp(-ft)$ extrapolates back to Γ_1 which is less than unity. As observed previously, this property is not possessed by the simpler models of molecular motion.

Itinerant Oscillator. The itinerant oscillator model of dielectric response was introduced by Hill.[41] It has considerable attraction as a model of the liquid state, and the name has become well known since Sears [42] used

[41] N. E. Hill, *Proc. Phys. Soc.*, 1963, **82**, 723.
[42] V. F. Sears, *Proc. Phys. Soc.*, 1965, **86**, 953.

it in an analogous discussion of translational motion in argon. Treatments of the model vary in detail,[43-45] but the essential idea is that a central molecule executes damped oscillations in a potential well which itself executes a random motion of, for example, damped but unconstrained rotation.

In one specific model the potential well arises from the interaction of the molecule with its nearest neighbours. These are taken to form a more or less rigid cage which rotates without surface slip in a continuum whose viscous, or viscoelastic, properties are those of the bulk fluid. In this sort of model we note that the exchange of momentum between molecule and cage does not permit the random motions, of molecule with respect to cage and of cage with respect to bulk medium, to be uncorrelated.

A crude calculation[45] of the spectral response of one version of this model gives absorption curves qualitatively resembling several experimental results, in that a resonance absorption appears in the frequency region above the familiar Debye absorption (which in this model corresponds to molecule and cage turning as a nearly rigid object) with a rapid fall-off of absorption at still higher frequencies. If the damping acts only at the outside of the cage (an unlikely circumstance) the frequency-dependent polarizability is in fact equal to

$$\frac{2\mu^2}{3I}\cdot\frac{j\omega\tau}{1+j\omega\tau}\cdot\frac{1-j\sigma\omega_0/\omega}{\omega_0^2-\omega^2+j\sigma\omega\omega_0}$$

where ω_0 is the frequency with which the molecule would oscillate if the cage were fixed, $\tau = \xi/2kT$ where the drag couple on the cage is $-\xi$ times its angular velocity, and $\sigma = I\omega_0/(\xi + jI_c\,\omega)$ where I_c is the moment of inertia of the cage. The relative simplicity of the expression depends once more on assuming $I\omega_0^2 \gg kT$.

This polarizability involves a set of characteristic times $\xi/2kT$, I_c/ξ, $\xi/I\omega_0^2$, and $(I_c/I\omega_0^2)^{\frac{1}{2}}$, between any two of which the correlation function may take a fairly simple form. However, this example indicates that for a linear system which fails to show 'normal mode' behaviour the frequency-dependent admittance may be much more straightforward than the corresponding correlation function.

There is no difficulty in the undamped plane motion which corresponds to the wagging of a rigid dipolar group elastically connected to the rest of a molecule which rotates freely as a whole, since in this case normal modes of motion exist, and are indeed obvious. For simplicity we take the mass centre of the polar group to be fixed in position within the molecule. If the angular position of the molecular body is given by α and that of the

[43] Y. Nakahara and H. Takahashi, *Proc. Phys. Soc.*, 1966, **89**, 747.
[44] P. S. Damle, A. Sjölander, and K. S. Singwi, *Phys. Rev.*, 1968, **165**, 277.
[45] G. Wyllie, *J. Phys.* (*C*), 1971, **4**, 564; *cf.* B. Lassier and C. Brot, *Discuss. Faraday Soc.*, 1969, **48**, 39.

dipole by θ, the constant total angular momentum of the molecule is

$$I\dot{\theta} + I_c\,\dot{\alpha} \equiv (I + I_c)\,\xi$$

The relative motion of body and polar group is given by

$$\ddot{\phi} + \omega_0^2\,\phi = 0$$

where $\phi = \theta - \alpha$, and the effective moment of inertia in this co-ordinate is $I_c I/(I + I_c)$.

In the thermal equilibrium distribution, ϕ and ξ are not correlated, so the correlation function for a linear combination $a\xi + b\phi$ is

$$\frac{a^2\langle\xi(0)\,\xi(t)\rangle + b^2\langle\phi(0)\,\phi(t)\rangle}{a^2\langle\xi^2\rangle + b^2\langle\phi^2\rangle}$$

Since $\theta = \xi + I_c\,\phi/(I + I_c)$, the dipole-correlation function $\langle\cos\theta\rangle$ is then

$$\gamma(t) = \exp\left\{-\frac{kT}{I + I_c}\left[\frac{I_c}{I\omega_0^2}(1 - \cos\omega_0\,t)\right]\right\}\exp\left[-\frac{kTt^2}{2(I + I_c)}\right]$$

if rotation is restricted to one plane. Provided $I\omega^2 \gg kT$, movement in two and three dimensions will change primarily the form of the second exponential factor in γ.

If we suppose that a suitable damping may be introduced without forcing strong correlation between the ξ and ϕ co-ordinates, which can of course be done self-consistently but by a very particular constraint on the damping forces, we obtain

$$\gamma = \gamma_1\gamma_2$$

where γ_1 represents a damped oscillation,

$$\gamma_1(t) = \exp\left\{-\frac{kT}{I + I_c}\cdot\frac{I_c}{I\omega_0^2}\left[1 - e^{-\beta t/2}\left(\cos\omega_1\,t + \frac{\beta}{2\omega_1}\sin\omega_1\,t\right)\right]\right\}$$

and γ_2 a damped rotation

$$\gamma_2(t) = \exp\left[\frac{kT}{(I + I_c)\,\beta^2}(1 - \beta t - e^{-\beta t})\right]$$

At long times, γ then behaves like

$$\gamma_1 \sim \exp\left[\frac{kT}{I + I_c}\left(\frac{1}{\beta^2} - \frac{I_c}{I\omega_0^2}\right)\right]\exp\left(-\frac{t}{\tau_r}\right)$$

with $\tau_r = (I + I_c)\,\beta/kT$. If this γ_1 is extrapolated back to $t = 0$, the value is more or less than unity according as $I\omega_0^2$ is more or less than $I_c\,\beta^2$.

Simple Forms of Correlation Function. It is worthwhile at this stage to summarize the forms of correlation function which have arisen in the discussion of simple models.

In situations such that inertial effects are unimportant, in particular those described by a diffusion equation with diffusivity independent of time, the correlation function is an exponential decay

$$\exp(-\alpha t) \tag{1}$$

or a superposition of such functions describable by a spectrum of relaxation times. On a time scale such that inertial effects are perceptible, the correlation function must have the form $(1 - ct^2)$ for sufficiently small t, where c depends usually on the relevant mean-square velocity in thermal equilibrium.

Typical correlation functions for linear Brownian motion described by a Langevin equation include, for free motion

$$\exp\left[-\ \alpha\beta^{-1}(\beta t + e^{-\beta t} - 1)\right] \tag{2}$$

and for a damped harmonic oscillator,

$$\exp(-\beta t/2) \times \begin{bmatrix} \cos\omega_1 t + (\beta/2\omega_1)\sin\omega_1 t \\ \text{or} \\ \cosh(\beta_1 t/2) + (\beta/\beta_1)\sinh(\beta_1 t/2) \end{bmatrix} \tag{3}$$

Correlation functions for free rotation include, for a plane rotor

$$\exp(-\ t^2 kT/2I) \tag{4}$$

for a linear molecule

$$1 - \frac{kT}{I} t \exp(-\ t^2 kT/2I) \int_0^t \exp(\tau^2 kT/2I)\, d\tau \tag{5}$$

and for a spherical molecule

$$\frac{1}{3} + \frac{2}{3}\left(1 - \frac{kT}{I}t^2\right)\exp\left(-\frac{kT}{2I}t^2\right) \tag{6}$$

For sufficiently damped rotation we obtain products of correlation functions of the linear-motion type, and for a plane rotation with an angular velocity which is itself a Gaussian process, the Kubo expression,

$$\gamma(t) = \exp\left[-\int_0^t (t - \tau)\langle\omega(0)\,\omega(\tau)\rangle\, d\tau\right] \tag{7}$$

yields in particular

$$\exp\left\{-\frac{kT}{I\omega_0^2}\left[1 - e^{-\beta t/2}\left(\cos\omega_1 t + \frac{\beta}{2\omega_1}\sin\omega_1 t\right)\right]\right\} \tag{8}$$

in the case of underdamped libration.

A polar group elastically connected to the rigid body of a freely rotating molecule gives

$$\exp\left[-\ \alpha(1 - \cos\omega_0 t) - \beta t^2\right] \tag{9}$$

where $\frac{1}{2}\omega_0^2\alpha + \beta = kT/2I$, I being the moment of inertia of the polar group itself. A more complex expression for the same motion with special damping has been given above.

These functions are all associated with well-defined mechanical models, but one may construct trial functions, to test against experiment, simply to give reasonable initial and final behaviour with a respectably smooth transition. One which is widely used [6-8] and gives a fair approximation to function (2), above, is

$$(\tau_1 - \tau_2)^{-1}\,[\tau_1 \exp(-\,t/\tau_1) - \tau_2 \exp(-\,t/\tau_2)] \qquad (10)$$

Another empirical trial function which is very thoroughly discussed by Birnbaum and Cohen,[46] who also give its Fourier transform, is

$$\exp\{\delta - [\delta^2 + (t/\tau_1)^2]^{\frac{1}{2}}\} \qquad (11)$$

The functions so far quoted in this section which conclude in a single exponential decay all extrapolate back along the exponential to values greater than unity at $t = 0$. The functions which extrapolate to smaller values derive from compound motions. We have obtained one from Shimizu's expression

$$\gamma = \gamma_1 \mathrm{e}^{-ft} + (1 - \mathrm{e}^{-ft})\,\gamma_2 \qquad (12)$$

using a γ_1 for which $\gamma_1(\infty) \neq 0$. The product form

$$\gamma = \gamma_1 \gamma_2 \qquad (13)$$

can exhibit similar behaviour if $\gamma_1(\infty) \neq 0$, $\gamma_2 \sim c \exp(-\,t/\tau)$, and $c\gamma_1(\infty) < 1$. It should, however, be remembered that Fixman and Rider found similar behaviour for a motion of a spherical top.

General Linearized Brownian Motion. Our discussion has not so far passed beyond the Langevin approximation, which must be modified for realistic description of the motion of molecules not vastly more massive than the molecules of the solvent medium. This is particularly necessary for the discussion of pure substances.

If one follows the motion of a molecule in time, its acceleration at one moment reflects the force which it is exerting on its neighbours, and the momentum it transfers to them influences their position, and the forces which they exert, at a later moment. The non-random part of the acceleration of the molecule at one moment is a function of the previous history of its motion, at the same instant that it is also an ordinary function of its position along with that of the surrounding molecules. Langevin took this non-random part to be simply proportional to the instantaneous velocity.

If we make the less restrictive assumption that it is a linear function of the preceding velocities we obtain [47-49] a modified Langevin equation of the form

$$\dot{u} = -\int_0^t K(t - \tau)\,u(\tau)\,\mathrm{d}\tau + F(t)$$

where $\langle F(t)\rangle = 0 = \langle u(0)\,F(t)\rangle$, and the kernel K can be shown to be $K(t) = \langle F(0)\,F(t)\rangle/\langle u(0)^2\rangle$. The frictional memory function is proportional to the correlation function of the residual force.

[46] G. Birnbaum and E. R. Cohen, *J. Chem. Phys.*, 1970, **53**, 2885.
[47] H. Mori, *Progr. Theor. Phys.*, 1965, **33**, 423.
[48] R. Kubo, *Rep. Progr. Phys.*, 1966, **29**, 255.
[49] P. Mazur and I. Oppenheim, *Physica*, 1970, **50**, 241.

The relation for γ which follows is, since $\langle u(0)\, F(t)\rangle = 0$,

$$\dot{\gamma} = -\int_0^t K(t-\tau)\,\gamma(\tau)\,\mathrm{d}\tau, \quad \text{where } \gamma = \frac{\langle u(0)\, u(t)\rangle}{\langle u^2\rangle}$$

and one may study the forms of γ arising from suitable forms of K. If K is a delta-function we regain the original Langevin equation. Since K and γ are both even functions of time and well-behaved near $t = 0$, we may expand both in even powers of t.

$$\gamma(t) = \sum_{n=0}^{\infty} \frac{(-1)^n}{(2n)!}\gamma_{2n}\, t^{2n}$$

$$K(t) = \sum_{n=0}^{\infty} \frac{(-n)^n}{(2n)!}\mu_{2n}\, t^{2n}$$

Then the γ_{2n} and μ_{2n} are necessarily connected. Berne and Harp [50] give the relations

$$\mu_0 = \gamma_2$$

$$\mu_2 = \gamma_4 - \gamma_2^2$$

$$\mu_4 = \gamma_6 - 2\gamma_4\gamma_2 + \gamma_2^3$$

One might reasonably explore the results of substituting for K a correlation function of any of the types hitherto considered, and solving the integral equation for γ either analytically or numerically. Little work has been done in this direction so far, but Berne and Harp report a numerical investigation with very striking results.

Correlations by Computation of Molecular Dynamics. The power of modern computing systems has made it possible to solve the dynamical equations of motion of a model system of several hundred molecules, with fairly realistic interaction potentials, and hence by direct calculation obtain correlation functions for linear velocity, angular velocity, dipole orientation, *etc.* Rahman's classic paper [51] on the motion of 864 atoms of model argon has stimulated a great amount of further work, of which we cite particularly that of Berne and Harp [50] on nitrogen and carbon monoxide, and that of Rahman himself and Stillinger on water.[52]

Berne and Harp followed the motion of 512 molecules of model CO, with a step interval of 5×10^{-15} s, at a temperature of 68 K and density 0.8558 kg dm^{-3}, with an interaction potential constructed from a Lennard-Jones interaction supplemented by dipole–dipole, dipole–quadrupole, and quadrupole–quadrupole terms. The correlation functions for linear velocity, angular velocity, and dipole orientation are published for times from zero to 10^{-12} s.

The corresponding memory functions were then computed, inverting the integral equation of the previous section, and verified by recomputing the correlation functions from the integral equation. The correlation

[50] B. J. Berne and G. D. Harp, *Adv. Chem. Phys.*, 1970, **17**, 63.
[51] A. Rahman, *Phys. Rev.*, 1964, **136A**, 405.
[52] A. Rahman and F. H. Stillinger, *J. Chem. Phys.*, 1971, **55**, 3336.

function values were very well reproduced, so the correspondence is reliable at least as far as 10^{-12} s, and the linear modified Langevin equation is applicable at least to this extent.

The memory functions for the self-correlations of velocity and of angular momentum show an initial fall to near zero along a very nearly Gaussian curve with a half-width much the same as for the correlation functions themselves. Berne and Harp therefore explored the result of assuming a Gaussian memory function with parameters agreeing with the first two terms in the expansion of the computed memory in even powers of t. This ensures automatically that correlation functions calculated on this basis have the first three terms in their own corresponding expansion correctly determined. However, the expansion terminated as a quadratic in t^2 deviates early and violently from the course of the function. The correlations calculated from the Gaussian memories do not stray far from the 'experimental' results until after their first minimum. This suggests that it might be fruitful to approximate the memory functions by Gram–Charlier series, as has already been done with some correlation functions.[53]

It is interesting that the normalized memory function for $\langle\mu(0)\cdot\mu(t)\rangle$, which has non-Gaussian structure, lies very close to the normalized correlation function $\langle J(0)\cdot J(t)\rangle/\langle J^2\rangle$ for the angular momentum vector, while the memory function for the angular momentum is very nearly Gaussian. The former correspondence is not surprising, since the motion of a linear molecule is equivalent to that of a particle on a sphere, so that the Kubo–Shimizu consideration of a plane rotation may apply with only the numerical change which makes $\langle J(0)\cdot J(t)\rangle$ the relevant correlation function in the exponent of Kubo's formula.

The closure of the exponent in the Kubo expression depends on the Gaussian behaviour of the process $\omega(t)$. Whether $J(t)$ is Gaussian in the numerical experiment can be checked directly since, if it were, higher correlations such as $\langle |J(0)|^2\cdot|J(t)|^2\rangle$ are known functions of $\langle J(0)\cdot J(t)\rangle$. Berne and Harp tested a number of these higher correlations and found that the Gaussian approximation obtained from the first correlation function gave too small an amplitude in the structure after 10^{-13} s, but very good agreement up to that time and the correct general shape.

The connection of dipole correlation with angular-momentum correlation cannot be quite so direct for a non-linear molecule. Rahman and Stillinger have published a detailed molecular-dynamics study of a water model containing 216 model molecules in a cube with periodic boundary conditions, at a density of 1 kg dm^{-3}. The short-time behaviour of the molecules is dominated by librations at 1—2 $\times$ 10^{13} Hz. They give Cole–Cole plots of the electric permittivity estimated by applying an analysis due to Nee and Zwanzig[17] to the computed molecular-correlation function. However, this application is uncertain in view of the

[53] V. F. Sears, *Canad. J. Phys.*, 1967, **45**, 237.

particular cut-off of the molecular interactions used by Rahman and Stillinger.

To avoid computation of the periodic electric field arising from the (in any case artificial) periodic boundary conditions, they count the interactions of each molecule with only those neighbours which lie within a sphere of radius about 9 Å. They show how to allow for this in order to estimate the Kirkwood static correlation factor g_K. The factor involved is far from unity, being $9\varepsilon_0/[(\varepsilon_0 + 2)(2\varepsilon_0 + 1)]$ with $\varepsilon_0 \sim 75$, and it is not evident what (smaller) change will be needed in Nee and Zwanzig's analysis.

Relaxation of Interacting Dipoles.—The elementary systems considered so far may, of course, be used directly as models for the dielectric relaxation of bulk material, with more or less success. It is much more interesting to treat them as realistic models of molecular behaviour, but then the relation between the relaxing orientation of one molecule and the relaxing polarization of a bulk sample requires further attention.

If the dipoles are so tightly correlated that all move together, we have essentially one molecule with an enormous dipole moment, as in the low-frequency response of a single ferroelectric domain. The response is likely to be strongly non-linear at moderate field intensities. Another example is a polypeptide molecule coiled into a solid polar bar, and in this case it may be possible to uncoil the molecule and study it in a less correlated motion.[54] In the most interesting situation of biological material, polar long-chain molecules are dispersed in an ionic fluid, and the total moment associated with one molecule is a sum of more or less correlated permanent dipoles with a certainly-correlated ionic atmosphere.

If dipoles are uncorrelated, the bulk response is simply the sum of the responses of the individual dipoles considered separately. This is clearly a sound approximation in gases, and in dilute solutions of polar molecules in inert solvents. In intermediate cases the correlation must be explicitly estimated, but one can frequently approximate by identifying a group of close neighbours, discussed in detail as a unit, while supposing that correlations outside the group, arising from short-range forces, may be neglected.* This is equivalent to saying that short-range forces give rise to short-range correlations, which may make sense in polymers but is certainly false in crystals. Even in long-chain molecules one may not always neglect the long-range correlation imposed by the inability of the chain to pass through itself.[55]

[54] G. Parry-Jones, M. Gregson, and M. Davies, *Chem. Phys. Letters*, 1969, **4**, 33; *Trans. Faraday Soc.*, 1971, **67**, 1630; *cf.* J. Seelig and G. Schwarz, *Biopolymers*, 1968, **6**, 1263.
[55] S. F. Edwards, *Discuss. Faraday Soc.*, 1970, No. 49, 43.

* This feature has frequently been introduced in various *ad hoc* empirical fashions in dielectric discussions. One analytically explicit form was used by Piekara (1939) to account for the non-linear electric field effects, *e.g.* in liquid nitrobenzene. The importance of these effects is in part precisely because they allow a sensibly direct appraisal of dipole inter-correlations (see Chapter 7).

In all cases there is one long-range correlation which cannot be neglected, arising from the electrostatic interaction of the dipoles. Professor Scaife's article in this Report (Chapter 1) gives a complete account of the macroscopic aspects of this interaction. Since it alters considerably the equation of motion of a polar molecule considered as a Brownian particle, some recent progress in the discussion must be reported here.

The shorter-range electrostatic correlation expressed in hydrogen-bond formation is only briefly noted in this chapter. Some new light is thrown on it by the numerical work of Rahman and Stillinger already referred to.

Long-range Dipole Correlations. For a realistic model of concentrated polar substances one must consider the reaction on a particular molecule of the fields arising from the displacement of other (possibly distant) molecules provoked by its own previous motion, *i.e.*, of the delayed reaction field due to the relaxing polarization of the surrounding medium.[56]

Onsager model—reaction field effects. In the simplest form of this model a chosen molecule is represented by a spherical cavity of suitable volume filled with fluid of relative permittivity ε_∞, containing a rigid dipole of value μ. This μ is chosen so that if μ_V is the measured vacuum dipole moment of the molecule, $\mu = (\varepsilon_\infty + 2)\,\mu_V/3$. Correct calculation of the orienting couple on the dipole due to a given external field leads to the Onsager relation

$$(\varepsilon_0 - \varepsilon_\infty)(2\varepsilon_0 + \varepsilon_\infty)/3\varepsilon_0 = 4\pi N\mu^2/3kT$$

where N is the numerical density of the molecules.

Several authors have extended the treatment of this model to determine its time-dependent response.[57–61] A particularly clear account, summarizing and extending earlier work on the subject, is given by Nee and Zwanzig.[17] They obtain

$$(\varepsilon - \varepsilon_\infty)(2\varepsilon + \varepsilon_\infty)/3\varepsilon = 4\pi N\mu^2\,\mathscr{L}(-\dot{\gamma})/3kT$$

or

$$\frac{\varepsilon_0(\varepsilon - \varepsilon_\infty)(2\varepsilon + \varepsilon_\infty)}{\varepsilon(\varepsilon_0 - \varepsilon_\infty)(2\varepsilon_0 + \varepsilon_\infty)} = \mathscr{L}(-\dot{\gamma}) \equiv \int_0^\infty \mathrm{d}t\, \mathrm{e}^{-\mathrm{j}\omega t}(-\dot{\gamma})$$

where $\gamma(t)$ once more denotes the normalized dipole-auto-correlation function referring to the rotation of the permanent dipole μ alone. Comparison with

$$\frac{\varepsilon - \varepsilon_\infty}{\varepsilon_0 - \varepsilon_\infty} = \int_0^\infty \mathrm{d}t\, \mathrm{e}^{-\mathrm{j}\omega t}(-\dot{\phi})$$

shows that ϕ and γ may differ significantly.

56 R. P. Mercier, *Physica*, 1969, **45**, 95.
57 R. H. Cole, *J. Chem. Phys.*, 1938, **6**, 385; 1965, **42**, 637.
58 F. Kneubühl and B. Keller, *Phys. Letters* (*A*), 1969, **29**, 605.
59 D. D. Klug, D. E. Kranbuehl, and W. E. Vaughan, *J. Chem. Phys.*, 1969, **50**, 3904.
60 B. K. P. Scaife, *Proc. Phys. Soc.*, 1964, **84**, 616.
61 (*a*) E. Fatuzzo and P. R. Mason, *Proc. Phys. Soc.*, 1967, **90**, 729; (*b*) see also an analysis of the relations in refs. 57, 58, 60, 61*a*: N. E. Hill, *J. Phys.* (*c*), 1972, **5**, 415.

The form of γ to be used in Nee and Zwanzig's equation is the true γ in the material, and itself depends on ε through the viscous torque experienced by a dipole rotating close to a lossy material. The torque N is related to the angular velocity Ω either by

$$N(t) = -\int_{-\infty}^{t} \mathrm{d}s\, \xi(t-s)\, \Omega(s)$$

or, in transformed variables, by

$$N(\omega) = -\xi(\omega)\, \Omega(\omega)$$

where $\Omega(\omega)$, for example, is that component of the angular velocity of the molecule which oscillates with angular frequency ω.

Nee and Zwanzig assume that ξ can be separated into mechanical and electrical parts, $\xi = \xi_0 + \xi_D$. The reaction field $\boldsymbol{R}$ on the permanent dipole is

$$\boldsymbol{R}(\omega) = \frac{8\pi N}{3} \cdot \frac{\varepsilon - \varepsilon_\infty}{2\varepsilon + \varepsilon_\infty} \cdot \frac{\mu(\omega)}{\varepsilon_\infty}$$

which leads, with a good deal of further work, to

$$\xi_D(\omega) = -\frac{2kT}{\mathrm{j}\omega} \frac{(\varepsilon_0 - \varepsilon_\infty)(\varepsilon - \varepsilon_0)}{\varepsilon_0(2\varepsilon + \varepsilon_\infty)}$$

A frequency-dependent diffusivity introduced by

$$D(\omega) \equiv kT/\xi(\omega) \equiv \int_0^\infty \mathrm{d}t\, \mathrm{e}^{-\mathrm{j}\omega t}\, D(t)$$

leads to a modified Langevin equation

$$\dot{\gamma} = -2\int_0^t \mathrm{d}s\, D(t-s)\, \gamma(s)$$

which yields on transformation

$$\mathscr{L}(-\dot{\gamma}) = \{1 + \mathrm{j}\omega[\xi(\omega)/2kT]\}^{-1}$$

which can be interpreted in terms of a frequency-dependent complex-relation time $\tau(\omega) \equiv \xi(\omega)/2kT$. There is no need for ξ_0 to be independent of frequency. If it is constant, the equation for ε reduces to the form previously given by Fatuzzo and Mason [61]

$$\frac{\varepsilon_0(\varepsilon - \varepsilon_\infty)(2\varepsilon + \varepsilon_\infty)}{\varepsilon(\varepsilon_0 - \varepsilon_\infty)(2\varepsilon_0 + \varepsilon_\infty)}\left[1 + \mathrm{j}\omega\tau_0 + \frac{(\varepsilon_0 - \varepsilon_\infty)(\varepsilon_0 - \varepsilon)}{\varepsilon_0(2\varepsilon + \varepsilon_\infty)}\right] = 1$$

where $\tau_0 = \xi_0/2kT$.

Onsager model with generalized mechanical behaviour. It is clear from the last four equations that if the true γ obeys a modified Langevin equation with memory function $2D(t)$, the essential transform

$$\mathscr{L}(-\dot{\gamma}) = [1 + \mathrm{j}\omega/2D(\omega)]^{-1} = \{1 + \mathrm{j}\omega[\xi(\omega)/2kT]\}^{-1}$$

can always be expressed in terms of some functions $\xi(\omega)$. In particular, all the inertial effects considered previously can be taken up under this notation, provided we can find the relevant D or ξ.

On the assumption of Nee and Zwanzig, ξ falls into two parts, one of which they determine as a known function of ε. The other is $2kT$ times the reciprocal of the transformed memory function for the dipole correlation function γ_0 of the motion in the absence of the electric dipole interactions, so

$$\mathscr{L}(-\dot{\gamma}) = \{[\mathscr{L}(-\dot{\gamma}_0)]^{-1} + \mathrm{j}\omega\xi_{\mathrm{D}}/(2kT)\}^{-1}$$

This additive separation is clearly valid if the interactions contributing to ξ_0 and ξ_{D} are uncorrelated. It is not clear that it is valid in the probable situation that ξ_0 and ξ_{D} are both largely contributed to by interactions with the same set of nearest neighbours. One would expect trouble of this sort to show strongly with water, so the question might usefully be tested on the Rahman–Stillinger computation.

Mandel[62] gives a clear general discussion of the problem presented when the short-range forces give rise, by the propagation of orientation correlation through chains of neighbours, to relatively long-range correlations. The extreme form of this situation is well understood in insulating crystals, where the dynamical correlations are just the lattice oscillations. At least at low temperatures, the polarization waves in such materials are coherent over large distances and the corresponding electrical response is a lightly damped high-frequency resonance.

Relaxation in Insulating Crystals. The low-frequency relaxation of ionic materials consists typically of a set of simple decays arising from the motion of defects.[63–65] Some of these have freedom of extensive motion and contribute to the conduction current; others are constrained to a few neighbouring sites. Except at high temperature, the defects are remnants of the previous thermal and mechanical history of the sample. Their movement is a thermally activated process and relaxation times normally vary with temperature as $\exp(A/kT)$. The experimental picture may be much complicated by the motion of electrons loosely bound to crystal defects.

In a crystal of polar molecules it may well be as easy to reverse a molecule in its site as to form any other defect. This reversal will cost less in energy the more disorder is already present, and this co-operative effect may give rise to a clear order–disorder transition.[65a] The ordered phase may in particular be ferro- or antiferro-electric. The ferroelectric transition in thiourea is essentially of this nature.

The long-range order in a crystal structure imposes correlation on the dipole moments of neighbouring structural units, and the normal modes of motion of the structure are running or standing waves of definite

[62] M. Mandel, *Physica*, 1972, **57**, 141.
[63] J. S. Dryden and R. J. Meakins, *Rev. Pure Appl. Chem.*, 1957, **7**, 43.
[64] R. Freymann and M. Soutif, 'La Spectroscopie Hertzienne', Dunod, Paris, 1962.
[65] A. S. Nowick, *Adv. Phys.*, 1965, **14**, 101; 1967, **16**, 1.
[65a] See N. E. Hill *et al.*, 'Dielectric Properties and Molecular Behaviour', Van Nostrand Reinhold, London, 1969, p. 393; M. Davies, *J. Chim. phys.*, 1966, **63**, 67; and especially C. Brot and I. Darmon, *J. Chem. Phys.*, 1970, **53**, 2271.

frequency and wavelength. The modes in which the oscillating displacement entails an electric polarization have a restoring force partly due to the electric field, and have in part the character of plasma oscillations, the whole range of wavelengths permitted to one branch of the spectrum corresponding to a range of frequencies with a non-zero lower bound, typically in the i.r.

The motion actually observed by i.r. absorption is a small part of the whole pattern, since for effective absorption one must match both frequency and wavelength in the electromagnetic and crystal oscillations. Absorption occurs essentially at the frequency of the longest waves present in the crystal. These waves are damped by anharmonic interaction with the numerous short-wave modes, except at very low temperatures where interaction with crystal defects and surfaces takes over. In spite of the complexity of the process one has to expect a simple lineshape and a fairly simple temperature dependence for a wave propagating in one crystal direction.

It must be noted that the dipole moment of a single structural unit has components from every mode in the branch. Its motion has components from every corresponding frequency, and the corresponding $\gamma(t)$ is a sum of terms such as $(\omega t)^{-1} \exp(-\alpha t) \cos \omega t$,[66] where the ω appearing are boundary frequencies of the band.

The effect of the strong correlations combined with the restriction to large wavelengths is to give a very narrow spectral response for the bulk substance compared with its structural units. (This may be broadened out in polycrystal specimens as a result of crystal anisotropy.) The crystal situation is in contrast with that in amorphous material, where the response of the units is regarded as only moderately perturbed by correlations which are largely accounted for by an internal-reaction field.

A complete internal-field theory should be able to show the transition from one situation to the other as the correlation range grows in comparison with the observing wavelength. This involves considering the propagated response in space and time described by an extended correlation function $\gamma(r, t)$ and the corresponding $\varepsilon(k, \omega)$,[67, 68] *i.e.* at molecular level considering the whole relaxation of polarization around a dipole as well as the motion of the dipole itself.

Relaxation near a ferroelectric transition. Ferroelectricity includes a range of diverse phenomena. There is a broad distinction between ferroelectrics in which the structural units are dipolar, but disordered, at temperatures well above the transition temperature, and those in which the units seem to be non-polar at such temperatures, the ordering and displacive classes, respectively.[69]

[66] R. O. Davies, in 'Fluctuation, Relaxation and Resonance in Magnetic Systems', ed. D. ter Haar, Oliver and Boyd, Edinburgh, 1962, p. 169.
[67] A. D. Buckingham, *Discuss. Faraday Soc.*, 1967, No. 43, 205.
[68] J. Stecki, *Discuss. Faraday Soc.*, 1967, No. 43, 212.
[69] F. Jona and G. Shirane, 'Ferroelectric Crystals', Pergamon, London, 1962.

Cochran has suggested that, particularly in displacive ferroelectrics, it should be possible to identify an optical mode of vibration in the crystal, whose frequency drops with temperature so that the mode becomes unstable at the transition.[70] If the restoring force in such a mode is due to the polarization field, screened by the macroscopic electric permittivity we should expect the frequency to go to zero inversely as the square root of the permittivity, which varies typically as $(T - T_c)^{-1}$. Thus the mode frequency should vary as $(T - T_c)^{\frac{1}{2}}$.

Measurements have been made of the audiofrequency noise due to critical fluctuations of polarization, which become very large as the restoring force disappears,[71, 72] but the 'soft modes' of ferroelectrics have been most clearly identified at relatively high frequency, in the sub-millimetre region. These modes are active in the i.r., but if clean crystals are available they are conveniently studied by Raman scattering.

Kaminow and Damen,[73] for example, observe depolarized Raman scattering from KH_2PO_4. The bandshape is fitted within experimental error at each temperature by a susceptibility

$$\chi(\omega) = \omega_0^2\, \chi(0)\, (\omega_0^2 - \omega^2 + j\omega 2\Gamma)^{-1}$$

Γ is nearly independent of temperature at 85 cm^{-1}, while ω_0 drops from equality with Γ, at 460 K, to 20 cm^{-1} near 120 K, tending towards zero at the transition just below. The soft mode thus appears to be overdamped through the whole range of observation.

This apparent overdamping, together with the odd constancy of Γ, suggests that we may be observing a band of incoherent, lightly damped oscillators whose frequency spread is determined by a locally varying static interaction. The fact that Γ is nearly equal to kT_c rather supports this interpretation. Against it is the expectation that inhomogeneous broadening of this sort does in the majority of cases (by grace of the central limit theorem) produce a Gaussian rather than a Lorentzian lineshape.

Recent theoretical treatments of the soft-mode behaviour include a detailed study by Onodera[74] using classical mechanics, and a theory of hydrogen-bond mechanics, including tunnelling effects, by Stamenkovic and Novakovic.[75] Onodera assumes a quartic potential function for his individual oscillators, with a bilinear interaction which reduces to $c\langle x\rangle x$, where x is the displacement, under the Weiss-molecular-field approximation. The model is soluble without further approximation (in series of elliptic functions), yielding the temperature variation of frequency and damping. If the quartic potential has a central hump larger than kT_c,

[70] W. Cochran, *Adv. Phys.*, 1960, **9**, 387; 1961, **10**, 401; 1969, **18**, 157.
[71] R. E. Burgess, *Canad. J. Phys.*, 1958, **36**, 1596.
[72] J. J. Brophy and S. L. Webb, *Phys. Rev.*, 1962, **128**, 584.
[73] I. P. Kaminov and T. C. Damen, *Phys. Rev. Letters*, 1968, **20**, 1105.
[74] Y. Onodera, *Progr. Theor. Phys.*, 1970, **44**, 1477.
[75] S. Stamenkovic and L. Novakovic, *Phys. Stat. Sol.* (*Germany*), 1970, **41**, 135.

the behaviour is that of an ordering ferroelectric; otherwise, with a small hump or none, it is that of a displacive material.

Suzuki and Kubo [76] have considered the dynamics of an Ising model near a critical point. They give the exact susceptibility $\chi(q, \omega)$ for a one-dimensional chain, and revert to the Weiss approximation for two and three dimensions to obtain

$$\chi(q, \omega) = (N\mu^2/kT)/[1 - (J/kT) + j\omega\tau]$$

where J and τ may depend on q but are not expected to vary rapidly with T in the neighbourhood of T_c.

Tani [77] discusses the dynamics of displacive ferroelectrics and obtains a damped-resonant response.

The high-frequency dielectric response of ferroelectrics seems so far to be fitted quite well by rather simple, mechanical models, at fixed temperature, and for that reason not to be very informative about the internal mechanism. Information about the temperature dependence of the mechanical parameters has to be supplemented by measurements of magnetic resonance and of X-ray and neutron scattering.[78]

Fluctuating Dipoles.—The varying dipoles in a crystalline material are described in terms of (more or less) anharmonic oscillations, though in some molecular crystals at higher temperatures this is a rather forced description of the movement of molecules in a rotator phase. In a fluid medium, with molecules in (more or less) regular rotation, the molecular dipole will not only fluctuate in magnitude, as well as direction, but these fluctuations interact with each other and with the fluctuations of other dipoles.

Simultaneous Rotation and Fluctuation. The scalar product $\boldsymbol{\mu}(0)\cdot\boldsymbol{\mu}(t)$ has the value $\mu(0)\,\mu(t)\cos\theta(t)$ so the dipole auto-correlation function is $\langle\mu(0)\,\mu(t)\cos\theta(t)\rangle$, where $\theta(t)$ is the angular displacement of the dipole axis and $\mu(0)$, $\mu(t)$ are the relevant magnitudes. The correlation function will separate into $\langle\mu(0)\,\mu(t)\rangle\,\langle\cos\theta(t)\rangle$ only if the fluctuations in μ and θ are uncorrelated, which is not quite true even for a molecule in perfectly free rotation. If the variation of dipole moment and molecular size is relatively small, the correlation between that and the rotation may be negligible. In that case, if $\langle\mu(0)\,\mu(t)\rangle$ is independently known, $\langle\cos\theta(t)\rangle$ may be obtained from an experimentally determined $\langle\boldsymbol{\mu}(0)\cdot\boldsymbol{\mu}(t)\rangle$.

Much of our previous argument has dealt with the situation where $\langle\mu(0)\,\mu(t)\rangle = \mu^2$, for an invariable magnitude μ. In the observation of the relaxation of a fixed permanent dipole, it is necessary to work through many decades of frequency, up to 10^{12} Hz or thereabout. If, on the other hand, $\mu(t)$ oscillates at a frequency of 10^{13}—10^{14} Hz, the whole range of interest for relaxation is a fraction of this frequency, and is in principle

76 M. Suzuki and R. Kubo, *J. Phys. Soc. Japan*, 1968, **24**, 51.
77 K. Tani, *J. Phys. Soc. Japan*, 1969, **26**, 93.
78 R. Blinc, S. Zumer, and G. Lahajnar, *Phys. Rev.* (*B*), 1970, **1**, 4456.

covered by a single technique of observation. Gordon, in particular, has published a valuable series of papers[1, 13, 34] on the estimation of rotary correlations from i.r. spectra. One result of such work is to throw emphasis on the evaluation of bandshapes towards the wings, where absorption is small, making renewed demands on experimental technique. It also becomes important to recall[2] that combination bands, hot bands, and bands arising from minor isotopic species may contribute heavily in the wings of a dominant band, while heterogeneous broadening from other sources may be very difficult to exclude.

Collision-induced Dipoles. Even in materials whose molecules possess no permanent dipole, molecular collisions, either of unlike molecules or of like molecules in skew attitudes, produce fluctuating dipoles whose time variation reflects the translational movement of the colliding molecules rather than any characteristic vibration of the molecules. This effect appears in its purest form in mixtures of rare gases.

Clearly when two identical spherical atoms collide their distortions are symmetrical and the lowest electric moment which can appear is quadrupole. The corresponding absorption is extremely weak. With dissimilar atoms a collision dipole is possible.

Marteau, Vu, and Vodar[79] have measured the absorption between 16 and 250 cm^{-1} for every pair except He–Ne from the set He, Ne, Ar, Kr, and Xe at various temperatures and high pressures. The absorption shows a single broad maximum in general shape not unlike the speed distribution of the gas molecules but with a longer tail to high frequencies. For the 19 spectra reported the frequency of the absorption maximum falls, well within an estimated error of about 6 cm^{-1}, on the line

$$\nu_{\max} = a + b[T/(m\sigma_0^2)]^{\frac{1}{2}}$$

where m is the reduced mass of the colliding pair and σ_0 the separation at zero value of the van der Waals 6 – 12 potential. No discontinuity in position of the absorption maximum was found on compressing to a gas–liquid mixture.

Kiss and Welsh[80] and Bosomworth and Gush[81] measured the absorption of the He–Ar and Ne–Ar mixtures. The latter authors verified the dominance of binary collisions (less surprising at their lower pressures). Their absorption curves, corrected for the induced emission which is important at low frequencies, yield a function, defined as

$$G(\bar{\nu}) \equiv \bar{\nu}^{-1}\,\alpha(\bar{\nu})\,[1 - \exp(-\,hc\bar{\nu}/kT)]^{-1}$$

where α is the absorption coefficient and $\bar{\nu}$ the wavenumber, which is nearly Gaussian at lower frequencies and exponential at higher. Sears has

[79] P. Marteau, H. Vu, and B. Vodar, *J. Quant. Spectroscopy Radiative Transfer*, 1970, **10**, 283.
[80] Z. J. Kiss and H. L. Welsh, *Phys. Rev. Letters*, 1959, **2**, 166.
[81] D. R. Bosomworth and H. P. Gush, *Canad. J. Phys.*, 1965, **43**, 751.

analysed these results [53, 82] and shows that the correlation function obtained from G is fitted within experimental error by

$$\gamma(t) = (1 + \varepsilon_4 \chi^4) \exp(-\chi^2)$$

where $\chi = t/\tau_0$.

Sears shows, incidentally, that the expansion he proposes in general

$$\gamma = \exp(-\chi^2) \sum \varepsilon_{2n} \chi^{2n}$$

includes, with a suitable scaling of the correlation time, that used by Levine and Birnbaum,[83] in the form $(1 + \chi^2/\alpha)^{-\alpha}$.

Collisional absorption intensities have been discussed in several papers,[83–86] mostly using the assumption that the total dipole induced in two molecules at separation r is $\mu(r) = \mu_0 \exp(-r/\xi)$. The absorption intensities found by Bosomworth and Gush fit well with the theory of Poll and van Kranendonk.[84]

Collisions of Polyatomic Molecules. An induced dipole will appear in an end-to-side collision event of two hydrogen molecules, partly by overlap interaction and partly by the induction of a dipole in each molecule by the field due to the permanent quadrupole of the other. The latter effect is relatively long range, varying as r^{-4}, and may well dominate the process.

For example, Bosomworth and Gush have measured induced absorption in H_2, N_2, and O_2,[81] confirming with improved precision measurements on N_2 made by Gebbie and his co-workers.[87] Poll had used these [85] to find a quadrupole moment for N_2 in good agreement with a wave-mechanical estimate, suggesting that quadrupole induction in fact predominates for N_2. The measurements on O_2, on the other hand, show a very broad absorption in spite of a low molecular quadrupole moment, implying dominance of overlap effects in this case. For both N_2 and O_2, rotation and translation absorptions overlap entirely, but in H_2 at 77 K rotation is well separated from the weak, low-frequency translation-absorption band.

The greater part of the work on induced absorption in gases is sufficiently accounted for in terms of the variation of the dipole in separate uncorrelated collisions. Recently van Kranendonk has shown [88] that a long-established anomaly in certain spectra is due to a negative correlation between the dipoles induced in successive collisions of a given molecule, arising since large collision-dipoles coincide with large momentum-exchanges. The effect is most evident for light molecules moving among heavier ones.

82 V. F. Sears, *Canad. J. Phys.*, 1968, **46**, 1163, 1501.
83 H. B. Levine and G. Birnbaum, *Phys. Rev.*, 1967, **154**, 86.
84 J. D. Poll and J. van Kranendonk, *Canad. J. Phys.*, 1961, **39**, 189.
85 J. D. Poll, *Phys. Letters*, 1963, **7**, 32.
86 S. L. Brenner and D. A. McQuarrie, *Canad. J. Phys.*, 1971, **49**, 837.
87 H. A. Gebbie, W. Stone, and D. Williams, *Mol. Phys.*, 1963, **6**, 215.
88 J. van Kranendonk, *Canad. J. Phys.*, 1968, **46**, 1173.

This 'intercollisional' correlation is also discussed by Sears. It must gain in importance with increasing density as the two characteristic times, the mean duration of a collision and the interval between collisions, converge with the approach to the liquid state.

Collision-induced Absorption in Liquids. High-frequency studies of liquids have shown the Poley absorption,[89] indicative of constrained reorientation,[41] for polar substances in the sub-millimetre band. A broad absorption band in the same region was discovered[90–97] for non-dipolar substances such as cyclohexane, benzene, carbon disulphide, and carbon tetrachloride. The order of magnitude of the absorption is compatible[98] with the induction of the fluctuating dipoles by the distribution of the atomic charge on neighbouring molecules.

The optical absorption $\alpha(\bar{\nu})$ is superficially not unlike a Poley absorption, though much less in intensity. This function of wavenumber must, however, be multiplied by

$$\bar{\nu}^{-1}[1 - \exp(-hc\bar{\nu}/kT)]^{-1}$$

to give the imaginary part of the transform of the (undifferentiated) dipole-correlation function. For $hc\bar{\nu} \ll kT$, this is equivalent to multiplication by $\bar{\nu}^{-2}(kT/hc)$ and the careful measurements of Pardoe show that for several polar and non-dipolar liquids this leads to a maximum certainly below 10 cm^{-1}. Even for benzene and cyclohexane, the tail of the dipole correlation function will have substantial components relaxing at 10^{11} s^{-1} or slower. These slow components are presumably associated with co-operative movement of several molecules, while the frequency band characteristic of molecular collisions is emphasized by the presentation of the absorption $\alpha(\nu)$.

In view of these results it is doubtful if the translational absorption in non-dipolar liquids is to be described by an 'oscillator in cage' model even with extremely high damping. On the other hand, even after allowing for translational absorption, the Poley absorption in polar liquids is a significant feature. Hill's recent success[99] in explaining the diminished absorption at the high-frequency border of this band suggests that itinerant oscillator models are useful for its discussion.

[89] J. Ph. Poley, *J. Appl. Sci. Res.* (*B*), 1955, **4**, 337; see ref. 3 and N. E. Hill *et al.*, ref. 65*a*.
[90] D. H. Whiffen, *Trans. Faraday Soc.*, 1950, **46**, 124.
[91] G. W. Chantry and H. A. Gebbie, *Nature*, 1965, **208**, 378.
[92] G. W. Chantry, H. A. Gebbie, B. Lassier, and G. Wyllie, *Nature*, 1967, **214**, 163.
[93] Y. Leroy, E. Constant, C. Abbar, and P. Desplanques, *Adv. Mol. Relaxation Processes*, 1968, **1**, 273.
[94] M. Davies, G. W. F. Pardoe, J. E. Chamberlain, and H. A. Gebbie, *Trans. Faraday Soc.*, 1968, **64**, 847; 1970, **66**, 273; *Chem. Phys. Letters*, 1968, **2**, 411.
[95] G. W. F. Pardoe, *Trans. Faraday Soc.*, 1970, **66**, 2699.
[96] S. G. Kroon and J. van der Elsken, *Chem. Phys. Letters*, 1967, **1**, 285.
[97] C. Brot, B. Lassier, G. W. Chantry, and H. A. Gebbie, *Spectrochim. Acta*, 1968, **24A**, 295.
[98] S. K. Garg, J. E. Bertie, H. Kilp, and C. P. Smyth, *J. Chem. Phys.*, 1968, **49**, 2551.
[99] N. E. Hill, *J. Phys.* (*C*), 1971, **4**, 2322.

Polymer Relaxation. This is too important to neglect, but much too complex to summarize fairly in brief compass. A great amount of experimental and theoretical work is still providing more questions than answers.[100–102]

Long-chain structures show varied behaviour according to the linkage of the dipole units of the molecule to the chain. The most general pattern arises when we have dipoles fixed to, but not all directed along, the polymer backbone, and dipoles fixed to side-groups which have (more or less) freedom to rotate about the axis connecting them to the main chain.

The motions of the molecule may be classified as: deformation of the backbone regarded as a purely linear, incompletely flexible object; torsion of segments of the backbone about its local mean line; and movement of side-groups relative to the backbone. All these, according to the chemical structure, may contribute to the polarization. The first two classes of motion must correlate; the third may in particular cases appear to be independent.

There is little evidence that inertial terms are important in the consideration of backbone motions. This facilitates analysis of the linear movement in terms of normal modes determined by the stiffness constants and the viscous drag. The corresponding polar response is a superposition of simple decays dominated by the lowest one or two modes.

A general account of the theory of polarization changes arising from segmental torsion has been given by Work and Fujita.[103] The experimental situation is, not surprisingly, complex. Many materials show two broad absorptions, named α and β, due to motions of this class. The α absorption is about one order of magnitude more intense than the β, which shows its peak at a substantially lower temperature than the α when measurements are made varying temperature at fixed frequency. The α absorption is more closely associated with the glass-transition temperature. Williams has recently shown[104–106] that a number of these α absorptions can be fitted using a correlation function of the form $\exp[-(t/\tau)^{\beta}]$, with values of the exponent β ranging from 0.38 to 0.56. There seems to be no very well defined microscopic description of the situation. The sensitivity of these absorptions to pressure suggests that the availability of free volume may be a rate-controlling factor.

Glarum[107] showed some time ago that if a dipole relaxes rapidly on the arrival of a mobile structural defect beside it, the dipole-correlation

[100] N. G. McCrum, B. E. Read, and G. Williams, 'Anelastic and Dielectric Effects in Polymeric Solids', Wiley, London, 1967.
[101] H. Block and A. M. North, *Adv. Mol. Relaxation Processes*, 1970, **1**, 309.
[102] E. Marchal, *Chem. Phys. Letters*, 1971, **12**, 9.
[103] R. M. Work and S. Fujita, *J. Chem. Phys.*, 1966, **45**, 3779.
[104] G. Williams and D. C. Watts, *Trans. Faraday Soc.*, 1970, **66**, 80.
[105] M. Cook, D. C. Watts, and G. Williams, *Trans. Faraday Soc.*, 1970, **66**, 2503.
[106] G. Williams, D. C. Watts, S. B. Dev, and A. M. North, *Trans. Faraday Soc.*, 1971, **67**, 1323.
[107] S. H. Glarum, *J. Chem. Phys.*, 1960, **33**, 369.

function depends on the (not usually exponential) distribution of first-passage times of defects through nearest-neighbour sites. This opens a way of generating new correlation functions which may prove of further interest. McDuffie and Litovitz[108] have pointed out limitations in Glarum's initial suggestion and insist that a more general model of co-operative relaxation is required. This leads to a field still more characterized by ingenious conjecture than well-founded argument. An interesting contribution has been made by Anderson and Ullman.[109]

The motion of side-chains at low frequencies will be much dependent on the nature of their connection to the backbone. At higher frequencies, particularly in solution, the damping will be largely characteristic of interaction with the solvent, while bonding to the main chain will be approximated by an additional elastic constraint. Something like a Poley absorption has therefore to be sought, and Chantry and co-workers[110] have recently demonstrated its presence in copolymers of ethylene and vinyl acetate, even in the absence of a solvent.

4 Conclusion

A review of this nature comes to a halt rather than a conclusion. The time-correlation-function technique is proving not merely fashionable but profitable, though there remain problems more easily discussed by analysis in the frequency variable. A fair vocabulary of functions giving consistent description of simple motions is now available, and we can proceed to compound them to describe more realistic situations. The solution of the internal field problem for the Onsager model opens a new field for calculation, and the massive computations of the molecular dynamicists offer well-defined systems for the testing of more speculative theories. This field of research is in a period of most interesting and fruitful development.

5 Appendix

The basic relation used in this article,

$$\frac{\varepsilon - \varepsilon_\infty}{\varepsilon_0 - \varepsilon_\infty} = - \int_0^\infty e^{-j\omega\tau} \left[\frac{\langle P(0)\, P(\tau)\rangle}{\langle P^2\rangle}\right] d\tau$$

is a special case of a fluctuation-dissipation theorem. The general theory is lucidly discussed in reviews by, for example, Zwanzig[18] and Kubo.[48] Callen[111] gives a very clear brief account.

Since one is interested, in the first instance, in linear response, there is no intrinsic error in using first-order quantum-mechanical perturbation

[108] G. E. McDuffie and T. A. Litovitz, *J. Chem. Phys.*, 1962, **37**, 1699; 1963, **39**, 729.
[109] J. E. Anderson and R. Ullman, *J. Chem. Phys.*, 1967, **47**, 2178.
[110] G. W. Chantry, J. W. Fleming, E. A. Nicol, H. A. Willis, and M. E. A. Cudby, unpublished results.
[111] H. B. Callen, in 'Fluctuation, Relaxation and Resonance in Magnetic Systems', ed. D. ter Haar, Oliver and Boyd, Edinburgh, 1962, p. 15.

theory. The absorption of energy is the difference between direct absorption and induced emission, so depends on the population difference between the states concerned, which in turn is determined by the temperature.

If co-ordinates Q_i of the responding system are driven by forces $F_i(t)$ through a coupling energy

$$\sum_i F_i(t)\, Q_i$$

the response, in first order, of Q_j is

$$\langle \Delta Q_j(t)\rangle = \sum_i \frac{1}{j\hbar}\int_{-t}^{\infty} F_i(t')\,\langle [Q_i(t'), Q_j(t)]_-\rangle\, dt'$$

where the expectation value in the integrand is taken over the unperturbed distribution. The corresponding complex susceptibility is then

$$\chi_{ij}(\omega) = (2\pi)^{-\frac{1}{2}}(j\hbar)^{-1}\int_{-\infty}^{\infty}\langle [Q_i(0), Q_j(t)]_-\rangle\, e^{-j\omega t}\, dt$$

$$= \frac{\exp(-\hbar\omega/kT) - 1}{\exp(-\hbar\omega/kT) + 1}\cdot (j\hbar)^{-1}\int_{-\infty}^{\infty}\langle Q_i\, Q_j(t) + Q_j(t)\, Q_i\rangle\, e^{-j\omega t}\, dt$$

use having been made of the Boltzmann distribution. In the classical limit, $\hbar \to 0$, this becomes

$$\chi_{ij}(\omega) = j\omega I_{ij}(\omega)$$

where I_{ij} is the spectral density corresponding to the cross-correlation $\langle Q_i\, Q_j(t)\rangle$. Planck's constant must be retained whenever we are interested in frequencies $\omega \gtrsim kT/\hbar$.

Some useful relations among the derivatives of a correlation function are obtained by assuming that any realization of the process $P(t)$ is analytic and expanding

$$P(t + \varepsilon)\, P(t + \tau + \delta)$$

as a product of Taylor expansions in ε and δ. If we then set $\varepsilon = \delta$ and use the stationarity condition

$$\langle P(t + \varepsilon)\, P(t + \tau + \varepsilon)\rangle = \langle P(t)\, P(t + \tau)\rangle$$

we obtain from successive terms

$$\langle \dot{P}(t)\, P(t + \tau)\rangle = -\langle P(t)\, \dot{P}(t + \tau)\rangle$$

$$\langle \dot{P}(t)\, \dot{P}(t + \tau)\rangle = -\tfrac{1}{2}[\langle \ddot{P}(t)\, P(t + \tau)\rangle + \langle P(t)\, \ddot{P}(t + \tau)\rangle]$$

and further simplification is obtained from the symmetry of the motion under time reversal, so that, for example,

$$\langle \ddot{P}(t)\, P(t + \tau)\rangle = \langle \ddot{P}(t)\, P(t - \tau)\rangle$$

$$= \langle \ddot{P}(t + \tau)\, P(t)\rangle$$

on another application of stationarity.

3
Dielectric Polarization in Gases

BY H. SUTTER*

1 Introduction

The Clausius–Mossotti function or total polarization, ${}_{\mathrm{T}}P$, of a gas can be expanded in inverse powers of the molar volume, V_{m}:

$$ {}_{\mathrm{T}}P = \frac{\varepsilon_{\mathrm{r}} - 1}{\varepsilon_{\mathrm{r}} + 2} V_{\mathrm{m}} = A_{\varepsilon} + \frac{B_{\varepsilon}}{V_{\mathrm{m}}} + \frac{C_{\varepsilon}}{V_{\mathrm{m}}^{2}} + \ldots \qquad (1) $$

where ε_{r} is the relative permittivity of the gas. The first dielectric virial coefficient A_{ε} represents the ideal-gas independent-molecule contribution to ${}_{\mathrm{T}}P$; the second dielectric virial coefficient B_{ε} represents the initial deviation of ${}_{\mathrm{T}}P$ from the ideal-gas value due to pairs of molecules; the third dielectric virial coefficient C_{ε} represents deviations due to triplets of molecules, and so on. Measurements of A_{ε} have long been used to derive accurate values of molecular polarizabilities and dipole moments, while studies of the higher coefficients have yielded values of other molecular multipole moments and provided information about intermolecular interactions. This article is intended to serve as an introduction to studies of the dielectric virial coefficients and to provide a summary of work done to date. The second section will deal with the statistical mechanical theory of the coefficients, the third with techniques of measurement, and the fourth with experimental results and their interpretation.

SI units and equations will be used throughout. Most of the original work, however, was done in the c.g.s.–e.s.u. system. Table 1 lists the SI units for the electromagnetic properties referred to in this article, as well as the conversion factors from the SI to the probably more-familiar e.s.u. system. To convert all equations to their c.g.s.–e.s.u. equivalents replace ε_0, the permittivity of free space (8.85×10^{-12} C^2 m^{-1} J^{-1}) by $(4\pi)^{-1}$.

2 Theory

The general theory of the dielectric virial coefficients has been examined by a number of workers, among them Kirkwood,[1] Buckingham and Pople,[2]

[1] F. G. Keyes and J. G. Kirkwood, *Phys. Rev.*, 1931, **37**, 202; J. G. Kirkwood, *J. Chem. Phys.*, 1936, **4**, 592.

[2] A. D. Buckingham and J. A. Pople, *Trans. Faraday Soc.*, 1955, **51**, 1029, 1173, 1179; *Discuss. Faraday Soc.*, 1956, **22**, 17.

* *Present address*: Union College, Schenectady, New York, U.S.A.

Hill,[3] Isihara,[4] and Kaufman and Watson.[5] Hill's method parallels Mayer's treatment of the compressibility factor of an imperfect gas. Kaufman and Watson and Isihara utilize cluster theory in an attempt to obtain a solution for ${}_{\mathrm{T}}P$ in closed form. We will follow the treatment of Buckingham and Pople, primarily for reasons of simplicity.

Table 1 *SI units and conversion factors to e.s.u.*

	Quantity	*S.I. unit*	*To change S.I. to e.s.u. multiply by*
α	(polarizability)	$C^2\ m^2\ J^{-1}$	0.89876×10^{16} (e.s.u.)
β	(first hyperpolarizability)	$C^3\ m^3\ J^{-2}$	2.6944×10^{20} (e.s.u.)
γ	(second hyperpolarizability)	$C^4\ m^4\ J^{-3}$	0.80777×10^{25} (e.s.u.)
A		$C^2\ m^3\ J^{-1}$	0.89876×10^{18} (e.s.u.)
B		$C^3\ m^4\ J^{-2}$	2.6944×10^{22} (e.s.u.)
μ	(dipole moment)	C m	2.9979×10^{11} (e.s.u.) 2.9979×10^{29} (Debye)
Θ	(quadrupole moment)	$C\ m^2$	2.9979×10^{13} (e.s.u.)
Ω	(octopole moment)	$C\ m^3$	2.9979×10^{15} (e.s.u.)
Φ	(hexadecapole moment)	$C\ m^4$	2.9979×10^{17} (e.s.u.)
E	(electric field)	$C^{-1}\ m^{-1}\ J\ (V\ m^{-1})$	3.3357×10^{-5} (e.s.u.)
E'	(electric field gradient)	$C^{-1}\ m^{-2}\ J\ (V\ m^{-2})$	3.3357×10^{-7} (e.s.u.)
A_ε, A_R	(first dielectric, refractivity virial coefficient)	$m^3\ mol^{-1}$	$10^6\ (cm^3\ mol^{-1})$
B_ε, B_R	(second dielectric, refractivity virial coefficient)	$m^6\ mol^{-2}$	$10^{12}\ (cm^6\ mol^{-2})$
C_ε, C_R	(third dielectric, refractivity virial coefficient)	$m^9\ mol^{-3}$	$10^{18}\ (cm^9\ mol^{-3})$

We first concern ourselves with deriving a general statistical expression for ${}_{\mathrm{T}}P$. For a homogeneous, isotropic, dielectric medium which reacts linearly to a uniform applied field, the specific polarization

$$\boldsymbol{P} = (\varepsilon_r - 1)\,\varepsilon_0\,\boldsymbol{E} \tag{2}$$

where $\boldsymbol{P}$ is the polarization of the dielectric (electric dipole moment per unit volume) and $\boldsymbol{E}$ is the average electric field in the dielectric. Although $\boldsymbol{E}$ is the field measured under most experimental conditions, the proper variable for statistical mechanics is $\boldsymbol{E}_0$, the field due to charges external to the sample. The relation between $\boldsymbol{E}$ and $\boldsymbol{E}_0$ depends on the shape of the dielectric. To simplify what follows we choose a spherical sample where

$$\boldsymbol{E} = \frac{3}{\varepsilon_r + 2}\,\boldsymbol{E}_0 \tag{3}$$

[3] T. L. Hill, *J. Chem. Phys.*, 1958, **28**, 61.

[4] A. Isihara, *Physica*, 1961, **27**, 857; *J. Chem. Phys.*, 1963, **38**, 2437; A. Isihara and R. V. Hanks, *J. Chem. Phys.*, 1962, **36**, 433.

[5] A. N. Kaufman and K. M. Watson, *Phys. Fluids*, 1961, **4**, 931; *J. Chem. Phys.*, 1962, **36**, 439.

Combination of equations (2) and (3), followed by multiplication by the molar volume, yields

$$\frac{\varepsilon_r - 1}{\varepsilon_r + 2} V_m = \frac{1}{3\varepsilon_0} \frac{\bar{M}(E_0)}{E_0} = \frac{1}{3\varepsilon_0} \left(\frac{\partial \bar{M}(E_0)}{\partial E_0} \right)_{E_0 = 0} \tag{4}$$

where $\bar{M}(E_0) = PV_m$ is the mean total macroscopic dipole moment of one mole of dielectric. The derivative on the right, taken in the limit of $E_0 = 0$, yields the linear term of the response to E_0.

The classical equilibrium statistical mechanical expression for $\bar{M}$ is

$$\bar{M}(E_0) = \frac{\int [M(\tau, E_0)\cdot e] \exp\{-[V(\tau) - M(\tau, E_0)\cdot E_0]/(kT)\}\,d\tau}{\int \exp\{-[V(\tau) - M(\tau, E_0)\cdot E_0]/(kT)\}\,d\tau} \tag{5}$$

where $M(\tau, E_0)$ is the molar moment for a configuration of molecules specified by the choice of molecular co-ordinates τ, V is the potential energy due to intermolecular forces and e is a unit vector in the direction of E_0. The integration is to be carried out over all values of the rotational and translational co-ordinates of the molecules. Substitution of equation (5) into equation (4), followed by differentiation, gives

$$\frac{\varepsilon_r - 1}{\varepsilon_r + 2} V_m = \frac{1}{3\varepsilon_0} \frac{\int \left[\left(\frac{\partial M}{\partial E_0} \right)_{E_0 = 0} \cdot e + \frac{1}{kT} (M\cdot e)^2 \right] \exp[-V/(kT)]\,d\tau}{\int \exp[-V/(kT)]\,d\tau}$$

$$= \frac{1}{3\varepsilon_0} \left[\left\langle \left(\frac{\partial M}{\partial E_0} \right)_{E_0 = 0} \cdot e \right\rangle + \frac{1}{3kT} \langle M^2 \rangle \right] \tag{6}$$

Note that the statistical averages in equation (6) are to be taken in the absence of a field. If we now assume M to be the sum of Avogadro's number, N_A, of individual molecular moments, μ, equation (6) can be written

$$\frac{\varepsilon_r - 1}{\varepsilon_r + 2} V_m = \frac{N_A}{3\varepsilon_0} \left[\left\langle \left(\frac{\partial \mu_1}{\partial E_0} \right)_{E_0 = 0} \cdot e \right\rangle + \frac{1}{3kT} \sum_{i=1}^{N_A} \langle \mu_1 \cdot \mu_i \rangle \right] \tag{7}$$

which is our general expression for ${}_TP$, the molar polarization.

We now turn our efforts to deriving statistical expressions for the dielectric virial coefficients from equation (7). We note from equation (1)

$$\lim_{V_m \to \infty} \left[\frac{\varepsilon_r - 1}{\varepsilon_r + 2} V_m \right] = A_\varepsilon \tag{8}$$

The above limit corresponds to treating the molecules individually by computing the averages in equation (7) with the intermolecular potential set equal to zero. On doing so, we obtain the classic result of Debye [6]

$$A_\varepsilon = \frac{N_A}{3\varepsilon_0} \left(\alpha_0 + \frac{\mu_0^2}{3kT} \right) \tag{9}$$

[6] P. Debye, 'Polar Molecules', Chemical Catalog Co., New York, 1929.

where α_0 and μ_0 are the mean polarizability and permanent dipole moment, respectively, of the isolated molecule.

The expression for B_ε, representing the pair contribution to ${}_{\mathrm{T}}P$, is obtained from equations (1) and (7) in a manner parallel to that used for A_ε.

$$B_\varepsilon = \lim_{V_m \to \infty} \left[V_m \left(\frac{\varepsilon_r - 1}{\varepsilon_r + 2} V_m - A_\varepsilon \right) \right] \tag{10}$$

$$= \lim_{V_m \to \infty} \left\{ \frac{N_A V_m}{3\varepsilon_0} \left[\left\langle \left(\frac{\partial \boldsymbol{\mu}_1}{\partial E_0} \right)_{E_0 = 0} \cdot \boldsymbol{e} \right\rangle - \alpha_0 \right] + \frac{1}{3kT} \left[\sum_{i=1}^{N_A} \langle \boldsymbol{\mu}_1 \cdot \boldsymbol{\mu}_i \rangle - \mu_0^2 \right] \right\}$$

For V_m sufficiently large the probability of a single molecule, i, other than molecule 1, having co-ordinates τ_i in a range $d\tau_i$ in the region of molecule 1 is

$$(\Omega V_m)^{-1} \exp[-u_{1i}/(kT)]\, d\tau_i$$

where u_{1i} is the intermolecular potential energy between molecules 1 and i, and

$$\Omega V_m = \int d\tau_i$$

Summing over the probabilities of $N_A - 1 \approx N_A$ equivalent molecules being in the region of molecule 1 yields

$$B_\varepsilon = \frac{N_A^2}{3\varepsilon_0 \Omega} \int \left[(\tfrac{1}{2}\alpha_{12} - \alpha_0) + \frac{1}{3kT} (\tfrac{1}{2}\boldsymbol{\mu}_{12}^2 - \mu_0^2) \right] \times \exp[-u_{12}/(kT)]\, d\tau \tag{11}$$

where $\boldsymbol{\mu}_{12} = \boldsymbol{\mu}_1 + \boldsymbol{\mu}_2$ and $\alpha_{12} = (\partial \boldsymbol{\mu}_{12}/\partial E_0)_{E_0=0} \cdot \boldsymbol{e}$ are the total dipole moment and polarizability, respectively, for the pair of interacting molecules. The derivations of expressions for C_ε and higher coefficients follow a similar course and lead to expressions involving the dipole moment, polarizability, and intermolecular potential for three or more molecules.

The statistical mechanical equations for ${}_{\mathrm{T}}P$ (7), A_ε (9), and B_ε (11) each consist of two terms, the first representing the polarization due to the field inducing moments in the molecules, and the second representing the polarization due to the statistical effect of dipoles tending to line up in the direction of the field. At some frequency, generally in the microwave or far i.r., orienting dipoles no longer equilibrate with the field, and we enter a dispersion region. In this region our equilibrium treatment [equation (5)] of ${}_{\mathrm{T}}P$ breaks down. On passing out of the high-frequency end of the dispersion, the treatment becomes valid again, provided only the induced-moment terms in equations (7), (9), and (11) are retained. Measurement of ${}_{\mathrm{T}}P$ on the high- and low-frequency sides of the dispersion thus allows the separation of the induced and orientation contributions to ${}_{\mathrm{T}}P$, A_ε, and B_ε. In the case of A_ε, where the induced polarization contribution is temperature independent, this separation can also be achieved by determining A_ε as a function of temperature. For B_ε and higher coefficients, however, the induced-polarization contributions depend on temperature through the Boltzmann factor.

The orientational contribution can be computed directly from the integrated line intensity of the absorption band associated with the dispersion by using the Kramers–Kronig relation in the form

$$\varepsilon(0) - 1 = \frac{2c}{\pi} \int_0^\infty \frac{\gamma(\omega)}{\omega^2} \mathrm{d}\omega \tag{12}$$

where $\varepsilon(0) - 1$ is the contribution to $\varepsilon_r - 1$ from the process involved in the absorption, $\gamma(\omega)$ is the absorption coefficient (a function of the angular frequency ω), and c is the velocity of light *in vacuo*. To compute the integral in equation (12), the absorption coefficient must either be measured through the whole band or be fitted to a theoretical lineshape, whose parameters are determined by measurements on part of the band. For absorptions due to pair and higher-order interactions, the breadth of the bands and the frequencies involved make the first method rather difficult to apply, and most work relies on the second. The measurement of the orientational contribution to B_ε, $B_{\varepsilon_{\text{or}}}$, for dipolar molecules is complicated by the absorption due to single non-interacting molecules, which masks that due to pairs. For non-dipolar molecules, the single-molecule absorption is absent, and the pair contributions are readily observed. Note that for $B_{\varepsilon_{\text{or}}}$ to equal zero the pair must have zero dipole moment for all possible orientations of molecule 1 relative to molecule 2. Since this condition is only satisfied for a pair made of two identical atoms, all gases except pure atomic gases will exhibit bimolecular pressure-induced absorption. At very high pressures, all gases will give evidence of absorption due to triplet and higher-order interactions. For non-interacting molecules, the energy absorbed goes into rotational motion. For pairs, the energy can go either into the rotational motion of one or both of the molecules or directly into translational motion. Absorption bands due to the latter mechanism tend to be very broad, extending from zero frequency through most of the microwave and far i.r. regions. Bands due to the former are much less broad, being similar in character to single-molecule rotational bands.

The optical analogue of equation (1), written in terms of n, the index of refraction, can be obtained through the relation $n^2 = \varepsilon\mu$. For a gas the magnetic permeability, μ, may be taken as equal to 1, giving

$$R_{\text{m}} = \frac{n^2 - 1}{n^2 + 2} V_{\text{m}} = A_R + \frac{B_R}{V_{\text{m}}} + \frac{C_R}{V_{\text{m}}^2} + \dots \tag{13}$$

R_{m} is the molar refraction or Lorentz–Lorenz function, and A_R, B_R, C_R, ... are the refractivity virial coefficients. The rigorous derivation of the statistical mechanical equations for R_{m}, A_R, and B_R, corresponding to equations (5)—(11), is complicated by the variation of the electric field in the dielectric with time and position. Bullough [7] has shown, however, that equations (5)—(11) are valid for R_m, A_R, and B_R, provided only the induced-moment terms are retained and the electric field of the light wave is relatively constant over the region of interest. This second condition is met in the visible, where the wavelength is *ca.* 5×10^2 nm while molecular diameters of simple molecules are *ca.* 0.5 nm and the range of inter-molecular forces involved in the calculation of B_R is *ca.* 3 nm. It should be pointed out, however, that the induced-moment terms in equations (5)—(11) are frequency dependent, and one cannot in general equate A_R and B_R to the low-frequency induced-moment contributions to A_ε and B_ε. More will be said on this in the following discussion of the quantities appearing in equations (9) and (11).

Polarizability of an Isolated Molecule.—In Cartesian tensor notation, the components of the molecular dipole moment $\boldsymbol{\mu}$, induced by an electric field $\boldsymbol{E}$, can be written as [8]

$$\mu_\alpha = \alpha_{0_{\alpha\beta}} E_\beta + \frac{1}{2!}\beta_{0_{\alpha\beta\gamma}} E_\beta E_\gamma + \frac{1}{3!}\gamma_{0_{\alpha\beta\gamma\delta}} E_\beta E_\gamma E_\delta + \ldots$$
$$+ \tfrac{1}{3}A_{0_{\alpha,\beta\gamma}} E'_{\beta\gamma} + \tfrac{1}{3}B_{0_{\alpha\beta,\gamma\delta}} E_\beta E'_{\gamma\delta} + \ldots \qquad (14)$$

The subscripts in equation (14) run over the co-ordinates x, y, z, and a repeated index denotes summation over co-ordinates. The second-rank tensor $\boldsymbol{\alpha}_0$ represents the polarizability, and the third-rank tensor $\boldsymbol{\beta}_0$ and fourth-rank tensor $\boldsymbol{\gamma}_0$ represent the first and second hyperpolarizabilities, respectively. The tensors $\boldsymbol{A}_0$ and $\boldsymbol{B}_0$ are polarizabilities describing the moment induced by a field gradient $\boldsymbol{E}'$. In a uniform field it is apparent from equation (14) that $(\partial\boldsymbol{\mu}_1/\partial\boldsymbol{E}_0)_{\boldsymbol{E}_0=0} = \boldsymbol{\alpha}_0$. The tensor $\boldsymbol{\alpha}_0$ is symmetric. ($\alpha_{0_{\alpha\beta}} = \alpha_{0_{\beta\alpha}}$) and can be diagonalized to give only three non-zero, independent components, $\alpha_{0_{xx}}, \alpha_{0_{yy}}, \alpha_{0_{zz}}$. The molecular constant α_0 of equation (9) is given by $\frac{1}{3}$(trace $\boldsymbol{\alpha}_0$). Symmetry considerations may reduce the number of independent components of $\boldsymbol{\alpha}_0$ to two or one and simplify the expression for α_0.[9] For example, in a spherical molecule $\alpha_{0_{xx}} = \alpha_{0_{yy}} = \alpha_{0_{zz}} = \alpha_0$, whereas in a linear or symmetric-top molecule, taking z to be the symmetry axis, $\alpha_{0_{xx}} = \alpha_{0_{yy}} = \alpha_{0_\perp}$, $\alpha_{0_{zz}} = \alpha_{0_\parallel}$, and $\alpha_0 = \frac{1}{3}(\alpha_{0_\parallel} + 2\alpha_{0_\perp})$.

The quantum mechanical expression for the ground-state polarizability, valid except in regions of electronic transition,

$$\alpha_{0_{\alpha\beta}}(\omega) = \frac{2}{\hbar} \sum_{\substack{\text{internal}\\ \text{states, } n\neq 0}} \frac{\omega_{n0}\langle 0|\mu_\alpha|n\rangle\langle n|\mu_\beta|0\rangle}{\omega_{n0}^2 - \omega^2} \qquad (15)$$

where μ_α is the dipole moment operator, $\sum_i e_i r_{i\alpha}$; ω_{n0} is the angular transition frequency between the ground state and state n, and $r_{i\alpha}$ is the α component of the vector $\boldsymbol{r}_i$ from the origin (usually the centre of mass)

[7] R. K. Bullough, *J. Phys. (A)*, 1970, **3**, 708, 726, 751.
[8] A. D. Buckingham, *Adv. Chem. Phys.*, 1967, **12**, 107.
[9] A. D. Buckingham and B. J. Orr, *Quart. Rev.*, 1967, **21**, 195.

to the element of charge e_i. The frequency-dependent denominator of equation (15) causes α_0 of most simple molecules to decrease by 2—20% on going from the blue end of the visible to the red. The frequency dependence decreases rapidly through the visible and, in the absence of other effects, the change in α_0 on going from the near i.r. to radio frequency is generally less than 2%. The other effects referred to relate to the contributions of nuclear motions to α_0. At optical frequencies, nuclear contributions are absent, causing the value of α_0 to be 5—15% below the low-frequency value for all species other than monoatomics and homonuclear diatomics. *Ab initio* calculations of α_0 using equation (15) have been carried out on a large number of molecules.[10] The disagreement between calculated and experimental values depends greatly on the accuracy of the wave functions used, and even for simple molecules can vary from 100% to < 1%.[11]

Multipole Moments of Isolated Molecules.—The permanent dipole moment μ_0, of equation (9) is the lead term in the multipole moment expansion of the charge distribution of an isolated molecule. The electric multipole moment of order n, $\xi^{(n)}$, is defined in tensor notation as [12]

$$\xi^{(n)}_{\alpha\beta\gamma\ldots\nu} = \sum_i e_i(-1)^n(n!)^{-1} r_i^{2n+1}\frac{\partial}{\partial r_{i\alpha}}\frac{\partial}{\partial r_{i\beta}}\frac{\partial}{\partial r_{i\gamma}}\cdots\frac{\partial}{\partial r_{i\nu}}(r_i^{-1}) \qquad (16)$$

The tensors $\xi^{(n)}$ are symmetric in all n indices, and vanish if contracted with respect to any pair of indices. In general, there are $(2n + 1)$ independent components, though in most cases we shall consider, symmetry will reduce this number to one.[12] Expressions for the first few moments are

dipole, $$\mu_\alpha = \sum_i e_i r_{i\alpha} \qquad (17a)$$

quadrupole, $$\Theta_{\alpha\beta} = \tfrac{1}{2}\sum_i e_i(3r_{i\alpha}r_{i\beta} - r_i^2\delta_{\alpha\beta}) \qquad (17b)$$

octopole, $$\Omega_{\alpha\beta\gamma} = \tfrac{1}{2}\sum_i e_i[5r_{i\alpha}r_{i\beta}r_{i\gamma} - r_i^2(r_{i\alpha}\delta_{\beta\gamma} + r_{i\beta}\delta_{\alpha\gamma} + r_{i\gamma}\delta_{\alpha\beta})] \qquad (17c)$$

hexadecapole, $$\Phi_{\alpha\beta\gamma\delta} = \tfrac{1}{8}\sum_i e_i[35r_{i\alpha}r_{i\beta}r_{i\gamma}r_{i\delta} - 5r_i^2(r_{i\gamma}r_{i\delta}\delta_{\alpha\beta} + r_{i\delta}r_{i\alpha}\delta_{\beta\gamma} + r_{i\alpha}r_{i\beta}\delta_{\gamma\delta} + r_{i\beta}r_{i\gamma}\delta_{\delta\alpha} + r_{i\beta}r_{i\delta}\delta_{\alpha\gamma} + r_{i\alpha}r_{i\gamma}\delta_{\beta\delta}) + r_i^4(\delta_{\alpha\beta}\delta_{\gamma\delta} + \delta_{\alpha\gamma}\delta_{\beta\delta} + \delta_{\alpha\delta}\delta_{\beta\gamma})] \qquad (17d)$$

where $\delta_{ij} = 0$ for $i \neq j$ and 1 for $i = j$. The quantum mechanical calculation of a multipole moment simply involves finding its expectation

[10] W. G. Richards, T. E. H. Walker, and R. K. Hinkley, 'A Bibliography of ab initio Molecular Wave Functions', O.U.P., London, 1971.

[11] H. J. Kolker and M. Karplus, *J. Chem. Phys.*, 1963, **39**, 2011; W. Kolos and L. Wolniewicz, *J. Chem. Phys.*, 1967, **46**, 1426.

[12] A. D. Buckingham, 'Physical Chemistry', Academic Press, New York, vol. 4, ch. 8,

value in the electronic state of interest. For example,

$$\boldsymbol{\mu}_0 = \left\langle 0 \left| \sum_i e_i \boldsymbol{r}_i \right| 0 \right\rangle \tag{18}$$

Values of the dipole and higher moments have been calculated for a large number of simple molecules.[10] The calculation of μ_0 is somewhat easier than that of α_0 because the former depends only on the ground-state wave function, whereas the latter depends on the excited-state wave functions as well. For simple molecules even relatively crude wave functions yield values of μ_0 within 20% of experimental values.

In what follows we shall have occasion to consider the field at a point p, $\boldsymbol{E}(p)$, due to a molecule. In the multipole moment formalism this can be written as[8]

$$E(p)_\alpha = T(p)_{\alpha\beta}\,\mu_\beta - \tfrac{1}{3}T(p)_{\alpha\beta\gamma}\,\Theta_{\beta\gamma} + \tfrac{1}{15}T(p)_{\alpha\beta\gamma\delta}\,\Omega_{\beta\gamma\delta} + \ldots \tag{19}$$

The $\boldsymbol{T}$ tensors are given by

$$T(p)_{\alpha\beta} = \frac{1}{4\pi\varepsilon_0}\nabla_\alpha\nabla_\beta R^{-1} = \frac{1}{4\pi\varepsilon_0}(3R_\alpha R_\beta - R^2\,\delta_{\alpha\beta})\,R^{-5} \tag{20a}$$

$$T(p)_{\alpha\beta\gamma} = \frac{1}{4\pi\varepsilon_0}\nabla_\alpha\nabla_\beta\nabla_\gamma R^{-1} \tag{20b}$$

$$= -\frac{3}{4\pi\varepsilon_0}[5R_\alpha R_\beta R_\gamma - R^2(R_\alpha\,\delta_{\beta\gamma} + R_\beta\,\delta_{\alpha\gamma} + R_\gamma\,\delta_{\alpha\beta})]\,R^{-7}$$

$$T(p)_{\alpha\beta\gamma\delta} = \frac{1}{4\pi\varepsilon_0}\nabla_\alpha\nabla_\beta\nabla_\gamma\nabla_\delta R^{-1} \tag{20c}$$

where $\boldsymbol{R}$ is the vector from the origin of the multipole expansion to p. The $\boldsymbol{T}$ tensors are symmetric in all indices and vanish if an index is repeated. The field gradient at p, $\boldsymbol{E}'(p)$, due to a molecule can be written[8]

$$E'(p)_{\alpha\beta} = T(p)_{\alpha\beta\gamma}\,\mu_\gamma - \tfrac{1}{3}T(p)_{\alpha\beta\gamma\delta}\,\Theta_{\gamma\delta} + \ldots \tag{21}$$

These expressions for $\boldsymbol{E}$ and $\boldsymbol{E}'$ amount to expansions in inverse powers of R which will converge provided the point p is outside the charge distribution represented by the expansion.

The Pair Polarizability $\boldsymbol{\alpha_{12}}$.—For an individual molecule the quantities α_0 and μ_0 can be directly related to the experimental value of A_ε. The corresponding terms for a pair of molecules, α_{12} and μ_{12}, are buried one layer deeper in the theory, appearing in the experimental quantity B_ε as averages over all relative orientations of molecules 1 and 2. A theory attempting to describe the variation of α_{12} and μ_{12} with orientation can only be tested by employing it along with a suitable intermolecular potential in equation (11), and comparing the calculated value of B_ε with experiment. Although it is unlikely that this method can lead to a unique choice of theory and potential, it has proved a stringent test of both.

Methods of deriving theoretical expressions for α_{12} generally fall into three categories; *ab initio*, semi-classical, and classical. *Ab initio* methods involve a complete quantum-mechanical calculation using equation (15), and an *ab initio* set of wave functions for the pair. This method is very difficult and time consuming since, even for the simplest case of a pair of atoms, it corresponds to a calculation of α_0 for a diatomic molecule at a large number of internuclear separations. Moreover, the calculations must be extremely accurate since the actual term in B_ε involves $\frac{1}{2}\alpha_{12} - \alpha_0$, *i.e.* the polarizability of molecule 1 in the presence of molecule 2 minus the polarizability of an isolated molecule, which is at least an order of magnitude less than either α_{12} or α_0. So far this method has only been applied to helium pairs.[13] The results indicate that the polarizability of a molecule is diminished ($\frac{1}{2}\alpha_{12} < \alpha_0$) at short range and increased ($\frac{1}{2}\alpha_{12} > \alpha_0$) at long range. Extension of the *ab initio* method to more complex systems will probably be very slow in coming, since calculations for the next simplest atomic gas involve five times as many electrons, while for the simplest diatomic case, H_2, the number of nuclei double and the number of orientational co-ordinates to consider increases from one to four.

Most classical expressions for α_{12} are derived using equations (14) and (17)—(21), with the assumption that the isolated molecule properties described by $\boldsymbol{\alpha}_0$, $\boldsymbol{\beta}_0$, $\boldsymbol{A}_0$, $\boldsymbol{\mu}_0$, *etc.* are not drastically changed by the presence of a second molecule. This assumption is equivalent to taking the wave function for the pair to be the product of two non-overlapping isolated-molecule wave functions. Although not a bad approximation at long range, this assumption most certainly fails at short range.

As an example of the classical calculation, consider the simplest case, a pair of spherical molecules. All components of $\boldsymbol{\beta}_0$, $\boldsymbol{A}_0$, and the multipole moments are zero for such molecules. In addition, fields due to moments induced by laboratory fields, typically of the order of 10^4 V m^{-1}, are so small that the terms in equation (14) involving $\boldsymbol{\gamma}_0$ and $\boldsymbol{B}_0$ may be neglected. Thus we retain only the lead terms in equations (14) and (19) and write

$$\mu_{1\alpha} = \alpha_{0_{\alpha\beta}}[E_{0\beta} + E_{2\beta}(1)] = \alpha_{0_{\alpha\beta}}[E_{0\beta} + T(1)_{\beta\gamma}\,\mu_{2\gamma}] \qquad (22)$$

where $E_{2\beta}(1)$ is the β component of the field at molecule 1 due to molecule 2, and $\mu_{1\alpha}$ is the α component of the moment of molecule 1. After noting $\mu_{1\gamma} = \mu_{2\gamma}$ by symmetry, differentiating with respect to $\boldsymbol{E}_0$, and averaging over all orientations of the pair relative to the field we obtain[2]

$$\tfrac{1}{2}\alpha_{12} = \alpha_0 + \frac{(4\pi\varepsilon_0)^{-2}\,2\alpha_0^3\,r_{12}^{-6}}{[1 + (4\pi\varepsilon_0)^{-1}\,\alpha_0\,r_{12}^{-3}]\,[1 - (4\pi\varepsilon_0)^{-1}\,2\alpha_0\,r_{12}^{-3}]} \qquad (23a)$$

where r_{12} is the distance from molecule 2 to molecule 1. Equation (23a) has an artificial singularity at $r_{12}^3 = 2\alpha_0$. This, however, corresponds to very small internuclear separations where the classical model is invalid. Furthermore, at such separations repulsive forces between the molecules are very large ($u_{12} \gg 0$), and the Boltzmann factor of equation (11) is vanishingly small. For most calculations this singularity is removed by

[13] T. K. Lim, B. Linder, and R. A. Kromhout, *J. Chem. Phys.*, 1970, **52**, 3831.

equating the denominator to one, a procedure again justified only at long range, to give

$$\tfrac{1}{2}\alpha_{12} = \alpha_0 + (4\pi\varepsilon_0)^{-2}\, 2\alpha_0^3\, r_{12}^{-6} \qquad (23b)$$

For molecules with anisotropic polarizabilities, where the assumption $\mu_{1\alpha} = \mu_{2\alpha}$ cannot be made, the second term on the right side of equation (23) is replaced by a series solution in the form α_0 times powers of $\alpha_0\, r_{12}^{-3}$.[1,2] The contribution to B_ε of the lead term in the series proportional to $\alpha_0^2\, r_{12}^{-3}$ vanishes on averaging over all orientations if the intermolecular potential is a function of r_{12} only. For isotropically polarizable molecules the lead term also vanishes, leaving the series equivalent to the second term on the right side of equation (23a). The lead term in this series, $(4\pi\varepsilon_0)^{-1}\, 2\alpha_0^3\, r_{12}^{-6}$, is exactly what is obtained by setting the denominator of the second term in equation (23a) equal to one.

Permanent dipole moments are a factor of 10^6 larger than moments induced by laboratory fields. Even at intermediate separations the field and field gradients of these moments are large enough so that the higher-order terms in equation (14) must be retained when computing μ_α and α_{12}. Buckingham[14] has shown that the retention of the β_0 and γ_0 terms leads to additional contributions to B_ε proportional to $\beta_0\, \mu_0 \langle r^{-3}\rangle$ and $\gamma_0\, \mu_0^2 \langle r^{-6}\rangle$, which are generally about the same size as those obtained using equation (23) and in some cases an order of magnitude larger.

At short range there is some question as to whether the multipole-moment representation of permanent and induced moments converges rapidly enough to provide an adequate description of the molecular charge distribution. Levine and McQuarrie[15] have derived an expression for α_{12} based on a model which represents the molecules as conducting metallic spheres. An exact solution of Laplace's equation is possible for this model giving α_{12} directly, without resorting to the multipole expansion. Values of $\tfrac{1}{2}\alpha_{12} - \alpha_0$ deduced from this model are positive at all separations, giving a value of B_ε about twice that obtained using equation (23b). This model is also invalid at very short range, having a singularity when the spheres come into contact.

The validity of either of these classical methods for calculating the induced moment contribution to B_ε can be judged by noting that both predict positive values of B_ε for atomic gases, in disagreement with the negative results reported for helium and neon. Comparison with *ab initio* calculations leaves little doubt that the inability of classical methods to deal with short-range interactions caused by electron overlap is responsible for the discrepancy.

Semi-classical methods of computing α_{12} generally involve the use of equation (15) with wave functions derived from model considerations. The accuracy and complexity of the calculation depend greatly on the

[14] A. D. Buckingham, *Trans. Faraday Soc.*, 1956, **52**, 747.
[15] H. B. Levine and D. A. McQuarrie, *J. Chem. Phys.*, 1968, **49**, 4181.

choice of model. In the theory of Van Vleck,[16] the pair wave function is initially represented as a product of two non-interacting harmonic-oscillator functions. Perturbation theory, with an interaction Hamiltonian consisting of the classical induced-dipole–induced-dipole energy, is then used to get higher-order wave functions. Van Vleck, and later Jansen and Mazur,[17] have shown that, to all orders of perturbation, expressions equivalent to the classical expression of equation (23) are obtained. Jansen *et al.*[17] have carried out a similar treatment using hydrogenic, Gaussian heavy-atom, and Slater helium functions. In all cases they obtained a positive value for $\frac{1}{2}\alpha_{12} - \alpha_0$ at all values of r_{12} and a value of B_ε about twice that obtained using equation (23b). Buckingham[18] has shown that the Jansen result can be approximated by retaining the γ_0 term in equation (14) and considering the field at molecule 1 to be $\boldsymbol{E}_0$ plus the field due to fluctuation moments in molecule 2. By relating the fluctuation field to the London dispersion energy, Buckingham obtains a term proportional to $\gamma_0\langle r^{-6}\rangle$, which is to be added to the right side of equation (23). This additional term increases the calculated value of B_ε for argon by about 150%. Heinrichs[19] has used a perturbation formalism, which includes the Lorentz local field in the zeroth-order Hamiltonian, to develop an approximate quantum-mechanical expression, valid in regions of negligible overlap, to replace the right side of equation (15). His calculations show that the value of α_{12} will be greater than that calculated from equation (23) for hydrogen atoms, where he considers transitions to many excited states, and less than the classical result for helium, where he considers transitions to only one excited state. These results are in accord with those of Linder and Kromhout[20] who find that α_{12} will be greater than its classical value if the electronic level density increases with increasing energy, less than the classical value if the level density decreases with increasing energy, and equal to the classical value if the level density remains constant. Except for the work of Heinrichs on helium, all the above semi-classical methods lead to positive values for the induced-moment contribution to B_ε which are too large when compared with experiment. They are all limited, as were the classical models, by their inability to deal with short-range overlap interactions.

DuPré and McTague[21] have used a Hirschfelder–Linnett wave function for the triplet state of the hydrogen molecule to approximate the wave function of a pair of helium atoms. On evaluating an approximate expression for the polarizability they find $\frac{1}{2}\alpha_{12} < \alpha_0$ at short range and $\frac{1}{2}\alpha_{12} > \alpha_0$ at long range.

[16] J. H. Van Vleck, *J. Chem. Phys.*, 1937, **5**, 320, 556, 991.
[17] L. Jansen and P. Mazur, *Physica*, 1955, **21**, 193, 208; L. Jansen and A. D. Solem, *Phys. Rev.*, 1956, **104**, 1291; L. Jansen, *Phys. Rev.*, 1958, **112**, 434.
[18] A. D. Buckingham, *Trans. Faraday Soc.*, 1956, **52**, 1035.
[19] J. Heinrichs, *Phys. Letters*, 1965, **18**, 251; *Chem. Phys. Letters*, 1967, **1**, 467, 560; 1969, **4**, 151.
[20] B. Linder and R. A. Kromhout, *J. Chem. Phys.*, 1970, **52**, 1615.
[21] D. B. DuPré and J. P. McTague, *J. Chem. Phys.*, 1969, **50**, 2024.

Michels *et al.*[22] and de Groot and ten Seldam[23] have attempted to represent the effect of short-range electron repulsion by a compressed-atom model. Although their model leads to negative corrections to $_{\mathrm{T}}P$ at very high pressures, it is not clear how these corrections relate to B_ε since the model considers single atoms compressed uniformly. Several authors[24–26] have developed expressions for the polarizability of a molecule in a condensed-phase environment, which indicate it to be less than α_0. These models, however, involve many-body interactions, and, as in the previous case, it is not clear how they can be related to B_ε.

To sum up, it seems clear from experiment that the polarizability of a molecule is increased by a second molecule at large intermolecular separations and decreased at small intermolecular separations. *Ab initio* methods reproduce this behaviour but are extremely difficult to apply even in the simplest of cases. Classical methods are much easier to use but are valid only in the long-range limit, failing completely at short range. Semi-classical methods seem to offer the most promise for obtaining a simple, general, yet accurate representation of α_{12}. To date, however, this promise has not been fulfilled.

The Pair Dipole Moment $\boldsymbol{\mu}_{12}$.—The dipole moment of a pair of molecules, $\boldsymbol{\mu}_{12}$, can be calculated quantum mechanically from equation (18), using a pair wave function. In practice this procedure is rarely attempted, since classical expressions for $\boldsymbol{\mu}_{12}$, derived using equations (14) and (19) and isolated-molecule properties, seem to work quite well. That the classical approach should fail for the first term in equation (11) and succeed for the second may seem surprising, but can be explained by noting that whereas the first term is zero to a first approximation, the second is not. Thus, an error of 10% in the classical expression for $\boldsymbol{\mu}_{12}$ leads to a similar error in the orientational contribution to B_ε, while an error of 10% in α_{12} can lead to a result of the wrong sign for the induced contribution.

The following example will serve to illustrate roughly the classical method. Consider two polarizable molecules possessing permanent dipole moments: on retaining only the lead terms in equations (14) and (19) we have

$$\mu_{1\alpha} = \mu_{0_{1\alpha}} + \alpha_{0_{1\alpha\beta}} T(1)_{\beta\gamma}\,\mu_{2\gamma} \tag{24}$$

Substitution of a similar result for $\mu_{2\gamma}$ leads to

$$\mu_{1\alpha} = \mu_{0_{1\alpha}} + \alpha_{0_{1\alpha\beta}} T(1)_{\beta\gamma}\,\mu_{0_{2\gamma}} + \alpha_{0_{1\alpha\beta}} T(1)_{\beta\gamma}\,\alpha_{0_{2\gamma\delta}} T(2)_{\delta\xi}\,\mu_{1\xi} \tag{25}$$

with a similar expression for $\mu_{2\alpha}$. The continued substitution of the whole expression for $\mu_{1\alpha}$ into the last term on the right of equation (25) leads to

[22] A. Michels, J. de Boer, and A. Bijl, *Physica*, 1937, **4**, 981.
[23] S. R. de Groot and C. A. ten Seldam, *Physica*, 1946, **12**, 669; 1947, **13**, 47; 1952, **18**, 891, 905, 910.
[24] J. Heinrichs, *Phys. Rev.*, 1969, **179**, 823; **188**, 1419.
[25] T. H. Keil, *J. Chem. Phys.*, 1967, **46**, 4404.
[26] R. M. Mazo, *J. Amer. Chem. Soc.*, 1964, **86**, 3470.

an infinite series of the form μ_0 times powers of $\alpha_0 r_{12}^{-3}$. The computation of $\frac{1}{2}\mu_{12}^2 - \mu_0^2 = \boldsymbol{\mu}_1 \cdot (\boldsymbol{\mu}_1 + \boldsymbol{\mu}_2) - \mu_0^2$ is then straightforward and yields a series of the form [2]

$$\mu_0^2[A_\mu + B_\mu \alpha_0 r_{12}^{-3} + C_\mu(\alpha_0 r_{12}^{-3})^2 + \ldots]$$

where the coefficients A_μ, B_μ, C_μ, ... are functions of the relative angular orientation of molecules 1 and 2. It is important to note that the existence of multipole moments of any order in molecule 2 will serve to induce a dipole moment in 1 and give a non-zero contribution to B_ε. For example, for molecules possessing only a permanent quadrupole moment one obtains an expression of the form

$$\Theta^2 \alpha_0^2 r_{12}^{-8}[A_\theta + B_\theta \alpha_0 r_{12}^{-3} + C_\theta(\alpha_0 r_{12}^{-3})^2 + \ldots]$$

while for octopoles one obtains

$$\Omega\alpha_0^2 r_{12}^{-10}[A_\omega + B_\omega \alpha_0 r_{12}^{-3} + C_\omega(\alpha_0 r_{12}^{-3})^2 + \ldots]$$

One check of the validity of the classical expression for $\boldsymbol{\mu}_{12}$ is the agreement between experimental and calculated values of B_ε. For dipolar molecules the agreement is poor, but this is generally believed to be due to inadequacies of the intermolecular potential rather than the expression for $\boldsymbol{\mu}_{12}$. For quadrupolar molecules agreement is very good, but for octupolar and hexadecapolar molecules calculation of B_ε is difficult owing to a lack of independently observed values for these moments. For atomic gas mixtures, the classical theory predicts that the orientational part of the mixed virial coefficient is zero. Kiss and Welch [27] and Bosomworth and Gush [28] have, however, observed far-i.r. pressure-induced absorption for Ne–He, Ar–He, and Ne–Ar pairs. Levine and Birnbaum [29] have applied their own lineshape theory to the data of Bosomworth and Gush and calculated values of $[\varepsilon(0) - 1]$ using equation (12). Values of $B_{\varepsilon_{or}}$ derived from these values are given in Table 5 (p. 99). As can be seen from the table, deviations from the classical result ($B_{\varepsilon_{or}} = 0$) are very small, lending strength to the belief that the classical approach is valid in most cases. More will be said on this in the discussion of the results of Bosomworth and Gush (Section 4).

The Intermolecular Potential u_{12}.—We will deal only briefly with the intermolecular potential. An excellent set of reviews on the subject can be found in the volume listed in ref. 8. For the computation of B_ε we are generally interested in a simple analytical expression for u_{12}. Since the list of possibilities is almost endless, emphasis will be placed on potentials that have been used in studies of B_ε. Generally these potentials may be broken into two parts: u_P, which is due to the permanent and induced moments of the molecules; and u_Q, which is due to the quantum nature of the molecule. The latter is, in turn, usually composed of an attractive

[27] Z. J. Kiss and H. L. Welsh, *Phys. Rev. Letters*, 1959, **2**, 166.
[28] D. R. Bosomworth and H. P. Gush, *Canad. J. Phys.*, 1965, **43**, 729, 751.
[29] H. B. Levine and G. Birnbaum, *Phys. Rev.*, 1967, **154**, 86.

long-range part to correspond to the London dispersion forces, and a repulsive short-range part to correspond to electronic-overlap forces. The form of u_Q most widely used in calculations of B_ε is the Lennard-Jones 6–12 potential,

$$u_{6-12} = 4\varepsilon\left[\left(\frac{r_0}{r_{12}}\right)^{12} - \left(\frac{r_0}{r_{12}}\right)^{6}\right] \tag{26}$$

The energy parameter ε equals the maximum depth of the potential well, and r_0 equals the value of r_{12} where the attractive 6-term and repulsive 12-term are equal. For non-spherical molecules, shape-dependent terms involving the relative angular orientations of the two molecules are sometimes added to the right side of equation (26). More will be said on this in Section 4.

The potential energy u_P can also be split into two parts: u_{elec}, due to permanent moments, and u_{ind}, due to induced moments. For a pair of molecules we have

$$u_{\text{elec}} = -\,[T(2)_{\alpha\beta}\,\mu_{0_{1\alpha}}\,\mu_{0_{2\beta}} + \tfrac{1}{3}T(2)_{\alpha\beta\gamma}(\mu_{0_{1\alpha}}\Theta_{0_{2\beta\gamma}} + \mu_{0_{2\alpha}}\Theta_{0_{1\beta\gamma}}) + \tfrac{1}{9}T(2)_{\alpha\beta\gamma\delta}(\Theta_{0_{1\alpha\beta}}\Theta_{0_{2\gamma\delta}} + \ldots) + \ldots] \tag{27}$$

$$u_{\text{ind}} = -\,\tfrac{1}{2}\alpha_{0_{1\alpha\beta}}\,E(1)_\alpha\,E(1)_\beta - \tfrac{1}{3}A_{0_{1\alpha\beta\gamma}}\,E(1)_\alpha\,E'(1)_{\beta\gamma} - \ldots \tag{28}$$

For triplet and higher-order potentials, needed to calculate C_ε and higher coefficients, the assumption has usually been made [4, 30, 31] that intermolecular forces are pairwise additive. The accuracy of this assumption is uncertain, and the lack of a reliable many-body potential is one of the major difficulties encountered in calculations of the higher-order coefficients.

For atomic gases the intermolecular potential most used in calculations of B_ε has been the Lennard-Jones 6–12, with the parameters ε and r_0 determined from gas viscosities and pressure virial coefficients. For non-dipolar gases possessing higher moments, most authors have used the 6–12 potential together with the appropriate terms from equations (27) and (28), while for dipolar gases some form of equations (26)—(28) is used with a shape-dependent term added to u_Q.

The integration of equation (11) is rather difficult for physically-realistic forms of u_{12}. Lennard-Jones,[32] followed by Buckingham and Pople,[2] showed that integrals of the form

$$\int_0^\infty r^{-n}\exp[-u_{6-12}/(kT)]\,\mathrm{d}r$$

can be computed in terms of the so called H functions, thus making the integration straightforward for spherical molecules. To deal with non-spherical molecules possessing multipole moments, where the potential

[30] D. A. McQuarrie and H. B. Levine, *Physica*, 1965, **31**, 749; *J. Chem. Phys.*, 1966, **44**, 3500.
[31] J. de Boer, F. van der Maesen, and C. A. ten Seldam, *Physica*, 1953, **19**, 265.
[32] J. E. Lennard-Jones, *Proc. Roy. Soc.*, 1924, **A106**, 441, 463.

can be written as $u_{6-12} + u'$, Pople[33] proposed expansion of $\exp(-u'/kT)$ in the series

$$1 - u'/kT + \frac{1}{2!}(u'/kT)^2 - \frac{1}{3!}(u'/kT)^3 + \ldots$$

followed by the Lennard-Jones method of integration. In his original paper, Pople was careful to point out that this expansion will converge rapidly only if $u' \ll kT$. For reasonable values of the dipole and quadrupole moments, this condition is by no means met over the whole range of the integral, the value of u' being 5—10 times kT for some orientations. Moreover, the largest contributions to the integral come somewhere in the region of r_0, precisely where u' is large and the expansion most slowly convergent. Kong[34] has found it necessary to retain terms through $(u'/kT)^8$ in his recent computations of B_ε for dipolar molecules. Kielich,[35] however, has generally retained only terms of order u'/kT, making his results somewhat suspect.

Retention of a large number of terms in the above series leads to such long and complicated expressions for B_ε that a number of workers have chosen to perform the integration numerically. For linear molecules, the use of the potentials described above necessitates integration over four dimensions, a job accomplished in a matter of minutes on a modern computer. Four dimensions also suffice for symmetric-top molecules, provided moments and polarizabilities described by tensors of rank greater than two are neglected. In the general case, however, six dimensions are needed, causing a vast increase in the amount of computer time required. Difficulties in computation have been the main reason that more accurate but more complex potentials have not been used in equation (11).

The integrations involved in calculations of C_ε and higher coefficients are almost impossible to carry out unless the intermolecular potentials used are solely radially dependent. McQuarrie and Levine[30] have calculated values of C_ε for linear quadrupolar molecules by first reducing the potential to a radial function by the method of Pople, and then carrying out the resulting three-dimensional integration numerically. They have tabulated the reduced integrals for a variety of potentials, including the 6–12 and Buckingham exp – 6. de Boer *et al.*[31] have carried out similar calculations for non-multipolar molecules using the 6–12 and Herzfeld square-well potentials.

3 Measurement Techniques

The order of accuracy required in measurements of the molar volume and the electric permittivity for dielectric virial coefficient determinations may

[33] J. A. Pople, *Proc. Roy. Soc.*, 1954, **A221**, 498, 508.
[34] C. L. Kong, *J. Chem. Phys.*, 1970, **53**, 1516, 1522; C. L. Kong and J. W. Larson, *ibid.*, 1971, **55**, 3051.
[35] S. Kielich, *Bull. Acad. polon. Sci., Ser. Sci. math. astron. phys.*, 1962, **10**, 485, 493, 657; *Physica*, 1962, **28**, 511, 1116; *Acta Phys. polon.*, 1962, **22**, 477; 1963, **24**, 389; 1964, **25**, 39; 1965, **27**, 305, 457; **28**, 95; *Mol. Phys.*, 1965, **9**, 549.

be illustrated by using the following approximate values for CO_2 at 10 atm and 323 K: $\varepsilon_r = 1.01$; $V_m = 2.5 \times 10^{-3}$ m³ mol⁻¹; $A_\varepsilon = 7.3 \times 10^{-6}$ m³ mol⁻¹; $B_\varepsilon = 50 \times 10^{-12}$ m⁶ mol⁻²; $B_\varepsilon/V_m = 2 \times 10^{-8}$ m³ mol⁻¹ (about 3 parts in 10^3 of A_ε). Hence, to determine B_ε to 10%, V_m must be measured to 1 part in 10^4 and ε_r to 3 parts in 10^6. In practice, the second requirement is relatively easy to achieve whereas the first is not.

Measurements of ε_r.—Stability under pressure is the prime requirement for capacitance cells used to determine dielectric virial coefficients. Cells of both parallel-plate and cylinder-within-a-cylinder design, stable to a few parts in 10^6 and usable to over 200 atm, have been described by a number of authors.[36–40] Cells for use in the microwave region[41–44] and cells for measuring refractivity[45, 46] have also been described. Early measurements at radio frequencies relied on the heterodyne beat method,[47–50] but more recent work[36–40, 51] has utilized the three-terminal transformer ratio-arm technique developed by Cole and Gross.[52] This second method eliminates difficulties due to stray capacitances and provides accuracies of better than 1 part in 10^6. For an excellent review of techniques at both radio and microwave frequencies see ch. 2 of ref. 53. For refractivity methods see refs. 45, 46, and 54.

Measurements of V_m.—Direct determination of V_m^{-1} using a gas density balance has in the past proved rather unsuccessful, leading to a value of B_ε for CHF_3 of the wrong sign.[55] Recently,[56] however, this method has been refined to the point where it may be useful in the future.

In almost all work done prior to 1960, V_m was determined indirectly from pressure measurements by means of an equation of state. However,

[36] D. R. Johnston, G. J. Oudemans, and R. H. Cole, *J. Chem. Phys.*, 1960, **33**, 1310.
[37] J. W. Stewart, *J. Chem. Phys.*, 1964, **40**, 3297.
[38] H. Sutter and R. H. Cole, *J. Chem. Phys.*, 1970, **52**, 132.
[39] E. C. Kerr and R. H. Sherman, *J. Low Temp. Phys.*, 1970, **3**, 451.
[40] A. N. M. Barnes, D. J. Turner, and L. E. Sutton, *Trans. Faraday Soc.*, 1971, **67**, 2902; A. N. M. Barnes and L. E. Sutton, *ibid.*, 2915, 2926.
[41] C. S. E. Philips, *J. Chem. Phys.*, 1955, **23**, 2388.
[42] M. G. Vallauri and P. W. Forsbergh, jun., *Rev. Sci. Instr.*, 1957, **28**, 198.
[43] W. Ho, I. A. Kaufman, and P. Thaddeus, *J. Chem. Phys.*, 1966, **45**, 877.
[44] H. W. deWijn and F. W. Heineken, *Physica*, 1959, **25**, 615.
[45] A. R. Blythe, J. D. Lambert, P. J. Petter, and H. Spoel, *Proc. Roy. Soc.*, 1960, **A255**, 427.
[46] D. H. Everett and R. J. Munn, *Trans. Faraday Soc.*, 1963, **59**, 2486.
[47] A. Michels and C. Michels, *Phil. Trans.*, 1933, **A231**, 409.
[48] I. E. Coop and L. E. Sutton, *J. Chem. Soc.*, 1938, 1269.
[49] J. G. Jelatis, *J. Appl. Phys.*, 1948, **19**, 419.
[50] J. Clay and F. van der Maesen, *Physica*, 1949, **15**, 467.
[51] R. H. Orcutt and R. H. Cole, *Physica*, 1965, **31**, 1779.
[52] R. H. Cole and P. M. Gross, jun., *Rev. Sci. Instr.*, 1949, **20**, 252.
[53] N. E. Hill, W. E. Vaughan, A. H. Price, and M. Davies, 'Dielectric Properties and Molecular Behaviour', Van Nostrand–Reinhold, London, 1969.
[54] P. Phillips, *Proc. Roy. Soc.*, 1920, **A97**, 225.
[55] A. D. Buckingham and R. E. Raab, *J. Chem. Soc.*, 1961, 5511.
[56] W. S. Haworth and L. E. Sutton, *Trans. Faraday Soc.*, 1971, **67**, 2907; *J. Phys. (E)*, 1970, **3**, 271.

as can be seen from the following argument, second and higher dielectric virial coefficients derived in this way are of limited accuracy. Consider the equation of state in the form

$$\frac{pV_m}{RT} = 1 + \frac{B_p}{V_m} + \frac{C_p}{V_m^2} + \ldots \tag{29}$$

where B_p, C_p, ... are the pressure virial coefficients. Elimination of V_m between equations (1) and (29) gives

$$\frac{\varepsilon_r - 1}{\varepsilon_r + 2}\frac{RT}{p} = A_\varepsilon + (B_\varepsilon - A_\varepsilon B_p)\frac{p}{RT} + \ldots \tag{30}$$

A plot of the quantity on the left *vs.* $p/(RT)$ will produce a line of intercept A_ε and slope $(B_\varepsilon - A_\varepsilon B_p)$. Values of A_ε determined by this method can be made reliable to a few parts in 10^4, being dependent only on the accuracy of the measured quantities p, T, and ε_r. The value of B_ε is more difficult to determine, appearing only in combination with $A_\varepsilon B_p$, which is generally an order of magnitude larger. Even for substances such as helium and carbon dioxide, where the equations of state are relatively well known, the disagreement between values of B_p obtained by various authors is about 5%.[57] Such an uncertainty in B_p corresponds to an uncertainty of about 100% in B_ε. The method is thus of limited worth for obtaining values of B_ε, except in the case of some dipolar gases where B_ε is the same order of magnitude as $A_\varepsilon B_p$.

Although equations of state are not known to the required accuracy over the density range where contributions to ${}_TP$ from B_ε and higher coefficients are appreciable, they become more accurate at low densities where the non-ideal terms of equation (29) are less significant. Several workers[36, 39, 58] have taken advantage of this by trapping a small known volume of gas at the density where ε_r is measured and allowing it to expand into a large calibrated volume. Since the ratio of volumes before and after expansion is known, measurement of the pressure after expansion followed by the use of an equation of state serves to determine V_m accurately at the initial density.

In an effort to determine the higher dielectric virial coefficients without determining V_m, Cole and co-workers[51, 59] have adapted the Burnett expansion technique[60] to dielectric measurements. A schematic figure of their apparatus is shown [Figure 1(a)]. Initially the system is filled with gas, the valve closed and ε_r measured. Volume B is then evacuated, the gas from cell A allowed to expand into B, the valve closed and ε_r measured again. The process is repeated a number of times giving a series of

[57] J. H. Dymond and E. B. Smith, 'The Virial Coefficients of Gases', Clarendon Press, Oxford, 1969.
[58] H. G. David, S. D. Hamann, and J. F. Pearse, *J. Chem. Phys.*, 1952, **20**, 969.
[59] D. R. Johnston and R. H. Cole, *J. Chem. Phys.*, 1962, **36**, 318.
[60] E. S. Burnett, *J. Appl. Mech.*, 1936, **3**, A136.

measurements of $f_i = (\varepsilon_r - 1)/(\varepsilon_r + 2)$ at a set of densities d_i related by

$$d_{i+k} = \left(\frac{V_A}{V_A + V_B}\right)^k d_i = N^k d_i \tag{31}$$

It can be shown that [59]

$$\frac{f_i}{f_{i+k}} = N^{-k} + (N^{-k} - 1)\frac{B_\varepsilon}{A_\varepsilon^2} f_i + (N^{-k} - N^k)\left(\frac{C_\varepsilon}{A_\varepsilon^3} - \frac{B_\varepsilon^2}{A_\varepsilon^4}\right) f_i^2 + \ldots \tag{32}$$

The quantities N^{-k} and $B_\varepsilon/A_\varepsilon^2$ can be obtained by plotting the left-hand side of equation (32) *vs.* f_i. Values of B_ε obtained in this manner are believed to be accurate to 5—20%.

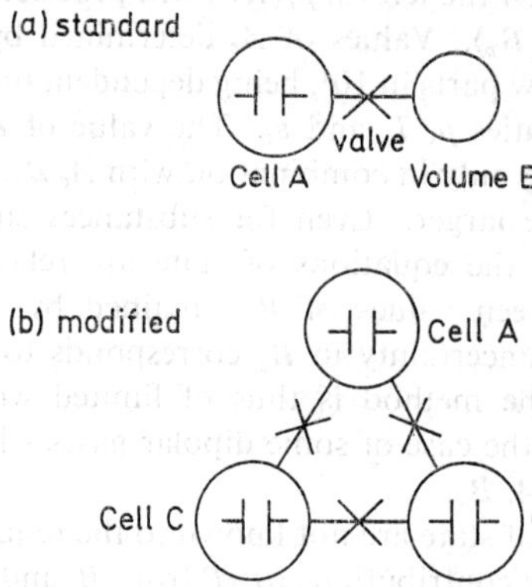

Figure 1 *Schematics of Burnett expansion techniques*

Cole [38, 61] has also developed a modification of the Burnett technique, involving three measuring cells with matched volumes and shown schematically in Figure 1(b). Initially all three cells are filled with gas, the valves closed, and ε_r is measured. Cell A is then evacuated, the gas from B and C allowed to expand into A, and another measurement of ε_r made. The procedure is repeated cyclically, evacuating cells B, C, A, B, C in turn. The values of f before and after a complete cycle of expansions are related by

$$f_i - R^{-1} f_{i+3} = \frac{B_\varepsilon}{A_\varepsilon^2}(1 - R) f_i^2 + \left[\frac{C_\varepsilon}{A_\varepsilon^3}(1 - R^2) - \frac{2B_\varepsilon^2}{A_\varepsilon^4}(1 - R)\right] f_i^3 + \ldots \tag{33}$$

Owing to the cyclic procedure, the density ratio R differs from $(2/3)^3$ only by terms which are quadratic or higher powers of the relative volume differences between cells. The two main advantages of this modification over the standard Burnett technique are: first, three values of ε_r are taken at each density; second, the volume ratio can be determined to a few parts in 10^6 independently of the dielectric measurements.[38] Values of B_ε obtained

[61] R. H. Orcutt and R. H. Cole, *J. Chem. Phys.*, 1967, **46**, 697.

using this method are believed to be a factor 2—5 times more accurate than those obtained from the standard Burnett technique.

Buckingham [62] has proposed another variation on the expansion procedure which consists of first measuring the sum of capacitances or optical pathlengths of two matched cells, one filled with gas and the other evacuated. The gas is then expanded to fill both cells and the change in the sum noted. The change, which is directly related to the non-ideality of the property under investigation, is then used to derive the virial coefficients. Preliminary measurements [62] indicate that the degree of accuracy of this method is similar to that of the modified Burnett technique. It does, however, possess the additional advantage that it automatically compensates for any linear variation with pressure of the amount of gas adsorbed on cell surfaces.

4 Results and Their Interpretations

Tables 2—4 (see Appendix, p. 95) list what the author believes are the most accurate values of the dielectric virial coefficients obtained to date. The values of the coefficients for a particular gas are all taken from a single paper, except in the case of a few dipolar gases where the authors published values of A_ε and B_ε in separate papers. Emphasis has been placed on accurate values of B_ε, and for some gases, particularly non-dipolar ones, more accurate values of A_ε can be found elsewhere in the literature (*e.g.* Table II of ref. 61). The uncertainty limits listed are those given by the authors. In most cases involving expansion techniques these reflect deviations of the experimental points from the computed least-squares curve, and contain no estimate of possible systematic errors. The papers listed under other references contain data which will lead to values of the virial coefficients considered less accurate than those given in the tables. In many cases these papers contain no actual values of the coefficients, but rather data from which values can be obtained.

A_ε.—The values listed in the tables were obtained by measuring ε_r as a function of T and p and extrapolating to $p = 0$ ($V_m \rightarrow \infty$) using equation (30). Values of α_0 and μ_0 can then be obtained using equation (9). The dipole moment values extracted in this fashion compare favourably with those determined by other methods.[38, 40]

B_ε of Atomic Gases.—For these gases only the first term in equation (11) contributes to B_ε. Orcutt and Cole [61] have calculated values of B_ε at 322 K using equation (23b) for α_{12} and a Lennard-Jones 6–12 intermolecular potential. In units of 10^{-12} m^6 mol^{-2}, they obtain the following values for B_ε: He, 0.03; Ne, 0.05; Ar, 2.0; Kr, 6.3. In all cases the calculated values are too large when compared with experiment, the relative disagreement decreasing on going from He to Kr. Ely and McQuarrie [63] have shown

[62] A. D. Buckingham, R. H. Cole, and H. Sutter, *J. Chem. Phys.*, 1970, **52**, 5960.
[63] J. F. Ely and D. A. McQuarrie, *J. Chem. Phys.*, 1971, **54**, 2885.

that quantum corrections to equation (11) will change the calculated value for He by less than 2% and those for Ne, Ar, and Kr by less than 0.5%. Since the Lennard-Jones parameters for these gases are well known, and no choice of intermolecular potential can explain the negative values observed for He and Ne, it seems clear that equation (23b) is a poor representation for α_{12}. A similar conclusion has been reached by Gray and Ralph and by Birnbaum and co-workers [64] in their interpretation of collision-induced depolarization ratios of atomic gases. In addition, it is well verified experimentally [65] that ${}_{\mathrm{T}}P$ and R_{m} of these substances are less in condensed phases than in the gas, indicating that short-range interactions decrease the polarizability, contrary to what would be expected on the basis of equation (23).

The Hartree–Fock calculations of Lim *et al.*[13] on He pairs show a decrease in polarizability at short range. This effect is outweighed, however, by an increase in polarizability at long range, and the value of B_ε they calculate is still positive. Calculations of B_ε are, of course, sensitive to the value of α_0 used, and there is some question as to the reliability of the procedure used by Lim *et al.* to obtain their value. Of the semi-classical calculations of α_{12} only Heinrich's two-state calculation [19] seems compatible with the experimental results. It is questionable, however, whether the two-state model is applicable to rare-gas atoms which have a number of electronic states close to the first excited state. Furthermore, Heinrich's theory takes no account of electron-overlap repulsion at short range. In summary, it seems fair to say that although the effects contributing to B_ε for the atomic gases are qualitatively understood, no satisfactory quantitative calculation of these effects has been made.

B_ε of Non-dipolar, Molecular Gases.—For these gases both terms in equation (11) contribute to B_ε. Attempts at separating the contributions have proceeded in three ways: (*i*) by measuring the Lorentz–Lorenz function at optical frequencies and assuming that the induced contribution to B_ε, $B_{\varepsilon\mathrm{ind}}$, is equal to B_R; (*ii*) by assuming that the temperature dependence introduced by the Boltzmann factor is small and may be ignored, leaving the induced term temperature independent; (*iii*) by using equation (12) to calculate $B_{\varepsilon\mathrm{or}}$ from pressure-induced absorption measurements. The first method has proved ineffective because all values of B_R reported in the literature to date were obtained using equation-of-state data to compute V_{m} and are therefore unreliable. Furthermore, the assumption $B_R = B_{\varepsilon\mathrm{ind}}$ is open to question, particularly for complicated molecules such as SF_6. Method (*ii*) is equivalent to assuming $u_{12} \ll kT$ and is strictly applicable

[64] C. G. Gray and H. I. Ralph, *Phys. Letters (A)*, 1970, **33**, 165; J. P. McTague and G. Birnbaum, *Phys. Rev. (A)*, 1971, **3**, 1376; H. B. Levine and G. Birnbaum, *J. Chem. Phys.*, 1971, **55**, 2914.

[65] R. L. Amey and R. H. Cole, *J. Chem. Phys.*, 1964, **40**, 146; C. P. Abbis, C. M. Knobler, R. K. Teague, and C. J. Pings, *J. Chem. Phys.*, 1965, **42**, 4145; C. M. Knobler, C. P. Abbiss, and C. J. Pings, *J. Chem. Phys.*, 1964, **41**, 2200; A. C. Sinnock and B. L. Smith, *Phys. Letters (A)*, 1968, **28**, 22.

only at high reduced temperatures. At the present time method (*iii*) seems to be the most reliable, but as can be seen from Table 5 (see Appendix, p. 99) it has been carried out for only a few gases.

Assuming that the separation can be accomplished, primary interest has been focussed on determining multipole moments from $B_{\varepsilon_{or}}$. In this case u_Q is determined from some other property (*e.g.* B_p or gas viscosity) and u_P and μ_{12}^2 are written in terms of the leading non-vanishing moment. The value of this moment is then adjusted until agreement is obtained between the calculated and experimental values of $B_{\varepsilon_{or}}$. As noted earlier, the expressions for μ_{12}^2 and u_P involve the square of the moment, making the calculated value of $B_{\varepsilon_{or}}$ very sensitive to the magnitude of the moment, but independent of its sign. This method has proved to be one of the most useful for determining the higher moments.

The values of the quadrupole moments of H_2 ($\pm$ 2.0 $\times$ 10^{-40} C m^2) [66] and N_2 ($\pm$ 5 $\times$ 10^{-40} C m^2),[67] obtained from pressure-induced absorption measurements, agree with the best values determined by other methods.[68] The values of $B_{\varepsilon_{or}}$ at 323 K, calculated using these moments, are 1.5 $\times$ 10^{-12} m^6 mol^{-2} for N_2 and 0.018 $\times$ 10^{-2} m^6 mol^{-2} for H_2, while the values of $B_{\varepsilon_{ind}}$ calculated using equation (23b), $u_Q = u_{6-12}$ and $u_P = 0$ are 1.8 $\times$ 10^{-12} m^6 mol^{-2} for N_2 and 0.036 $\times$ 10^{-12} m^6 mol^{-2} for H_2.[61] In both cases the total calculated B_ε is larger than the observed value, the most probable explanation being the inadequacy of the classical expression for α_{12}.

For CO_2 the value of $B_{\varepsilon_{ind}}$ obtained by method (iii) is negative ($\approx$ $-$ 1 $\times$ 10^{-12} m^6 mol^{-2} at 323 K), and although highly uncertain ($\pm$ 6 $\times$ 10^{-12} m^6 mol^{-2}) is definitely less than the value of 9.3 $\times$ 10^{-12} m^6 mol^{-2} calculated using equation (23b). Bose and Cole [69] have performed extensive calculations of B_ε considering such effects as the anisotropy of polarizability and the anisotropy of both the attractive and repulsive parts of u_Q. They found that the calculated value of $B_{\varepsilon_{or}}$ changes drastically (60%) if u_{elec} is ignored but only slightly (8%) if u_{ind} or the anisotropy of polarizability is ignored. Using a value of 4.0 $\times$ 10^{-12} m^6 mol^{-2} for $B_{\varepsilon_{ind}}$ (based on early pressure-induced absorption measurements of Birnbaum [70]), they obtained a value of $\pm$ 14.4 $\times$ 10^{-40} C m^2 for Θ, while Birnbaum *et al.*[71] have found that a value $\pm$ 15.0 $\times$ 10^{-40} C m^2 fits their more recent pressure-induced absorption data at all temperatures. Both values are in excellent agreement with the value of $-$ 14.3 $\times$ 10^{-4} C m^2 obtained by the direct measurement of Buckingham *et al.*[72]

66 J. P. Colpa and J. A. A. Ketelaar, *Mol. Phys.*, 1958, **1**, 343.
67 J. A. A. Ketelaar and R. P. H. Rettschnick, *Mol. Phys.*, 1963, **7**, 191; J. D. Poll, *Phys. Letters*, 1963, **7**, 32.
68 D. E. Stogryn and A. P. Stogryn, *Mol. Phys.*, 1966, **11**, 371.
69 T. K. Bose and R. H. Cole, *J. Chem. Phys.*, 1970, **52**, 140.
70 G. Birnbaum, unpublished results.
71 W. Ho, G. Birnbaum, and A. Rosenberg, *J. Chem. Phys.*, 1971, **55**, 1028, 1039.
72 A. D. Buckingham, R. L. Disch, and D. A. Dunmur, *J. Amer. Chem. Soc.*, 1968, **90**, 3104.

Ethylene has proved a most interesting gas, both experimentally and theoretically. Early measurements [51, 59] showed B_ε increasing with increasing temperature. Later measurements,[62, 73] however, indicated that this apparent anomaly was due to experimental error. The separation of $B_{\varepsilon_{or}}$ and $B_{\varepsilon_{ind}}$ is hampered by a lack of experimental data. Birnbaum and Maryott [74] have derived a value of 9×10^{-12} m^6 mol^{-2} for $B_{\varepsilon_{or}}$, based on a theoretical lineshape and their measurement of the dielectric loss at 24 GHz. This in turn gives a value of over 40×10^{-12} m^6 mol^{-2} for $B_{\varepsilon_{ind}}$, a figure far larger than one would expect from comparison with CO_2, and about three times the value (16×10^{-12} m^6 mol^{-2}) calculated using equation (23b). In deriving their theoretical lineshape, Birnbaum and Maryott neglect contributions to the intermolecular forces from u_{elec} and u_{ind}, assume the molecules are linear, and approximate the resonant contributions to the loss by a non-resonant formula. In the case of C_2H_4 these three assumptions seem questionable, the first seeming especially drastic in view of the strong dependence of $B_{\varepsilon_{or}}$ of CO_2 on u_{elec}. Bose and Cole [73] have used method (*ii*) to obtain a value of 6.0×10^{-12} m^6 mol^{-2} for $B_{\varepsilon_{ind}}$, which seems roughly in line with the values for other gases. The applicability of this method is also open to question, however, since ε/k is 230 K for C_2H_4. Ethylene has two quadrupole moments which are in the plane of the molecule, one parallel and the other perpendicular to the carbon—carbon bond. Although their separate contributions to $B_{\varepsilon_{or}}$ may be resolved in pressure-induced rotational spectra,[75] no values exist for the individual moments at present. By assuming a 6–12 intermolecular potential (neglecting u_P) Bose and Cole were able to obtain an approximate value of 13.0×10^{-40} C m^2 for $\bar{\Theta} = |\Theta_1 - \Theta_2|$.

Birnbaum and Rosenberg [76] have determined values of $B_{\varepsilon_{or}}$ for the tetrahedral molecules CH_4, CD_4, and CF_4. Using a 6–12 intermolecular potential, they obtained the following values for the octopole moments: CH_4, $\pm 8.3 \times 10^{-50}$ C m^3; CD_4, $\pm 8.0 \times 10^{-50}$ C m^3; CF_4, $\pm 15.0 \times 10^{-50}$ C m^3. For CH_4, use of their value of $B_{\varepsilon_{or}}$ and Johnson and Cole's [59] value of B_ε gives a value of 6.3×10^{-12} m^6 mol^{-2} for $B_{\varepsilon_{ind}}$, compared with the value of 13.2×10^{-12} m^6 mol^{-2} found using equation (23b).

For SF_6, the measurements of Rosenberg and Birnbaum [77] give a very small value of $B_{\varepsilon_{or}}$, comparable to those obtained from atomic gas mixtures. By neglecting possible contributions from dipoles caused by overlap forces, and attributing all of $B_{\varepsilon_{or}}$ to the effects of the hexadecapole moment, Rosenberg and Birnbaum calculate an upper limit of $\pm 17 \times 10^{-60}$ C m^4 for Φ. In view of the accuracy of both the measurement of $B_{\varepsilon_{or}}$ and the

[73] T. K. Bose and R. H. Cole, *J. Chem. Phys.*, 1971, **54**, 3829.
[74] G. Birnbaum and A. A. Maryott, *J. Chem. Phys.*, 1962, **36**, 2026, 2032.
[75] J. Zwicker, unpublished results.
[76] G. Birnbaum and A. Rosenberg, *Phys. Letters* (*A*), 1968, **27**, 272; *J. Chem. Phys.*, 1968, **48**, 1396.
[77] A. Rosenberg and G. Birnbaum, *J. Chem. Phys.*, 1970, **52**, 683.

value of B_ε obtained by Nelson and Cole,[78] it seems unlikely that the large value of $B_{\varepsilon \text{ind}}$ ($\approx 64 \times 10^{-12}$ m^6 mol^{-2}), which is roughly twice the value calculated using equation (23b), is anything but real. A possible explanation may lie in a large nuclear contribution to $B_{\varepsilon \text{ind}}$. This is suggested by two facts: first, SF_6 has an unusually large nuclear contribution to A_ε of 30%; second, it possesses a number of vibrational modes which are not i.r. active in the isolated molecule but which would become so in the presence of a second molecule. An examination of the pressure-induced vibrational spectrum might thus prove of interest.

B_ε of Dipolar Gases.—A glance at Table 3 (p. 97) shows that, for dipolar gases, both positive and negative values of B_ε are found, which are generally orders of magnitude larger than those for non-dipolar gases. Although both terms of equation (11) contribute to B_ε for these gases, the usual assumption made is that $B_{\varepsilon \text{ind}}$ is negligibly small compared with $B_{\varepsilon \text{or}}$. On the basis of the values of $B_{\varepsilon \text{ind}}$ found for non-dipolar molecules, this assumption seems reasonable enough, but there are some indications that it may not be generally valid. As mentioned earlier, Buckingham [14] has shown that, owing to contributions proportional to $\beta_0\,\mu_0\langle r^{-3}\rangle$, the value of $B_{\varepsilon \text{ind}}$ calculated for dipolar molecules using classical expressions for α_{12} can be of the order of 100—200 $\times 10^{-12}$ m^6 mol^{-2}. Everett and Munn [46] have measured B_R for methanol, diethylamine, and triethylamine and found values of 25 300 $\pm$ 1000 $\times 10^{-12}$ m^6 mol^{-2} (303.4 K), 83 100 $\pm$ 1000 $\times 10^{-12}$ m^6 mol^{-2} (303.7 K), and 49 500 $\pm$ 1000 $\times 10^{-12}$ m^6 mol^{-2} (313.6 K), respectively. Although these results were obtained using an equation of state for V_m, the values of B_R are so large ($> A_R B_p$) that the uncertainty caused is relatively small. The high values for the first two gases may possibly be due to hydrogen bonding which substantially alters the electronic structure of the molecules, but such effects should not be present in triethylamine. An attempt to interpret the value for triethylamine in terms of the β_0 hyperpolarizability leads to a value for β_0 three orders of magnitude larger than has been found in other molecules. Blythe *et al.*[45] have measured B_R for the gases CH_3F, CH_3Cl, NH_3, and SO_2, using equation-of-state data for V_m, and obtained values of 6.6 $\times 10^{-12}$ m^6 mol^{-2} (298 K), $-$ 23 $\times 10^{-12}$ m^6 mol^{-2} (295 K), $-$ 124 $\times 10^{-12}$ m^6 mol^{-2} (298 K), and $-$ 90 $\times 10^{-12}$ m^6 mol^{-2} (298 K), respectively. Although these values are rather uncertain ($\pm$ 100 or more), they do give some reassurance that, for these molecules at least, $B_{\varepsilon \text{ind}}$ is not too large.

One set of refractivity measurements which are rather difficult to interpret are those of Beaume and Coulon [79] on the Lorentz–Lorenz function of NH_3 as a function of frequency and density. Plots of their data indicate that B_R is very dependent on frequency, having the values 1130 $\pm$ 100 $\times 10^{-12}$ m^6 mol^{-2}, 680 $\pm$ 100 $\times 10^{-12}$ m^6 mol^{-2}, 450 $\pm$ 100 $\times 10^{-12}$ m^6 mol^{-2}, and 220 $\pm$ 100 $\times 10^{-12}$ m^6 mol^{-2} at the

78 R. D. Nelson, jun. and R. H. Cole, *J. Chem. Phys.*, 1971, **54**, 4033.
79 R. Beaume and R. Coulon, *Compt. rend.*, 1967, **265**, *B*, 309.

wave lengths 447.1 nm, 501.5 nm, 587.6 nm, and 667.8 nm, respectively. The difference between their value and that of Blythe *et al.* (taken at 546.0 nm) may possibly be due to the use of a different equation of state. The frequency dependence of B_R is, however, independent of the equation of state and most unexpected. Although a possible explanation might be a low-lying electronic absorption band for the pair, causing the denominator of equation (15) to be frequency dependent, a rough extrapolation of B_R^{-1} *vs.* λ^{-2} indicates that this transition should lie in the 300—400 nm region, where both the gas at low pressures and the solid are transparent.

If the possibility of $B_{\varepsilon_{ind}}$ contributing significantly to B_ε is ignored and the classical expression for μ_{12}^2 accepted, it is immediately clear that the potentials of equations (26)—(28) are inadequate, for they predict positive values only for B_ε, *i.e.* the potential-energy minimum at a given internuclear distance is for a collinear, head-to-tail alignment of dipoles. It is for this reason that attempts at interpreting B_ε for dipolar gases have generally focussed on finding an adequate intermolecular potential.

Perhaps the simplest sort of modification that can be made to the potential is to include terms in u_{elec} and u_{ind} due to the higher moments. Until recently such modification was hampered by the lack of accurate values for these moments. Although recent measurements of the Zeeman effect [80] and pressure-induced rotational absorption [81] have begun to yield values for the quadrupole moments of dipolar molecules, to date no attempts have been made to incorporate these values into calculations of B_ε.

Smith and co-workers [82] have proposed moving the dipole moment a distance a away from the origin of u_Q. This 'off-centre dipole' model is equivalent to having the dipole plus an infinite series of higher multipole moments of value $na^{n-1}\,\mu_0$, at the origin, where n is the order of the moment and takes on the values 2, 3, 4, ... The model thus includes one additional adjustable parameter, a, but eliminates the need for specific values of the higher moments. As shown in Figure 2(a), such a model provides for negative values of B_ε by favouring an antiparallel alignment of dipoles. By judicious choice of r_0, a, and ε, a value of B_ε more positive than that found with the dipole at the origin can also be obtained.

The Lennard-Jones 6–12 potential of equation (26) is angle independent and, strictly speaking, valid only for spherically symmetric molecules. Since dipolar molecules cannot be spherically symmetric, a number of authors have proposed angle-dependent forms of u_Q designed to reflect the shape of the molecule. Although a large number of such potentials have been proposed, difficulties of integration severely limit the usefulness of most, and to date only two have been applied to studies of B_ε.

[80] W. H. Flygare and R. C. Benson, *Mol. Phys.*, 1971, **20**, 225.
[81] S. Weiss and R. H. Cole, *J. Chem. Phys.*, 1967, **46**, 644; A. I. Baise, *Chem. Phys. Letters*, 1971, **9**, 627.
[82] K. P. Lawley and E. B. Smith, *Trans. Faraday Soc.*, 1963, **59**, 301; J. H. Dymond and E. B. Smith, *ibid.*, 1964, **60**, 1378.

Buckingham and Pople[2] have proposed adding an angle-dependent term to the repulsive part of the 6–12 potential. The term contains an adjustable 'shape factor', D, which has the effect of giving an ellipsoidal shape to the potential. As shown in Figure 2(b), a positive value of D gives an ellipsoid whose long axis is parallel to the dipole (cigar shape) and favours antiparallel alignment of dipoles at all internuclear separations, whereas a negative D places the long axis perpendicular to the dipole (plate shape)

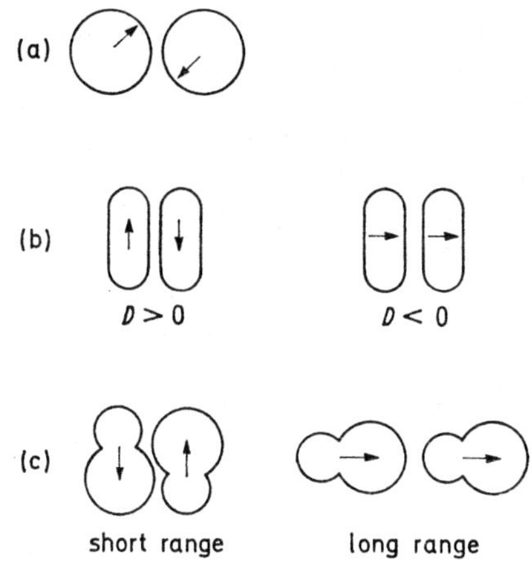

Figure 2 *Intermolecular potentials for dipolar gases showing favoured orientations*

and favours a parallel, head-to-tail alignment at all separations. A sufficiently positive value of D thus produces a negative value of B_ε, whereas a negative value of D produces a large positive value of B_ε.

Sweet and Steele[83] and Kong[34] have chosen to divide the molecule into two constituent groups and centre a 6–12 potential on each. The potential u_Q is then written as the sum of all pair interactions between the constituent groups of the two molecules. After a number of approximations, Kong obtains a potential consisting of a single 6–12 term with adjustable parameters r_0 and ε, and directional terms whose parameters are fixed by the 6–12 functions of the constituent groups. The latter are determined empirically from pressure-virial coefficients. This model differs from the Buckingham and Pople model in two respects: firstly, it has one less adjustable parameter; secondly, it predicts that all molecules prefer $\rightarrow\ \rightarrow$ orientations at long range and $\uparrow\ \downarrow$ at short range, the difference between cigar- and plate-shaped molecules simply being that the latter favour the $\rightarrow\ \rightarrow$ orientation to much smaller separations than the former.

Sutter and Cole[38, 84] have applied the Buckingham and Pople model to the halogenomethanes CH_3F, CHF_3, CH_3Cl, $CHClF_2$, and $CClF_3$.

[83] J. R. Sweet and W. A. Steele, *J. Chem. Phys.*, 1967, **47**, 3022, 3029; *ibid.*, 1969, **50**, 668.
[84] H. Sutter and R. H. Cole, *J. Chem. Phys.*, 1971, **54**, 4988.

Pressure virial coefficients, which are insensitive to the value of the shape factor, were used to evaluate r_0 and ε, and these in turn were used to extract a value of D from B_ε. They found that the model roughly reproduces the temperature dependence of B_ε and predicts the expected shapes for CH_3F (cigar), CH_3Cl (cigar), $CHClF_2$ (plate), and $CClF_3$ (spherical). Contrary to expectations, however, the model yields a cigar shape for CHF_3. A recently-discovered error in their numerical integration program may change the reported values for D slightly, but is expected to leave their findings qualitatively unchanged. Nelson and Cole [78] have shown that the experimental data for $CClF_3$ can also be fitted by neglecting shape considerations, retaining octopole terms in μ_{12} and u_P, and assuming a value of 33×10^{-50} C m^3 for the octopole moment.

Barnes and Sutton [40] have attempted to fit their values of B_ε for simple dipolar gases with both the Buckingham and Pople and off-centre-dipole models. They concluded, in agreement with Sutter and Cole, that the Buckingham and Pople model requires values of D at odds with expected molecular shapes. The off-centre-dipole calculations could not be made to agree with experiment unless the shape parameter D was used as well. Kong and Larson,[34] using the two-centre model and the Pople expansion method of integration, have published calculated values of B_ε in qualitative agreement with the experimental results of Sutter and Cole. However, recent calculations of Brookmeyer and Cole,[85] using numerical integration techniques, show that this agreement is due to a fortuitous combination of the errors arising from the expansion of the two-centre potential and the Pople integration technique.

In summary, it seems clear that B_ε of dipolar molecules is very sensitive to both molecular shape and higher-order moments. At the present time, however, no single model has been found which produces agreement between calculation and experiment.

B_ε for Mixtures.—The values of the dielectric virial coefficients of a mixture are determined in the same way as those of a pure gas. For a two-component mixture we can, in the absence of reaction, write

$$A_{\varepsilon_{\text{obs}}} = \sum_{i=1}^{2} x_i A_{\varepsilon_i} = x_1 A_{\varepsilon_1} + (1 - x_1) A_{\varepsilon_2} \tag{35a}$$

$$B_{\varepsilon_{\text{obs}}} = \sum_{ij=1}^{2} x_i x_j B_{\varepsilon_{ij}} = x_1^2 B_{\varepsilon_{11}} + 2x_1(1 - x_1) B_{\varepsilon_{12}} + (1 - x_1)^2 B_{\varepsilon_{22}} \tag{35b}$$

$$C_{\varepsilon_{\text{obs}}} = \sum_{ijk=1}^{2} x_i x_j x_k C_{\varepsilon_{ijk}} = x_1^3 C_{\varepsilon_{111}} + 3x_1^2(1 - x_1) C_{\varepsilon_{112}} + 3x_1(1 - x_1)^2 C_{\varepsilon_{122}} + (1 - x_1)^3 C_{\varepsilon_{222}} \tag{35c}$$

85 B. Brookmeyer and R. H. Cole, personal communication.

where $A_{\varepsilon_{obs}}$, $B_{\varepsilon_{obs}}$, *etc.* are the measured dielectric virial coefficients for the mixture and x_i is the mole fraction of component i. Analysis of experimental results usually involves the use of equation (35a) along with known values of the pure-gas first virial coefficients, A_{ε_1} and A_{ε_2}, to determine x_1. The value of x_1 is then combined with known values of the pure-gas second virial coefficients, $B_{\varepsilon_{11}}$ and $B_{\varepsilon_{22}}$, to obtain a value for the mixed second virial coefficient $B_{\varepsilon_{12}}$. Note that there are two mixed third virial coefficients, $C_{\varepsilon_{112}}$ and $C_{\varepsilon_{122}}$, and that their determination requires values of $C_{\varepsilon_{obs}}$ for a minimum of two mixtures with different compositions.

Although accurate values of mixed coefficients are slightly more difficult to obtain than pure-gas values, they are attractive theoretically for two reasons. Firstly, by careful choice of components the interaction terms in μ_{12}^2 and u_P can be simplified, and secondly, mixed virials provide a stiff test of the intermolecular potential parameters determined from pure-gas virials.

The classical expressions for α_{12}, μ_{12}^2, and u_P are similar to the pure-gas expressions with the slight added complication of having two sets of isolated-molecule parameters: α_{0_1}, α_{0_2}, μ_{0_1}, μ_{0_2}, *etc.* The determination of the mixed-pair parameters involved in u_Q is much more difficult. The determination of these parameters from mixed viscosities and pressure virial coefficients is hampered by a shortage of experimental data, and the usual procedure is to employ a set of empirical combining rules which relate the mixed parameters to those of the pure gases. A number of such rules have been proposed for the 6–12 potential,[86] the most widely used for dielectric calculations being

$$r_{0_{12}} = \tfrac{1}{2}(r_{0_1} + r_{0_2}), \quad \varepsilon_{12} = (\varepsilon_1 \varepsilon_2)^{\frac{1}{2}} \tag{36}$$

where r_{0_1} and ε_1 are the pure-gas parameters. The very lack of experimental data that makes the use of these rules necessary also makes their accuracy difficult to assess. DiPippo and Kestin [86] have shown that, for mixtures involving atomic gases, CO_2, and N_2, differences between mixed-parameter values found from equation (36) and those obtained directly from mixed viscosities are as large as 5% for $r_{0_{12}}$ and 100% for ε_{12}. Dorsch and De Rocco [86] have found that the use of equation (36) for the mixtures $CH_4 + SF_6$, $CH_4 + (Me)_4C$, and $CH_4 + (Me)_4Si$ leads to calculated values of B_p which differ by over 20% from those obtained experimentally. It thus appears that, although there usually exists little choice other than to employ combining rules, values of $B_{\varepsilon_{12}}$ so calculated must be viewed with caution.

The work of Bose and Cole on mixtures of CO_2[69] and C_2H_4[73] with Ar illustrates the simplifications possible in calculations of mixed virials. Since Ar has no multipole moments, $u_{12\text{elec}}$ is zero and μ_{12}^2 is also simplified.

[86] H. M. Lin and R. L. Robinson, jun., *J. Chem. Phys.*, 1971, **54**, 52; R. DiPippo and J. Kestin, *J. Chem. Phys.*, 1968, **49**, 2192; R. R. Dorsch and A. G. De Rocco, *J. Chem. Phys.*, 1971, **54**, 2727.

Moreover, Bose and Cole were able to show that values of $B_{p_{12}}$ calculated using equation (36) were within 10% of their experimentally determined values, indicating that the combining rules are a good approximation for these mixtures. Analysis of $B_{\varepsilon_{12}}$ of the CO_2 + Ar mixtures, for the quadrupole moment of CO_2 yielded a value of $\pm$ 13.7 $\times$ 10^{-40} C m^2, in good agreement with the value obtained from B_ε of the pure gas. In their analysis of $B_{\varepsilon_{12}}$ for C_2H_4 + Ar, Bose and Cole were forced to assume, as they did for pure C_2H_4, that $u_P = u_{elec} + u_{ind} = 0$. Since this assumption is more nearly valid for the mixed virial, the value of $\bar{\Theta}$ obtained from $B_{\varepsilon_{12}}$ (14.2 $\times$ 10^{-40} C m^2) is probably more reliable than the value obtained from the pure-gas coefficient (13.0 $\times$ 10^{-40} C m^2).

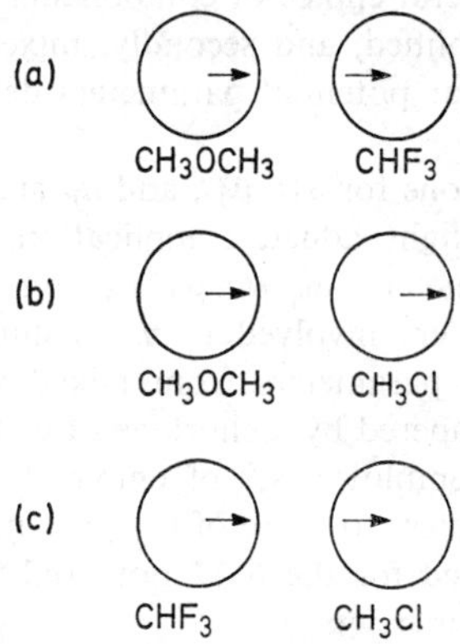

Figure 3 *Off-centre-dipole model for mixtures*

The work of Barnes and Sutton [40] on polar gas mixtures illustrates the use of $B_{\varepsilon_{12}}$ to test the intermolecular potential. As Barnes and Sutton point out, one of the hoped-for advantages of the off-centre-dipole model would be its ability to deal with values of $B_{\varepsilon_{12}}$ where $B_{\varepsilon_{12}}$ does not lie between $B_{\varepsilon_{11}}$ and $B_{\varepsilon_{22}}$. Unfortunately, the fact that the model fails even qualitatively can be seen by considering Figure 3.[40] Both CH_3OCH_3 and CHF_3 have pure-gas values of B_ε of the order of 1—2 $\times$ 10^{-9} m^6 mol^{-2}. The value of $B_{\varepsilon_{12}}$ for these gases is, however, of the order of 20 $\times$ 10^{-9} m^6 mol^{-2}. This can be explained by simply assuming that the dipole of one points out from the centre whereas the other points into the centre, thus favouring a head-to-tail alignment in the mixture [Figure 3(a)]. For CH_3OCH_3 + CH_3Cl, the value of $B_{\varepsilon_{12}}$ is about the same size as for pure CH_3OCH_3, leading to the conclusion that the dipole of CH_3Cl should be pointing out from the centre as in CH_3OCH_3 [Figure 3 (b)]. A mixture of CHF_3 and CH_3Cl would thus be expected to have a large positive value of $B_{\varepsilon_{12}}$ since, as illustrated in Figure 3(c), head-to-tail alignment should be favoured. Unfortunately, the negative value of $B_{\varepsilon_{12}}$ found experimentally is completely at odds with this prediction.

Studies of the pressure-induced absorption of atomic-gas mixtures offer an excellent opportunity to examine non-classical contributions to μ_{12}. As can be seen from Table 5 (p. 99), the value of $B_{\varepsilon_{or}}$ for these mixtures, due solely to 'non-classical' dipole moments, is very small. Similar small contributions to $B_{\varepsilon_{or}}$ for other gases would be difficult to observe in the presence of the generally much larger contributions from multipole moments. As Buckingham [87] has shown, the dipole-moment function of a pair of spherical atoms of different dimensions is composed of two terms, a short-range overlap term, similar to the repulsive term of the intermolecular potential, and a long-range electron-correlation term, similar to the London dispersion term of the potential. The long-range correlation term can be expanded as a multipole series which, for atoms in S states, begins

$$\mu_{C_{12}}(r) = \frac{D_7}{r_{12}^7} + \frac{D_9}{r_{12}^9} + \ldots \tag{37a}$$

The functional form of the short-range overlap term is much more complicated but has usually been simplified to

$$\mu_{S_{12}}(r) = A_S \exp(-\lambda_S r_{12}) \tag{37b}$$

Buckingham has calculated both long- and short-range contributions to μ_{12} for the He + H pair and finds that they are opposite in sign, with the D_7 term predominating at all distances but especially at short range, where it increases sharply. Presumably, this sharp increase is due to the failure of the perturbation method which leads to the D_7 term and assumes no overlapping of wave functions. Matcha and Nesbet [88] have calculated the overlap dipole for He–Ne, He–Ar, and Ne–Ar pairs using Hartree–Fock theory. They have found that their results can be fitted by equation (37b), and that the overlap dipoles are very small, being $< 0.033 \times 10^{-30}$ C m (0.01 D) at $r_{0_{12}}$. Byers Brown and Whisnant [89] have derived formulae for the calculation of D_7 for two S-state atoms using isolated-molecule parameters. Although they give no values for D_7, they do find that the contributions to μ_{12} from the long-range term are opposite in sign to the short-range contributions calculated by Matcha and Nesbet. Brenner and McQuarrie [90] have attempted to fit the data of Bosomworth and Gush [28] with a dipole-moment function composed of the D_7 term of equation (37a) plus the exponential term of equation (37b). They find that there is no value of D_7 which can be combined with the values of A_S and λ_S calculated by Matcha and Nesbet to reproduce the experimental spectrum. Thus it appears, as was the case for α_{12}, that although the qualitative behaviour of non-classical contributions to μ_{12} is understood, no calculations exist that are in accord with experiment.

[87] A. D. Buckingham, *Colloq. Internat. Centre natl. Rech. sci. (Paris)*, 1959, **77**, 57.
[88] R. L. Matcha and R. K. Nesbet, *Phys. Rev.*, 1967, **160**, 72.
[89] W. Byers Brown and D. M. Whisnant, *Chem. Phys. Letters*, 1970, **7**, 329.
[90] S. L. Brenner and D. A. McQuarrie, *Canad. J. Phys.*, 1971, **49**, 837.

C_ε and Higher Coefficients.—As can be seen from Tables 2—4, few reliable values of C_ε exist, while for D_ε and higher coefficients no values have been reported at all. As mentioned earlier, calculations of C_ε are difficult to carry out owing to difficulties of integration and a lack of knowledge about three-body intermolecular interactions. Ho *et al.*[71] have used the integrals tabulated by McQuarrie and Levine[30] to calculate $C_{\varepsilon_{or}}$ for CO_2 and found a value of -8.2×10^{-14} m^9 mol^{-3} at 296 K. Since the calculations depend on the pairwise additivity of intermolecular forces and assume $u_P = 0$, they must be considered to be in good agreement with the value of -14.4×10^{-14} m^9 mol^{-3}, which Ho *et al.* derive from their pressure-induced absorption measurements.

Summary.—To date, studies of dielectric virial coefficients have added much to our knowledge of charge distributions, polarizabilities, and interactions of molecules. These studies have also pointed out deficiencies in a number of present theories, most notably those attempting to describe the polarizability of a pair of molecules and the angular dependence of intermolecular forces. In the future much theoretical work needs to be done to remove these deficiencies. Much experimental work also needs to be done. Values of the second and higher dielectric virial coefficients have been determined for relatively few gases over a limited temperature range, and few, if any, reliable values exist for the second and higher refractivity virial coefficients.

The author wishes to thank Prof. R. H. Cole and Prof. A. D. Buckingham for their advice and suggestions concerning this manuscript, and Dr. C. Graham for many interesting and stimulating conversations.

Appendix: Selected Values of Dielectric Virial Coefficients

Table 2 *Selected values of dielectric virial coefficients of non-dipolar gases*[a]

Gas	*Temp.* (K)	$A_\varepsilon \times 10^6$ ($m^3\ mol^{-1}$)	$B_\varepsilon \times 10^{12}$ ($m^6\ mol^{-2}$)	$C_\varepsilon \times 10^{18}$ ($m^9\ mol^{-3}$)	*Ref.*	*Other Refs.*
^3He	3.8	0.51695 ± 0.00009	− 0.030 ± 0.004	—	39	
He	322.2	0.519 ± 0.001	− 0.06 ± 0.04	—	61	36, 50, 91, 92
Ne	322.2	0.998 ± 0.001	− 0.30 ± 0.10	—	61	
Ar	322.2	4.142 ± 0.002	0.39 ± 0.20	—	61	36, 44, 50, 51, 92—94. LL[b]: 95—97
	322.9	4.1397 ± 0.0006	0.72 ± 0.12	—	69	
Kr	322.2	6.267 ± 0.003	5.6 ± 0.3	—	61	
H_2	322.2	2.033 ± 0.001	0.03 ± 0.10	—	61	37, 91, 99—102
N_2	322.2	4.390 ± 0.002	0.60 ± 0.2	—	61	36, 44, 91, 99, 104—106. LL[b]: 96—98, 107, 108

[a] Methods of measurement: Burnett expansion, refs. 51, 59; modified Burnett expansion, refs. 61, 69, 73, 78; Buckingham expansion, ref. 62; equation of state, refs. 37, 44, 47, 50, 54, 91—118; expansion plus equation of state, refs. 36, 39; [b] LL, Lorentz–Lorenz function measured.

91 F. G. Keyes and J. L. Oncley, *Chem. Rev.*, 1936, **19**, 195.
92 G. J. Oudemans and R. H. Cole, *J. Chem. Phys.*, 1959, **31**, 843.
93 A. Michels, C. A. ten Seldam, and S. D. J. Overdijk, *Physica*, 1951, **17**, 781.
94 A. Cops, M. Caubergh, and W. van Dael, in 'Proceedings of the 3rd International Cryogenic Engineering Conference', Berlin, 1970, p. 504.
95 A. Michels and A. Botzen, *Physica*, 1949, **15**, 769.
96 C. E. Bennett, *Phys. Rev.*, 1940, **58**, 263.
97 C. E. Bennett, *Phys. Rev.*, 1935, **47**, 790.
98 C. E. Bennett, *Phys. Rev.*, 1934, **45**, 200.
99 H. H. Uhlig, J. G. Kirkwood, and F. G. Keyes, *J. Chem. Phys.*, 1933, **1**, 155.
100 R. McNabney, W. Moulton, and W. L. Beuschlein, *Phys. Rev.*, 1935, **47**, 695.
101 A. van Itterbeek and K. de Clippeleir, *Physica*, 1948, **14**, 349.
102 A. Michels, P. Sanders, and A. Schipper, *Physica*, 1935, **2**, 753.
103 R. Nacasch and R. Coulon, *J. Phys.* (*Radium*), 1966, **27**, 735.
104 H. L. Andrews, *Physics*, 1931, **1**, 366.
105 J. W. Broxon, *Phys. Rev.*, 1931, **38**, 2049.
106 A. Michels, A. Jaspers, and P. Saunders, *Physica*, 1934, **1**, 627.
107 A. Michels, H. and L. Lebesque, and S. R. de Groot, *Physica*, 1947, **13**, 337.
108 C. E. Bennett, *Phys. Rev.*, 1931, **37**, 263.

Table 2 (*cont.*)

Gas	*Temp.* (K)	$A_\varepsilon \times 10^6$ ($m^3\ mol^{-1}$)	$B_\varepsilon \times 10^{12}$ ($m^6\ mol^{-2}$)	$C_\varepsilon \times 10^{18}$ ($m^9\ mol^{-3}$)	*Ref.*	*Other Refs.*
CO_2	302.6	7.358 ± 0.002	57.6 ± 0.9	− 2764 ± 177	69	44, 47, 50, 51, 59, 91, 99, 109—113
	322.9	7.350 ± 0.006	50.7 ± 0.9	− 2515 ± 235	69	LL[b]: 54, 96, 114, 115
	369.5		36 ± 3	—	62	
CH_4	242	6.518 ± 0.001	9.0 ± 0.4	—	36	91, 99
	315	6.543 ± 0.001	7.3 ± 1.8	—	36	
C_2H_4	323	10.757 ± 0.005	47.2 ± 1.4	− 3028 ± 556	73	51, 59, 62, 116. LL[b]: 117
	373	10.730 ± 0.004	41.7 ± 1.2	− 2797 ± 489	73	
	423	10.727 ± 0.004	37.6 ± 2.4	− 2502 ± 959	73	
SF_6	323	16.478 ± 0.009	64.1 ± 0.4	− 6400 ± 320	78	
O_2	—	—	—	—		112. LL[b]: 118
CF_4	—	—	—	—	—	LL[b]: 45
C_3H_8	—	—	—	—		91

[b] LL, Lorentz–Lorenz function measured.

109 F. G. Keyes and J. G. Kirkwood, *Phys. Rev.*, 1930, **36**, 754.
110 A. Michels and L. Kleerekoper, *Physica*, 1939, **6**, 586.
111 A. Michels and C. Michels, *Phil. Mag.*, 1932, **13**, 1192.
112 A. van Itterbeek and J. Spaepen, *Physica*, 1944, **11**, 43.
113 A. van Itterbeek and K. de Clippeleir, *Physica*, 1947, **13**, 459.
114 A. Michels and J. Hamers, *Physica*, 1937, **4**, 995.
115 M. Opladen, *Z. Physik*, 1927, **42**, 160.
116 H. G. David, S. D. Hamann, and J. F. Pearse, *J. Chem. Phys.*, 1951, **19**, 1491.
117 A. Michels, A. Botzen, and S. R. de Groot, *Physica*, 1947, **13**, 343.
118 H. I. Eadie and J. Satterly, *Trans. Roy. Soc. Canada, Sec. II*, 1922, **16**, 63.

Table 3 *Selected values of dielectric virial coefficients of dipolar gases*[a]

Gas	*Temp.* (K)	$A_\varepsilon^6 \times 10^6$ ($m^3\ mol^{-1}$)	$B_\varepsilon \times 10^{12}$ ($m^6\ mol^{-2}$)	$C_\varepsilon \times 10^{18}$ ($m^9\ mol^{-3}$)	*Ref.*	*Other Refs.*
HCl	292.5	31.50	4000 ± 1000	—	119	103
	312.8	29.89	3600 ± 1000	—	119	
SO_2	292.7	65.78	1700 ± 1000	—	119	LL[b]: 45
	313.2	62.17	1700 ± 1000	—	119	
	334.7	58.98	1400 ± 1000	—	119	
CH_3F	323.2	72.13 ± 0.01	− 1307 ± 37	—	38	LL[b]: 45
	369.5	64.03 ± 0.01	− 606 ± 30	—	38	
	416.5	57.64 ± 0.02	− 331 ± 66	—	38	
CH_3Cl	303.2	84.17 ± 0.04	− 6650 ± 1200	—	40	LL[b]: 45
	323.2	79.55 ± 0.01	− 4470 ± 200	—	38	
	369.5	71.04 ± 0.03	− 2517 ± 50	—	38	
	404.8	65.87 ± 0.02	− 1696 ± 60	—	38	
CHF_3	303.2	63.62 ± 0.08	1315 ± 100	—	40	55, 58, 62, 82, 119—121
	323.2	60.35 ± 0.02	1125 ± 52	− 3·5 ± 0.5 × 10^5	38	
	369.5	53.87 ± 0.01	903 ± 20	− 2.6 ± 0.6 × 10^5	38	
	416.5	48.81 ± 0.01	704 ± 10	− 1.4 ± 0.4 × 10^5	38	
CH_3CN	353.79	273.68 ± 0.33	− 1.01 ± 0.11 × 10^6	—	55	
	373.63	258.50 ± 0.17	− 0.61 ± 0.03 × 10^6	—	55	
	405.24	239.26 ± 0.15	− 0.34 ± 0.06 × 10^6	—	55	
	433.15	224.01 ± 0.11	− 0.17 ± 0.01 × 10^6	—	55	
$CHClF_2$	303.2	56.88 ± 0.04	2600 ± 200	—	40	
	323.1	54.16 ± 0.03	2551 ± 63	—	84	
	373.4	48.72 ± 0.05	1676 ± 61	—	84	
	425.0	44.50 ± 0.01	1235 ± 93	—	84	

[a] Methods of measurement: Burnett expansion, ref. 40; modified Burnett expansion, refs. 38, 78, 84, 121; Buckingham expansion, ref. 62; equation of state, refs. 45, 46, 55, 79, 82, 91, 99, 101, 103, 119, 122; expansion plus equation of state, refs. 58, 120; density balance, ref. 55; [b] LL, Lorentz–Lorenz function measured.

119 K. P. Lawley and L. E. Sutton, *Trans. Faraday Soc.*, 1963, **59**, 2680.
120 S. D. Hamann, *J. Chem. Phys.*, 1970, **53**, 475.
121 H. Sutter and R. H. Cole, *J. Chem. Phys.*, 1967, **46**, 2014.

Table 3 (*cont.*)

Gas	*Temp.* (K)	$A_\varepsilon \times 10^6$ ($m^3\ mol^{-1}$)	$B_\varepsilon \times 10^{12}$ ($m^6\ mol^{-2}$)	$C_\varepsilon \times 10^{18}$ ($m^9\ mol^{-3}$)	*Ref.*	*Other Refs.*
$CClF_3$	323	19.167 ± 0.009	58.5 ± 0.4	—	78	40
	369.5	18.566 ± 0.003	53.5 ± 5.0	—	38	
	416.5	—	46.8 ± 6.0	—	38	
CCl_2F_2	323.2	24.90 ± 0.04	235 ± 140	—	40	
CCl_3F	369·5	27.42 ± 0.01	215 ± 30	—	38	
CF_2CH_2	303.2	50.99 ± 0.06	− 723 ± 100	—	40	
	323.2	48.47 ± 0.04	− 447 ± 80	—	40	
	343.2	46.36 ± 0.04	− 240 ± 100	—	40	
CH_3OCH_3	303.2	48.25 ± 0.04	2800 ± 1000	—	40	119
	323.2	46.22 ± 0.04	1600 ± 400	—	40	
	343.2	44.35 ± 0.04	1600 ± 400	—	40	
AsF_3	—	—	—	—		55
NH_3	—	—	—	—		91, 99, 101, 122. LL[b]: 45, 79
$CHCl_3$	—	—	—	—		55
CH_3OH	—	—	—	—		LL[b]: 46
$NH(Et)_2$	—	—	—	—		LL[b]: 46
$N(Et)_3$	—	—	—	—		LL[b]: 46

[b] LL, Lorentz–Lorenz function measured.

[122] F. G. Keyes and J. G. Kirkwood, *Phys. Rev.*, 1930, **36**, 1570.

Table 4 *Values of dielectric virial coefficients of mixtures*[a]

Mixture	*Temp.* (K)	$B_{\varepsilon_{12}} \times 10^{12}$ (m^6 mol^{-2})	*Ref.*
Ar + CO_2	322.9	14.3 ± 1.4	69
Ar + C_2H_4	323	14.4 ± 1.5	73
Ar + CHF_3	369.5	53 ± 8	62
SF_6 + $CClF_3$	323	−73.8 ± 0.4	78
CHF_3 + CHF_2Cl	323.2	950 ± 400	40
	343.2	150 ± 600	40
CHF_3 + CH_3Cl	343.2	− 800 ± 600	40
CHF_3 + CF_2CH_2	323.2	950 ± 200	40
CHF_3 + CH_3OCH_3	292.7	2.70 ± 0.2 × 10^4	119
	323.2	1.68 ± 0.15 × 10^4	40
	343.2	1.18 ± 0.08 × 10^4	40
CH_3OCH_3 + HCl	292.7	4.47 ± 0.05 × 10^5	119
	312.8	2.45 ± 0.03 × 10^5	119
CH_3OCH_3 + SO_2	291.7	4.6 ± 0.3 × 10^4	119
	313.2	2.7 ± 0.2 × 10^4	119
	334.7	1.6 ± 0.2 × 10^4	119
CH_3OCH_3 + CH_3Cl	323	1100 ± 1000	40
	343	950 ± 500	40
CH_3OCH_3 + $CHClF_2$	303	2.4 ± 0.2 × 10^4	40
	323	1.8 ± 0.12 × 10^4	40
	343	1.25 ± 0.07 × 10^4	40
CH_3OCH_3 + CF_2CH_2	343	1400 ± 700	40

[a] Methods of measurement: Burnett expansion, ref. 40; modified Burnett expansion, refs. 69, 73, 78; Buckingham expansion, ref. 62; equation of state, ref. 119.

Table 5 *Values of $B_{\varepsilon_{or}}$ calculated from pressure-induced absorption*

Gas	*Temp.* (K)	$B_{\varepsilon_{or}} \times 10^{12}$ (m^6 mol^{-2})	*Ref.*
CO_2	233	106 ± 11	71
	273	73 ± 7	71
	296	62 ± 6	71
	333	48 ± 5	71
CH_4	298	1.2	76
CD_4	298	1.1	76
CF_4	298	2.3	76
SF_6	298	0.44	77
He + Ne	295	0.0378	28, 29
He + Ar	295	0.202	28, 29
Ne + Ar	295	0.301	28, 29

4
Time Domain Methods

BY A. SUGGETT

1 Introduction

Relaxation methods, in general, monitor the rate at which equilibrium is re-attained after the application of a perturbation which may be either a step or a periodic function. In the past, dielectric relaxation methods have almost invariably employed the latter alternative. Until quite recently, for example, molecular reorientations with relaxation times less than about 10^{-4} s were investigated purely in the frequency domain because of the total unavailability of equivalent time domain techniques. Previous reviews[1, 2] which include transient dielectric methods cover most of the relevant information up to 1969. For the purposes of comparison with the present 'state of the art' it is nevertheless instructive to reconsider the position 10 years ago, and Figure 1 shows the area of contribution of transient, *i.e.* step function, methods at that time. Vaughan[2] in 1969 considered that transient methods could be used to cover the frequency range 10^{-3}—10^{+3} Hz (equivalent time range, about 10^{+2}—10^{-4} s). The recent resurgence in time domain dielectric measurement may be judged by the expansion in the available time range – now about 10^{+3}—10^{-11} s (equivalent frequency range 10^{-4}—10^{+10} Hz, *i.e.* the total range of Figure 1).

Especially at the short times, the use of time domain methods, as opposed to their point-by-point frequency domain equivalents, is advantageous in a number of ways. They can, for example, be considered as truly spectroscopic techniques because of their broad-band nature and their capacity to generate dielectric properties as a continuous function of time or, by appropriate transformation, frequency. In the past few years, time domain methods have received fresh impetus from advances in two different types of method: firstly, the d.c. step response technique as used by Reddish[3] and Williams[4] has been up-graded in sensitivity and bandwidth through

[1] W. Reddish, *Pure Appl. Chem.*, 1962, **5**, 723.
[2] W. E. Vaughan, in 'Dielectric Properties and Molecular Behaviour', ed. N. E. Hill, W. E. Vaughan, A. H. Price, and M. Davies. Van Nostrand, London, 1969.
[3] (*a*) W. Reddish, *Soc. Chem. Ind. Monograph* No. 5, 1959, p. 138; (*b*) W. Reddish, *J. Polymer Sci. Part C, Symposia*, 1966, 123.
[4] (*a*) G. Williams, *Polymer*, 1963, **4**, 27; (*b*) G. Williams and D. C. Watts, *Trans. Faraday Soc.*, 1970, **66**, 80; (*c*) G. Williams, D. C. Watts, S. B. Dev, and A. M. North, *ibid.*, 1971, **67**, 1323.

improvements in instrumentation and data handling; secondly, the availability of high-frequency sampling oscilloscope systems has facilitated the development of pulse techniques known collectively as time domain spectroscopy (t.d.s.). The earliest t.d.s. measurements indicated striking advantages over frequency domain measurements in terms of bandwidth and speed of measurement, and subsequent theoretical, experimental, and

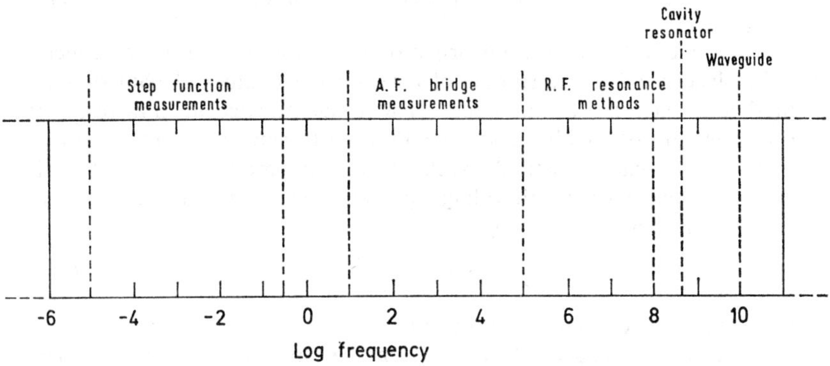

Figure 1 *Frequency spectrum of dielectric measurements*
(Reproduced by permission from *Pure Appl. Chem.*, 1962, **5**, 723)

instrumental improvements have raised t.d.s. to an equivalent level of precision.

For the convenience of the reader, an arbitrary separation has been made into Slow Response Techniques (time range 10^{+3}—10^{-7} s), and Fast Response Techniques (time range 10^{-7}—10^{-11} s). The former are, in part, well documented elsewhere [1–4] and are not discussed here in any great detail, but the experimental methods associated with fast response measurements are largely unfamiliar and are therefore more comprehensively described.

2 Slow Response Techniques

Experimental Methods.—The initial fleeting excursions from frequency domain into time domain (for example, ref. 5) appear to have been made because, at that time, steady-state measurements at very low frequencies ($\sim 10^{-1}$ Hz) were unsatisfactory. Step-up, step-down, and ramp voltages were variously applied to capacitors containing dielectric samples, and the transient current $i(t)$, or charge $q(t)$, responses monitored over a wide range of times; such approaches have been reviewed.[2] Although it is now quite feasible to make steady-state measurements at very low frequencies,

[5] D. W. Davidson, R. P. Auty, and R. H. Cole, *Rev. Sci. Instr.*, 1951, **22**, 678.

the value of slow response transient methods is increasing as their convenience (for rapid evaluation of complex permittivity over a wide range of frequencies) increases.

The current response $i(t)$, to a unit voltage step of a unit vacuum capacitor filled with a dielectric material, is related to the complex relative permittivity (ε^*) of the dielectric by the Fourier transform

$$\varepsilon^*(\omega) = \int_0^\infty i(t) \exp(-j\omega t)\, dt \tag{1}$$

Direct numerical Fourier transformation when a wide range of frequencies is involved requires the storage of a very large number of data samples [$i(t)$] and extensive computation. The usual practice in slow-response time-domain methods has consequently been to employ[3,4] more approximate transformation methods, such as the Hamon approximation.[6] This relates the current per unit voltage at time t, $i(t)$, to the dielectric loss ε'' at the frequency $\nu = 1/10t$ by

$$\varepsilon''(\omega) = \frac{i(1/10\nu)}{2\nu C_0} \tag{2}$$

where C_0 is the capacitance of the air-filled cell.

An alternative and much more acceptable approach has recently been reported by Hyde.[7] He pointed out that as in practice the charge q varies in an approximately linear manner with $\log t$, then sampling $q(t)$ at equal intervals of $\log_2 t$ could provide a conveniently small number of samples for a Fast Fourier transform programme. The transforms used by Hyde[7] have the forms

$$\varepsilon'(\omega) = \sum_{p=-\infty}^{\infty} \Delta q(n)\, x(n) \tag{3}$$

$$\varepsilon''(\omega) = \sum_{p=-\infty}^{\infty} \Delta q(n)\, y(n) \tag{4}$$

where $n = 2^p$, and $x(n)$, $y(n)$ are constant coefficients. The number of terms on the right-hand sides of equations (3) and (4) was limited to 1, 3, or 5, although in principle it is possible to extend this method to include further terms. The above sampling and Fourier transformation approach is embodied in the spectrometer design (Figure 2). In Figure 2 the measuring circuit is essentially as described earlier by Reddish,[3b] and the processor is a specially designed small computer for conversion of voltage–time data into $\varepsilon^*(\omega)$ by means of the approximate transform procedures. Two such spectrometers were constructed. The first, which measures $q(t)$ over the time range 10^{+3}—10^{-3} s, uses the classical method of applying a single pulse and sampling the resultant decay at pre-selected intervals. The second, with a time range of 10^{-1}—10^{-7} s, uses a series of

[6] B. V. Hamon, *Proc. Inst. Elec. Eng.*, 1952, **99**, 151.

[7] (*a*) P. J. Hyde, *Proc. Inst. Elec. Eng.*, 1970, **117**, 1891; (*b*) P. J. Hyde, Ph.D. Thesis, Imperial College, University of London, 1971.

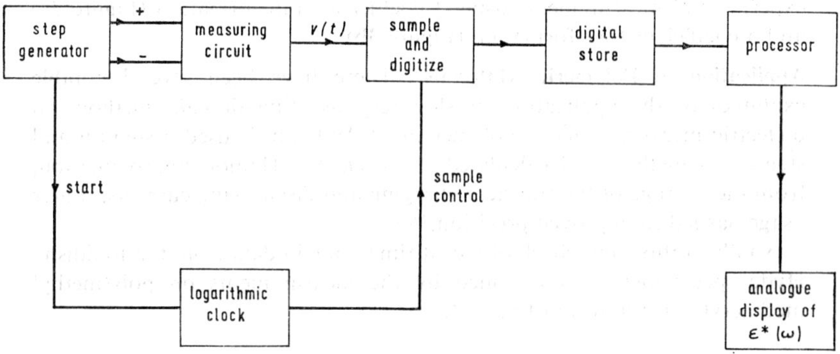

Figure 2 *Spectrometer block diagram*
(Reproduced by permission from *Proc. Inst. Electrical and Electronic Eng.*, 1970, **117**, 1891)

equal length pulses and takes successive $q(t)$ samples on successive pulses. The application of these 'dielectric spectrometers' to measurement of a lumped circuit having a unique relaxation time of 10^{-4} s is shown in Figure 3. The solid lines are the theoretical curves, and the points are the

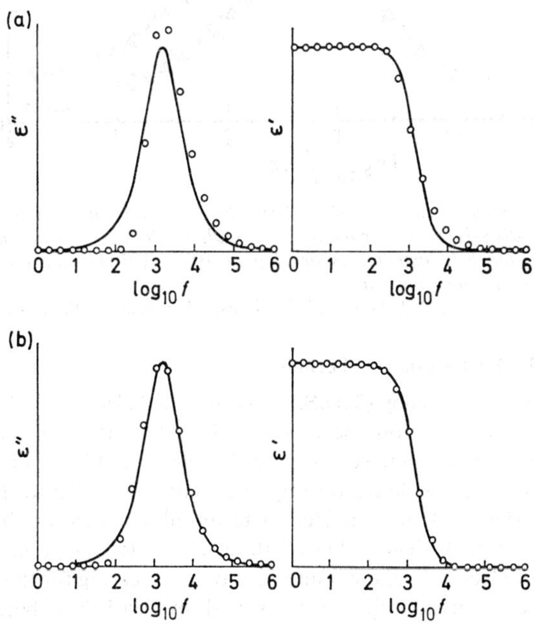

Figure 3 *Experimental results for Debye lumped circuit with* $\tau_0 = 10^{-4}$ s (a) *using* 1-*coefficient transform*, (b) *using* 5-*coefficient transform*
(Reproduced by permission from *Proc. Inst. Electrical and Electronic Eng.*, 1970, **117**, 1891)

experimental measurements using 1-coefficient transformation (Figure 3a) and 5-coefficient transformation (Figure 3b).

Application to Polymeric Materials.—There have been several notable examples of the application of slow-response time-domain methods to dielectric measurements of polymers.[1, 3, 4] Williams[4a] used a step-up and step-down method and calculated $\varepsilon''(\omega)$, *via* the Hamon approximation, from the average of the transient charging and discharging currents. Later usage has led to improved precision.[4b, 4c]

An illustration and check of recent time domain data from the Reddish–Hyde spectrometer[7] is provided by the measurements on poly(methyl methacrylate)[7b] shown in Figure 4.

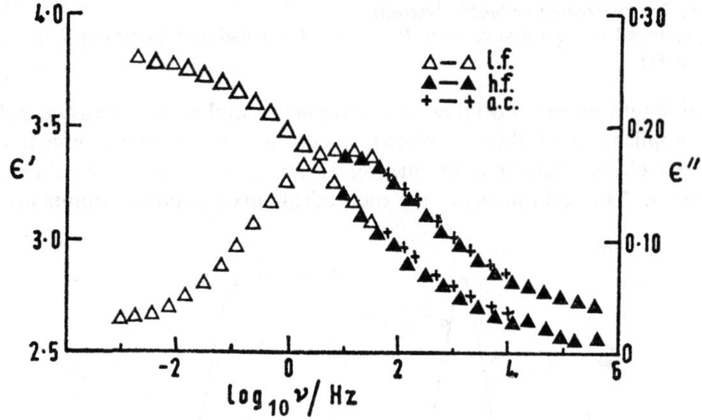

Figure 4 *Dielectric parameters of poly(methyl methacrylate) obtained from the Reddish–Hyde spectrometer; ε′ – scale on left, ε″ – scale on right. The l.f. and h.f. instruments overlap at* $10^1 — 10^2$ Hz; *the a.c. results are careful bridge measurements*
(Based on the results of P. J. Hyde, Ph.D. Thesis, Imperial College, London, 1971)

3 Fast Response Techniques

Time Domain Spectroscopy (T.D.S.).—Transient dielectric methods for the study of molecular motions occurring in less than 10^{-7} s are a relatively recent addition to the chemist's armoury.* The availability of tunnel diode pulse-generators and wide-band sampling oscilloscopes led to the development in the 1960's of pulse reflection techniques known as time domain reflectometry (t.d.r.). The value of these methods was soon recognized in the fields of electronic and communication engineering for the qualitative analysis of transmission line systems and by 1965 had been used for

* [Senior Reporter's note]. It should be emphasized that all these methods presenting dielectric spectra from time domain observations can equally well provide the time dependence of any process (physical or chemical) whose progress can be presented as a capacitance change.

quantitative measurements [8] (network analysis). The first application of t.d.r. to the measurement of dielectric properties was reported in 1969 by Fellner-Feldegg.[9] Subsequent improvements in theory and techniques have been made by a number of workers.[10–16] The term 'time domain spectroscopy' now embraces a range of alternative and sometimes complementary time domain methods.

Basic Theory. In the most commonly used t.d.s. methods, a voltage step-pulse whose Fourier spectrum comprises a wide range of frequencies is propagated in a low-loss coaxial line. Such a pulse remains unaltered as long as the propagation characteristics of the line are unchanged, but any discontinuity in the line, such as a transition to a section with a different characteristic impedance, will give rise to a partial reflection and partial transmission of the pulse. The nature of the discontinuity will be indicated by the resulting pulse shapes while its location will be shown by the time relationship between incident and reflected pulses.

For a coaxial line of negligible resistance, filled with dielectric, the characteristic impedance is given by

$$Z = Z_0/\sqrt{[\varepsilon^*(\omega)]} \qquad (5)$$

where Z_0 is the characteristic impedance of the air-filled line, and ε^*, the complex relative permittivity, is given by

$$\varepsilon^* = \varepsilon' - j\varepsilon'' - j\sigma/\omega\boldsymbol{\varepsilon} \qquad (6)$$

σ is the specific electric conductance and $\boldsymbol{\varepsilon}$ the permittivity of free space. Considering a wave propagated in this line of impedance Z_0, for the transition to a region of impedance Z, the voltage reflection and transmission coefficients ρ, τ are given by

$$\rho = \frac{Z - Z_0}{Z + Z_0} \qquad (7)$$

$$\tau = \frac{2Z}{Z + Z_0} \qquad (8)$$

For a dielectric sample Z, ρ, and τ will be complex frequency-dependent quantities, and their frequency-dependence will be shown in differences between the time domain waveforms (pulse shapes) of the incident signal

[8] F. Davis and H. W. Loeb, *Proc. Inst. Electrical and Electronic Eng.*, 1965, **53**, 1649.
[9] H. Fellner-Feldegg, *J. Phys. Chem.*, 1969, **73**, 616.
[10] A. M. Nicolson and G. F. Ross, *IEEE Trans. Instrument. Measur.*, 1970, **IM-90**, 377.
[11] H. Fellner-Feldegg and E. F. Barnett, *J. Phys. Chem.*, 1970, **74**, 1962.
[12] A. Suggett, P. A. Mackness, M. J. Tait, H. W. Loeb, and G. M. Young, *Nature*, 1970, **228**, 456.
[13] H. W. Loeb, G. M. Young, P. A. Quickenden, and A. Suggett, *Ber. Bunsengesellschaft Phys. Chem.*, 1971, **75**, 1155.
[14] M. J. Tait, A. Suggett, F. Franks, P. A. Quickenden, and S. Ablett, *J. Solution Chem.*, in press.
[15] M. J. C. van Gemert, *J. Phys. Chem.*, 1971, **75**, 1323.
[16] M. J. C. van Gemert and J. G. de Graan, *Appl. Sci. Res.*, in press.

and of the responses, *i.e.* reflected and transmitted signals. The behaviour of a dielectric in the frequency domain may be found by measuring a suitable incident pulse and the corresponding responded pulse† and calculating the ratio of the Fourier transforms.[12] The basis of the transformation is that any pulse in the time domain, $f(t)$, is related to a complex spectrum $F(\omega)$ in the frequency domain by the Fourier transform

$$F(\omega) = \int_{-\infty}^{+\infty} f(t) \exp(-j\omega t)\, dt \tag{9}$$

Thus if $G(\omega)$ is the frequency response of the dielectric system and $f_i(t)$ and $f_r(t)$ are the incident and responded pulses, then

$$G(\omega) = \frac{F_r(\omega)}{F_i(\omega)} = \frac{\int_{-\infty}^{+\infty} f_r(t) \exp(-j\omega t)\, dt}{\int_{-\infty}^{+\infty} f_i(t) \exp(-j\omega t)\, dt} \tag{10}$$

$G(\omega)$ is thus defined for all frequencies contained in the incident pulse, *i.e*, it is defined for all ω where $F_i(\omega) \neq 0$. The only major restriction in the use of this Fourier transform approach is that the pulse form must be known for all values of t or, in other words, the decay of any relaxation process must be monitored effectively to completion. The Samulon[17] modification of the Shannon sampling theorem[18] has been shown to be an accurate and convenient method of Fourier-transforming step pulses. The relevant relation is

$$F(\omega) = \frac{T \exp(j\omega T/2)}{2j \sin \omega T/2} \sum_{n=-\infty}^{+\infty} [f(nT) - f(nT - T)] \exp(-jn\omega T) \tag{11}$$

provided that $F(\omega) = 0$ for $\omega > \pi/T$, where T is the time interval between the digitized data samples.

Experimental Methods. There are many possible experimental configurations for t.d.s. measurements. Figure 5, for example, shows in diagrammatic form a basic system for making t.d.s. reflection measurements. The tunnel diode produces a fast rise (*i.e.* about 30 ps) pulse which is split into two by the matched 'tee'. One part is used to provide the incident signal for the reflection measurement and the other to provide a time marker pulse. The signal pulse travels past the sampling point at A to the air–dielectric interface B where it is partially reflected and partially transmitted. The reflected pulse is monitored at A while the transmitted portion undergoes further reflections generating 'echo' pulses corresponding to the pathways ABCBA, ABCBCBA, *etc.* In Figure 6 an idealized signal form is shown

17 H. A. Samulon, *Proc. Inst. Radiation Eng.*, 1951, **39**, 175.
18 C. Shannon, *Proc. Inst. Radiation Eng.*, 1949, **37**, 10.

† 'Responded pulse' is the term used for reflected or transmitted pulses or for a combination of these.

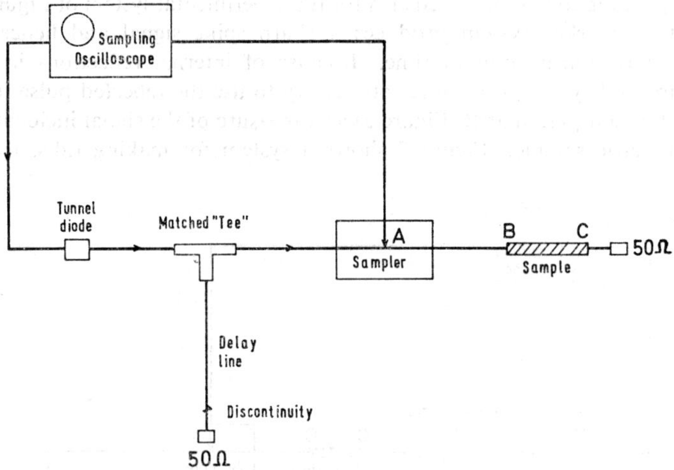

Figure 5 *Basic system for t.d.s. reflection measurements*
(Reproduced by permission from *Ber. Bunsengesellschaft Phys. Chem.*, 1971, **75**, 1155)

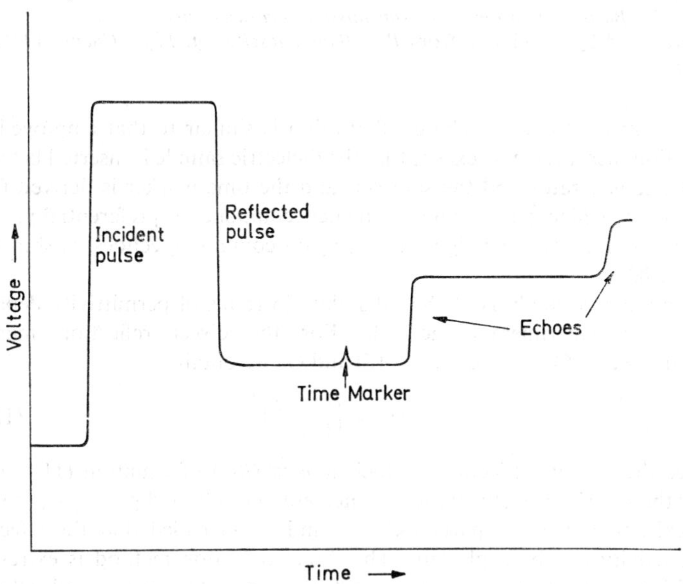

Figure 6 *Idealized waveform from the arrangement of Figure* 5
(Reproduced by permission from *Ber. Bunsengesellschaft Phys. Chem.*, 1971, **75**, 1155)

corresponding to that observed at A for the experimental system of Figure 5. The time marker system produces a sharp spike signal and hence an accurate reference point in time. Because of internal reflections in the experimental system, it is more satisfactory to use the reflected pulse from a short circuit placed at B (Figure 5) as a measure of the signal incident on the dielectric sample. Figure 7 shows a system for making t.d.s. trans-

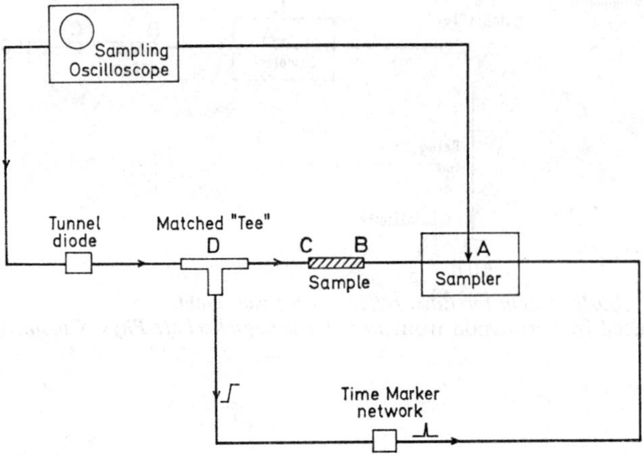

Figure 7 *Basic system for t.d.s. transmission measurements*
(Reproduced by permission from *Ber. Bunsengesellschaft Phys. Chem.*, 1971, **75**, 1155)

mission measurements. The configuration is similar to that employed for reflection measurements except that the dielectric sample is inserted between the pulse generator and the sampler, and the time marker is derived from a network which passes high-frequency components preferentially. The corresponding idealized signal form, again containing echoes, is shown in Figure 8.

The required analysis of the t.d.s. data in terms of permittivity depends upon the experimental method. For the direct reflection method (Figures 5 and 6) from equations (5) and (7) we obtain

$$\varepsilon^* = \left(\frac{1 - \rho}{1 + \rho}\right)^2 \qquad (12)$$

where the voltage reflection coefficient $\rho \equiv G(\omega)$ of equation (11). Thus after the Fourier transformation of incident and reflected pulses, separation of real and imaginary parts yields ε' and ε'', provided that the effect of conductivity can be neglected. The direct reflection method is extremely sensitive to any imprecision in time referencing of the incident and reflected pulses. This can introduce serious errors in the calculated ε^* at frequencies greater than about 1 GHz. When, however, methods can be devised to minimize these errors[13] then acceptable dielectric information can be obtained at frequencies in excess of 10 GHz. The low-frequency limit of the method is determined by the appearance of the echoes (Figure 6) at a time, t, determined by the length of the cell and the permittivity of the sample. Thus for aqueous solutions and a cell length of 30 cm (about the maximum commercially available), $t = \sim 18$ ns. A direct time domain

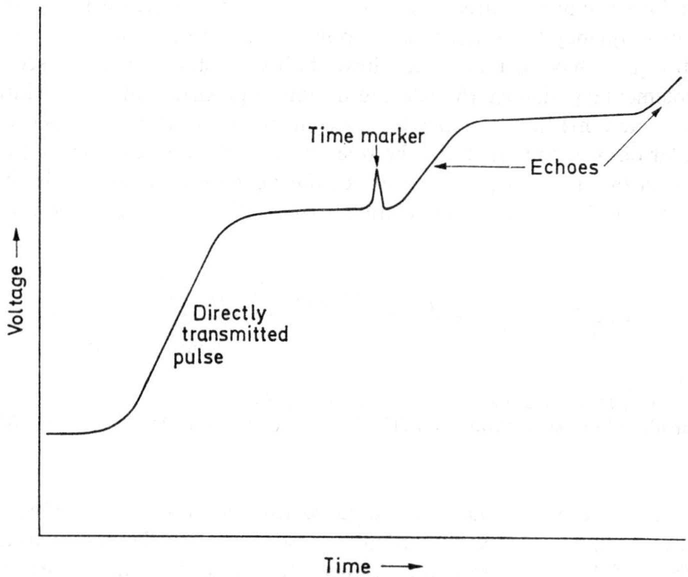

Figure 8 *Idealized waveform from the arrangement of Figure 7*
(Reproduced by permission from *Ber. Bunsengesellschaft Phys. Chem.*, 1971, **75**, 1155)

conversion of t.d.s. reflection data, originally suggested by Fellner-Feldegg,[9] assumed equation (13) to hold when the complex quantities ε^* and ρ were replaced by time-dependent quantities $\varepsilon(t)$ and $\rho(t)$ respectively. Although this was later shown to be mathematically invalid,[19] the convenience of the method is such that it would be valuable to tabulate the magnitude of the error in a measured relaxation time as a function of ε_0, ε_∞, τ, *etc.*, for various dielectric models.

Based upon the experimental configuration of Figure 7, a method has recently been described[13] in which the transmitted pulse through a given length of dielectric sample was compared with that through a different length of sample. It was shown that if T_a and T_b are the transmission coefficients through dielectric lengths x_a and x_b then

$$T_a/T_b = \exp\left[-\,j\omega\,\sqrt{(\varepsilon^*)}\,(x_a - x_b)/\mathbf{c}\right] \tag{13}$$

[19] T. A. Whittingham, *J. Phys. Chem.*, 1970, **74**, 1824.

where c is the velocity of light *in vacuo*; expressing T_a/T_b as $s\exp(j\phi)$ then

$$\varepsilon^* = \frac{c^2}{\omega^2(x_a - x_b)^2}[\phi^2 - (\ln s)^2 - 2j\phi \ln s] \tag{14}$$

This method (transmission coefficient ratios) does not suffer from sensitivity to time referencing errors, although the available frequency range is somewhat lower than the direct reflection method—the decrease in the upper cut-off frequency being dependent upon the length of dielectric-filled line. Loeb *et al.*[13] have demonstrated how the two methods can be used in a complementary manner to yield the optimum precision and bandwidth.

A related method suitable for measuring the electrical properties of annular discs of solid material, between 0.4 and 10 GHz, has been reported by Nicolson and Ross;[10] a diagram of their experimental system is shown in Figure 9. In this work the relative permeability of the dielectric was not

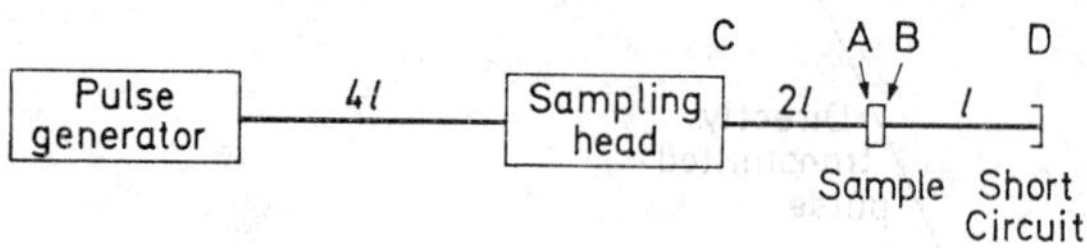

Figure 9 *Measurement system of Nicolson and Ross*
(Reproduced by permission from *IEEE Trans. Instrument. Measur.*, 1970, **IM-90**, 377)

assumed to be unity, but was measured together with permittivity. The various lengths in this system are chosen so that the reflected pulse (pathway CAC) and the doubly transmitted pulse (pathway CABDBAC) are well separated in time: they each have a 'time window' of $2l/c$ seconds. The transmission and reflection scattering coefficients $S_{21}(\omega)$ and $S_{11}(\omega)$, which are obtained by Fourier transformation, are related to the voltage reflection coefficient ρ [equation (7)] and z, the transmission coefficient between the faces A and B of the dielectric sample by

$$S_{21}(\omega) = \frac{(1-\rho^2)\,z}{1-\rho^2 z^2} \tag{15}$$

$$S_{11}(\omega) = \frac{(1-z^2)\,\rho}{1-\rho^2 z^2} \tag{16}$$

where ρ and z are more generally defined in terms of the relative permittivity ε^*, the relative permeability μ^*, and the thickness d of the disc, as

$$\rho = \frac{\sqrt{(\mu^*/\varepsilon^*)} - 1}{\sqrt{(\mu^*/\varepsilon^*)} + 1} \tag{17}$$

and

$$z = \exp\left[-\frac{j\omega}{c}\sqrt{(\mu^*\,\varepsilon^*)}.d\right] \tag{18}$$

From these relations one obtains

$$\frac{\mu^*}{\varepsilon^*} = \left(\frac{1 + \rho}{1 - \rho}\right)^2 \tag{19}$$

$$\mu^* \varepsilon^* = - \left[\frac{c}{\omega d} \ln \left(\frac{1}{z}\right)\right]^2 \tag{20}$$

from which μ^* and ε^* may be separately determined.

Fast-response time-domain measurements suffer from error sources which in general are more commonly associated with spectroscopic techniques (*e.g.* n.m.r.) than with other dielectric methods. The existence of these sources of error must be recognized, particularly when selecting the most appropriate method for the collection and handling of experimental data. It is convenient to separate error sources into four major categories:

(*a*) In approximating to a continuous Fourier transform by a discrete Fourier transform, errors may occur owing to 'truncation and aliasing' (terminology of ref. 20). They can be minimized by suitable choice of the time interval between digitized samples.

(*b*) Imprecision in referring two transient responses to the same point in time can create serious difficulties in the measurement of permittivity at frequencies greater than a few GHz. This is especially so for high-permittivity samples, and for aqueous systems an indirect method using transmission measurements to produce a time reference for reflection measurements has been employed.[13] Timing errors are discussed in more detail elsewhere.[10, 13]

(*c*) As a consequence of the relatively low spectral intensity of the tunnel diode pulse (only a few hundred millivolts in amplitude), the ratio of signal to random noise (from, for example, amplifier circuitry) falls off rapidly as the frequency increases. Some form of signal averaging can considerably reduce the resulting errors. Also under the heading of noise-like errors, we include the effect of high-frequency timing jitter (*i.e.* random fluctuations in the time-location of the pulse) and a low-frequency (of the order of minutes) 'drift' in the time-location of the pulse. Two alternative experimental approaches to the minimization of such errors have been reported; Loeb *et al.*[13] used a signal averager effectively to eliminate random noise, but at the same time scanned very quickly (about 60 ms per scan) to minimize this effect of pulse drift. A more sophisticated method using a small computer was suggested by Nicolson and Ross[10] based upon a three-point scanning method. The computer is used to step the time base of the sampling oscilloscope to discrete positions and repeated measurements are made at three points (Figure 10). At the first fixed point the signal waveform slope is zero, at the second fixed point the slope is finite, and the third position incrementally scans the time window. Changes in

[20] A. M. Nicolson, *IEEE Trans. Instrument. Measur.*, 1968, **IM-17**, 395.

the difference voltage, f_{21}, can be used by the computer to generate correction voltages in the scanning circuitry to return point 2, and hence the time window, to its original position. Successive values of f_{31} are stored as the third point scans across the window. The same small computer can be used to obtain the average signal for a small number of such scans and to perform the required transformations.

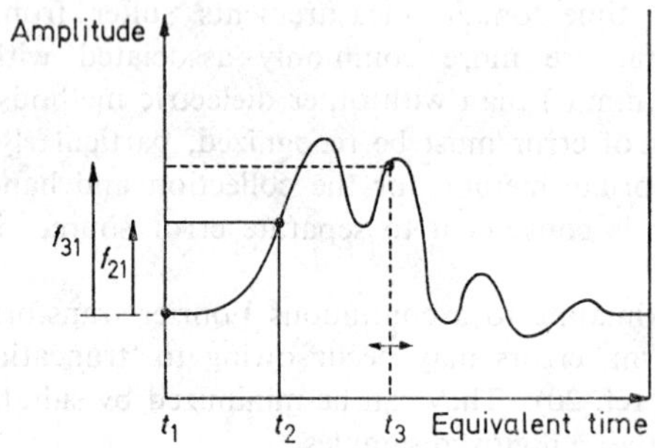

Figure 10 *Principle of three-point scanning of a sampling oscilloscope waveform* (Reproduced by permission from *IEEE Trans. Instrument. Measur.*, 1970, **IM-90**, 377)

(*d*) Because of impedance mismatches between components in the t.d.s. system, as well as the required reflected and/or transmitted pulses, the observed signal waveforms also show the presence of a great number of multiple internal reflections. One sees, as it were, the spectrum of the sample plus the spectrum of the spectrometer. It has been demonstrated [13] that by judicious choice of coaxial line distribution in the experimental system, one can often displace many of the unwanted reflections away from the time window of interest.

Application to Dielectric Measurements.—*Polar Liquids.* At this early stage in the development of fast-response time-domain methods, much of the experimental application has been directed towards dielectric systems whose behaviour was already reasonably well established by conventional frequency domain methods. Fellner-Feldegg, for example, in his original application [9] of direct reflection t.d.s. to dielectric measurements, selected a number of aliphatic alcohols for this very reason. The values so obtained for the 'static' relative permittivity (ε_0) of the alcohols were in acceptable agreement with literature data, although for reasons given earlier the relaxation times were not necessarily so. The more correct Fourier transform approach was later applied [12] to alcohol relaxation processes, and data for ethanol, propan-1-ol, and butan-1-ol at 298 K are shown on Figure 11. This clearly illustrates the broad-band nature of the technique. Although relaxation parameters calculated from this work are in much better agreement with those from classical methods than were the first t.d.s. estimates (see Table), there was still some discrepancy in the reported relaxation time values for propanol and butanol. This may have been due in part to the method of data recording used in these early studies (X–Y plotter) which is likely to be affected by error sources of all four types discussed earlier.

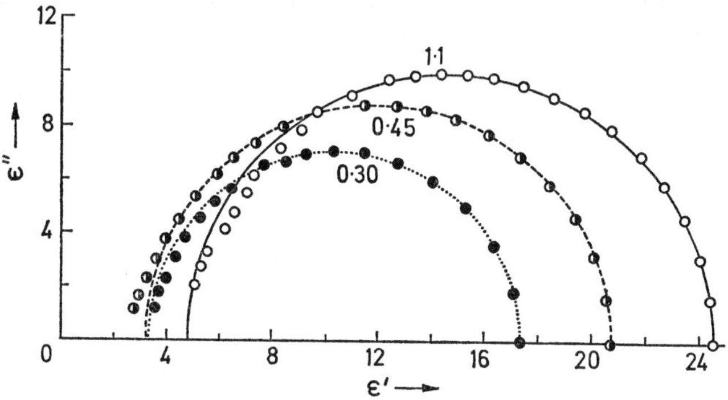

Figure 11 *Complex plane representation of t.d.s. data for ethanol, propan-1-ol, and butan-1-ol at 298 K; ○—○, ethanol; ◑---◑, propan-1-ol; ●·····●, butan-1-ol. Indicated frequencies are in GHz.*
(Reproduced by permission from *Nature*, 1970, **228**, 456)

However, very recent measurements[21] obtained under conditions where such errors are minimized[13] are substantially in agreement with those of Figure 11 and the wide-band frequency domain studies of Sagal[22] (see

Table *Comparison of the results obtained by different techniques for the major dielectric relaxation process of ethanol, propan-1-ol, and butan-1-ol at* 298 K

Parameter	*T.D.S.*[a]	*T.D.S.*[b]	*T.D.S.*[ch]	*Frequency domain methods*[h]
τ(EtOH)/ps	130	140	170	170[d], 150[e]
τ(Pr^{n}OH)/ps	220	330	325	385[f]
τ(Bu^{n}OH)/ps	330	480	470	575[f], 488[d]
ε_0(EtOH)	24.3	24.5	24.6	24.35[d], 24.4[e]
ε_0(Pr^{n}OH)	20.5	20.6	20.6	20.5[f]
ε_0(Bu^{n}OH)	16.8	17.3	17.1	17.0[f], 17.4[d]
ε_∞(EtOH)	8.3	4.8	4.6	4.5[d], 4.4[e]
ε_∞(Pr^{n}OH)	7.2	3.0	4.2	3.3[f]
ε_∞(Bu^{n}OH)	4.2	3.3	3.3	3.2[g], 3.3[d]

[a] Ref. 9; [b] Ref. 12; [c] Ref. 21; [d] Ref. 22; [e] J. A. Saxton, R. A. Bond, G. T. Coats, and R. M. Dickinson, *J. Chem. Phys.*, 1962, **37**, 2132; [f] Ref. 23; [g] W. Dannhauser and R. H. Cole, *J. Chem. Phys.*, 1955, **23**, 1762; [h] Corrected to 298 K where necessary.

[21] M. J. C. van Gemert, P. A. Quickenden, and A. Suggett, unpublished data.
[22] M. W. Sagal, *J. Chem. Phys.*, 1962, **36**, 2437.

Table). The rather long relaxation times estimated by Garg and Smyth [23] may be a consequence of their restricted frequency range.

Water and Aqueous Solutions. The high permittivity and high relaxation frequency of water presents a formidable challenge to t.d.s. methods, as virtually all error percentages increase as sample permittivity and measuring frequency increase. At the same time, it is in the microwave region

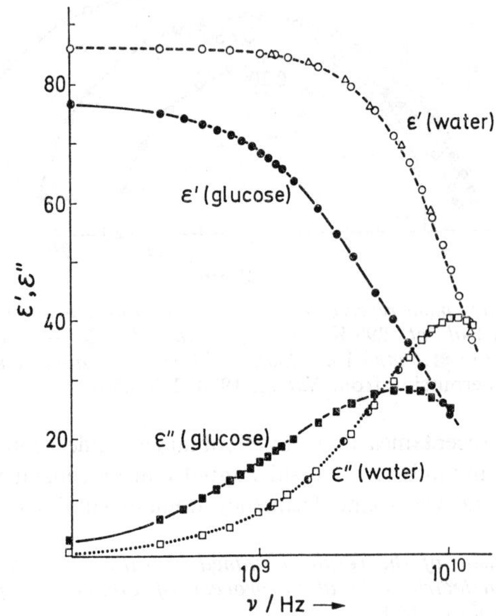

Figure 12 *The complex permittivity of water and* 2.8 *molal aqueous glucose solution at* 278 K; ○, *t.d.s measurement of* ε′(*water*); □, *t.d.s. measurement of* ε″(*water*); ————, ..., *data interpolated from ref.* 24; △, ◑, *data from ref.* 25; —●—, —■—, *t.d.s. measurements on* 2.8 *molal glucose*

(Reproduced by permission from *Ber. Bunsengesellschaft Phys. Chem.*, 1971, **75**, 1155)

that some of the main advantages of time domain methods are most apparent; for example, the limited number of frequency points normally covered in a frequency domain microwave study is not adequate for analysis of overlapping relaxations in aqueous solutions. Time domain methods are of course not restricted in this way.

The high-frequency precision of t.d.s. methods has been tested [13] by making measurements on water itself. The data for water at 278 K are shown on Figure 12, together with a representative sample of frequency domain measurements. On the basis of most of the previous dielectric

[23] S. K. Garg and C. P. Smyth, *J. Phys. Chem.*, 1965, **69**, 1254.

measurements on water, Mason and Hasted[24] derived the most likely relaxation parameters, and an interpolation of their results to 278 K is shown by the dotted and dashed lines in Figure 12, together with some recent measurements made by Pottel and co-workers.[25]

T.d.s. has very recently been applied, together with ^{17}O n.m.r. relaxation, to studies of the hydration properties of monosaccharides, the ^{17}O n.m.r. providing the time-averaged properties of the water in the system, and t.d.s. the frequency spectrum which was resolvable in terms of the different types of relaxing species present. In Figure 12, the complex permittivity of a 2.8 molal glucose solution is shown compared with that of water. The analysis of the t.d.s. data for the monosaccharide solution, both by the approximate time domain method[9] and by a minimization analysis of the Fourier transformed information, indicated a minimum of three relaxation processes (provided that the bulk water relaxation was restricted to its usual narrow distribution of relaxation times). The two smaller (in amplitude) lower frequency relaxations could satisfactorily be represented by Debye processes. For glucose solutions, the dependence of the separated relaxation times upon the solution viscosity is shown in Figure 13, where processes 1, 2, and 3 are ascribed to the bulk water, glucose, and hydration water respectively.† This work has yielded estimates for the extent of hydration of a number of monosaccharides, and has shown that the interactions between water and chemically similar 'hydrophilic' solutes such as glucose and ribose are significantly different and depend critically upon the stereochemistry of the solute molecules. Why certain stereochemical arrangements of hydroxy-groups are hydrated to greater or lesser extents, it is suggested, can be understood in terms of current ideas on the long-range order in water.

Polymers. Nicolson and Ross,[10] using the methods developed for solid discs of material, examined the permittivity and permeability of a number of polymeric substances, and the results are given in Figure 14 for a disc of maple wood, the only one of their samples to exhibit any detectable loss in the frequency range 0.4—10 GHz.

Some Theoretical Considerations.—Since the first application of fast-response time-domain methods to dielectric measurement, the value of a direct time domain conversion of the data has been universally recognized. It is after all a strange paradox that, in order to obtain information about time-dependent molecular motions from the time-dependent reflection coefficient, $P(t)$, one must go through the stage of the frequency-dependent reflection coefficient $\rho(j\omega)$. Yet so far, no mathematically rigorous treatment for the direct conversion of $P(t)$ to the required relaxation parameters has been forthcoming.

[24] P. R. Mason and J. B. Hasted, see J. B. Hasted in 'Water—A Comprehensive Treatise', ed. F. Franks, 'The Physics and Physical Chemistry of Water,' Plenum Press, New York, 1972, Vol. 1.

[25] R. Pottel, personal communication in ref. 24.

†See Chapter 5, pp. 161, 162.

In a theoretical treatment of time domain responses, van Gemert and de Graan[16] used a reverse procedure, calculating $P(t)$ from a known $\varepsilon(j\omega)$ behaviour. This treatment is idealized because it was necessary to assume a perfect step pulse with zero rise-time. Using Laplace transform

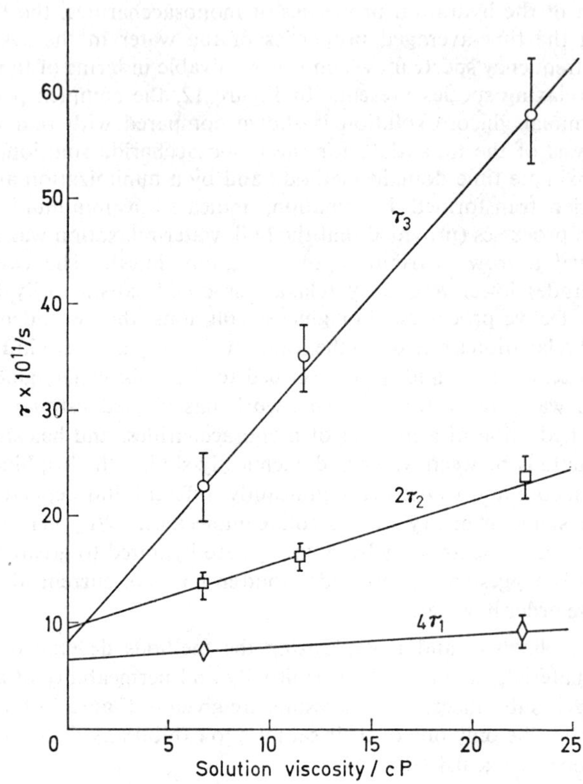

Figure 13 *Aqueous glucose solutions. Dependence of the relaxation times upon solution viscosity.* ○, τ_3; □, $2\tau_2$; ◇, $4\tau_1$
(Reproduced by permission from *J. Solution Chem.*, in press)

notation, where the quantity $j\omega$ is replaced by s where necessary ($s = \gamma + j\omega$, where γ is real) they derive the relations

$$P(t) = \mathscr{L}^{-1}\left\{\frac{\rho(s)}{s}\right\} \tag{21}$$

and

$$\rho(s) = \frac{\sqrt{s} - \sqrt{[s.\varepsilon(s) + \sigma/\varepsilon]}}{\sqrt{s} + \sqrt{[s.\varepsilon(s) + \sigma/\varepsilon]}} \tag{22}$$

They thus calculate $P(t)$ in the absence of significant conductivity for three forms of $\varepsilon(s)$ behaviour, Debye, Cole–Cole, and Davidson–Cole types. In

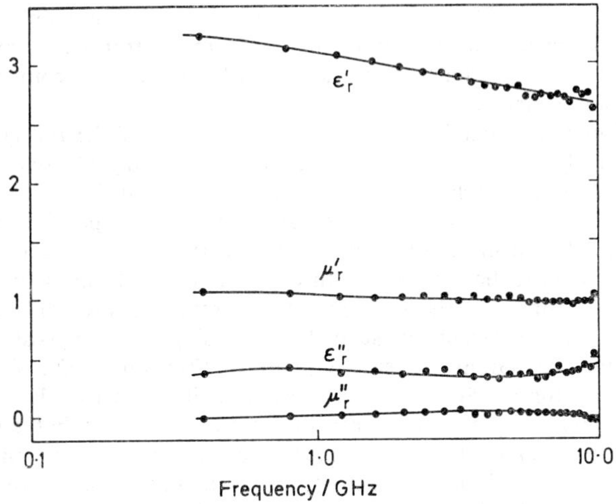

Figure 14 *Components of the complex relative permittivity and relative permeability of a maple wood disc*
(Reproduced by permission from *IEEE Trans. Instrument. Measur.*, 1970, **IM-90**, 377)

Figure 15 for example, their computed responses for Cole–Cole dielectrics with $\varepsilon_0 = 80$, $\varepsilon_\infty = 8$, and $\alpha = 0$ (*i.e.* Debye) and $\alpha = 0.5$ are given. It is concluded that for Debye and Davidson–Cole behaviour, $P(t)$ effectively reaches its asymptotic value by a time $t \simeq 2\tau_0$. In contrast, the corresponding times for Cole–Cole dielectrics with $\alpha = 0.2$ and 0.5 are $\sim 10\tau_0$ and

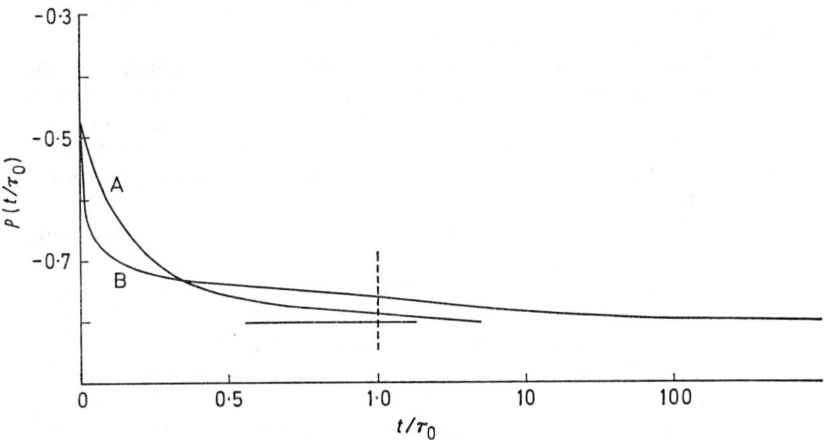

Figure 15 *Step response of the reflection coefficient for a Cole–Cole dielectric with* $\varepsilon_0 = 80$, $\varepsilon_\infty = 8$; A, $\alpha = 0$; B, $\alpha = 0.5$. *For* $t/\tau_0 > 1$ *a logarithmic time scale is used*
(Reproduced by permission from *Appl. Sci. Res.* in press)

$300\tau_0$ respectively. These workers conclude that it is impossible to determine the dielectric relaxation time directly from the transient response, as the type of dielectric model involved cannot be recognized from the shape of the response curves.

When ionic conductivity is predominant, instead of the reflected pulse in Figure 15 levelling off at a value of ρ corresponding [*cf.* equation (13)] to the low-frequency relative permittivity, ε_0, it instead decays to the level of a short circuit ($\rho = -1$). The suggestion [9] that the slope of the reflected pulse at $t = 0$ is a linear function of the low-frequency conductivity was later [15, 16] shown to be erroneous. In any case for real pulses, because of the finite rise-time and the presence of a number of internal reflections, the precise measurement of such limiting slopes is impossible. Van Gemert and de Graan,[16] using equations (21) and (22), calculated $P(t)$ allowing for varying conductivities. Despite such calculations, it is clearly apparent that, as yet, the measurement of ε^* by t.d.s. in the presence of appreciable conductivity is a hazardous affair: in this context the definition of 'appreciable conductivity' is itself uncertain. More work is required in order to characterize, for example, the effects of Fourier transforming conductivity-perturbed transient responses, and the consequent errors in the computed values of ε' and ε''.

Future Developments.—One of the most appealing aspects of t.d.s. is the number of possible variations in experimental method that one has available with the same basic equipment, although only a few of the possibilities have been reported. It is perhaps most likely that the major advances will come from the extension of the technique to longer times (lower frequencies) and thus into the area of macromolecular relaxation processes. The factor in the present methods which determines the length of the t.d.s. time window is the appearance of the echo pulses, and bearing this in mind there are at least three possibilities to be considered:

(*a*) The construction of very long sample cells, so that much more of the directly reflected pulse (see Figure 6) can be monitored before the onset of echoes. Fellner-Feldegg [9] has discussed this problem and has described a spiral coaxial cell which, empty, has a return trip time of about 50 ns. This approach, however, is only feasible when the dielectric samples are available in abundant supply.

(*b*) The analysis of the reflected or transmitted signal waveform including the echoes. Depending upon whether the line is terminated with a short circuit, open termination, or 50 Ω termination, equations can be derived relating $G(\omega)$, of equation (10), to ε^*.

(*c*) Perhaps the most promising approach involves the use of very thin sample cells, so that the echo pulses are totally enclosed within the rise (or fall) times of the directly transmitted or reflected pulses. This method has been reported informally by Fellner-Feldegg,[26] and it has the significant advantage that only very small amounts of sample are required.

[26] H. Fellner-Feldegg, presented at the Dielectrics Discussion Group, Bedford College, London, 1971.

The reflected pulse from a thin sample of, for example, an aliphatic alcohol, has a form similar to that shown in Figure 16. Because of the elimination of the echo pulses, this waveform can be directly interpreted in terms of the relaxation characteristics of the sample, and for a time

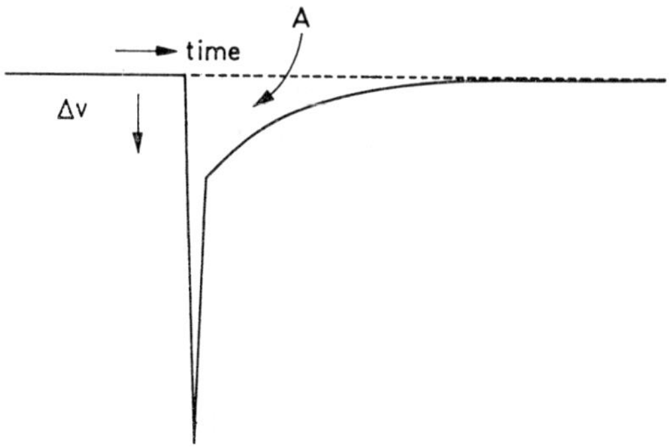

Figure 16 *Typical waveform observed for the reflection of a fast-rise voltage pulse from a thin dielectric sample enclosed in a coaxial line*

which depends only upon the length of the applied pulse: if the cell thickness, x, is such that the condition

$$\omega x \sqrt{(\varepsilon^*)}/c \ll 1$$

is satisfied (usually $0.1 \text{ mm} \leqslant x \leqslant 1 \text{ mm}$), then the sample permittivity $\varepsilon^*(\omega)$ and the measured reflection coefficient, $R(\omega)$, are related[27] by

$$R(\omega) = \frac{j\omega x}{2c}[1 - \varepsilon^*(\omega)] \tag{23}$$

The use of a 'standard dielectric' substance as a reference for the sample under test (*e.g.* water for aqueous solutions) may be convenient, as the ratio of the Fourier transforms of the reflected pulses from the sample and 'standard', $F(\omega)$ and $F'(\omega)$ respectively, gives

$$\frac{F(\omega)}{F'(\omega)} = \frac{R_{\text{sample}}}{R_{\text{standard}}} = \frac{1 - \varepsilon^*_{\text{sample}}}{1 - \varepsilon^*_{\text{standard}}} \tag{24}$$

and no accurate measurement of the sample thickness is required.

For the case of a dielectric having a single, discrete relaxation time, τ, the reflection of a perfect step pulse from the thin cell can be treated purely in the time domain, in which case the decaying tail of the waveform (Figure 16) is an exponential ($\propto \exp[-t/\tau]$), and the total area, A, under the curve is given by

$$A = \frac{x}{2c}(1 - \varepsilon_0) \tag{25}$$

where ε_0 is the static relative permittivity.

[27] H. Fellner-Feldegg, Hewlett-Packard Application Note 153, *J. Phys. Chem.*, in press.

Fellner-Feldegg,[27] using a multiple scanning technique incorporating the three-point method of Nicolson and Ross [10] (Figure 10), has recently demonstrated thin-sample t.d.s. by dielectric measurements on alcohols. Figure 17 shows the data for butan-1-ol at 298 K plotted on the complex

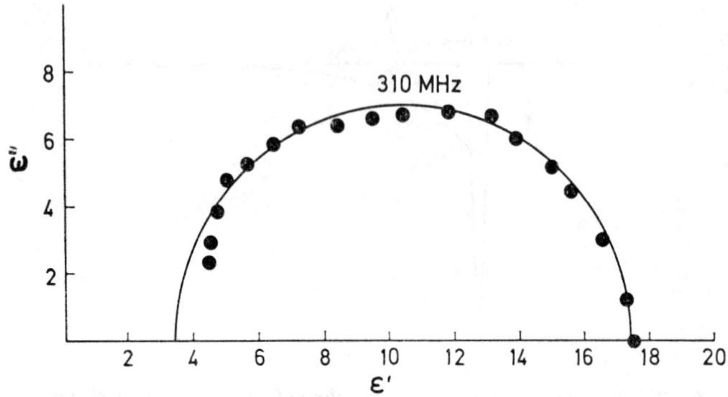

Figure 17 *Components of the complex permittivity of butan-*1*-ol at* 298 K *obtained by thin-sample t.d.s.*
(Based on the results of H. Fellner-Feldegg, ref. 27)

plane diagram. Such results can be obtained in a few minutes, clearly indicating the remarkable speed and simplicity of t.d.s. measurements, when data acquisition and processing are automated using a small, on-line computer.

5

Dielectric Properties of Water and of Aqueous Solutions

BY J. B. HASTED

1 Introduction

In recent years complex permittivity measurements have continued to be made on aqueous systems; the range of frequency coverage and the accuracy of measurement have improved since the last reviews of the subject,[1, 2] but it is in the interpretation of these data, and the light they throw on the structure and properties of water and aqueous solutions, that the more important advances have been made.

It is likely that the next few years will see the extension of experiments in this field in two ways: (i) the exploitation of time-domain techniques will undoubtedly produce data of wider scope than exist at present; (ii) the submillimetre band of frequencies (300—5000 GHz), in which some measurements have already been made, will be exploited in numerous, and hopefully systematic, measurements. Probably the theories that are discussed in this article cannot be fully tested until these new extensions of the experimental data come about. The intention of the present discussion is to review the ideas that have already arisen, in order that the need for the extension of data may be properly assessed.

There have been great advances in the study of liquid water by techniques such as *X*-ray diffraction,[3] far-i.r. absorption,[4] molecular orbital calculation,[5] and computerized molecular dynamics.[6] The dielectric relaxation processes in the liquid must now be examined in the light of these advances.

There has also been some speculative research, of importance in the study of surface H_2O, which will not be considered here. This research includes: (i) investigation of the possibility of the existence of 'kinks', or structures, in the temperature dependence of various physical properties,

1 J. B. Hasted, *Progr. Dielectrics*, 1961, **3**, 103.
2 D. Eisenberg and W. Kauzmann, 'The Structure and Properties of Water', Clarendon Press, Oxford, 1969.
3 A. H. Narten, M. D. Danford, and H. A. Levy, *Discuss. Faraday Soc.*, 1967, No. 43, p. 97.
4 G. W. Chantry, 'Submillimetre Spectroscopy', Academic Press, New York, 1972; see also ref. 30.
5 Janet Del Bene and J. A. Pople, *J. Chem. Phys.*, 1970, **52**, 4858.
6 A. Rahman and F. H. Stillinger, *J. Chem. Phys.*, 1971, **55**, 3336.

and (ii) appraisal of evidence for the existence of (impure) 'polywater', formed by condensation of water vapour inside very fine capillary tubes.[7]

As regards aqueous solutions of electrolytes, there has been some significant advance in the understanding of the dielectric behaviour of the hydration sheath around the ions. In the field of non-electrolyte solutions, no claim can yet be made that there is good understanding. It is obvious that the data are still insufficient; this is the field where measurements are most needed.

In applied science, there has been great stimulus to study the dielectric properties of surface water; this stimulus comes from two distinctly different applications, *i.e.* moisture sensing and determination, and diathermic heating. Although there is insufficient space for detailed discussion of these problems, it is important to be able to assess their importance in relation to related branches of technology, and a few words will not be out of place.

Industrial applications of moisture measurement by microwave absorption are numerous; they are relative rather than absolute measurements, strongly dependent upon calibration; they must compete with other techniques, such as neutron absorption. A popular frequency band is 3—10 GHz; at lower frequencies the electrolytic conductivity can mask the absorption due to dipolar relaxation, so that the technique becomes inconveniently electrolyte-sensitive; at higher frequencies the penetration is usually insufficient; the technique is normally carried out in transmission, but measurement by reflection is also possible. Kilohertz frequencies are also popular in studies of systems containing bound water.

Reflection[8] and ground-wave propagation[9] are the normal modes used in geological remote sensing of moisture. When obtained from an aeroplane, radar echoes from ground water are studied in terms of their range, scattering cross-section (*i.e.* dependence of reflection on angle), and Doppler shift. In temperate conditions the penetration of a radar pulse into moist soil is usually insufficient for the technique to be of much value in hydrological studies of ground water, but in deeply frozen earth and in hot desert sand the penetration could well be sufficient for remote sensing to be very important.

It has been mentioned that the use of radar propagation in studies of ground water must be based upon extensive laboratory measurements of dielectric properties. However, it is not clear that, at the time of writing, sufficiently extensive laboratory work has yet been done.

The diathermic applications of dielectric studies are not solely biological, although the most important are medicinal and culinary. There also exist industrial diathermic applications, concerned mostly with fast drying and

[7] J. B. Hasted, *Contemporary Physics*, 1971, **12**, 133.
[8] R. K. Moore, in 'Proceedings of 4th Conference on Remote Sensing of the Environment', 1966, p. 339.
[9] 'Electromagnetic Probing in Geophysics', ed. J. R. Wait, Golem, Colorado, U.S.A., 1971.

the setting of plastics. Applications in the liquid phase require that the microwave dielectric properties of aqueous solutions be intensively studied and well understood.

2 Current Ideas on the Structure of Water

Although the molecules in a liquid are in continual motion, there exists at any instant a configuration which can be regarded as the instantaneous structure. The time average of these configurations can still be regarded as

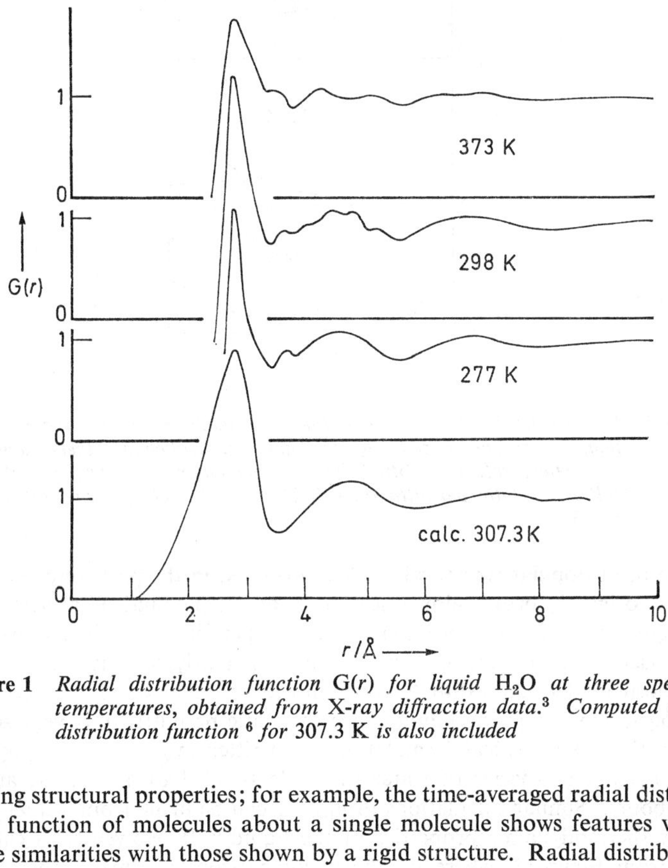

Figure 1 *Radial distribution function* G(*r*) *for liquid* H_2O *at three specified temperatures, obtained from* X-*ray diffraction data.*[3] *Computed radial distribution function*[6] *for* 307.3 K *is also included*

having structural properties; for example, the time-averaged radial distribution function of molecules about a single molecule shows features which have similarities with those shown by a rigid structure. Radial distribution functions which have been deduced from *X*-ray diffraction measurements[3] on water at various temperatures are shown in Figure 1. The broad maxima at certain nuclear separations tell us that the arrangement of the molecules is neither random (as it would be in a gas) nor regular (as it would be in a crystal). However, it is not possible to deduce just what is the typical instantaneous structure from this information alone.

It is only on the basis of a correct knowledge of the most probable instantaneous structure that an accurate calculation of the static electric permittivity can be made. It is indeed likely that the comparison of experimental and calculated temperature variation of the static electric permittivity can serve as a test of the adequacy of the structural model proposed.

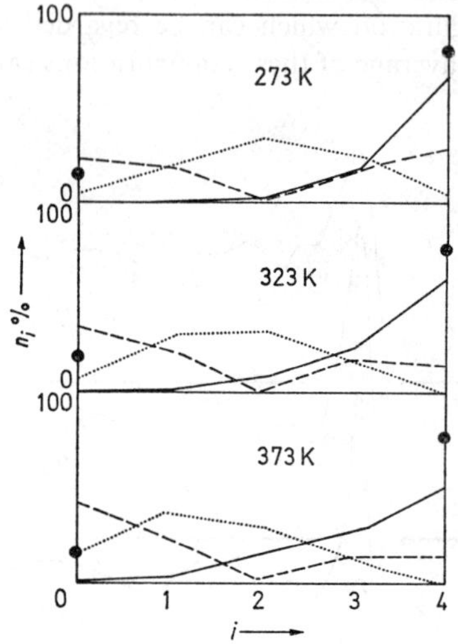

Figure 2 *Contrasting bond state populations for liquid* H_2O *at three specified temperatures, proposed by different authors. Broken line, thermodynamic properties, ref.* 13; *dotted line, thermodynamic properties, ref.* 14; *full line, static permittivity, ref.* 11; *closed circles,* X-*ray diffraction, ref.* 3

The most popular structural models proposed in the past were versions of the original 'broken-down ice' idea of Bernal and Fowler.[10] The attempts to apply such versions to static permittivity calculations[11, 12] were successful, but did not provide a particularly sensitive test of the structural model. An estimate could, however, be made of the extent of hydrogen-bonding in the liquid, since the static permittivity is very sensitive to this. A statistical formulation was written in terms of the populations $n_{4,3,2,1,0}$ of species of water molecule bonded to 4, 3, 2, 1, and 0 neighbours. Similar calculations of $n_{4,3,2,1,0}$ can be made from a treatment of the thermodynamic properties of the liquid.[13, 14] However, as can be seen from Figure 2, where graphical comparisons are made, there is a considerable discrepancy between the extent of hydrogen-bonding that is

[10] J. D. Bernal and R. H. Fowler, *J. Chem. Phys.*, 1933, **1**, 515.
[11] G. H. Haggis, J. B. Hasted, and T. S. Buchanan, *J. Chem. Phys.*, 1952, **20**, 1452.
[12] J. Lennard-Jones and J. A. Pople, *Proc. Roy. Soc.*, 1951, **A205**, 155.
[13] G. Nemethy and H. A. Scheraga, *J. Chem. Phys.*, 1962, **36**, 3382; 1964, **41**, 680.
[14] K. Buijs and G. R. Choppin, *J. Chem. Phys.*, 1963, **39**, 2035; 1964, **40**, 3120.

required to interpret the different types of phenomenon. It cannot be claimed that the permittivity calculations determine the structure uniquely.

The broken-down ice structure is usually applied, with oxygen atoms possessing the ability to make any number up to four bonds to other oxygen atoms. Thus it is possible for an open network to be formed in three dimensions. Distorted rather than fragmented networks have also been considered. An entirely different type of model would have to be envisaged if branching were unlikely; only chains and rings would then be found.

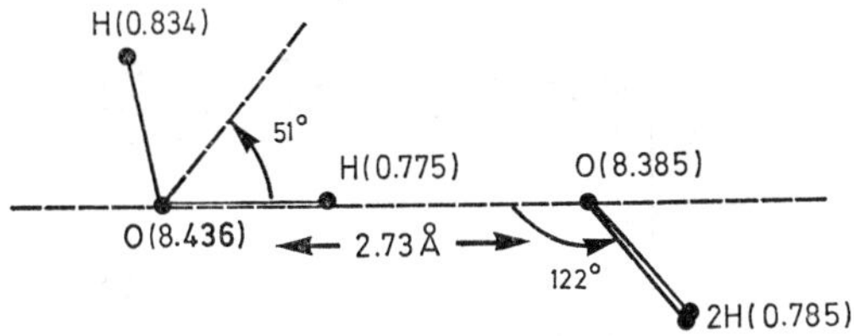

Figure 3 *Predicted minimum energy configuration of the* $(H_2O)_2$ *dimer* [5]

Such a model has seldom been considered in recent years, but the impact of molecular orbital theory now makes it necessary to do so.

Del Bene and Pople [5] have published extensive *ab initio* minimal basis LCAO–SCF–MO calculations for groups of up to six water molecules. It is well known that such calculations are not particularly accurate for absolute values of energy levels, but they are highly successful for the comparison of the energies of different geometrical configurations of molecules. For example, the equilibrium nuclear separations of diatomic molecules calculated by minimizing the MO energies are usually correct to within ± 1%.

This technique of minimization has been applied to the calculation of the most stable configuration of the water dimer; the second molecule was systematically moved around, both in orientation and distance from the first molecule, until the minimum energy was achieved; the equilibrium configuration of the dimer was thereby predicted to be as illustrated in Figure 3. Its energy is tabulated, together with those of other species, in Table 1. The stiffness of the bonds was also calculated, and it was found that only rotation about the axis of the hydrogen bond is relatively easy.

In seeking to calculate the structure of the most stable trimer, it is at once seen that configurations (1)—(3) are possible. The third configuration

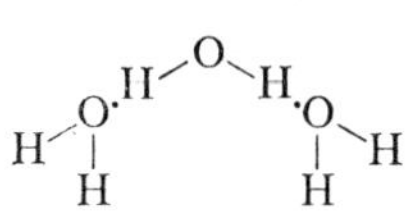

(1)

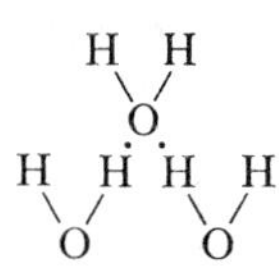

(2)

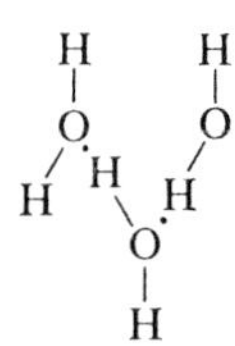

(3)

Table 1 *Assignments, calculated energies, correlation factors, and populations of ring-and-chain species of* H_2O *multimer*

Species	q	*Notation*	*Calculated energies* E_i/kJ mol^{-1}	*Scaled energies* /kJ mole^{-1}	*Calculated* g_i	*Values of* n_i *at temperature* T: T = 273 K	T = 373 K	T = 473 K	T = 573 K
Free molecule	1	1c	0	0	1	0.100	0.214	0.314	0.390
Chain	2	2c	− 25.49	− 4.109	0.658	0.020	0.056	0.091	0.117
Chain	3	3c	− 61.42	− 9.892	0.770	0.009	0.026	0.041	0.051
Chain	4	4c	− 102.7	− 16.56	0.734	0.006	0.016	0.025	0.028
Chain	5	5c	− 173.2	− 27.94	0.745	0.022	0.036	0.036	0.031
C_n ring	3	3r	− 70.40	− 11.36	≃ 2.03	0.013	0.032	0.047	0.054
C_n ring	4	4r	− 158.7	− 25.60	≃ 2.71	0.171	0.171	0.140	0.107
C_n ring	5	5r	− 222.4	− 35.88	≃ 3.39	0.356	0.226	0.135	0.081
C_n ring	6	6r	− 270.0	− 43.52	≃ 4.07	0.266	0.146	0.075	0.040
S_n ring	3	3s	− 77.56	− 8.714	≃ 0.68	0.004	0.015	0.027	0.035
S_n ring	4	4s	− 177.0	− 19.86	0	0.017	0.034	0.041	0.041
S_n ring	5	5s	− 243.5	− 27.36	≃ 0.68	0.011	0.020	0.021	0.018
S_n ring	6	6s	− 301.2	− 33.83	0	0.005	0.008	0.008	0.007

is found to be the most stable, and, uniquely in this configuration, the hydrogen bonds are each stronger than that in the dimer, by approximately 4 kJ mol^{-1}. Moreover, this configuration is susceptible to lengthening in the same manner, forming chains similar to those proposed to exist in liquid alcohols by Dannhauser and Cole.[15]

The most stable chain tetramer is that formed in an approximately *cis*-planar configuration. The binding energy, as will be seen from Table 1, is considerably greater than three of the hydrogen-bonds of the dimer, and actually slightly greater than four such bonds.

The most stable pentamer is in an approximately cyclic configuration (4). Although the ring is not precisely closed, the binding energy of this

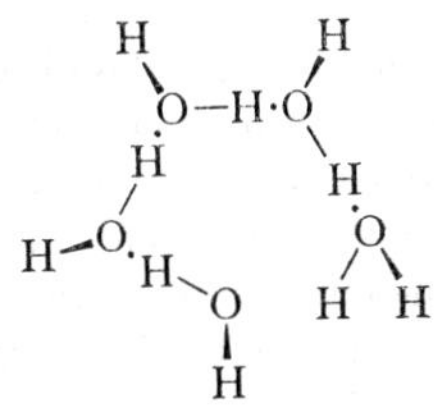

(4)

pentamer (Table 1) is almost as large as that of seven dimer hydrogen-bonds.

Thus there is evidence from MO theory that a 'pseudo-co-operative effect' is active in water hydrogen-bonding; bonds become stronger when more molecules are added to the chain. This is not to say that all bonds are of equal strength in a chain; in fact it is very likely that they are not equal. This strengthening of hydrogen-bonds is evidence that chains may be present in liquid water; the higher the temperature, the shorter the chains.

It was also found by MO calculations that ring species were even more stable than chains; since this applies to the five-membered ring, it is unlikely that many chains with more than five members (*e.g.* helical chains) would be found in the liquid, since most units of five members would close into rings.

The calculated energies of the three-, four-, five-, and six-membered rings are shown in Table 1. The structures can possess either a C_n or S_n symmetry axis, where n is the number of units in the multimer. The multimer is described in terms of the oxygen–oxygen distance (R) between nearest neighbours and the three angles α, β, and γ. In C_n symmetry, the α rotation moves each external proton out of the oxygen plane, the β then moves the hydrogen-bonded protons out of the plane, and the γ rotation allows for the third dimension of the hydrogen-bond, bending away from the oxygen–oxygen line. In C_n symmetry all the external hydrogen nuclei are on the same side of the oxygen plane, but in S_n symmetry alternate external hydrogens are on opposite sides of the plane

[15] W. Dannhauser and R. H. Cole, *J. Chem. Phys.*, 1955, **23**, 1762.

(with one exception, where n is odd). In the MO calculations the angles α, β, and γ have been determined by minimization of the energy; although they are not shown in Table 1, their values will be used below in calculations of the dielectric correlation factor.

It is clear from the calculated energies given in Table 1 that the cyclic multimers are in all cases more stable than the chain multimers having identical values of n. This is an unexpected conclusion for the three- and four-membered rings, since it might be expected that these would be full of strain and therefore less stable. It would appear that a multimer species might exist in either its cyclic S_n ground state, or in metastable cyclic C_n or chain configurations.

An explanation of the high stability of rings can be given on the basis of calculated electron densities of the oxygen and hydrogen atoms. The local negative charge at the oxygen atom increases with growth in multimer size, while the bonding hydrogen acquires larger positive charge. Thus each molecule is polarized by its neighbours. This mutual polarization enhances the strength of the hydrogen-bonds, and increases the stability of the ring structures. Because of the symmetry of the structures, each bond is equal in strength in C_n symmetry, and in S_n symmetry where n is even.

MO calculations were made on puckered six-membered rings of the type found in the ice I lattice, but it was found that this ring has less intermolecular binding (by 29.62 kJ mol^{-1}) than the cyclic structure of S_6. Thus the puckering of a six-membered ring, which might be required in order to satisfy some local order, needs energy; it would not occur in an isolated ring. Local order might be brought about by the formation of an additional hydrogen-bond to one member of the ring. The energy of such a bond has not been calculated, but an attempt has been made to calculate the energy of the simplest form of branched structure, namely the pentamer in which one central molecule is bonded to four neighbours, acting as proton donor to two and proton acceptor to two more. This pentamer has an energy of 125.0 kJ mol^{-1}, which is only slightly more than than that of four isolated hydrogen-bonds. Thus there is relatively little additional binding energy arising from the branching effect at one centre. Although similar calculations have not been made for ring structures with additional branching, it may be concluded that, contrary to previous supposition, and particularly at high temperatures, branching structures are energetically unfavourable in liquid water, although they appear in ice I. This revolutionary conclusion, although open to criticism on various grounds, must be taken seriously; we must consider in detail the effects of assuming a ring-and-chain structure upon the variation of the physical properties of the liquid with temperature. In the next section, which concerns the static permittivity, some progress in this direction is reported.

A different approach to the structural problem is provided by the technique of computerized molecular dynamics. It is now possible to set up in a computer program a microcanonical ensemble of perhaps two hundred molecules in an initially random geometrical arrangement and work out their ensuing motions under classical interactions. Over a chosen period, the time averages of the important physical parameters (radial distribution function, dielectric correlation factor, *etc.*) can be extracted from the program, together with the appropriate molecular motion parameter, which yields the temperature; ageing of the program under time-varying 'matched' boundary conditions allows a simulation of different temperatures.

This approach stems from the older model-building techniques of Bernal, but the computer has made it enormously more powerful. The first application of the approach to water has been made by Rahman and Stillinger,[6] who have treated the H_2O molecule as a rigid tetrahedron, at whose corners are two positive (screened) charges 0.19*e* and two equal negative charges; there is also a van der Waals force, spherically symmetrical about the central oxygen. In this way the hydrogen-bonding feature, with bending (and possibly even breakage) of bonds, is simulated. The hydrogen-bond energy for this model is 27.05 kJ mol^{-1}, but there is no co-operative effect such as has been discussed above in the MO calculations.

The radial distribution function (or pair correlation function) of the oxygen nuclei that is produced by the computer program is included in Figure 1, and it will be seen that the broad features of the function are reproduced. An attempt was made to divide this distribution function into solid angle fractions, and it was found that, of the nearest neighbours, approximately 5% are "necessarily so seriously misaligned that no hope for a hydrogen-bond between them exists. One possible explanation would be that an 'interstitial' molecule were involved in such a pair, surrounded by, but only weakly interacting with, an enclosing network of hydrogen-bonds. Alternatively, two molecules, each incompletely hydrogen-bonded to their surroundings, could by chance simply back into one another". This is in effect something of a justification of bond-breaking statistics.

Both static and time-dependent dielectric parameters have been calculated from the computer molecular dynamical model, and they will be discussed in the ensuing sections.

3 Calculations of Static Electric Permittivity

In order to apply the Kirkwood–Fröhlich theory[16, 17] to the calculation of the static permittivity ε_0 of water at temperature *T*, the approximation is sometimes made that the liquid contains a number of different molecular species. It is necessary to know:

(i) the populations n_i of different species *i* present in the liquid;

(ii) the 'free molecule' dipole moment μ_0 that is appropriate to each species;

[16] J. G. Kirkwood, *J. Chem. Phys.*, 1939, 7, 911.

[17] H. Fröhlich, 'Theory of Dielectrics', Oxford University Press, 1949.

(iii) the correlation parameter g_i appropriate to each species i: g is defined as μ^*/μ_0, where μ^* is the average dipole moment of a sphere embedded in its own medium, when the dipole at the centre is held in a given orientation;

(iv) the 'infinite-frequency' permittivity ε_∞, which contains contributions only from intramolecular processes, induced polarization, and molecular vibrations.

The static or zero-frequency permittivity ε_0 is then given by equation (1),

$$\frac{(\varepsilon_0 - \varepsilon_\infty)(2\varepsilon_0 + \varepsilon_\infty)}{\varepsilon_0(\varepsilon_\infty + 2)^2} = \frac{4\pi N_0}{9kTv}\sum_i n_i g_i \mu_0^2 \tag{1}$$

where N_0 is the Avogadro number, k the Boltzmann constant, and v the molar volume.

It is, of course, not a realistic approximation to separate the molecules in the liquid into a small number of different species, each with characteristic environmental and orientational properties. There is likely to be a continuous gradation between the environmental properties of the water molecules in the liquid. Nevertheless it is possible that the 'limited number of species approximation' can achieve some success. It was originally applied [11] to a broken-ice network in such a way that the temperature variation of ε_0 between 273 and 633 K was accurately fitted. The first step in this application was the definition of species and the calculation, using statistics, of their population from the latent heat of evaporation L.

Since in tridymite ice I the H_2O molecule can form four hydrogen-bonds to its neighbours, there are two bonds for each H_2O. The proportion p of bonds broken on fusion is calculated from equation (2), where the

$$p = 1 - [(L - W)/2\overline{\Delta H}] \tag{2}$$

energy W is that due to van der Waals forces, and $\overline{\Delta H}$ is the average energy of breakage of the hydrogen-bond. Comparison with other reported values of species populations calculated from the variation with temperature of static permittivity shows the measure of their success. It will be seen from Figure 2 that when the simplest assumptions are made about molecular species (*i.e.* molecules bonded to four, three, two, one, or zero neighbours), the results of calculations from thermodynamic properties and static permittivity are not in concord.

We now consider how the approximation can be applied to the ring-and-chain species proposed in the last section. Following the methods of Haggis *et al.*,[11] the probabilities of breaking and making of each type of bond are equated [equation (3)], and for $j = 68$ such equations can be written

$$m_{i1} f(T) \exp(-\Delta E_j/kT) = n_{i2}\, n_{i3} f_j(T, p) \tag{3}$$

for the breakage of a number m_{i1} of bonds of species 1, yielding numbers n_{i2} and n_{i3} of two fragment species 2 and 3. An energy $\Delta E_j = E_{i2} + E_{i3} - E_{i1}$ is involved in each process. $f(T)$ is a function of temperature only, but

$f_j(T, p)$ is a function both of temperature and of the condition of the surrounding water molecules. For $i = 13$ species, the detail of the $j = 68$ equations is given in Table 2.

Table 2 *Kinetic processes for interconversions of ring-and-chain species of* H_2O *polymers*

Assigned value of j	*Notation* *i*1	*i*2	*i*3	*Assigned value of j*	*Notation* *i*1	*i*2	*i*3
1	6r	5r	1c	35	6s	5s	1c
2	6r	4r	2c	36	6s	4s	2c
3	6r	3r	3c	37	6s	3s	3c
4	5r	4r	1c	38	5s	4s	1c
5	5r	3r	2c	39	5s	3s	2c
6	4r	3r	1c	40	4s	3s	1c
7	6r	5c	1c	41	6s	5c	1c
8	6r	4c	2c	42	6s	4c	2c
9	6r	3c	3c	43	6s	3c	3c
10	5r	4c	1c	44	5s	4c	1c
11	5r	3c	2c	45	5s	3c	2c
12	4r	3c	1c	46	4s	3c	1c
13	4r	2c	2c	47	4s	2c	2c
14	3r	2c	1c	48	3s	2c	1c
15	6r	3r	3r	49	6s	3s	3s
16	5c	4c	1c	50	5c	4s	1c
17	5c	3c	2c	51	5c	3s	2c
18	4c	3c	1c	52	4c	3s	1c
19	4c	2c	2c	53	6s	5r	1c
20	3c	2c	1c	54	6s	4r	2c
21	2c	1c	1c	55	6s	3r	3c
22	5c	4r	1c	56	5s	4r	1c
23	4c	3r	2c	57	5s	3r	2c
24	4c	3r	1c	58	4s	3r	1c
25	5r	5c	—	59	6r	5s	1c
26	4r	4c	—	60	6r	4s	2c
27	3r	3c	—	61	6r	3s	3c
28	5s	5c	—	62	5r	4s	1c
29	4s	4c	—	63	5r	3s	2c
30	3s	3c	—	64	4r	3s	1c
31	3r	3s	—	65	6s	3s	3r
32	4r	4s	—	66	6r	3s	3r
33	5r	5s	—	67	6r	3s	3s
34	6r	6s	—	68	6s	3r	3r

The populations n_i are normalized so that $\sum_i n_i = 1$. Simple sum rules of the type $qn_i = m_i$ can be written for ring species r and s, and $(q - 1) n_i = m_i$ for chain species c: q is the number of molecules in the species. From the $j = 68$ equations, the 13 populations n_i can be calculated, together with all ratios $r_j = f_j(T, p)/f(T)$. In the calculations reported here, the ratios for the breakage processes of each species have been assumed identical at any single temperature (*e.g.* $r_1 = r_2 = r_3 = r_7 = r_8$

$= r_9 = r_{15} = r_{34} = r_{59} = r_{60} = r_{61} = r_{66} = r_{67}$). The solutions show all r to be equal to q for ring species, and to $(q - 1)$ for chain species, which is a vindication of the simple Saha-type equation $n_{i1} \exp(-\Delta E_j/kT) = n_{i2}\, n_{i3}$.

The calculation of the populations n_i must yield results which are consistent with the temperature variation of the latent heat of vaporization. The proportion p of bonds broken is given by equation (4).

$$\sum_i m_i = 2(1 - p) \qquad (4)$$

The average energy of breakage of the hydrogen-bond is related to the energies of the species by equation (5), where $q_0 = q$ for ring species and

$$\overline{\Delta H} = -\sum_i n_i E_i q_0 \qquad (5)$$

$q_0 = (q - 1)$ for chain species. Combining equations (2), (4), and (5) with the sum rules, we have equation (6).

$$-\sum_i n_i q_0 \cdot \sum_i n_i E_i \Big/ q_0 = L - W \qquad (6)$$

There are therefore two independent methods of testing the kinetics of species populations: realistic and temperature-independent van der Waals energy W, as calculated from equation (6), and realistic and correctly temperature-varying $\bar{g} = \sum_i n_i g_i$.

Calculations using the unchanged (Table 1) species energies calculated by Del Bene and Pople yield strongly negative values of W. The smallest uniform scaling factor by which these would have to be reduced for W to be positive is 7. With uniform scaling factor 8, the values of W are fairly acceptable (9.52 kJ mol^{-1} at 273 K, 9.36 at 373 K, 7.86 at 473 K, 1.21 at 573 K), but the weighted correlation parameter $\bar{g} = \sum_i n_i g_i$ is then very unrealistic (0.82 at 273 K, 1.01 at 373 K, 1.10 at 473 K, 1.14 at 573 K). This is because the most stable rings of S_n symmetry are relatively non-polar; the dielectric evidence suggests that these species must be forbidden. When their scaling factor is raised to 8.9, whilst that of the remaining species is held at 6.2 (actual energies as in column 5, Table 1), the temperature variation of W is smaller, and the magnitude and temperature variation of $\bar{g}$ are much more acceptable. For this optimized calculation the values of W and $\bar{g}$ are shown in Table 3 and the species populations in Table 1. The values of g have been calculated from $\varepsilon_0(T)$, using $\varepsilon_\infty = 1.8$, the value at 400 cm^{-1}, which is probably low because of the occurrence of librational processes below this frequency.* On the evidence of the static permittivity, the ring-and-chain model would be acceptable provided only that those extremely large scaling factors are not held to be improbable.

* [Senior Reporter's note]. The difficulty in accepting such a value is that much of the dispersion between 50 and 400 cm^{-1} is due to translational modes, whereas much significant librational absorption occurs above 400 cm^{-1} (see ref. 30).

The reconciliation of the activation energy of dielectric relaxation with the energies of bond breakage is only possible when the MO energies are scaled down in magnitude. The most frequently encountered processes, such as 6 ring → 5 ring + 1, possess energy defects of about 29 kJ mol^{-1}, and this figure is in excess of the range of activation energies encountered in the relaxation process† (18.7—21.0 kJ mol^{-1}). The scaling-down factor

Table 3 *Comparison of calculations for ring-and-chain species with experimental data*

T/K	L (*experimental*) /kJ mol^{-1}	W/kJ mol^{-1} *from equation* (6)	g (*calc. from* ε_0 *exptl.*)	$\bar{g} = \sum_i n_i g_i$
273	44.7	14.05	2.87	2.93
373	40.5	18.4	2.56	2.22
473	35.1	19.4	2.24	1.67
573	25.1	13.5	1.64	1.42

necessary to achieve consistent population statistics would probably suit the activation energy of relaxation, which is calculated using equation (7), where the average energy is defined as shown in equation (8). The symbol

$$1/\tau \propto \exp(-\overline{\Delta E}/RT) \tag{7}$$

$$\overline{\Delta E} = \frac{1}{(j)} \sum_{(j)} \Delta E_j \cdot m_{i1} \tag{8}$$

(j) denotes that only those processes j are included which exhibit the smallest ΔE_j available to any bond species i. Computations of $\overline{\Delta E}$ are in progress.

We now turn to discussion of the computer dynamical procedure, which shows great promise for calculations of static permittivity. The model of Rahman and Stillinger [6] contains sufficient information in the computer program to enable the mean dipole moment $\overline{M}$ of a sphere of known radius to be calculated at a known temperature. This moment is related to the Kirkwood correlation parameter g by equations (9) and (10), with g defined as in equation (10).

$$\overline{\boldsymbol{M}} = \boldsymbol{\mu}^* + \int \boldsymbol{P}\,\mathrm{d}v \tag{9}$$

$$g = \mu^*/\mu_0 \tag{10}$$

The polarization volume integral is solved using the spatial variation of potential V [equation (11)], which finally yields equation (12). At 307.5 K,

$$4\pi\boldsymbol{P} = -\varepsilon(\varepsilon_0 - 1)\nabla V \tag{11}$$

$$\overline{\boldsymbol{M}} = 9\varepsilon_0\,\mu^*/(\varepsilon_0 + 2)(2\varepsilon_0 + 1) \tag{12}$$

† The energies are of the same order for heavy water.[18]

[18] C. H. Collie, J. B. Hasted, and D. M. Ritson, *Proc. Phys. Soc. London*, 1948, 60, 145.

the computations yield $g = 2.96$, but since the intermolecular forces chosen for the model are probably too strong, the authors propose a lower value, $g = 2.72$.

However, the most recent discussions [19] favour these high values of g although values of the order of 20% lower had originally been favoured. This is because the Fröhlich equation [equation (1)] differs from the earlier version of Kirkwood, and treats the inner field in a more nearly correct manner. It is no longer necessary to make a calculation of the H_2O dipole moment in its surroundings in the liquid, as had been necessary in the application of the Kirkwood equation. The dipole moment of the free molecule, $\mu = 1.84$ D, is used in equation (1), together with $\varepsilon_\infty = 1.80$ at 293 K. This leads to a value of $g = 2.82$, which is sufficiently close to that calculated from the computer dynamics model to warrant optimism for future calculations. The exact choice of ε_∞ will continue to present difficulties until the far-i.r. data are complete over a wide range of temperature.

The marked fall of g with increasing temperature (at 373 K, $g = 2.56$; at 473 K, $g = 2.24$; at 573 K, $g = 1.64$, assuming a temperature-invariant value of ε_∞) should, of course, be a sensitive test of any assumptions used in the computer dynamics. The success of the early statistical calculations,[11] in which molecules are classified according to the number of hydrogen-bonds they form, lies principally in the ability to reproduce the rapidly falling values of g.

Interpretation of pressure effects can be a profitable field for the application of dipole correlation theory. Some studies of the effects upon the static electric permittivity have appeared recently. Owen *et al.*[20] made measurements in the frequency range 40—480 MHz, which yielded effectively ε_0; the temperature range was 273—343 K. Density measurements under pressure,[21] and refractive index measurements under pressure [22] have enabled the Kirkwood–Fröhlich equation to be applied,[23] but whereas the dependence of ε_0 upon density, $d \ln \varepsilon_0 / d \ln \rho$, is $+ 0.8852$ at 273 K and $+ 1.1627$ at 343 K, the dependence of $g\mu_0^2$ upon density is of the opposite sign: $d \ln g\mu_0^2 / d \ln \rho$ is $- 0.51$ at 273 K and $- 0.24$ at 343 K. Thus the fall of $g\mu_0^2$ which is induced by raising temperature is also induced by raising pressure. The origin or origins of this effect can only be surmised: the hydrogen-bonded structure may become broken, or perhaps, in addition, the dipole moment may be lowered.

[19] J. B. Hasted, in 'Water – A Comprehensive Treatise', ed. F. Franks, Vol. 1, 'The Physics and Physical Chemistry of Water', Plenum Press, New York, 1972, Ch. 8.
[20] B. B. Owen, R. C. Miller, C. E. Milner, and H. L. Cogan, *J. Phys. Chem.*, 1961, **65**, 2065.
[21] G. S. Kell and E. Whalley, *Phil. Trans.*, 1965, **258A**, 565.
[22] R. M. Waxler and C. E. Weir, *J. Res. Nat. Bur. Stand., Sect. A*, 1963, **67**, 163; 1965, **69**, 325.
[23] E. Whalley, *Adv. High Pressure Res.*, 1966, **1**, 172.

4 The Dielectric Properties of Water at Very High Frequencies

Few new microwave measurements of the dielectric properties of water have appeared in recent years,[19] but submillimetre-wave measurements of both refractive index and absorption coefficient have been reported.

Microwave measurements have established with some high degree of probability that a small spread of relaxation times is exhibited by pure water. According to the Cole–Cole equation, the region between the static permittivity ε_0 and the infinite frequency permittivity ε_∞ is spanned by the complex permittivity ε^*, as defined in equation (13), where τ is the principal relaxation time and α the spread parameter; $\omega = 2\pi\nu$ is the radial

$$\varepsilon^* = (\varepsilon_0 - \varepsilon_\infty)/[1 + (\mathrm{j}\omega\tau)^{1-\alpha}] \qquad (13)$$

frequency. A value of $\alpha = 0.015$ at 293 K was reported by Grant and Shack.[24]

Now that regression programs for data analysis can be written for a computer, it is possible to place correctly determined limits of reliability upon ε_∞ and α at all temperatures. A program has been written by Mason *et al.*[25] which obtains ε_∞, τ, and α, and calculates the limits of confidence with which they can be regarded. This program has been applied to ten sets of measurements [19] which have been assumed to have equal percentage errors; the best values of α, τ, and ε_∞ and their confidence limits are obtained, together with the errors in each measurement. Inspection of these errors shows that, with the exception of one set of measurements at one frequency which must be discarded because of systematic error, all the measurements are of the same order of accuracy; therefore the decision to weight them equally has been correct. Measurements of the value of ε_0 are weighted by a factor ten times more accurate than the microwave measurements; no attempt is made to deduce ε_0 from this program.

The values of ε_∞, τ, and α and some limits of uncertainty are listed in Table 4. Interesting features are as follows:

(i) The principal relaxation times are close to those originally reported;[16] the curvature of the activation energy plot of log τ *vs.* $1/(T/\mathrm{K})$ is confirmed. The statistical interpretation of this has been discussed in Section 3 above.

(ii) There is likewise very little doubt (> 99% certainty) that the spread of relaxation times is real. Moreover, as well as can be seen, the spread parameter α is temperature-independent. Its interpretation is still open to question.

(iii) More accurate values of ε_∞ have been obtained than has been possible hitherto. The temperature variation and the limits of uncertainty are shown in Figure 4. It is the submillimetre-wave data (890 GHz laser) [26] which contribute by far the largest weight to these ε_∞ values. This is

[24] E. H. Grant and R. Shack, *Brit. J. Appl. Phys.*, 1967, **18**, 1807; E. H. Grant, *J. Phys. Chem.*, 1969, **73**, 4386.

[25] P. R. Mason, J. B. Hasted, and L. Moore, to be published, 1972.

[26] J. E. Chamberlain, G. W. Chantry, H. A. Gebbie, N. W. B. Stone, T. B. Taylor, and G. Wyllie, *Nature*, 1966, **210**, 790.

because the microwave frequencies, on which reliance has hitherto been placed, are far too low to give accurate values of ε_∞. The values of ε' measured at 890 GHz are within 4% of ε_∞, and therefore the accuracy in

Table 4 *Microwave dielectric data and relaxation parameters deduced from constraint fitting program*

T/K	ε_∞	$\tau/10^{-11}$ s	α
273	4.46	1.79	0.014
283	4.10	1.26	0.014
293	4.23	0.931	0.013
303	4.20	0.72	0.012
313	4.16	0.58	0.009
323	4.13	0.48	0.013
333	4.21	0.39	0.011
348	4.49	0.32	—
99% *confidence limits at* 293 K	$\Delta\varepsilon_\infty$ ± 0.192	$\Delta\tau$ ± 0.108	$\Delta\alpha$ ± 0.0067

ε_∞ is very close to the accuracy of the measurement of ε'. This is, of course, not the case for microwave measurements. It should be noted that at 283 K and 273 K the submillimetre-wave data are lacking and the 99% confidence limits are much larger than the values given for 293 K, which

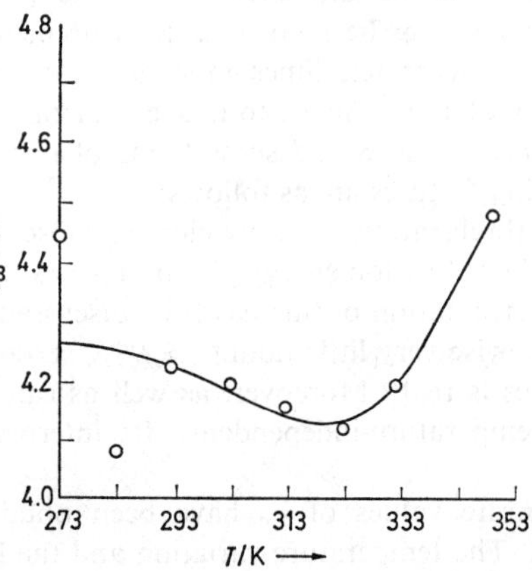

Figure 4 *Temperature variation of* ε_∞ *for principal relaxation process of* H_2O

also apply to higher temperatures. The smooth line drawn through the 348—293 K data points is extrapolated to 273 K without regarding the fluctuation in the two lowest points as being significant. Consideration suggests the rise at $T > 323$ K is real. This detail is being investigated in further experiments at the National Physical Laboratory, which at the time of writing confirm the high value.

The regression analysis program has been written for the principal relaxation of water, with its spread; no allowance is made for any second process, although we can see from recent measurements at higher frequencies (see Figure 5) that there is such a process, which must produce

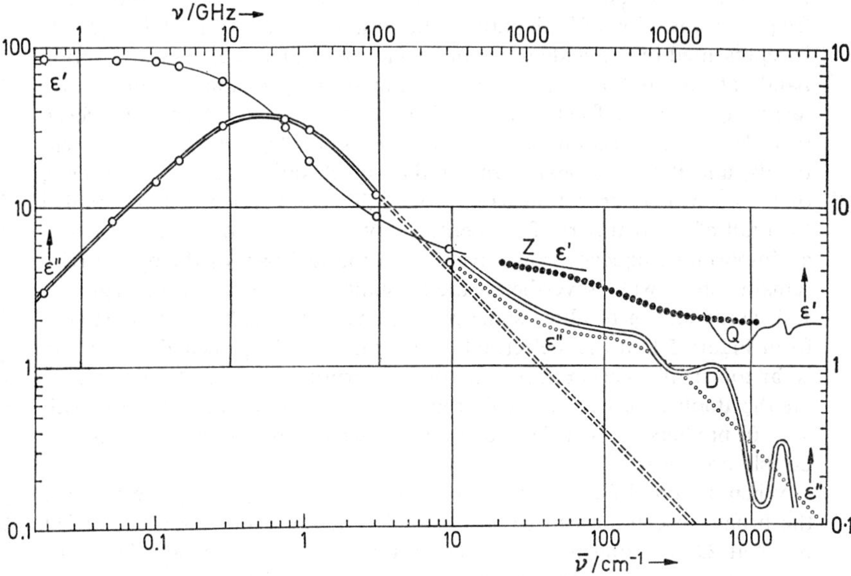

Figure 5 *Fully logarithmic plot of ε′ (single line) and ε″ (double line) against frequency ν and wavenumber ν̄, for liquid* H_2O *at 293 K. Included are microwave data (large open circles), and submillimetre-wave data* (D, *ref.* 27; Z, *ref.* 28; Q, *ref.* 29). *Broken line shows ε″ expected on the basis of a single relaxation process. Small circles, respectively closed and open, show calculations of ε′ and ε″ from proposed second relaxation process*

a correction, at present difficult to estimate; a modified program is being written to allow for this process, but is not as yet available.

The opening up of the submillimetre region of the spectrum is an event of some importance for aqueous materials science.[4] It rests upon the availability of three experimental techniques:

(i) Laser sources of radiation, such as the HCN and the H_2O lasers.

(ii) Fourier absorption and reflection spectroscopy, by means of which an interferogram obtained using a white radiation source (such as a mercury vapour lamp) can be mathematically converted into a frequency function of the optical constants.

(iii) Sufficiently quiet and sensitive detectors, some of which need to be cooled by liquid helium.

Absorption measurements on water at frequencies $\geqslant$ 300 GHz [$\bar{\nu} \geqslant 10$ wavenumbers (cm^{-1})] have been available for some years,[27] over a range of temperature. Fixed-frequency laser measurements of refractive index over a range of temperature are also available, and interferometer measurements of refractive index in the range 30—100 cm^{-1} have been made over a range of temperature.[28] At higher wave numbers still, $\bar{\nu} > 400$ cm^{-1}, i.r. measurements of both refractive index and absorption are available.[29] The frequency variation of both parts of the complex permittivity at $T = 298$ K is represented over a wide range of frequency in Figure 5. The absorption bands at 600 and 1600 cm^{-1} are accompanied by regions of anomalous dispersion in the refractive index. The absorption at 600 cm^{-1} has been regarded as a vibration of the water molecules 'against' the hydrogen-bonds, and it is a resonance rather than a relaxation process. However, there is a broad absorption band around 200 cm^{-1} which is accompanied by a fall of the real part of the permittivity. It is unfortunate that there is an important frequency range over which the real part of the permittivity remains unknown. Nevertheless the presumption of a fall in this region is sufficiently strong to allow some predictions to be made. It may be seen from Figure 5 that the dielectric loss arising from the principal absorption is far too small to account for the total absorption in the 200 cm^{-1} region; on the double logarithmic scale the Debye equation [equation (13) with $\alpha = 0$] predicts a linear 45° slope fall-off, which appears in Figure 5 as a double broken line.

When a second Debye relaxation process is added to the principal one, the broad absorption can be approximately fitted. This has been carried out[19] at 298 K using the following parameters for the second relaxation: $\varepsilon_0 = 4.1$, $\varepsilon_\infty = 1.8$, relaxation wavelength $2\pi c\tau = 0.008$ cm. The resulting values $\varepsilon'(\nu)$ and $\varepsilon''(\nu)$, which are not intended to be accurate, are shown in Figure 5 as circles, open for loss and closed for the permittivity.

To obtain the best possible fit, a full regression analysis will be necessary, and results of this are not yet available. The outstanding question to be answered is whether the broad absorption is to be regarded as a relaxation process, or whether it is a resonance process, a molecular libration (partial rotation within a potential well) such as is responsible for submillimetre-wavelength absorption bands in many liquids; it is possible that both processes contribute.[30] A resonance process would produce an effect in the real permittivity similar to an anomalous dispersion; a relaxation process merely produces a monotonic fall in the real permittivity.

[27] J. E. Chamberlain, G. W. Chantry, H. A. Gebbie, N. W. B. Stone, T. B. Taylor, and G. Wyllie, *Nature*, 1966, **210**, 790; D. A. Draegert, N. W. B. Stone, B. Curnutte, and D. Williams, *J. Opt. Soc. Amer.*, 1966, **56**, 64; V. M. Zolotarev, B. A. Mikhailov, L. I. Alperovich, and S. I. Popov, *Optics and Spectroscopy*, 1969, **27**, 430.

[28] J. E. Chamberlain, M. S. Zafar, and J. B. Hasted, in course of publication.

[29] M. R. Querry, B. Curnutte, and D. Williams, *J. Opt. Soc. Amer.*, 1969, **59**, 1299.

[30] M. Davies, *Ann. Reports* (*A*), 1970, **67**, 65. The summary there (p. 76) suggests that several different processes contribute to the broad absorption from 50 to 400 cm^{-1} and higher wavenumbers.

An intramolecular process would contribute [31] to the value of ε_∞ employed in the Fröhlich–Kirkwood equation [equation (1)] for the calculation of the static permittivity. A dipole relaxation process would not contribute, so that the required correlation factors g would be much larger. It has been contended [31] that the large ε_∞ values, together with $g = 1$, could between them explain the large static permittivity of water; the required ε_∞ would, however, have to fall monotonically with increasing temperature, unlike the behaviour shown in Figure 4. In the calculation of Section 3, the value $\varepsilon_\infty = n^2$ (far-i.r.) has been used.

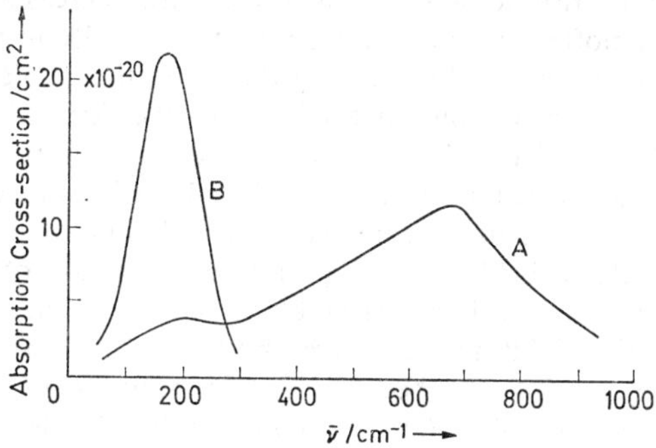

Figure 6 *Absorption cross-section*/cm² *for* 0.01% *solution of water in carbon tetrachloride* (B) *contrasted with that for pure water* (A)

The interpretation of the peak at 200 cm^{-1} as a dielectric process must ultimately depend on a detailed study of its variation, in magnitude and position, with temperature. Both the absorption and the elevation of the real part of the permittivity may be due to orientation in the applied field of those molecules which are free to do so without breakage of a hydrogen-bond. The magnitudes of the absorption and permittivity elevation should depend upon the populations of these species; at high temperatures the populations will be much enhanced, and ε_∞ will be increased. The activation energy for the 200 cm^{-1} process should be very small, but some frequency dependence of the peak may possibly arise from a change of the rate of formation with temperature of the species which dominates the relaxation time. This species might crudely be regarded as those water molecules which are not tightly bound by hydrogen-bonds. For water dissolved in non-polar solvents, the proportions of these species will be greatly enhanced. Measurements by Pardoe and Gebbie [32] in carbon tetrachloride show a greatly enhanced absorption peak, which is shown in Figure 6, thus confirming the interpretation of the second relaxation process as arising from 'unbonded' molecules. In the statistical interpretation these

[31] N. E. Hill, *Trans. Faraday Soc.*, 1962, **59**, 344.
[32] G. W. F. Pardoe and H. A. Gebbie, 'Symposium on Submillimetre Waves', Polytechnic Inst. Brooklyn, 1970, p. 643.

would be characterized as molecules bonded to one or zero neighbours, which might rotate without breakage of a hydrogen-bond.

One interesting feature of the absorption data of Figure 5 is that at *ca.* 1000 cm^{-1} the loss falls well below the calculated values (small open circles). However, this is a well-known anomaly (the dipole inertia effect) in the Debye theory, in which it is assumed that the polarization induced by an applied field tends exponentially towards its equilibrium value. It follows that an abrupt change in the field would produce an instantaneous change in the polarization (zero inertia), and that the absorption per centimetre would rise to a constant value with increasing frequency. Nevertheless, radiofrequency relaxations do not give rise to finite absorption coefficients for visible light; therefore the absorption at high frequencies must behave differently from that of the Debye predictions, and the anomaly at 1000 cm^{-1} could be nothing more than an example of this. Various attempts to account for inertial effects have been made.[33, 34]

The response of a dipole, surrounded by neighbours, to an applied field can in fact be related in detail to the frequency variation of the complex permittivity. A theoretical treatment of the effect on the dipole of the 'lag in the reaction field' has been given by Scaife[36] and by Fatuzzo and Mason.[37] Equation (14) has been derived by Klug *et al.*[38] and also by Nee and Zwanzig,[35] in which $\gamma(t)$, defined in equation (15), is the normalized correlation function of the permanent dipole moment μ. The function

$$\varepsilon_0 \frac{[\varepsilon^*(\omega) - \varepsilon_\infty][2\varepsilon^*(\omega) + \varepsilon_\infty]}{\varepsilon^*(\omega)[\varepsilon_0 - \varepsilon_\infty][2\varepsilon_0 + \varepsilon_\infty]} = \frac{4\pi\rho\mu^2}{3kT}\int - \frac{d\gamma}{dt}\exp(i\omega t)\,dt \qquad (14)$$

$$\gamma(t) = \langle \boldsymbol{\mu}(0)\cdot\boldsymbol{\mu}(t)\rangle / \boldsymbol{\mu}(0)\cdot\boldsymbol{\mu}(0) \qquad (15)$$

$\gamma(t)$ can, of course, be calculated from a structural model, provided that it is a dynamical one. The molecular dynamical computer program of Rahman and Stillinger,[6] discussed above, has provided a function $\gamma(t)$ appropriate to a temperature of 307.5 K; the function is illustrated in Figure 7. When it is applied in equation (14) it yields a frequency dependence of complex permittivity which can be represented in the Cole–Cole form, as in Figure 8. A comparison is made with experimental data[19] at 293 K in this Figure. It is apparent not only that the principal relaxation is close to the Debye form, but that the magnitude of the relaxation time is of the correct order. Also, a subsidiary 'curlecue' relaxation is predicted at higher frequencies, although it is rather larger and of different form from that which is observed experimentally.

[33] G. Birnbaum and E. R. Cohen, *J. Chem. Phys.*, 1970, **53**, 2885.
[34] Y. Leroy, E. Constant, C. Abbar, and P. Desplanques, *Adv. Mol. Relax. Processes*, 1967—8, **1**, 237; see also ref. 30, p. 67.
[35] Tsu-Wei Nee and R. Zwanzig, *J. Chem. Phys.*, 1970, **52**, 6353.
[36] B. K. Scaife, *Proc. Phys. Soc.*, 1964, **84**, 616.
[37] E. Fatuzzo and P. R. Mason, *Proc. Phys. Soc.*, 1967, **90**, 729.
[38] D. D. Klug, D. E. Kranbuehl, and W. E. Vaughn, *J. Chem. Phys.*, 1969, **50**, 3904.

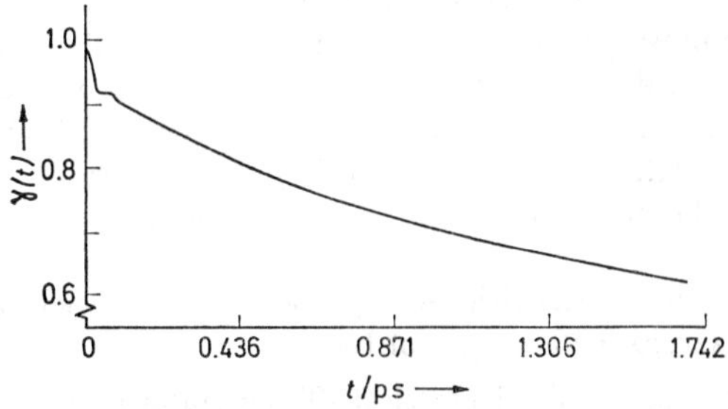

Figure 7 *Computed dipole correlation function $\gamma(t)$ for water at* 307.3 K

The subsidiary relaxation process is clearly attributable to the initial behaviour of $\gamma(t)$; the curlecue behaviour, which implies an 'anomalous dispersion' feature in the refractive index, has been described as a 'librational motion'. Librational absorptions in the submillimetre-wave region are found in polar liquids; however, the absence of 'anomalous dispersion' in the data of Figure 5 favours the interpretation in terms of the relaxation of those molecules which can re-orient without breaking a hydrogen-bond.

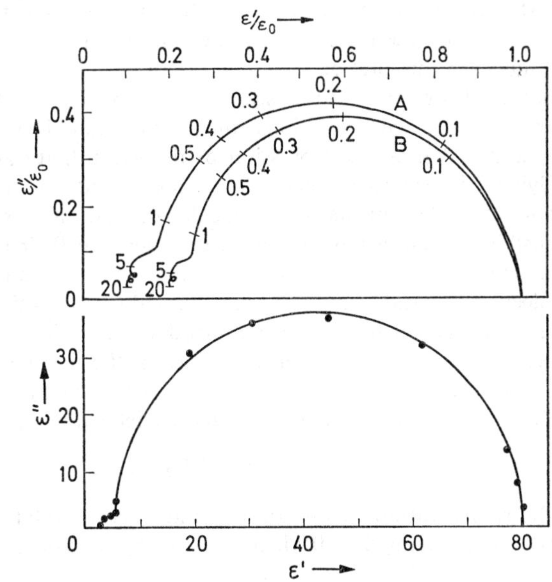

Figure 8 *Top: Cole–Cole plot computed from $\gamma(t)$ for chosen values of $\varepsilon_\infty/\varepsilon_0$, equal to* 0.05 (A) *and* 0.15 (B). *Frequency in units of* 10^{12} Hz *is indicated along the curves. Below: The Cole–Cole plot for water at* 293 K *shown on the same scale*

It is the responsibility of experimentalists to see their task in terms of producing the most complete correlation function possible, for comparison with that obtained from dynamical calculations. It is the goal of the theoreticians to seek to combine the techniques of MO theory with dynamical calculations; calculations using quantum statistical mechanics are made possible by the existence of large computers.

5 Static Permittivity of Aqueous Electrolyte Solutions: the Hydration Numbers of Ions

We shall now discuss the depression of the static permittivity of water by the addition of electrolyte solutes, which is a phenomenon of some importance in the understanding of the hydration sheath of the ions. It is essentially a dielectric saturation phenomenon; the strong electric fields in the neighbourhood of the ions produce a non-linear polarization, which renders the local water molecules ineffective as regards orientation in the applied field. It is possible to make estimates of the extent of hydration, or 'hydration number', of water molecules considered to be 'bound irrotationally' to the average ion; these estimates are in reasonable agreement with hydration numbers estimated on the basis of activity coefficients, entropies, mobilities, and viscosities. The hydration number must be distinguished from the number of water molecules actually adjacent to the ion in the first or second layers of hydration (the hydration sheath); it does not follow that all of these molecules can be considered to be attached to the ion as it moves in the solution.

The Onsager theory was first successfully applied to the static permittivities of electrolyte solutions by Ritson and Hasted,[39] who showed that almost the entire depression of the permittivity arises from the region lying between two spheres, of radius 2 and 4 Å, centred on the ion. Dielectric decrements were calculated, and in a subsequent paper [11] plausible hydration numbers were obtained. In their calculation a discontinuous model was used, with the assumption that the first sheath of water molecules was fully saturated, *i.e.* oriented to a (positive) ion.

Considerable refinement has now been introduced by Glueckauf,[40] who obtained hydration numbers which lie between those obtained from activity coefficients and those from entropies. The calculation has been carried out in terms of the volume variable $d\phi$, as defined in equation (16),

$$d\phi = 2n_0(4\pi r^2\, dr) \tag{16}$$

where r is the radius variable from an ion, and n_0 is the total electrolyte ion number density. Taking $\phi = 0$ when $r = 0$, integration of equation (16)

$$\phi = \tfrac{8}{3}n_0\, \pi r^2 \tag{17}$$

[39] D. M. Ritson and J. B. Hasted, *J. Chem. Phys.*, 1948, **16**, 11.
[40] E. Glueckauf, *Trans. Faraday Soc.*, 1964, **60**, 1637.

yields equation (17). The mean value $\bar{\varepsilon}$ of the continuously variable permittivity is related to the local value $\varepsilon(\phi)$ by equation (18). Equation (18)

$$\frac{d\bar{\varepsilon}}{d\ln(1-\phi)} = \frac{3\bar{\varepsilon}[\varepsilon(\phi)-\bar{\varepsilon}]}{2\bar{\varepsilon}+\varepsilon(\phi)} \tag{18}$$

is combined with the saturation equation for the dependence of ε upon field strength E, given by Booth,[41] as equation (19), where n is the refractive

$$\varepsilon = n^2 + (\varepsilon_0 - n^2)\frac{3}{u}\left(\frac{1}{\tanh u} - \frac{1}{u}\right) \tag{19}$$

index, $u = \beta E$, and the field E is defined in equation (20), for $r < a$, and in equation (21) for $r > a$. The value of β is defined in equation (22)

$$E = -\, ez/\varepsilon r^2 \tag{20}$$

$$E = -\frac{ez}{er^2}\left\{\frac{\exp(\kappa a)\cdot(1+\kappa r)}{(1+\kappa a)\exp(\kappa r)}\right\} \tag{21}$$

$$\beta = \frac{\sqrt{73}}{6}(n^2+2)\,\mu_0/kT \tag{22}$$

for $r > a$, and a is the distance of closest approach of ions, and κ the coefficient (proportional to the square root of the concentration) of the Debye–Hückel theory of electrolytes; z is the ionic valency.

The continuous variation of permittivity given by equation (18) can be integrated to give the permittivity ε_{soln} of a mixture of two components of volumes ϕ_1, ϕ_2 and permittivities ε_1, ε_2. The result of such integration is equation (23). It will be seen from Figure 9 that the comparison of calcula-

$$\varepsilon_{soln} = \frac{\varepsilon_2[2(\varepsilon_2-1)(1-\phi_2)+3\varepsilon_1(1-\phi_1)]}{3\varepsilon_2(1-\phi_1)-(\varepsilon_2-\varepsilon_1)(1-\phi_2)} \tag{23}$$

tions from equation (23) with early data[42] at concentrations $\leqslant 2\,\text{mol}\,l^{-1}$ is encouraging. To some extent the curvature of the $\varepsilon_{soln}(c)$ relation is reproduced; although in the limit of low concentration the dependence is linear, the curvature becomes more marked when data for higher concentrations[43] are considered. Comparisons between ions of different valency can be made by means of the relationship between the concentrations c and $\alpha c = z^+[1 + (z^+)^{\frac{1}{2}}/2]$, for which the permittivities respectively of a $z^+ : 1$ electrolyte and a $1 : 1$ electrolyte are equal.

This continuous model, although valuable, is inadequate to interpret either the differences which are found in data for higher concentrations,[44] or the anomalously large values of ε_{soln} for Li^+ and H^+; nor can adequate estimation of hydration numbers be made without invoking a discontinuous model. For Glueckauf's continuous and discontinuous models the variations of ionic radii and electric permittivity are shown in Figure 10. Values

[41] F. Booth, *J. Chem. Phys.*, 1951, **19**, 391, 1327, 1615.
[42] J. B. Hasted, C. H. Collie, and D. M. Ritson, *J. Chem. Phys.*, 1948, **16**, 1.
[43] R. M. Noyes, *J. Amer. Chem. Soc.*, 1952, **84**, 515.
[44] F. E. Harris and C. T. O'Konski, *J. Phys. Chem.*, 1956, **61**, 310.

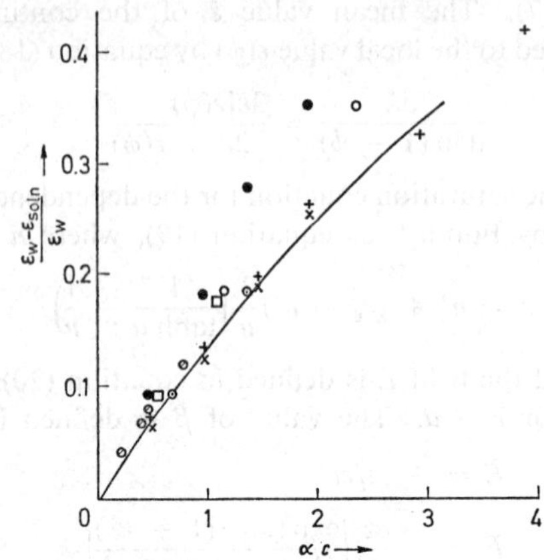

Figure 9 *Comparison of measured variation with concentration of permittivity of electrolytic solutions* [42] *with calculations from Glueckauf's continuous model,*[40] *which gives the full line*

Key: ● LiCl
\+ NaCl
× KCl and RbCl
□ $MgCl_2$
○ $BaCl_2$
⊙ $LaCl_3$
⊘ NaF and KF

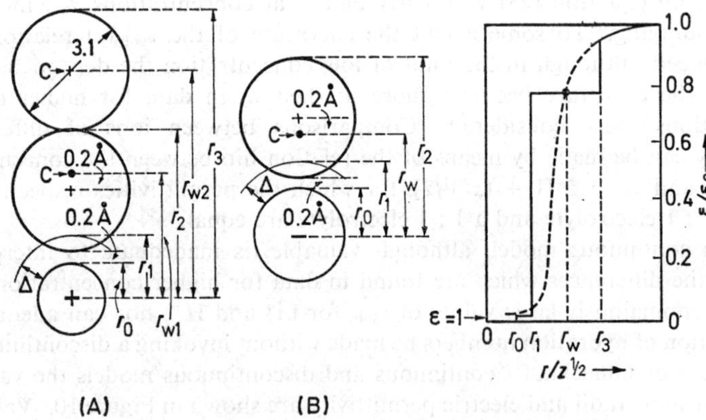

Figure 10 *Parameters used in Glueckauf's discontinuous model for electrolytes, and their comparison with static permittivity calculated as a function of distance from the ion centre.* (A) *cations;* (B) *anions;* C *is the centre of the water dipole;* Δ = *the range where* ε = 1

of Δ (defined in Figure 10) are those previously chosen for fitting thermodynamic data: $\Delta = 0.35$ Å for Cl^- and F^-, 0.56 Å for I^-, and 0.64 Å for all cations. The water radius $r_W = 1.55$ Å is slightly larger than one half of the mean oxygen–oxygen distance in pure water. For H^+, $\Delta = 0$, but $r_0 = 0.25$ Å throughout. This is equivalent to fitting the hydrogen ion into the central space of four water molecules in closest tetrahedral packing.

In Table 5 a successful comparison is made between the calculated (discontinuous model) and the experimental fractional dielectric decrements δ [$\delta = (\varepsilon_{soln} - \varepsilon_W)/c\varepsilon_W$] and δ_+, δ_-, using the convention $\delta = (\delta_+ + \delta_-)/2$.

The mean hydration number h_ε is calculated by assuming that the calculated relative permittivity change of any water molecule is a measure of its partial immobilization [equation (24)], where $\bar{n}_p$ is the number of

$$h_\varepsilon = \sum_p \bar{n}_p \left\{1 - \frac{\varepsilon_{Wp} - n^2}{\varepsilon_W - n^2}\right\} \tag{24}$$

water molecules which can be fitted into the pth layer and ε_{Wp} is the mean permittivity of the water molecules of that layer; n is the refractive index of water. The second layer makes a significant contribution only for H^+, Li^+, and multivalent ions. In Table 6 calculated hydration numbers h_ε are compared with those (h_a) calculated from activity coefficients [45] and those (h_s) calculated from entropies,[43] and will be seen to lie between the two. Thus the discontinuous model is consistent with other electrolyte solution theories.

It may be remarked that large polyatomic ions will exhibit only very small dielectric saturation effects, but will contribute to the dielectric decrement by reason of their large volume. Such ions, of volume fraction v and zero dipole moment, have been treated [46] by the mixture relation of

$$\varepsilon_{soln} = \frac{\varepsilon_W\left[1 + \dfrac{n^2}{2\varepsilon_W} - v\left\{1 - \dfrac{n^2}{2\varepsilon_W}\left(3 - \dfrac{\varepsilon_W}{\varepsilon_{soln}}\right)\right\}\right]}{1 + \dfrac{n^2}{2\varepsilon_W} + \dfrac{v}{2}\left(1 + \dfrac{n^2}{\varepsilon_{soln}}\right)} \tag{25}$$

equation (25), which has been found quantitatively satisfactory for tetra-alkylammonium bromide solutions; naturally, this relation is inapplicable to small ions.

Some unsolved problems of research into electrolyte static permittivity may be outlined. Extension of calculations using the Glueckauf continuous model for the high concentration range should be carried out as a preliminary to studying the extreme non-linearity of $\varepsilon_s(c)$ functions at high concentrations. For solutions in the region 10—12 mol l^{-1}, ε_s hardly varies with concentration. The use of the Debye–Hückel electric field [equation (22)] contributes

[45] E. Glueckauf, *Trans. Faraday Soc.*, 1955, **51**, 1241.
[46] R. Pottel and U. Kaatze, *Ber. Bunsengesellschaft phys. Chem.*, 1969, **73**, 437.

Table 5 *Comparison of dielectric decrements calculated on Glueckauf's discontinuous model with (bracketed) experimental values* [42]

	δ_-	H^+	Li^+	Na^+	K^+	Rb^+	Cs^+	Mg^{2+}	Ba^{2+}	La^{3+}
δ_+		0.306	0.166	0.148	0.132	0.132	0.154	0.437	0.410	0.613
F^-	0.146[a]	—	0.16	0.15 (0.16)	0.14 (0.17)	0.14 (0.17)	0.15	—	—	—
Cl^- [a]	0.160	0.23 (0.24)	0.16 (0.18)	0.15 (0.15)	0.15 (0.13)	0.15 (0.13)	0.16	0.41 (0.39)	0.39 (0.39)	0.61 (0.54)
I^-	0.212	—	0.19	0.18 (0.22)	0.17 (0.22)	0.17 (0.22)	0.18	—	—	—

[a] $z = 1$, $\delta_- = 0.146$; $z = 2$, $\delta_- = 0.186$; $z = 3$, $\delta_- = 0.205$.

Table 6 *Comparison of ion hydration numbers h_ε calculated from Glueckauf's treatment of static dielectric constant with values calculated from activity coefficients (h_a) and entropies (h_s)*

Ion:	H^+	Li^+	Na^+	K^+	Rb^+	Cs^+	Ba^{2+}	Mg^{2+}	La^{3+}	Al^{3+}	F^-	Cl^-	I^-
h_a	3.9	3.4	2.0	0.6	0	0	3	5	7.5	12	1.8	0.9	0.9
h_ε	3.8	3.2	2.1	1.1	1.1	1.1	6	9	12	16	2.2	1.4	1.1
h_s	—	4.2	2.9	1.6	1.1	1.0	6	11	15	19	—	—	—

to this non-linearity. It will be recalled that Debye and Falkenhagen[47] predicted a *positive* contribution to the permittivity of electrolytes ($\propto c^{\frac{1}{2}}$), arising from the asymmetry produced in an ion atmosphere by the application of an external field. Although claimed to have been found experimentally at very low concentration, the effect is not substantiated, because of the difficulties inherent in low-frequency measurements. At these concentrations the ionic atmosphere is expected to relax in the MHz region, therefore the effect will not be present at higher frequencies. Time-domain measurements may possibly demonstrate a relaxation in this region. An increase in the low-frequency conductivity is expected to accompany this relaxation, and some recent measurements are reported to show anomalies.[48]

6 Relaxation of Electrolyte Solutions

The broad features of the dielectric relaxation of electrolyte solutions may be summarized as follows:

(i) All electrolytes in dilute solution show a principal relaxation time which decreases with increasing concentration, but at molarities greater than *ca.* 5, the $\tau(c)$ curve has a minimum, and the principal relaxation time can exceed that of pure water. $\tau(c)$ curves for small ions turn round at lower concentrations. There is disagreement between the results obtained by different workers.

(ii) The spread parameters of the principal relaxation times increase with increasing concentration, becoming several times larger than those observed for pure water.

(iii) There are considerable differences between the proton magnetic and the dielectric relaxation times. Up to the present there is no satisfactory quantitative interpretation of dielectric relaxation times of electrolytes.

The original measurements[42] were made at only two or sometimes three frequencies and not very large concentrations; this procedure was insufficient to allow the detection of any spread of relaxation times. The shortened relaxation times were observed, low-frequency conductivity corrections being applied to the ε''. The relaxation of conductivity predicted by Debye and Falkenhagen has no observable effect, for reasons discussed in the original papers.[42] However, an ionic atmospheric resonance of this type can be detected in sodium–ammonia solutions.[49]

Measurements on electrolytic solutions were subsequently made by Harris and O'Konski[44] at three frequencies and larger concentrations; spread parameters were reported, and the rising principal relaxation times were found at higher concentrations. Recent measurements have been

[47] P. Debye and H. Falkenhagen, *Z. Physik.*, 1928, **28**, 121, 401; H. Falkenhagen, M. Leist, and G. Kelbig, *Ann. Physik*, 1952, **11**, 15.
[48] P. Maurel, Thesis, Université des Sciénces, Montpellier, 1971.
[49] J. B. Hasted and S. H. Tirmazi, *J. Chem. Phys.*, 1969, **50**, 4116.

reported by Pottel and colleagues,[50–52] who sometimes used up to six or seven frequencies; the data are analysed in terms of equation (26), which is an empirical relaxation time-spread equation. The Cole–Cole semi-

$$\varepsilon^* = \varepsilon_\infty + (\varepsilon_0 - \varepsilon_\infty)/[1 + i^{1-\alpha}(2\pi\nu\tau)^{1-\beta}] \qquad (26)$$

circular arcs are plotted, and as usual the angle between the ε' axis and the line connecting the centre of the circle to the point $(0, \varepsilon_\infty)$ is $\alpha\pi/2$; typical data are shown in Figure 11.

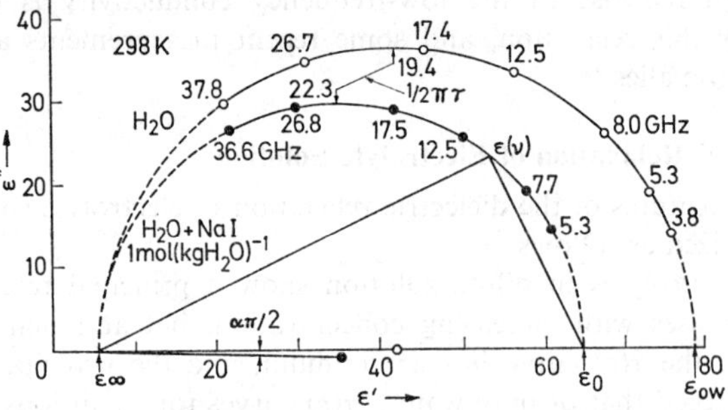

Figure 11 *Cole–Cole diagrams for water and* 1 mol (kg H_2O)$^{-1}$NaI *solution* [53]

The parameter β is given by equation (27),

$$\beta = 1 + \frac{1}{\varepsilon''}\left\{\frac{d\varepsilon'}{d(2\pi\nu\tau)}\right\}_{2\pi\nu\tau=1} . \cos\left(\frac{\alpha\pi}{2}\right) \qquad (27)$$

$$\approx -\frac{\log_{10} e}{2\pi\tau} . \frac{\Delta q}{\Delta(\log \nu)} \qquad (28)$$

where $q = (1/\nu)(|\varepsilon_0 - \varepsilon^*|)/(|\varepsilon^* - \varepsilon_\infty|)$. The lengths of the chords $\varepsilon_0 - \varepsilon^*(\nu)$ and $\varepsilon^*(\nu) - \varepsilon_\infty$ are used to calculate q, and the parameter β is calculated from the slope of the plot of q *vs.* $\log \nu$. In the Cole–Cole relaxation spread equation [equation (13)], $\beta = \alpha$, but this was not found by Pottel and colleagues, who used not only the graphical techniques but also computer regression analysis. Limits of uncertainty were not reported.

Typical dependences of spread parameters on concentration are shown in Figure 12. Certain large discrepancies in relaxation times between the results obtained by the Harris group and the Pottel and Hasted groups are shown in Figure 13.

Although the general trends of relaxation times of electrolyte solutions can be regarded as established, one must consider the possible reasons for

[50] K. Giese, U. Kaatze, and R. Pottel, *J. Phys. Chem.*, 1970, **74**, 3718.
[51] R. Pottel and O. Lossen, *Ber. Bunsengesellschaft phys. Chem.*, 1967, **71**, 135.
[52] D. Aldolph, quoted by R. Pottel, in 'Water – A Comprehensive Treatise,' ed. F. Franks, Vol. II, 'Aqueous Solutions of Simple Molecules and Ions,' Plenum Press, New York, Ch. 17, in course of publication.

discrepancies both in interpretation and in magnitudes of principal relaxation times. There are three possible lines of approach:

(i) Investigation of the accuracy of the assumption that the conductivity (σ) contribution [$\varepsilon' = \varepsilon'' + (2\sigma/\nu)$] is well approximated by using the low-frequency conductivity.

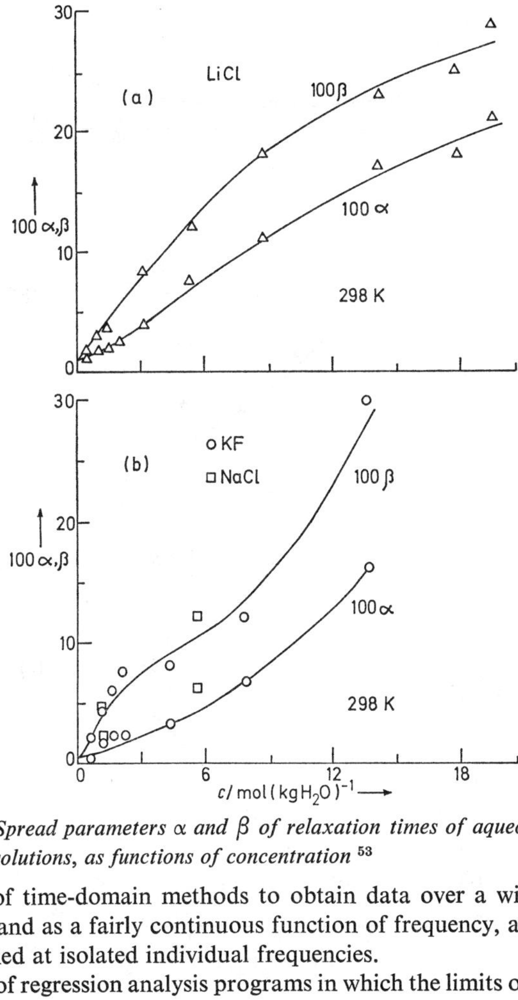

Figure 12 *Spread parameters α and β of relaxation times of aqueous electrolyte solutions, as functions of concentration* [53]

(ii) Use of time-domain methods to obtain data over a wider range of frequency, and as a fairly continuous function of frequency, as opposed to data obtained at isolated individual frequencies.

(iii) Use of regression analysis programs in which the limits of uncertainty are also calculated.

The basis of the comparison of dielectric and nuclear magnetic relaxation times is as follows: the relative dielectric relaxation molar difference,

[53] R. Pottel, in 'Water – A Comprehensive Treatise', ed. F. Franks, Vol. II, 'Aqueous Solutions of Simple Molecules and Ions', Plenum Press, New York, Ch. 17, in course of publication.

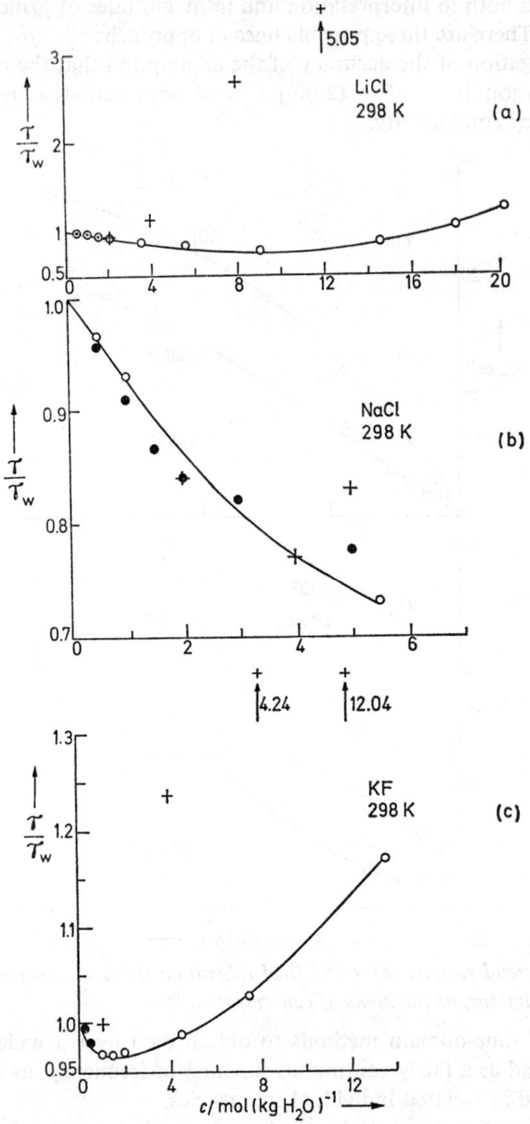

Figure 13 *Principal relaxation times τ of some aqueous electrolyte solutions (divided by relaxation time of water) as a function of concentrations.*

Key: ● *data of Hasted* et al.[42]
\+ *data of Harris and O'Konski*[43]
○ *data of Pottel*[53]

defined as in equation (29), may be compared with the relative molar

$$B_d = \frac{1}{\tau_w}\left(\frac{d\tau}{dc}\right)_{c\to 0} \tag{29}$$

shift of the intramolecular proton magnetic relaxation rate $1/T_1$, using the quantity B_m, defined in equation (30).

$$B_m = T_{1w}\left(\frac{d(1/T_1)}{dc}\right)_{c\to 0} \tag{30}$$

Data given by Pottel[53] are compared for electrolytes of different sums of ionic radii in Figure 14. It is apparent that, for all except the largest ions,

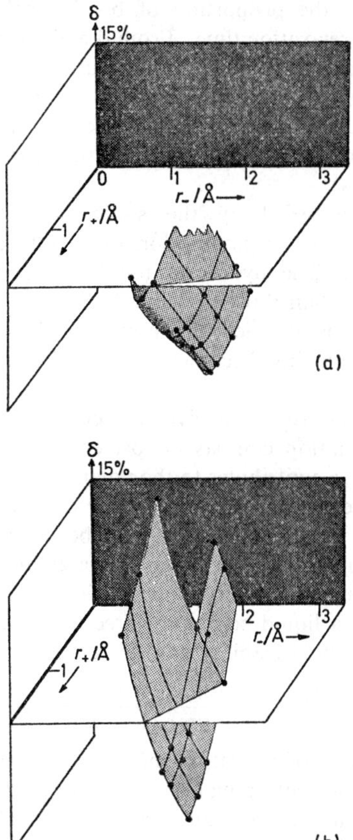

Figure 14 *Measured relative molar shifts* (a) $\delta = B_d$, *the dielectric relaxation time, and* (b) $\delta = B_m$, *the intramolecular proton magnetic relaxation rate, of aqueous alkali-metal halide solutions; these are plotted on the vertical axis against negative ion radii on the horizontal axis and against positive ion radii on the third axis*

the dielectric and the proton magnetic relaxation times contain different information about the influence of the ions on the motion of water molecules. Apparently, the coulombic charge can influence the rate of change of the electron density in the region of the protons in a different way from that in which it influences the rate of change of orientation of the dipole.

The quantitative interpretation of the dielectric relaxation times is still not on a satisfactory basis. The earliest attempt in this direction [54] was made on the basis of an ion-oriented hydration sheath, for the formation of which a calculated number of hydrogen-bonds must be broken. This breakage changes the equilibrium of species in the liquid, and statistical relationships connect the proportion of bonds broken, the equilibrium populations, and the relaxation time. From the observed shift of relaxation time one can calculate the number of molecules in the sheath, and show that for temperatures between 276 and 298 K it is approximately the same as the number calculated from the depression of the static permittivity (comparison in Figure 4 of ref. 54). This treatment is open to criticism on the following grounds:

(i) The evidence afforded by the static permittivity treatment of Glueckauf suggests that the assumption of a fully charge-oriented first sheath is unrealistic. The hydration numbers calculated by the early techniques are higher than those of Table 5.

(ii) No explanation is afforded of the reason why in very concentrated solutions the relaxation time becomes larger and can exceed that of pure water.

It is this second feature which gives the clue to the real situation. A very concentrated solution consists almost entirely of hydration sheath; such a region of course contributes to the polarization; its principal relaxa-time time τ_h could be characteristically longer than, or at least as long as, that of pure water, although it might still be concentration-dependent. There is also the question of exchange of water molecules between free liquid and hydration sheath; Pottel and Giese [55] have shown that the relaxation time τ of a liquid containing free liquid and hydration liquid without orientational polarizability can be represented by equation (31),

$$\tau = \tau_w\left\{1 + \frac{n_h\,c/(1 - n_h\,c)}{1 - [n_h\,c/(1 - n_h\,c)]}\left(\frac{\tau_w}{\tau_{ex}}\right)\right\}^{-1} \tag{31}$$

where n_h is the number of hydration molecules and τ_{ex} is the mean time for which a water molecule remains within the hydration sphere. This time τ is clearly less than τ_w at small concentrations, although at very high concentrations it might be greater.

It is true that Pottel includes in this theory an attempt to divide the anionic and cationic contributions; such divisions are very difficult at the

[54] J. B. Hasted and G. W. Roderick, *J. Chem. Phys.*, 1958, **29**, 17.
[55] K. Giese, quoted in ref. 53.

present time, but this does not destroy the validity of the concept. A complete theory would require the use of τ_w, τ_h, and τ_{ex}; τ_h could possibly be related to bond-breakage statistics. There is clearly some sort of 'structural mismatch' between hydration sheath and free liquid, but whether this is a sufficiently large effect to account for the entire shift of principal relaxation time is another question.

7 Static Permittivity of Aqueous Solutions of Non-electrolytes

The static permittivities of mixtures of organic liquids with water may be interpreted in terms of polarization theory. We shall see that water can be used as a solvent for the determination of dipole moments of highly polar molecules, but for less polar molecules the Kirkwood–Fröhlich equation has been used as a method of demonstrating molecular interactions between water and solute.

The polarization P_m of a mixture having molar fractions x_1 and x_2 respectively of water and solute was given by Oster[56] as equation (32),

$$P_m = x_1 P_1 + x_2 P_2 \tag{32}$$

whence, in the original form of the Kirkwood equation, the value of the static permittivity of the mixture, ε_{0m}, is related to those of the components, ε_{01} and ε_{02}, by equation (33), where the mean molar volume v_m is defined in equation (34). Rearranging equation (33) provides equation (35). In

$$\frac{(\varepsilon_{0m} - 1)(2\varepsilon_{0m} + 1)v_m}{9\varepsilon_{0m}} = \frac{(\varepsilon_{01} - 1)(2\varepsilon_{01} + 1)x_1 v_1}{9\varepsilon_{01}} + \frac{(\varepsilon_{02} - 1)(2\varepsilon_{02} + 1)x_2 v_2}{9\varepsilon_{02}} \tag{33}$$

$$v_m = x_1 v_1 + x_2 v_2 \tag{34}$$

$$\varepsilon_{0m} \approx \varepsilon_{01} + \left\{\frac{(\varepsilon_{02} - 1)(2\varepsilon_{02} + 1)}{2\varepsilon_{02}} - (\varepsilon_{01-1})\right\}\frac{x_2 v_2}{v_m} \tag{35}$$

the limit of low solute concentration, ε_{0m} is taken to be a linear function of concentration [equation (36)], where c_2 is the concentration of solute

$$\varepsilon_{0m} = \varepsilon_{01} + \delta c_2 \tag{36}$$

in moles per litre of solution. The dielectric increment δ may be calculated from equation (37). A comparison of experimental with calculated dielectric

$$\delta = \left\{\frac{(\varepsilon_{02} - 1)(2\varepsilon_{02} + 1)}{2\varepsilon_{02}} - (\varepsilon_{01} - 1)\right\}\frac{v_2}{1000} \tag{37}$$

increments is given in Table 7. The agreement is surprisingly good, but it is obvious that for a molecule such as 1,4-dioxan the additivity of polarizations is a poor assumption; there is apparently some form of very intimate association of the dioxan and water molecules.

[56] G. Oster, *J. Amer. Chem. Soc.*, 1946, **68**, 2036.

A method which has recently been applied[57] to the study of completely miscible liquids of molar fractions x_1 and x_2 is the examination of the behaviour of the correlation factor g throughout the entire range of concentration. The polarization is taken, as above, to be the sum of the

Table 7 *Comparison of dielectric increments* (T = 298 K) *observed and calculated from equation* (37)

Solute	$-\delta$ (*calculated*)	$-\delta$ (*observed*)
Methyl alcohol	1.79	1.4
Ethyl alcohol	3.11	2.6
n-Propyl alcohol	4.22	4.0
t-Butyl alcohol	6.26	6.3
Acetone	4.13	3.2
Diethyl ether	7.57	7.1
Ethylene glycol	2.04	1.8
Aniline	6.40	7.6
Methyl acetate	5.65	5
Pyridine	5.27	4.2
Acetonitrile	2.08	1.7
Nitromethane	2.11	2.0
1,4-Dioxan	6.45	8.3

contributions from each set of dipoles, using the original form of the Kirkwood equation [equations (38) and (38a)]. For the interpretation of this

$$P_s = \left\{\frac{(\varepsilon_{0m} - n^2)(2\varepsilon_{0m} + n^2)}{9\varepsilon_{0m}}\right\} v \qquad (38)$$

$$P_s = \frac{4\pi N_0}{3kT}(x_1 g_1 \mu_1^2 + x_2 g_2 \mu_2^2) \qquad (38a)$$

equation, the refractive index is interpolated linearly between those of the two liquids, as shown in equation (39). The dipole moments are calculated from

$$n = n_1 + (n_1 - n_2) x_2 \qquad (39)$$

those of the free molecules, μ_{01} and μ_{02}, using for internal consistency the Onsager relation in the form of equations (40) and (41). Since it is impossible

$$\mu_1 = (n^2 + 2)\,\mu_{01}/3 \qquad (40)$$

$$\mu_2 = (n^2 + 2)\,\mu_{02}/3 \qquad (41)$$

to separate two correlation factors g_1 and g_2 from a single value of P_s without some assumption, we make the simplest one possible, namely that $g_1 = g_2 = g$, so that g may be defined by equation (42). The assump-

$$g = \frac{3P_s kT}{4\pi N_0}(x_1 \mu_1^2 + x_2 \mu_2^2) \qquad (42)$$

[57] J. B. Hasted, in 'Water – A Comprehensive Treatise', ed. F. Franks, Vol. II, 'Aqueous Solutions of Simple Molecules and Ions', Plenum Press, New York, Ch. 7, in course of publication.

tion is most plausible for dilute solutions, when P_s is dominated by the more common dipole; the single parameter g then approximates to a quantitative measure of the effect of the addition of the other liquid upon the dipole correlation of the dominant solvent. More accurate interpretations of the behaviour of the mixture would have to involve cross-correlation terms, between one dipole and its behaviour in the field due to dipoles of the other species.

Figures 15(a)—(d) show the results of calculations [57] of the g parameters from early data on mixtures.

The simplest aqueous mixtures are those containing liquids with small correlation factors (acetone) and those which are non-polar (1,4-dioxan is non-dipolar although there are two opposed moments within the molecule). For these there is a steady monotonic fall of correlation factor as the molar fraction of liquid increases. The liquids are all efficient at 'breaking the structure' of water, *i.e.* causing breakage of the hydrogen-bonds.

The temperature dependence of the structure-breaking effect is of some interest in the t-butyl alcohol–water mixture. It is apparent that the temperature variation of correlation factor is much more marked in mixtures of high t-butyl alcohol content than it is in water. This reflects the structural interpretations discussed earlier. The temperature variation for t-butyl alcohol given by Dannhauser and Cole [15] is actually more marked still, partly because of differences arising from the use of ε_∞ instead of n^2 in the Kirkwood formula. Whereas the temperature variation of g for t-butyl alcohol is supposed to arise from variation in chain length ($33 > \Delta H^\ddagger > 25$ kJ mol^{-1}), that for water is supposed to arise from a change in the proportions of highly bonded and less highly bonded water molecules; when the the structure is a three-dimensional network rather than a coiled chain, the statistics of hydrogen-bond breakage result in a weaker temperature variation of the correlation factor.

The polyhydric alcohols ethylene glycol and glycerol also possess correlation factors much smaller than that of water, and mixtures of these with water also show an almost monotonic decrease of correlation factor with decreasing molar fraction of water. However, at the lower temperature of Figure 15(c) there is a region of low alcohol concentration where the correlation factor is slightly higher than that of pure water. This effect is also present in ethyl alcohol–water mixtures. The correlation factors of the two liquids do not differ greatly, but there is a tendency for small proportions of ethyl alcohol to increase the hydrogen-bonding of water; there is also a tendency for small proportions of water to decrease the hydrogen-bonding in alcohol.

In methyl alcohol–water mixtures there is a structure-making effect, maximizing the correlation factor at 40 molar per cent of alcohol. By contrast, both n-propyl and isopropyl alcohols, at *ca.* 50 molar percent,[2] have a marked structure-breaking effect, the correlation factors of these mixtures being well below those both of the alcohols and of water. The temperature variation suggests that the structure of the 50% mixture is rather more alcohol-like than water-like.

The addition of water to pure n-propyl alcohol, isopropyl alcohol, or t-butyl alcohol results in a lowering of the correlation factor; the com-

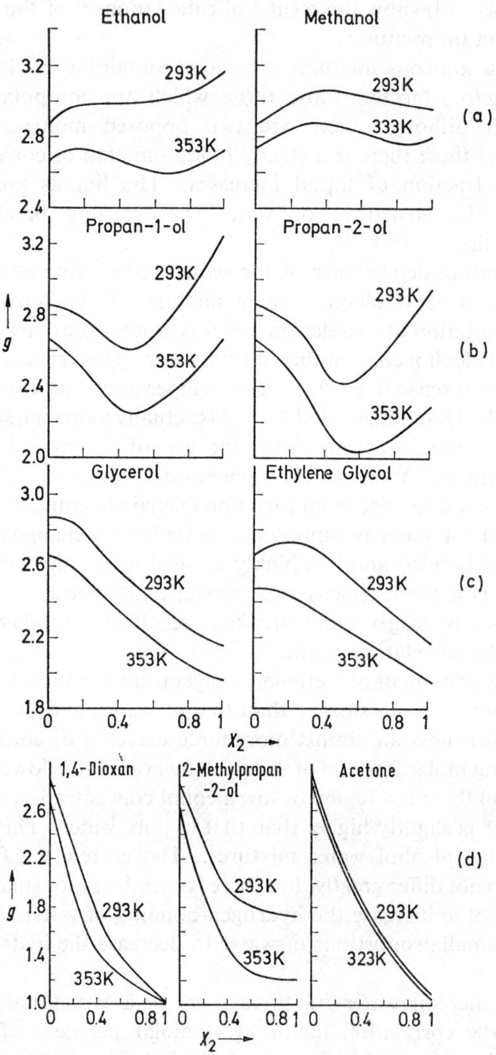

Figure 15 *Variation of correlation factors with molar fractions; specified organic liquids mixed with water*

paratively rare water molecules are apparently able to cause breakage of the alcohol chains.

8 Calculation of Dipole Moments from Static Permittivities of Solutions

One of the purposes for which static permittivities of aqueous non-electrolyte solutions have been determined, by Franks and his colleagues, is for calculation of the dipole moments of molecules. There are molecules for which no studies can be made in the vapour phase or in solution in non-polar solvents; however, sufficiently good comparisons of determinations of dipole moments in aqueous solution with other determinations can sometimes be made to allow an estimate of the limitations of the method.

The theory of Buckingham[58] gives an expression [equation (43)] for the static permittivity ε_{0m} of a mixture of i components of molar fractions x_i and having dipole moments μ_{0i} in the free molecule, where P_{0i} is the

$$\frac{(2\varepsilon_{0m} + 1)(\varepsilon_{0m} - n^2)}{3(2\varepsilon_{0m} + n^2)} \sum_i \frac{x_i M_i}{d} = \frac{4\pi N}{9kT} \sum_i x_i \mu_{0i}^2 f_i g_i$$
$$= \sum_i x_i P_{0i} f_i g_i \qquad (43)$$

orientation polarization of the ith component, and M_i its molecular weight; d is the density of the mixture and N the total number of molecules per unit volume; n^2 is the square of the refractive index of the solution, which is sufficiently well approximated by linear interpolation between the squares of the refractive indices of the components.

In the absence of an applied field, the moment of a spherical specimen of the solvent containing a solute molecule of moment μ_{0i} is $f_i \mu_0$. The factor g_i is related to the energy w of a molecule of moment μ_{0i} by the expression (44) when a uniform field E exists in the solution. θ is the angle

$$w = - g_i \mu_{0i} E \cos\theta \qquad (44)$$

between E and the axis of the dipole. Employing the assumption that the molecules are ellipsoids of internal permittivity n^2, possessing permanent dipoles along one of the axes, surrounded by a continuous medium of permittivity ε_s, it can be shown that f_i and g_i are defined by the equations (45) and (46). The shape factors A_i, which depend upon the semi-axial ratios, have been calculated.[59]

$$f_i = \frac{(2\varepsilon_{0m} + 1)\{1 + (n_i^2 - 1) A_i\}}{3\{\varepsilon_{0m} + (n_i^2 - \varepsilon_s) A_i\}} \qquad (45)$$

$$g_i = \frac{\varepsilon_{0m}\{1 + (n_i^2 - 1) A_i\}}{\varepsilon_{0m}\{1 + (n_i^2 - \varepsilon_s) - A_i\}} \qquad (46)$$

[58] A. D. Buckingham, *Austral. J. Chem.*, 1953, **6**, 93, 323.
[59] J. A. Osborne, *Phys. Rev.*, 1945, **62**, 351.

For dilute solutions the molecular polarization of the solvent (water) may be assumed to vary linearly with concentration, and can be calculated from the experimental data for a well-behaved solution of a liquid of known orientation polarization, which does not have a marked effect on the structure of water. Acetone has been chosen for this purpose.

Table 8 lists dipole moments calculated from static permittivities of solutions, using equations (43)—(46). In a few cases comparisons can be

Table 8 *Dipole moments derived from static permittivities of aqueous solutions*

Molecule	μ_{0i}/D *derived*	μ_{0i}/D *Other source*	*Reference*
Acetone	3.7	2.85 (gas)	66
Pyridine	3.1	2.5 (gas)	58
Glycine	13.3	14.2 (calc.)	58
β-Alanine	17.5	17.4 (calc.)	58
Tetrahydrofuran	2.4	1.63 (gas)	66
Tetrahydrofurfuryl alcohol	3.5	2.3 (gas)	66
D-Galactose	5.3	—	66
L-Arabinose	4.3	—	67
D-Glucose	4.7	—	66
D-Mannose	4.8	—	66
D-Ribose	5.1	—	66
Myoinositol	5.0	—	66

made with dipole moments obtained by other means. It appears that the dipole moments of amino-acids ($\mu_0 \geqslant 10$ D) are usually well approximated by the values derived in this way. However, the smaller dipole moments, such as those of monosaccharides, may not be so well approximated. The conditions which must be satisfied for the model to be applicable are: (i) the molecule must be sufficiently rigid for the variation of geometry between individual molecules to be small. It should be understood that the theory predicts the root-mean-square dipole moment rather than the mean value of the dipole moment; (ii) the extent of interaction of the molecule with water molecules must be sufficiently small for the dielectric behaviour of the water to be unaffected, and for the dipole moment of the solute molecule to be unaffected by hydration.

It would seem, from an examination of the four cases pyridine, acetone, tetrahydrofuran, and tetrahydrofurfuryl alcohol, that values of dipole moments of the order 2—4 D are likely to be up to 50% too high when calculated from the permittivity of their aqueous solutions. However, some confidence in the internal consistency of the calculations is to be derived from the fact that, although the dielectric decrements are temperature-dependent, the calculated dipole moments are not. Thus the correlation between the dipoles of solute and water molecules does not change with temperature. Since the correlation between dipoles in pure liquids and pure alcohols changes quite markedly with temperature, the implication is that when there is correlation it must be temperature-dependent; therefore the correlation between solute molecule and water might be presumed to be small.

Where there are grounds for supposing that there is strong correlation between the two types of molecule the model will not be of great value. For example, its application to dioxan–water mixtures yields a negative dipole moment for dioxan. If the calculation is carried out assuming the existence of a dioxan hydrate, a positive moment can be derived.

9 Relaxation of Aqueous Solutions of Non-electrolytes

Much of the high-frequency work that has been done on aqueous non-electrolyte solutions has yielded data at too few frequencies for an analysis of them to be free from ambiguity. It is to be hoped that the new time-domain techniques will remedy this deficiency.

An analysis of data for nitrogen-containing heterocyclic compounds and tetramethylammonium bromide[60] has been made by Pottel and Kaatze[61] in terms of a hydration sheath surrounding the solute molecule.

The simplest model assumes that the N water molecules in each hydration sheath have a different relaxation time $\tau_h = p\tau_w$ from that of the pure water in the remaining volume of the solution, τ_w. The water in the hydration sheath contributes a certain fraction q of the entire static permittivity of the solution. The frequency variation of the complex permittivity is taken to be a sum of two Debye terms (subscripts w and h respectively for pure water and hydration sheath water) as shown in equation (47) where

$$\varepsilon^* - \varepsilon_\infty = \frac{(\varepsilon_{0w} - \varepsilon_\infty)(1 - q)}{(1 + j\omega\tau_w)} + \frac{(\varepsilon_{0h} - \varepsilon_\infty)q}{(1 + j\omega\tau_h)} \tag{47}$$

$q = Nx/x'$, x and x' being respectively the solute and solvent molarities. No attempt is made to apply any of the usual mixture relations, although some wide-frequency-band calculations of this type have been made elsewhere.[62] Equation (47) can only be a rough approximation to the measured $\varepsilon^*(\omega)$. The observed linear variation of principal relaxation time τ_s with solute concentration is not consistent with the results calculated using equation (47). Nevertheless some interesting deductions concerning the nature of the postulated hydration sheath can be made with its aid.

Using the fact that the observed principal relaxation time τ_s is determined by the condition $\left(\frac{d\varepsilon''}{d\omega}\right)_{\omega\tau_s = 1} = 0$, one obtains equation (48), where $y = \tau_s/\tau_w$.

$$\frac{q}{(1 - q)} = \frac{(y^2 - 1)(p^2 + y^2)^2}{p(y^2 + 1)^2(p^2 - y^2)} \tag{48}$$

[60] R. L. Kay, T. Vituccio, C. Zawoyski, and D. F. Evans, *J. Phys. Chem.*, 1966, **70**, 2336.
[61] R. Pottel and U. Kaatze, *Ber. Bunsengesellschaft phys. Chem.*, 1969, **73**, 437.
[62] J. P. de Loor, Thesis, University of Leiden, 1956.

From the Cole–Cole equation [equation (13)] one may derive equation (49), and when equations (48) and (49) are combined the result is equation (50). Therefore the observed quantities τ_s and α lead to unique

$$\alpha = 1 + \left\{\frac{1}{\varepsilon''}\cdot\left(\frac{d\varepsilon'}{d(\omega\tau_s)}\right)_{\omega\tau_s=1}\right\}\cos\left(\frac{\alpha\pi}{2}\right) \tag{49}$$

$$\frac{p}{y} = \left\{y - \frac{(1-\alpha)}{[\cos(\alpha\pi/2)]}\right\}\{y(1-\alpha)/[\cos(\alpha\pi/2) - 1]\}^{-1} \tag{50}$$

values of τ_h and N. Some values of $(p - 1)$ are presented in Figure 16. They demonstrate dependence upon the ratio $R = v_{solute}/v_{solvent}$ of solute

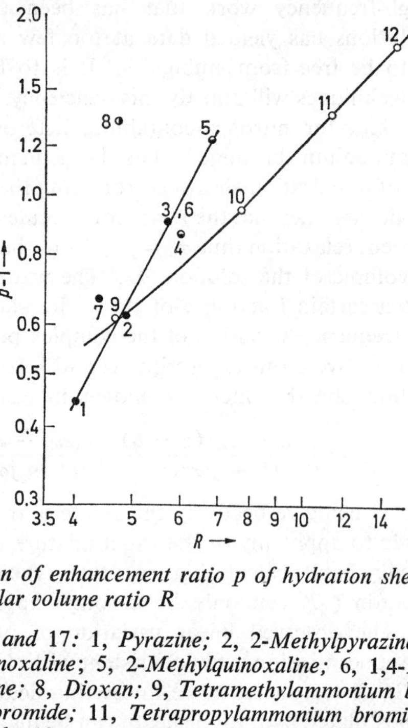

Figure 16 *Variation of enhancement ratio p of hydration sheath relaxation time with molar volume ratio R*

Key to Figures 16 *and* 17: 1, *Pyrazine;* 2, 2-*Methylpyrazine;* 3, 2,6-*Dimethylpyrazine;* 4, *Quinoxaline*; 5, 2-*Methylquinoxaline;* 6, 1,4-*Diazabicyclo*[2,2,2]-*octane;* 7, *Pyridine;* 8, *Dioxan;* 9, *Tetramethylammonium bromide;* 10, *Tetraethylammonium bromide;* 11, *Tetrapropylammonium bromide;* 12, *Tetrabutylammonium bromide;* 13, *n-propanol.*

to solvent molar volumes. Some calculated values of N are displayed in Figure 17. Two guide lines are shown in this Figure, representing sheath thickness in terms of the number N_1 of water molecules in the inner hydration sheath, *i.e.* with surface directly adjacent to the solute molecule. The value of N_1 is calculated using a model of spherical solute molecules and cubic water molecules to give equation (51). The conclusion to be drawn

$$N_1 = (36\pi)^{1/3}\left(\frac{v_{solute}}{v_{solvent}}\right)^{2/3} \tag{51}$$

is that the effect on the relaxation time does not penetrate much beyond the inner sheath. However, on this model the contribution made by the solute molecule to the permittivity is neglected. The approximation is not always satisfactory but it may be of value when the permittivity of the pure solute is $\gtrsim$ 0.1 times that of pure water.

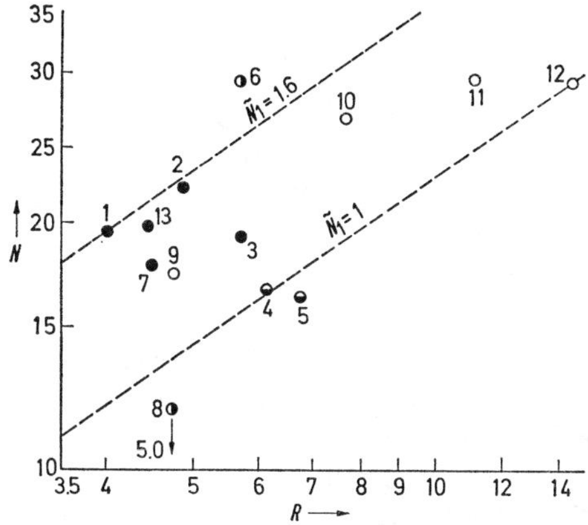

Figure 17 *Variation of hydration sheath population N with molar volume ratio R*

10 Hydration Sheaths of Long Lifetime

Until recently, measurements of relaxation times of aqueous solutions of non-electrolytes had not been extended to sufficiently low frequencies to allow the calculation of any but the principal relaxation time of the water. We have seen that this relaxation time is not identical with that of pure water, but relaxation times widely separated from that of pure water have not been reported.

Measurements on monosaccharides, by Tait *et al.*,[63] using time-domain techniques, have extended over the wide frequency range 10^7—10^{10} Hz. Using both the original direct time conversion of the data [64] and the full Fourier transform analysis,[65] three separate relaxation times were found. Their values at 278 K and the activation energies are as follows:

	$10^{11} \times \tau n_0$	$\Delta H^{\ddagger}$/kJ mol^{-1}
Process 1	1.7 ± 0.2	(18.8)
Process 2	5.2 ± 0.9	23 ± 4
Process 3	11.0 ± 4.5	38 ± 8

[63] M. J. Tait, A. Suggett, F. Franks, S. Ablett, and P. A. Quickenden, personal communication, to be published. See H. W. Loeb, G. M. Young, P. A. Quickenden, and A. Suggett, *Ber. Bunsengesellschaft phys. Chem.*, 1971, **75**, 1155.

[64] H. Fellner Feldegg, *J. Phys. Chem.*, 1969, **73**, 616.

[65] A. Suggett, P. A. Mackness, M. J. Tait, H. W. Loeb, and G. M. Young, *Nature*, 1970, **228**, 456; see also Chapter 4.

Figure 18 illustrates a typical $\varepsilon''(\nu)$ function and its breakdown into three components. The shortest relaxation time is that of the bulk of the liquid, but its activation energy cannot be deduced quantitatively from these measurements. On the basis of the contribution to the total static permittivity, and other evidence, it is concluded that the *second* process corresponds

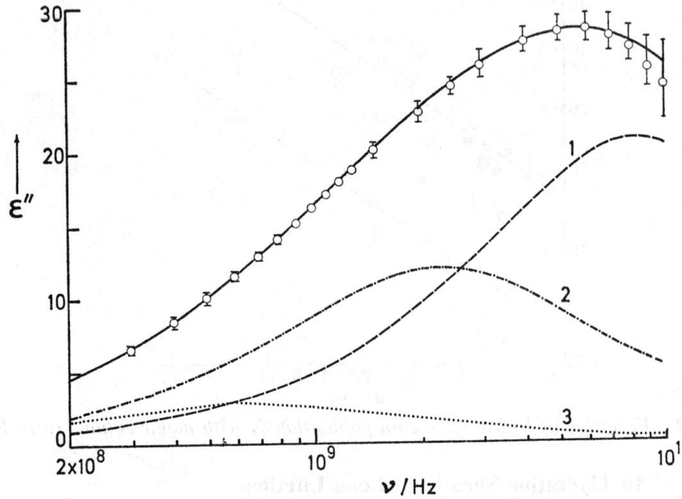

Figure 18 *The total dielectric loss and its resolution into three components for an aqueous solution of glucose,* 2.8*-molar, at* 278 K. (*After Suggett* et al.: cf. *ref.* 65—67)

to the reorientation of the glucose molecule. The *third* process may therefore correspond to the relaxation of the immediate hydration sheath of the molecule. It might, on the other hand, be regarded as a measure of the rate of exchange of molecules between the hydration sheath and the bulk liquid state. The high energy of activation presumably corresponds to the 'breakage' of more than one hydrogen-bond.

This proposal represents a distinct advance from the previous state of our knowledge of aqueous solutions of non-electrolytes. The multiplicity of relaxation times may be present in other solutions; the frequency ranges studied are inadequate to enable generalizations to be made.

Mr. L. Moore and Mr. K. Gray are responsible for the matrix algebra solutions of equation (3) and the computer programming leading to equation (6). I am grateful to Dr. P. R. Mason and Dr. F. Franks for permission to include unpublished data.

[66] B. de Nooijer, P. A. Mackness, M. J. Tait, H. W. Loeb, and G. M. Young, Unilever Report, 1970.

[67] F. Oehme and M. Feinauer, *Chem.-Ztg.*, 1962, **86**, 71.

6

Dielectric Polarization Phenomena in Biomolecular Systems

BY G. SCHWARZ

1 Introduction

The processes of molecular biology generally display a high degree of specificity and efficiency unknown for simple physico-chemical phenomena. These peculiar properties can be attributed to appropriately evolved control and regulation mechanisms in which certain polymeric substances play a vital role. This refers particularly to nucleic acids, proteins, and membranes. Such macromolecular compounds usually have definite conformations (*i.e.* overall structures) which are of crucial significance for their biological function. Therefore, it will be of the greatest importance to elucidate the nature of pertinent structures and their functional potentialities. This is the main scope of biophysical chemistry.

Because of the question as to the way in which the response of biological systems to outside stimulation is brought about, it will also be of particular interest to examine how the structure and function of biopolymers can be affected by the variation of external parameters. In this regard the effects of the interaction of an electric field with biomolecular systems are of considerable significance, since electric fields are known to occur in biological cells (especially at membranes) and to take part in regulation processes as well as in information transfer (as reflected, *e.g.*, by nerve pulses).

Unfortunately, most real biological systems are fairly complex, so that substantial difficulties are encountered in the application of suitable experimental techniques and the evaluation and interpretation of data. Therefore, more clearly defined and reasonably uncomplicated model systems have frequently been studied in order to gather some quantitative understanding of relevant phenomena. Synthetic polypeptides, for instance, have been widely used as simple model compounds for proteins.

Studying dielectric polarization effects is one of the ways of obtaining useful information on biophysico-chemical properties. A particularly large number of such investigations have been carried out with proteins (*e.g.* refs. 1—4). In comparison with the solvent, which is usually water, fairly

[1] (*a*) J. L. Oncley, *Chem. Rev.*, 1942, **30**, 433; (*b*) J. L. Oncley, 'Proteins, Amino Acids and Peptides', ed. Cohn and Edsall, Reinhold, New York, 1945, Ch. 22.
[2] R. Lumry and R. Hue, *J. Phys. Chem.*, 1965, **69**, 1162.
[3] P. Moser, P. G. Squire, and C. T. O'Konski, *J. Phys. Chem.*, 1966, **70**, 744.
[4] H. Hendrickx, R. Verbruggen, M. Y. Rosseneu-Motreff, V. Blaton, and H. Peeters, *Biochem. J.*, 1968, **110**, 419.

large dielectric increments have generally been observed. These data have mostly been interpreted in terms of rotational orientation of permanent dipoles which are inherent in the macromolecular structure. Such a mechanism is well-known from studies of small molecules and doubtless represents also an essential factor in describing dielectric properties of biopolymers. There are, however, other mechanisms of dielectric polarization and relaxation which are not encountered with small molecules but which must be taken into account if one deals with macromolecular substances.

Instead of reviewing the extensive experimental work on the dielectric behaviour of biomolecular systems, this chapter will be mainly concerned with a discussion of the basic aspects of various mechanisms which may yield relevant dielectric polarization. These result from the following effects of an applied external electric field:

(i) displacements of atomic and electronic charges within a molecule and rotational orientation of permanent molecular dipoles as proposed by classical theories;
(ii) polarization of the charge distribution at the interface between macromolecular particles and the surrounding medium;
(iii) perturbation of the equilibrium of a (fast) chemical reaction process which leads to changes in the overall dipole moment.

Studying dielectric phenomena caused by these effects provides information about a number of interesting properties, namely structural data such as dipole moments and rotational relaxation times which can be interpreted in terms of size and shape of the particles under consideration, parameters describing counterion atmospheres (*e.g.* ion mobilities and densities), and finally kinetic quantities, *i.e.* chemical relaxation times and rate constants of underlying chemical reactions. Furthermore, direct effects of an applied electric field can be analysed. This concerns especially the field-induced displacement of a chemical equilibrium, which is of significance with regard to the problem of whether electric signals in biological systems could directly control metabolic processes by affecting the reactivity of biopolymers (*e.g.* the catalytic activity of an enzyme).

2 Principles and the Classical Approach [5,6]

Fundamentals.—Applying an electric field of strength $\boldsymbol{E}$ to an isotropic medium of volume V results in the formation of an overall dipole moment which is parallel to the field and generally described by the relation

$$M = \varepsilon(\varepsilon - 1) VE \tag{1}$$

where $\boldsymbol{\varepsilon} = 8.85 \times 10^{-12}$ F m^{-1} [or $(4\pi)^{-1}$ in electrostatic units] stands for the absolute permittivity of free space. The relative permittivity of the

[5] C. F. J. Böttcher, 'Theory of Electric Polarization', Elsevier, Amsterdam, 1952.
[6] C. P. Smyth, 'Dielectric Behavior and Structure', McGraw-Hill, New York, 1955.

medium is denoted by ε. It is also called the dielectric constant if ε does not depend on $\boldsymbol{E}$. This is usually the case for small field strengths.

Owing to the finite rate of polarization, the actual magnitude of $\boldsymbol{M}$ will lag behind its equilibrium value if the field is changed fast enough. This occurs for a periodic field of sufficiently high frequency. Under such circumstances a phase shift between $\boldsymbol{M}$ and $\boldsymbol{E}$ will be encountered which results in an energy absorption known as dielectric loss. Simultaneously, the amplitude of $\boldsymbol{M}$ is decreased. Using complex notation, this dielectric dispersion can be adequately expressed in quantitative terms by writing $\boldsymbol{E} = \boldsymbol{E}_0 \cdot \exp\,(\mathrm{j}\omega t)$ and introducing a complex relative permittivity (or dielectric constant)†

$$\varepsilon^* = \varepsilon' - \mathrm{j}\varepsilon'' \tag{2}$$

($\boldsymbol{E}_0$ = amplitude of the field strength, $\mathrm{j} = \sqrt{-1}$, $\omega = 2\pi\nu$, being the angular frequency of the field, and t = time). The real permittivity ε' and the dielectric-loss parameter ε'' become functions of ω within a certain frequency range where ω is comparable to the reciprocal time-constant of polarization. Below and above this dispersion region ε' assumes constant values ε_0 and ε_∞, respectively, while ε'' vanishes. Both quantities, ε' as well as ε'' can be determined experimentally by means of electric impedance measurements.[7] For the capacitance C we have the relation

$$C = C_0\,\varepsilon' + C_1 \tag{3}$$

where C_0 and C_1 are specific constants of the measuring cell which must be determined by calibration. On the other hand, ε'' is proportional to the electric conductivity κ of the system according to

$$\kappa = \boldsymbol{\varepsilon}\omega\varepsilon'' \tag{4}$$

This quantity is linearly related to the electric conductance of the measuring cell, *i.e.* the reciprocal resistance of the latter can be described by

$$\frac{1}{R} = B_0\,\kappa + B_1 \tag{5}$$

with B_0 and B_1 being constants.

Ordinarily the value of κ evaluated by means of (5) includes a background conductivity κ_0 which is due to transport of free-charge carriers like ions and/or electrons. Since κ_0 corresponds to a finite value of κ at $\omega \to 0$ we then have apparently $\varepsilon'' \to \infty$ in the limit of low frequencies. In most studies of dielectric-loss effects such complication is avoided by discussing only the 'dynamic' conductivity $\kappa - \kappa_0$.

[7] H. P. Schwan, 'Physical Techniques in Biological Research', ed. W. L. Nastuk, Academic Press, New York, 1963, vol. VI, Ch. 6; N. E. Hill, W. E. Vaughan, A. H. Price, and M. Davies, 'Dielectric Properties and Molecular Behaviour', Van-Nostrand–Reinhold, London, 1969, Ch. 2.

† There are good reasons for abandoning the term 'dielectric constant' for a factor most of whose molecular interest centres on its variation—with concentration, temperature, frequency, field strength, *etc.* [Senior Reporter].

Generally, all potential electric and dielectric behaviour of a medium is taken into account by writing

$$\varepsilon^* = \varepsilon' - \mathrm{j}\frac{\kappa}{\varepsilon\omega} \tag{6}$$

where ε' and κ represent the overall relative permittivity and conductivity, respectively, of the system. The ε' may even include complex contributions provided the corresponding ε'' is not already included in the κ-term.

Most dielectric effects are caused by relaxation mechanisms involved in the formation of the overall dipole moment. Accordingly, there may be one or more contributions to $\boldsymbol{M}$ which approach their equilibrium values with a rate proportional to the instantaneous deviation from the latter. Thus, if $\boldsymbol{M}_i$ is such a part of $\boldsymbol{M}$ we have

$$\frac{\mathrm{d}\boldsymbol{M}_i}{\mathrm{d}t} = -\frac{\boldsymbol{M}_i - \bar{\boldsymbol{M}}_i}{\tau_i} \tag{7}$$

with the bar indicating the equilibrium value and τ_i being the relaxation time of the process. In the ordinary case of an $\boldsymbol{M}_i$ produced by a harmonically oscillating field we put of course $\boldsymbol{M}_i = \boldsymbol{M}_{i0} \exp(\mathrm{j}\omega t)$ where $\boldsymbol{M}_{i0}$ denotes the time-independent amplitude. This yields

$$\boldsymbol{M}_i = \frac{\bar{\boldsymbol{M}}_i}{1 + \mathrm{j}\omega\tau_i} \tag{8}$$

and an analogous expression for the corresponding contribution to ε^*. Frequently, a more or less extended distribution of individual relaxation processes is encountered. Such a phenomenon is described by a relaxation spectrum, *i.e.* the ensemble of relaxation times together with their respective weight (or amplitude) factors.

In cases where the individual relaxation mechanisms are closely related to each other the corresponding overall dielectric dispersion can be quantitatively fairly well represented by the Cole–Cole empirical equation [8]

$$\varepsilon^* = \varepsilon_\infty + \frac{\Delta\varepsilon_0}{1 + (\mathrm{j}\omega\tau_0)^{1-h}} \tag{9}$$

The quantity $\Delta\varepsilon_0 = \varepsilon_0 - \varepsilon_\infty$ stands for the (total) dielectric increment of the dispersion effect. The relaxation spectrum is characterized by the parameter h which satisfies the condition $1 > h \geqslant 0$ and by a special average of the relaxation times, τ_0. According to (9) a plot of ε'' versus ε' results in a circular curve. This proves to be very useful for extrapolation purposes. The centre of the circle is found below the ε'-axis for $h > 0$ but falls on that axis for $h = 0$. In that latter case we obtain a complete semicircle. This corresponds to a pure Debye dispersion [9] which involves

[8] K. S. Cole and R. S. Cole, *J. Chem. Phys.*, 1941, **9**, 341.

[9] (*a*) P. Debye, 'Polar Molecules', Chemical Catalog, New York, 1929; (*b*) *cf.* the abnormally large and low-frequency (*ca.* 20 GHz) dispersion due to the atomic-displacement term in a helical polypeptide. See G. Parry-Jones, M. Gregson, and M. Davies, *Trans. Faraday Soc.*, 1971, **67**, 1630; M. Davies, P. Maurel, and A. H. Price, *J.C.S. Faraday I*, 1972, in the press.

only a single relaxation time τ. It permits comparatively simple formulations of ε' and ε'', namely

$$\varepsilon' = \varepsilon_\infty + \frac{\Delta\varepsilon_0}{1 + \omega^2\tau^2}; \quad \varepsilon'' = \frac{\omega\tau}{1 + \omega^2\tau^2}\Delta\varepsilon_0 \quad \text{(10a and b)}$$

More complicated dielectric properties may be described by superimposing such terms with different relaxation times and the corresponding values of $\Delta\varepsilon_0$. This always results in a flattening of the dispersion curves (*i.e.* ε' and ε'' plotted *vs.* $\log \omega$) in comparison with the basic Debye functions of (10).

Induced and Orientational Polarization.—In principle, the applied field may produce an overall dipole moment in two different ways, namely by means of

(*a*) induced polarization, *i.e.* the formation of new dipoles due to charge separation within the molecular structures of the system;

(*b*) orientational polarization, *i.e.* preferential orientation of originally randomly aligned dipoles.

In general, mechanisms of both kinds are superimposed.

The classical effect of induced polarization is caused by displacing atomic and electronic charges within the individual molecules. The induced dipole moment can be expressed as

$$p = \alpha E' \quad (11)$$

where E' is the effective field strength and α the field-independent polarizability of the molecule. We may assume that there are N_s molecules of species s having a polarizability α_s. Then the total induced dipole moment becomes

$$M_{\text{in}} = \sum_s N_s \alpha_s F_s \quad (12)$$

The local field F_s at the position of a polarizable molecule of type s may be different from the applied field strength E. Since E is defined as the field produced by external means (*e.g.* a charged condenser) within the polarized material, it includes the field of the induced molecular dipole under consideration. The latter must of course not be taken into account for the effective field strength which induces that dipole.

In the case of spherical symmetry of the molecule and the absence of oriented permanent dipoles of species s, the well-known relation

$$F_s = \frac{\varepsilon + 2}{3} E \quad (13)$$

can be established. It must be emphasized that for non-spherical molecules considerable deviations from (13) may occur. Parallel to the axis of a long rod F_s becomes equal to E, whereas perpendicular to it we obtain $F_s = \varepsilon E$.

The rate of displacing atoms and electrons within a molecule corresponds to optical frequencies, and hence no dielectric dispersion of M_{in} is normally encountered in or below the microwave range ($\nu \lesssim 10$ GHz). Since all other polarization effects are not that fast, the polarizabilities α_s determine exclusively the electric permittivity at optical frequencies,† ε_{op}. Thus at sufficiently high frequencies the relations (12) and (13) may be combined with (1) to evaluate α_s. For the pure species s consisting of isotropic spherical molecules, α_s can be given by the refractive index n_s if Maxwell's relation

$$\varepsilon_{\text{op}} = n^2 \tag{14}$$

(which is valid in the absence of absorption) is used. The result reads

$$\alpha_s = \frac{3\varepsilon}{N}\frac{(M_r)_s}{\rho_s}\frac{n_s^2 - 1}{n_s^2 + 2} = \frac{3\varepsilon_0}{N} R_s \tag{15}$$

where R_s is defined as the molar refraction of species s [N = Avogadro's number, ρ_s = density, $(M_r)_s$ = molecular weight, and $(M_r)_s/(N\rho_s) = v_s$ equals the volume per molecule]. Inserting (15) into (12) leads to

$$\frac{n^2 - 1}{n^2 + 2} = \sum_s c_s R_s \tag{16}$$

n being the refractive index of the mixture and c_s denoting the concentration of an individual component as expressed by its mole number N_s/N per unit volume. Of course, equations (15) and (16) must be appropriately modified when non-spherical molecules are concerned.

Orientational polarization results from the fact that randomly oriented dipoles tend to be aligned parallel to the applied field, in which position they have minimum electric energy. This tendency, however, is usually counteracted by rotational diffusion, *i.e.* thermal movements of dipole axes which tend to restore a random distribution. Ordinarily we deal with small electric field strengths satisfying the condition

$$\mu_s E \ll kT \tag{17}$$

(μ_s = permanent dipole moment of particle s, k = Boltzmann's constant, and T = absolute temperature). In this case statistical mechanics leads to the following classical relationship for the overall dipole moment caused by the oriental mechanism:

$$M_{\text{or}} = \sum_s \frac{1}{1 + \text{j}\omega(\tau_{\text{r}})_s} N_s \cdot \frac{\mu_s^2}{3kT} (E_{\text{d}})_s \tag{18}$$

where

$$(\tau_{\text{r}})_s = \frac{1}{2(D_{\text{r}})_s} \tag{19}$$

represents the rotational relaxation time of μ_s and $(D_{\text{r}})_s$ its rotational diffusion coefficient. The field strength $(E_{\text{d}})_s$ refers to the average directing

† Including the i.r., *i.e.* ε_{op} here $= n^2$ at, say, 100 cm^{-1} [Senior Reporter].

field acting on the dipole. It will be generally different from $\boldsymbol{F}_s$ and $\boldsymbol{E}$. This is due to the fact that the inherent field of the dipole μ_s causes a polarization of the surrounding medium which generates a reaction field at the position of the dipolar molecule. For reasons of symmetry this field must have the same direction as μ_s, so that it cannot exert an orienting torque on the dipole. It contributes, however, to the local field $\boldsymbol{F}_s$. The general theory [5] yields

$$\boldsymbol{F}_s = \left[1 + \frac{f_s}{1 - f_s\,\alpha_s}\cdot\frac{1}{1 + j\omega(\tau_r)_s}\cdot\frac{\mu_s^2}{3kT}\right](\boldsymbol{E}_{\mathrm{d}})_s \tag{20}$$

where α_s denotes the polarizability in the direction of μ_s (when there is anisotropic behaviour) and f_s equals the factor by which a dipole moment must be multiplied in order to obtain its reaction field. For an ellipsoid, one finds parallel to its semi-axis a_j (j = 1, 2, or 3), according to electrostatic theory,[10]

$$f_{(j)} = \frac{3}{a_1\,a_2\,a_3}\cdot\frac{A_j(1 - A_j)\,(\varepsilon - 1)}{\varepsilon + (1 - \varepsilon)\,A_j} \tag{21}$$

where

$$A_j = \frac{a_1\,a_2\,a_3}{2}\int_0^\infty \frac{\mathrm{d}u}{(u + a_j^2)\,[(u + a_1^2)\,(u + a_2^2)\,(u + a_3^2)]^{\frac{1}{2}}} \tag{22}$$

For a sphere of radius a we obtain $A_1 = A_2 = A_3 = \frac{1}{3}$ so that

$$f = \frac{1}{a^3}\cdot\frac{2(\varepsilon - 1)}{2\varepsilon + 1} = \frac{4\pi}{3v}\cdot\frac{2(\varepsilon - 1)}{2\varepsilon + 1} \tag{23}$$

(v = volume of the particle).

Many biomolecules are not spherical but can be approximated by an ellipsoid of revolution with $a_1 \gtrless a_2 = a_3$. In these cases the elliptic integral (22) can be expressed in a closed form.[11] The most interesting shape would be a very elongated (*i.e.* rod-like) ellipsoid with $a_1 \gg a_2 = a_3$. Then one has simply

$$f_{(1)} \to 0; \quad f_{(2)} = f_{(3)} \to \frac{1}{2v}\cdot\frac{\varepsilon - 1}{\varepsilon + 1} \tag{24a and b}$$

The directing field $\boldsymbol{E}_{\mathrm{d}}$ likewise depends on the shape of the molecular dipole in question. It may be calculated from the field strength $\boldsymbol{G}^*$ which is generated by the applied field within a correspondingly shaped cavity in the surrounding medium of permittivity ε. Since the directing field must not include a contribution due to the permanent dipole which was originally present in that cavity, it will be composed of $\boldsymbol{G}^*$ plus the reaction field of the dipole which is induced by the directing field, *i.e.* we have

$$\boldsymbol{E}_{\mathrm{d}} = \boldsymbol{G}^* + f\alpha\boldsymbol{E}_{\mathrm{d}} \tag{25}$$

yielding

$$\boldsymbol{E}_{\mathrm{d}} = \frac{1}{1 - f\alpha}\boldsymbol{G}^* \tag{26}$$

[10] Th. G. Scholte, *Physica*, 1949, **15**, 437.
[11] J. A. Stratton, 'Electromagnetic Theory', McGraw-Hill, New York, 1941.

In the case of a spherical dipolar molecule this results in

$$(E_d)_s = \frac{\varepsilon(2 + n_s^2)}{2\varepsilon + n_s^2} E \tag{27}$$

whereas for a long rod-like particle with a dipole parallel to its long axis

$$(E_d)_s \rightarrow E \tag{28}$$

Rotational Relaxation.—As far as small molecules having dimensions of a few angströms are concerned, the rotational relaxation time τ_r under ordinary conditions (particularly in aqueous solution at room temperature) has the order of magnitude of 10^{-10}—10^{-11} s. This corresponds to dielectric dispersion in the microwave range of 1—10 GHz. Macromolecular particles display dielectric relaxation far below these frequencies. Such behaviour is readily explained by the strong dependence of τ_r on the length of the dipolar axis of the molecule. By calculation of the rotational diffusion coefficient D_r according to hydrodynamic theory and inserting the result in (19) one obtains for spherically shaped dipoles [9]

$$\tau_r = \frac{4\pi\eta a^3}{kT} \tag{29}$$

(η = viscosity of the solvent). In the case of ellipsoidal particles different rotational diffusion coefficients apply to rotations about any of the three axes. They have been calculated theoretically [12, 13] so that corresponding values of τ_r can be specified. For the most interesting special case of a prolate ellipsoid of revolution with $a_1 \gg a_2 = a_3$ we obtain

$$\tau_r = \frac{8\pi\eta a_1^3}{3kT(2 \ln 2p - 1)} \tag{30}$$

($p = a_1/a_2 \gg 1$) regarding rotation about the small axis. Accordingly, rotational relaxation of globular proteins in aqueous solution falls in the frequency range of about 0.1—10 MHz. Rod-like polymers may easily display such dispersion effects even at much lower frequencies.

Application to Polar Biopolymers.—On the basis of the above general relationships, the classical dielectric polarization of any biomolecular system can be evaluated. In the particularly interesting case of a dilute solution of polar biopolymers with a uniform rotational diffusion coefficient a comparatively simple relation can be derived because of the fact that orientational polarization of the solute occurs far below the relaxation range of the solvent. The complex permittivity (without the contribution of background conductivity) turns out to be

$$\varepsilon^* = \varepsilon_\infty + \frac{\Delta\varepsilon_0}{1 + j\omega\tau_r} \tag{31}$$

[12] R. Gans, *Ann. Physik.*, 1928, **86** (IV. Folge), 652.
[13] F. Perrin, *J. Phys. Radium*, 1934, **7**, 497.

where

$$\Delta\varepsilon_0 = g\frac{N\mu^2}{3\varepsilon kT}c_0 \tag{32}$$

(c_0 = concentration of the solute in moles per volume) equals the static dielectric increment with regard to ε_∞, *i.e.* the value of ε at sufficiently high frequencies, where orientational polarization of the macromolecular dipole moment μ no longer occurs. The factor g comes in because of the reaction field. It is found that

$$g = \frac{\varepsilon(n^2 + 2)^2}{(2\varepsilon + n^2)^2}\cdot\frac{2\varepsilon + 1}{3} \tag{33}$$

for spherical particles. (Note that ε refers to the solvent whereas n is to be taken for the solute.) For biological macromolecules appropriate values of n are difficult to ascertain by direct measurements. As far as globular proteins in aqueous solution are concerned, an empirical value of $g = 1.93$ (corresponding to $n^2 = 1.41$) has been used.[1] If we deal with long rod-like molecular dipoles (*e.g.* helical biopolymers) we may simply put $g = 1$. Consequently, it becomes very important to take into account the shape of the biopolymer molecules when measured values of $\Delta\varepsilon_0$ are to be evaluated.

Owing to their definite structures, most biomolecules have an appreciable permanent dipole moment which must lead to dielectric polarization *via* the rotational mechanism of preferential orientation. Thus pertinent experimental investigation permits a direct determination of the molecular dipole moments and rotational relaxation times (or rotational diffusion coefficients, respectively). These are characteristic factors for many macromolecules and give valuable information regarding structural properties such as length, shape, and mass.

The dielectric behaviour of proteins in aqueous solutions was first extensively studied by Oncley and co-workers.[1] They interpreted the data in terms of rotational polarization of permanent dipole moments. The latter were found to be in the range of 100—1000 D (1 D = 10^{-18} e.s.u.) while the relaxation times came out at *ca.* 10^{-7} s. Despite some criticism,[14, 15] the preferential-orientation effect must still be considered the principal dielectric-polarization mechanism of proteins.[3, 4] This view is also supported by dielectric dispersion studies of various proteins in solvents of different viscosity.[16, 17] The measured relaxation times were indeed proportional to η as predicted by (29) and (30). Nevertheless, for very large molecules ($M_r > 10^5$) indications of other mechanisms, whose relaxation does not depend on the bulk viscosity of the solvent, have been observed.[16]

[14] J. G. Kirkwood and J. B. Shumaker, *Proc. Nat. Acad. Sci. U.S.A.*, 1952, **38**, 855.
[15] C. T. O'Konski, *J. Phys. Chem.*, 1960, **64**, 605.
[16] S. Takashima, *J. Polymer Sci., Part A-1, Polymer Chem.*, 1962, **56**, 257.
[17] A. Minataka, *Biochim. Biophys. Acta*, 1966, **126**, 570.

Actually there are only a few proteins which can be described by a single rotational relaxation time, as would be expected for a practically spherical macromolecule. Such behaviour was found for instance with myoglobin.[18] From the relaxation time a radius of 21 Å could be calculated according to (29).[2] This is in very good agreement with the value determined from the known structure as established by means of *X*-ray diffraction.[19]

In most other cases the dispersion curves are flattened in comparison with pure Debye functions as presented in (10). This indicates the existence of more than one relaxation time. A rather obvious explanation is based on an ellipsoidal shape of the protein molecule with components of the permanent dipole along each of the three principal axes. In general, this should result in three corresponding relaxation times.[13] Applying this concept to practical examples, it turns out that the assumption of an ellipsoid of revolution with only two relaxation times is quite sufficient for a quantitative fit of the experimental curves. In such a way, the ratio of the axes can be determined.[1] It must be emphasized, however, that the data really do not permit an unambiguous discrimination between two or more individual relaxation times.

Helical polypeptides constitute a simpler biopolymer system which also displays a distinct rotational mechanism of dielectric dispersion.[20] The α-helix structure implies an appreciable permanent dipole moment parallel to its long axis and a relaxation frequency which may become very low because of the considerable length of the particles. Since the latter cannot be made uniform for synthetically prepared samples a more or less extended relaxation spectrum is inevitable.

Conformational transformations of biopolymers generally lead to substantial changes of the dielectric properties which reflect the accompanying changes of charge distribution and overall structure. In the case of proteins, the dielectric increments have been found to be increased drastically by denaturation.[21] Besides such intramolecular transitions, binding or aggregation phenomena can usually also be followed by means of dielectric measurements. A peculiar high-frequency dispersion (around 100 MHz) of protein solutions has been discussed in terms of bound water molecules which are to some extent immobilized in comparison with free water.[22, 23]

Double-stranded helical nucleic acids in aqueous solution likewise exhibit very pronounced dielectric dispersion behaviour at fairly low frequencies. This was originally also interpreted on the basis of rotational orientation

[18] H. O. Marcy and J. Wyman, *J. Amer. Chem. Soc.*, 1942, **64**, 638.

[19] J. C. Kendrew, R. E. Dickerson, B. E. Strandberg, R. G. Holt, D. R. Davies, D. C. Philips, and V. C. Shore, *Nature*, 1960, **185**, 422.

[20] H. Watanabe, K. Yoshioka, and A. Wada, *Biopolymers*, 1964, **2**, 91; M. Sharp, *J. Chem. Soc.* (*A*), 1970, 1596. See also ref. 9(*b*).

[21] S. Takashima, *Biochim. Biophys. Acta*, 1964, **79**, 531.

[22] H. P. Schwan and K. Li, *Ann. New York Acad. Sci.*, 1965, **125**, 344.

[23] (a) E. H. Grant, *Ann. New York Acad. Sci.*, 1965, **125**, 418; (b) E. H. Grant, *J. Mol. Biol.*, 1966, **19**, 133; *cf.* Chapter 5, pp. 161, 162.

of permanent dipoles.[24, 25] Such an explanation, however, is contradicted by the fact that the two strands of the Watson–Crick double helix[26] have opposite directions. Therefore, the fixed-charge distribution must be completely symmetrical, so that practically no permanent dipole moment can exist. It turned out that in this case the effect can be attributed to the polyionic nature of the nucleic acid molecule. As will be pointed out in the next section, polarization of the counterion atmosphere of a polyion by an applied electric field results in substantial induced dipole moments.

3 Induced Polarization in Macromolecular Systems

If an electric field is applied to macromolecular or colloidal systems as encountered in biological samples, a number of peculiar polarization phenomena may occur which are unknown in solutions of small molecules. These are due to the fact that we actually deal with microscopically heterogeneous systems involving material of different dielectric properties. There are defined interfaces where charges can be accumulated and separated. Some aspects can be discussed from a purely phenomenological point of view, observing only the difference of the bulk dielectric parameters of the material involved. The most conspicuous dielectric dispersion behaviour of biological material, however, cannot be explained in this straightforward way without taking into account the molecular structure and dynamics of the macromolecular interfaces.

The Phenomenological Approach to Orientation.—We may consider the system as a number of macromolecular particles which are embedded in a fluid medium of isotropic and frequency-independent electric permittivity and conductivity. Upon application of an electric field the system will tend to assume a state of minimum electric energy. This can ordinarily be expected to result in some kind of polarization (*e.g.* pearl chain formation[27]). A general equation for pertinent calculations has been derived by means of quasi-electrostatic theory.[28] Let us consider an arbitrarily shaped body consisting of material with a complex electric permittivity ε_b^* which includes the values ε and κ according to (6). Immersing it in a medium of complex electric permittivity ε_a^* which is subjected to an electric field strength $\boldsymbol{E}_0$ requires a certain amount of work. The time average of the latter measures the electric energy of the body under the conditions stated. It comes out as the real part of a complex expression, namely

$$w = \mathrm{Re}\frac{1}{2}\left[\boldsymbol{\varepsilon}\frac{\hat{\varepsilon}_a^*}{\varepsilon_a^*}\int_{V_b}\boldsymbol{E}_0(\varepsilon_a^* - \varepsilon_b^*)\,\boldsymbol{E}_b^*\cdot\mathrm{d}V\right] \tag{34}$$

[24] L. G. Allgen, *Acta Physiol. Scand.*, 1950, **22** (suppl.), 76.
[25] S. Takashima, *J. Mol. Biol.*, 1963, **7**, 455.
[26] J. D. Watson and F. H. C. Crick, *Nature*, 1953, **171**, 737, 964.
[27] M. Saito and H. P. Schwan, 'Proceedings of the Fourth Annual Tri-service Conference on Biological Effects of Microwave Radiation', Plenum Press, New York, 1954, vol. I, p. 85.
[28] G. Schwarz, *J. Chem. Phys.*, 1963, **39**, 2387.

The integration has to be performed over the volume of the body with E_b^* being the effective complex field-strength built up inside the body (note that there may be a certain phase shift with regard to E_0!). The caret on ε_a^* indicates its complex conjugate.

It can be generally shown that usually, for any particle, one clearly defined direction of orientation with regard to the applied field exists (leaving aside completely symmetrical particles). This direction depends on the frequency of the field.[20] More detailed quantitative evaluation requires explicit knowledge of E_b^*. The problem has been generally solved for ellipsoidal bodies.[29] On the basis of the results obtained, most particles of practical interest can at least be discussed in first approximation.

It is well known that in isotropic and purely dielectric systems (*i.e.* in the absence of conductivity and dielectric loss) ellipsoidal particles always tend to be aligned with their longest axis parallel to the applied field.[11] This is in principle not true, however, if the conductivity of the material involved must also be taken into account. Nevertheless, it can be shown that still one of the principal axes of the ellipsoid is oriented parallel to the field in the state of minimum energy (setting aside very abnormal cases where the complex dielectric constant of the particle is described by a tensor whose principal axes do not coincide with those of the ellipsoid).

If we consider the simple case of ellipsoids with isotropic and frequency-independent electric permittivity and conductivity, we find that the electric energy w_i for axis i parallel to the field can be represented as

$$\frac{w_i}{V_b E_0^2} = v_{i,\infty} + \frac{\Delta v_{i,0}}{1 + (\omega\tau_i)^2} = V_i \tag{35}$$

The parameters $v_{i,\infty}$, $\Delta v_{i,0}$, and τ_i are functions of the bulk permittivities and conductivities of the particle and the surrounding medium. Apparently, the v_i undergo a dispersion of the ordinary Debye-type displayed by the real electric permittivity according to (10a). It may now easily occur that the frequency curves of two different v_i intersect each other at a certain frequency value. Provided the corresponding energy is smaller than that for the third axis there will evidently be a sudden switch of orientation if this point is passed on changing the frequency. Since this concerns two axes of the ellipsoid a turn-over through an angle of 90° occurs. There may be up to three such turn-over frequencies for the simple system described by (35).[29]

A biological cell may be approximated by a membrane-covered ellipsoid and can also be treated by the general theory.[29, 30] It turns out that at lower frequencies the shortest axis is aligned parallel to the field, whereas after passing one or more turn-over frequencies the longest axis will be so oriented at sufficiently high frequencies (see the example illustrated in Figure 1). Assuming reasonable values for the parameters involved, the

[29] G. Schwarz, M. Saito, and H. P. Schwan, *J. Chem. Phys.*, 1965, **43**, 3562.
[30] M. Saito, H. P. Schwan, and G. Schwarz, *Biophys. J.*, 1966, **6**, 313.

switch of orientation can be predicted for frequencies of the order of 10—1000 MHz and field intensities well above 10 V cm^{-1}. Such an effect has indeed been observed experimentally with living cells.[31] Its disappearance after the cells have been killed can easily be explained by the fact

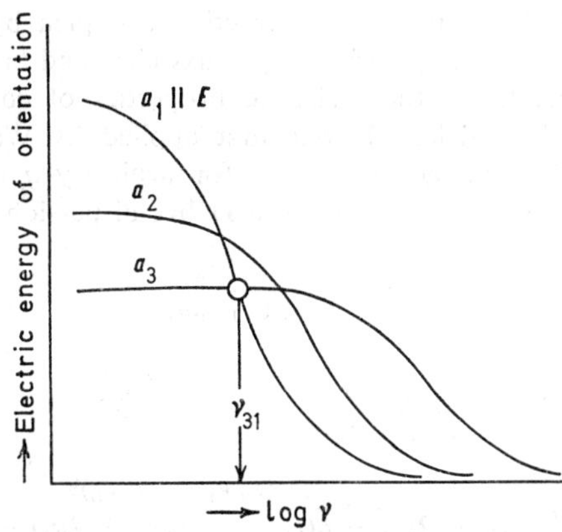

Figure 1 *Schematic sketch of the frequency-dependent electrical energy of an ellipsoid which is oriented with a principal axis a_i parallel to the applied field **E** ($a_1 > a_2 > a_3$). The case refers to a simple membrane-covered ellipsoid with equal internal and external electric permittivity and conductivity, respectively. There is one turnover frequency ν_{31} corresponding to the intersection point of the curves regarding a_2 and a_3. Apparently $a_3 \parallel \mathbf{E}$ is the stable state for frequencies $\nu < \nu_{31}$ whereas for $\nu > \nu_{31}$ this applies to $a_1 \parallel \mathbf{E}$ (provided of course that one has sufficient field strength to overcome thermal movements)*

that the cell membrane has become too porous to maintain its insulating function.

Maxwell–Wagner Dispersion.—Macroscopic heterogeneities of the complex permittivity must always result in an apparent overall dielectric dispersion, even if the above-discussed orientation effect cannot occur. We may investigate this for the case of spherical particles of radius a and complex permittivity ε_b^* which are suspended in a medium with complex permittivity ε_a^*. It follows from electrostatic theory that the presence of one such sphere is equivalent to a dipole moment given by

$$p = \frac{\varepsilon_b^* - \varepsilon_a^*}{\varepsilon_b^* + 2\varepsilon_a^*} a^3 \mathbf{E}_0 \tag{36}$$

A larger spherically shaped range of the suspension having a radius R and containing N particles would thus be equivalent to a dipole moment

$$\frac{\varepsilon^* - \varepsilon_a^*}{\varepsilon^* + 2\varepsilon_a^*} R^3 \mathbf{E}_0 = Np \tag{37}$$

31 A. A. Teixeira-Pinto, L. L. Nejelski, J. L. Cutler, and J. H. Heller, *Exp. Cell Research*, 1960, **20**, 548.

where ε^* describes the effective complex permittivity of the suspension. Clearly it follows from (36) and (37)

$$\frac{\varepsilon^* - \varepsilon_a^*}{\varepsilon^* + 2\varepsilon_a^*} = q\frac{\varepsilon_b^* - \varepsilon_a^*}{\varepsilon_b^* + 2\varepsilon_a^*} \tag{38}$$

with q ($= Na^3/R^3$) being the volume fraction occupied by the particles. This equation was originally derived by Maxwell[32] and has proved to be very useful in calculating the dielectric properties of colloidal systems involving spherical particles. It was first applied by Wagner to small volume fractions.[33] The general result for arbitrary q and frequency-independent values of ε and κ leads to a Debye dispersion described by[34]

$$\varepsilon^* = \varepsilon_\infty + \frac{\Delta\varepsilon_0}{1 + j\omega\tau} \tag{39}$$

where

$$\varepsilon_\infty = \varepsilon_a \frac{\varepsilon_b + 2\varepsilon_a + 2q(\varepsilon_b - \varepsilon_a)}{\varepsilon_b + 2\varepsilon_a - q(\varepsilon_b - \varepsilon_a)} \tag{40}$$

$$\Delta\varepsilon_0 = 9q(1 - q)\frac{(\varepsilon_a\kappa_b - \kappa_a\varepsilon_b)^2}{[(\varepsilon_b + 2\varepsilon_a) - q(\varepsilon_b - \varepsilon_a)]\,[\kappa_b + 2\kappa_a) - q(\kappa_b - \kappa_a)]^2} \tag{41}$$

$$\tau = \frac{\varepsilon_b + 2\varepsilon_a - q(\varepsilon_b - \varepsilon_a)}{\kappa_b + 2\kappa_a - q(\kappa_b - \kappa_a)}\varepsilon_0 \tag{42}$$

For a suspension of particles with small ε_b and κ_b in an aqueous electrolyte solution the dielectric increment $\Delta\varepsilon_0$ turns out to be very small (< 0.01) whereas the dispersion frequency should be around 10—100 MHz. Accordingly, such Maxwell–Wagner effects cannot have significance for the interpretation of dielectric behaviour of spherically shaped biological particles (*e.g.* globular proteins) as long as they are formally assumed to consist of homogeneous material having comparatively low-bulk conductivity and permittivity. The same applies to an extension of the theory with regard to rod-like particles (*e.g.* helical DNA).[35]

The situation becomes more interesting if we consider particles which are surrounded by a shell having different dielectric properties.[36] Such a three-phase system can be treated in an analogous way as the above two-phase system.[34] Assuming the shell to be a practically non-conductive membrane and the external and internal medium to be an aqueous electrolyte, a dielectric dispersion can be derived which explains nicely the large effect observed in the MHz range with membrane-covered biological cells. It can be qualitatively interpreted by a membrane

[32] J. C. Maxwell, 'Treatise on Electricity and Magnetism', Oxford, 1892, vol. I.
[33] K. W. Wagner, *Arch. Elektrotech.*, 1914, **3**, 83.
[34] H. Pauly and H. P. Schwan, *Z. Naturforsch.*, 1959, **14b**, 125.
[35] S. Takashima, *Adv. in Chem. Ser.*, 1967, No. 63, 232.
[36] H. P. Schwan, G. Schwarz, J. Maczuk, and H. Pauly, *J. Phys. Chem.*, 1962, **66**, 2626.

capacitance (of about 1 μF cm^{-2}) which is short-circuited at higher frequencies of the field.[37]

Another model of experimental interest concerns the case of a highly conductive shell around practically non-conductive material. It may be applied to macromolecules or colloidal particles in electrolyte solution which usually have counterion atmospheres so that the field may displace freely movable ionic charges on their surfaces. The resulting dielectric effect turns out to be equivalent to a simple Maxwell–Wagner dispersion of particles having an apparent bulk conductivity of

$$\kappa_b = \kappa_b^0 + 2\frac{\lambda_0}{a} \tag{43}$$

(κ_b^0 = bulk conductivity of the material of the particle). The quantity λ_0 is given by

$$\lambda_0 = \sum_i e_i^2 \sigma_i u_i \tag{41}$$

(e_i, σ_i, and u_i are the electric charge, surface density, and mechanical mobility of a surface-charge carrier of type *i*) and defines the (frequency-independent) surface conductivity. A complete theory of dielectric dispersion effects due to such surface conductivity, which applies also to ellipsoidal particles has been given by O'Konski.[15] Under certain conditions dielectric increments of appreciable magnitude can be predicted. However, the very large electric permittivities and extremely low dispersion frequencies which have been observed with biological cells and colloidal suspensions cannot be explained in this way.[36, 37] In order to understand these phenomena, it becomes necessary to analyse the effect of an external field on the distribution of counterions in the interface between the particle and the surrounding fluid medium.

Effects Due to Polarization of Counterion Distribution.—We consider first spherical particles (of radius *a*) with a layer of counterions which are freely mobile on the surface, but which cannot escape into the solvent because of the strong attractive forces of the highly charged particle. Without external manipulations, a symmetrical distribution of counterions can be assumed. After application of an external field, a displacement of counterions will occur which leads to some asymmetry of the distribution. The final equilibrium is determined by the opposing effects of the field and the diffusion of counterions, which tends to restore a random distribution. It can be shown that the phenomenon may be quantitatively described by a frequency-dependent complex surface conductivity[38]

$$\lambda = \frac{j\omega\tau}{1 + j\omega\tau}\lambda_0 \tag{45}$$

[37] H. P. Schwan, in 'Advances in Biological and Medical Physics', ed. J. H. Lawrence and C. A. Tobias, Academic Press, New York, 1957, vol. 5, p. 148.
[38] G. Schwarz, *J. Phys. Chem.*, 1962, **66**, 2636.

where λ_0 is equal to the quantity introduced by equation (44) as applied to the counterions being the charge carriers. The relaxation time

$$\tau = \frac{a^2}{2ukT} \tag{46}$$

is essentially determined by the (surface) diffusion coefficient of the counterions, $D = ukT$. It turns out that this is equivalent to an apparent complex bulk permittivity of the particle of

$$\varepsilon_b^* = [\varepsilon_b^*]^0 + \frac{1}{1 + j\omega\tau} \cdot \frac{e_0^2 a \sigma_0}{kT} \tag{47}$$

($[\varepsilon_b^*]^0$ refers to the material of the particle without the special properties of the surface; e_0 and σ_0 are the electric charge and surface density of the counterions, respectively.) Inserting this ε_b^* in (38) permits an evaluation of the dielectric properties of the overall system. The enormous dielectric increments of colloidal polystyrene spheres in an electrolyte solution of fairly high conductivity could be very well described by

$$\Delta\varepsilon_0 = \frac{9}{4} \frac{q}{(1 + \frac{1}{2}q)^2} \cdot \frac{e^2 a \sigma_0}{\varepsilon kT} \tag{48}$$

For a particle diameter of about 1 μm, $\sigma_0 \approx 2 \times 10^{13}$ cm^{-2}, and $q \approx 0.2$, a $\Delta\varepsilon_0$ of *ca.* 35 000 is obtained at room temperature. The dispersion occurs in the low-frequency range of 1—10 kHz, in quantitative agreement with the relaxation time given by (46).[38] There can be little doubt that the still more remarkable low-frequency dielectric dispersion of biological cells and tissue is also due to such counterion-polarization effects.

The same mechanism must in principle apply to globular proteins. However, because of the small dimensions of these biopolymers the dielectric increments are of comparatively small magnitude. Furthermore, their dispersion falls in the same frequency range as that of the rotational polarization mechanism of permanent dipole moments. Since the latter is apparently predominant, it would be difficult to distinguish the counterion effect.

Another, but to some extent similar mechanism has been proposed by Kirkwood and Shumaker with special regard to protein molecules.[14] These authors pointed out that at appropriate pH values protons may fluctuate between the various acidic and basic groups of the amino-acid side-chains. This was shown to result in a non-vanishing mean square of the overall dipole moment of the molecule, namely

$$\Delta\mu^2 = e^2 a^2 f^2 \sum_i \frac{y_i}{(1 + y_i)^2} \tag{49}$$

where e denotes the charge of a proton and f stands for a geometric factor depending on deviation from spherical shape (a = effective radius). The running index i refers to all proton-donating groups. With the respective pK value indicated by pK_i we have

$$\log y_i = \mathrm{p}K_i - \mathrm{pH} \tag{50}$$

Numerical calculations were carried out for several proteins. They yielded dipole moments of about the same magnitude as actually measured. The $\Delta\mu^2$ apparently becomes fairly large when the pH is close to the pK range of the frequently occurring carboxy- and amino-groups, respectively. Therefore, sharp maximum values of the dipole must be expected at about pH 4 and 10. Experimental tests of this characteristic feature have failed, however.[39, 40] The discrepancy between the theory and these measurements are so obvious that one must conclude that protons are not sufficiently mobile to contribute significantly to the dielectric polarization of protein solutions.

Actually a very crucial kinetic aspect has apparently always been overlooked so far. The original assumption of a proton mobility comparable to that in free solution which allows for dispersion frequencies in the MHz range must not be made in this special case. It is only applicable to the diffusion-controlled polarization of counterion atmospheres as discussed before. In contrast to those counterions the protons of the Kirkwood–Shumaker model are chemically bound at specific sites and must dissociate before they can jump to another site. Thus the lifetime of protons at a given site has to be taken into account with regard to the relaxation time of the overall fluctuation process. Its effect can be readily estimated on the basis of the rate constant of protolytic dissociation

$$k_\mathrm{D} = k_\mathrm{R}\, K_\mathrm{dis} \tag{51}$$

The rate constant of recombination k_R has an upper bound of about $10^{10}\ \mathrm{l\ mol^{-1}\ s^{-1}}$ corresponding to a diffusion-controlled mechanism. K_dis denotes the dissociation constant. Since the mechanism requires p$K \approx$ pH in order to yield appreciable dipole moments we may set $K_\mathrm{dis} \approx c_{\mathrm{H}^+}$ (= hydrogen-ion concentration). Hence the lifetime of bound effective protons can be estimated to be

$$\tau_\mathrm{p} = \frac{1}{k_\mathrm{D}} \approx 10^{-10}\ \mathrm{l^{-1}\ mol\ s}/c_{\mathrm{H}^+} \tag{52}$$

which at pH 7 leads to $\tau_\mathrm{p} \approx 10^{-3}$ s so that the corresponding dispersion frequency cannot be expected to be above *ca.* 100 Hz. Relaxation in the MHz range would be possible only at very low pH values which are usually not accessible in pertinent experiments.

In contrast to the situation with globular (bio)polymers, polarization of counterion atmospheres can be much better differentiated for very elongated ellipsoidal or rod-like polymers. In many cases there is no interfering effect due to permanent dipoles. Even if these exist, they usually have sufficiently

[39] S. Takashima, *J. Phys. Chem.*, 1965, **69**, 2281.
[40] W. Goebel and H. Vogel, *Z. Naturforsch.*, 1964, **19b**, 292.

slow rotational relaxation times, owing to the strong dependence on the length as displayed by equation (30). Polarization of the counterion cloud by an external field, however, will be comparatively fast and produce extremely large induced dipole moments parallel to the long axis. On the basis of a simple model the polarizability can be obtained as [41]

$$\alpha = \frac{\alpha_0}{1 + j\omega\tau} \tag{53}$$

where

$$\alpha_0 = \tfrac{1}{2}\pi\varepsilon\varepsilon_{\text{eff}}\,\beta L^3 \tag{54}$$

and

$$\tau = \frac{\pi\varepsilon\varepsilon_{\text{eff}}}{2uze_0^2}\,L^2 \tag{55}$$

[ε_{eff} = effective permittivity in the ionic cloud, β = fraction of counterions on the polyion surface (= degree of association), L = length of the particle, u = mechanical mobility of the counterion along the axis, and z = number of associated counterions per unit length of the polyion]. Rotational orientation of these induced dipoles leads to anisotropy of electrolytic conductivity [41, 42] and to birefringence.[43] Dielectric polarization of polyelectrolytes has also been attributed to this mechanism.[44] Two different relaxation processes must be considered. The first one is controlled by the rate of displacing counterions as described by the relaxation time of (55). This is fairly fast since τ turns out to be of the order of magnitude of 10^{-6}—10^{-5} s for a particle length of several thousand ångströms. The second process refers to the rotational orientation of the induced dipoles, which is described by a relaxation time of one-third of that specified by equation (30) (with $a_1 = L/2$).[41, 42] According to this, rotational relaxation will be so slow that by the usual dielectric measuring techniques only the first relaxation process would be observed.

It should be emphasized that one fundamental difference must be taken into account when the polarization of the counterion clouds of rods is compared with that of spheres. In the latter case, it is only diffusion which counteracts the displacements of counterions caused by the field. On the surface of rod-like polyions, counterions must also be displaced against a component of the attractive forces exerted by the fixed charges. Without observing this, the term $ze_0^2/4\pi\varepsilon\varepsilon_{\text{eff}}$ in equation (55) will be substituted by kT.[35] Such a result, however, should only be applied if the electric energy of displacing a counterion against the attraction of the polyion along its axis is negligible in comparison with the thermal energy,

[41] G. Schwarz, *Z. phys. Chem.* (*Frankfurt*), 1959, **19**, 286.
[42] (a) M. Eigen and G. Schwarz, *J. Colloid Sci.*, 1957, **12**, 181; (b) M. Eigen and G. Schwarz in 'Electrolytes', Pergamon Press, Oxford, 1962, p. 309.
[43] C. T. O'Konski and A. J. Haltner, *J. Amer. Chem. Soc.*, 1957, **79**, 5634.
[44] M. Mandel and A. Jenard, *Trans. Faraday Soc.*, 1963, **59**, 2158, 2170.

i.e. if

$$\frac{ze_0^2}{4\pi\varepsilon\varepsilon_{\mathrm{eff}}} \ll kT \tag{56}$$

This does not hold true for polyions of ordinary charge density. One would rather encounter the opposite situation so that equation (55) is usually more realistic. Unfortunately, the alternatives are difficult to distinguish experimentally, since both of them exhibit a dependence on the square of the length of the particle, and z, as well as $\varepsilon_{\mathrm{eff}}$, can only be roughly estimated from other data. In certain systems both energy terms of (56) may even be of about the same order of magnitude so that the values of τ calculated from either equation do not differ very much from each other anyway.

Appropriate biopolymer systems which should display dielectric polarization of the kind under consideration are linear polyelectrolytes like polypeptides or nucleic acids in aqueous solution. As was mentioned earlier, polyamino-acids in the α-helix conformation apparently carry considerable permanent dipole moments which ought to vanish when a transition to the randomly coiled state occurs. In fact, this is actually observed when the dielectric behaviour of helix–coil transitions in non-aqueous solutions is measured.[20] In the case of poly(glutamic acid) in aqueous solution a different picture was observed.[45] The dielectric increments registered as a function of the pH values are much larger for the coiled state (pH $\gtrsim$ 6) than for the helical one (pH $\lesssim$ 4). The relaxation times are about the same in either state. There is nevertheless a distinct minimum of both quantities in the transition range. Such behaviour can readily be interpreted in the following way: the rotational mechanism of permanent dipole orientation applies to the helical state, where all the carboxy-groups are neutralized so that no counterion cloud exists, whereas, on the other hand, polarization of the ionic atmosphere occurs in the coiled state where the polymer behaves as a fairly stretched polyion owing to dissociation of the carboxy-groups. Under the latter circumstances most of the permanent dipole moment disappears, since the individual dipoles of the chain elements cancel each other to a great extent because their orientation is quite random. Since the macromolecule assumes a comparatively compact shape in the transition state, the accompanying minima of the dielectric increment and the dielectric relaxation time are not surprising.

The counterion mechanism of dielectric polarization is also in reasonable agreement with the data collected for DNA in the double-helical state as well as in the coiled denatured form.[35] For the latter, much smaller dielectric increments and relaxation times are observed, indicating contraction of the polymer upon denaturation. The rotational relaxation time for helices as measured by means of flow birefringence was generally found

[45] S. Takashima, *Biopolymers*, 1963, **1**, 171.

to be distinctly greater than the actually observed dielectric relaxation time which indeed followed the L^2-dependence predicted for the counterion mechanism. Polarization of the counterion cloud of DNA is also reflected by the occurrence of the orientation field effect.[46]

4 Dielectric Phenomena Caused by Chemical Rate Processes

In addition to the physical mechanisms of dielectric polarization discussed above, there is also a potential chemical basis of such effects. It is in principle encountered with systems which involve chemical reactions resulting in changes of the overall dielectric properties.

Thermodynamic Foundation.[47, 48]—A single-step chemical reaction process can in a general way be formulated as

$$\nu_1' \mathrm{R}_1 + \nu_2' \mathrm{R}_2 + \ldots \rightleftharpoons \nu_1'' \mathrm{R}_1 + \nu_2'' \mathrm{R}_2 + \ldots \tag{57}$$

where the R_i ($i = 1, 2, \ldots$) denote the reaction partners and the respective ν_i' and ν_i'' stand for the stoicheiometric coefficients on the left- and right-hand sides of the reaction scheme (in most cases either ν_i' or ν_i'' will of course be zero). In a closed system mass conservation requires that any changes of mole numbers n_i of a given R_i are described by

$$\delta n_i = \nu_i\, \delta\xi \tag{58}$$

with $\nu_i = (\nu_i'' - \nu_i')$. The extent of reaction ξ is the only independent variable measuring variations of the amounts of the various species involved in the reaction process. The Gibbs free energy of the system, G, can be considered as a function of the absolute temperature T, pressure p, electric field intensity $\boldsymbol{E}$, and ξ. Its total differential reads

$$\mathrm{d}G = -\,S\,\mathrm{d}T + V\,\mathrm{d}p - \boldsymbol{M}\,\mathrm{d}\boldsymbol{E} - A\,\mathrm{d}\xi \tag{59}$$

where S denotes the entropy and $\boldsymbol{M}$ the overall dipole moment parallel to $\boldsymbol{E}$. The affinity

$$A = -\sum_i \nu_i \mu_i \tag{60}$$

(μ_i = chemical potential of R_i) may be written as

$$A = RT\Big[\ln K - \sum_i \nu_i \ln c_i - \sum_i \nu_i \ln f_i\Big] \tag{61}$$

where K equals the thermodynamic equilibrium constant, R is the gas constant, and c_i and f_i are the concentration and activity coefficient of R_i, respectively. Since (59) represents a total differential we have generally

$$\left(\frac{\partial A}{\partial \boldsymbol{E}}\right)_{T,p,\xi} = \left(\frac{\partial \boldsymbol{M}}{\partial \xi}\right)_{T,p,\boldsymbol{E}} \equiv \Delta \boldsymbol{M} \tag{62}$$

[46] N. Ise, M. Eigen, and G. Schwarz, *Biopolymers*, 1963, **1**, 343.
[47] K. Bergmann, M. Eigen, and L. DeMaeyer, *Ber. Bunsengesellschaft phys. Chem.*, 1963, **67**, 819; M. Davies, *Acta Phys. Polon.*, 1971, **40A**, 561.
[48] G. Schwarz, *J. Phys. Chem.*, 1967, **71**, 4021.

Accordingly, $\Delta \boldsymbol{M}$ describes the change of the overall dipole moment during the course of the reaction under the condition that the intensive variables T, p, and $\boldsymbol{E}$ are kept constant. For the sake of simplicity we may assume that the f_i are not changed by the field. Then by inserting (61) into (62) we obtain

$$\left(\frac{\partial \ln K}{\partial \boldsymbol{E}}\right)_{T,p} = \frac{\Delta \boldsymbol{M}}{RT} \tag{63}$$

This means that an electric field tends to displace a chemical equilibrium in the direction of greater polarization. There will be a corresponding dielectric relaxation controlled by the kinetics of the chemical reaction steps.

By means of equation (1) the total dielectric increment due to the chemical effect may be expressed as

$$\Delta\varepsilon_{\text{ch}} = \frac{1}{\boldsymbol{\varepsilon} V}\cdot\frac{\partial \boldsymbol{M}}{\partial \boldsymbol{E}} = \frac{1}{\boldsymbol{\varepsilon} V}\cdot\frac{\partial \boldsymbol{M}}{\partial \bar{\xi}}\cdot\frac{\partial \bar{\xi}}{\partial \ln K}\cdot\frac{\partial \ln K}{\partial \boldsymbol{E}} \tag{64}$$

(assuming no electrostriction, *i.e.* $\partial V/\partial \boldsymbol{E} = 0$). The bar refers to the equilibrium value of the extent of reaction. With

$$\Gamma = \frac{\partial \bar{\xi}}{\partial \ln K} \tag{65}$$

as a factor which is determined by the relation between the equilibrium concentrations and K we find by employing (62) and (63)

$$\Delta\varepsilon_{\text{ch}} = \frac{\Gamma}{\boldsymbol{\varepsilon} V}\cdot\frac{(\Delta \boldsymbol{M})^2}{RT} \tag{66}$$

For ordinary systems which can be described by equation (1) we have to use

$$\Delta \boldsymbol{M} = \boldsymbol{\varepsilon}\frac{\partial \varepsilon}{\partial \xi} V\boldsymbol{E} \tag{67}$$

Accordingly, $\Delta\varepsilon_{\text{ch}}$ becomes proportional to E^2. Therefore the chemical effect of dielectric polarization cannot usually be detected experimentally at small field intensities. It should only be measurable when extremely high static fields are applied. By imposing 200 kV cm^{-1} on the dimerization process of 6-hexanolactam the effect has actually been observed.[47]

It must be emphasized, however, that a $\Delta \boldsymbol{M}$ which is directly proportional to $\boldsymbol{E}$ as specified by (67) rests on the implicit assumption that the rate of polarization is much faster than the rate of chemical relaxation. Such an assumption is very well justified for small molecular dipoles which can freely rotate in a liquid. Then we have a rotational relaxation time $\tau_r \approx 10^{-10}$—10^{-11} s which is small compared with pertinent chemical relaxation times τ_{ch} in almost all cases of practical interest. Under these circumstances, chemically induced dielectric behaviour cannot be possible for small $\boldsymbol{E}$. It can occur, on the other hand, in the reverse case of $\tau_r \gg \tau_{\text{ch}}$. This condition may actually be encountered in macromolecular reaction systems, *e.g.* those involving biopolymers, since rotational relaxation is comparatively slow here. A quantitative discussion of simple model systems is presented to illustrate the essential aspects of the phenomenon.

Simple Model Processes and General Aspects.—Let us consider an intramolecular reaction process of the type

$$A_1 \underset{k_{21}}{\overset{k_{12}}{\rightleftharpoons}} A_2 \tag{68}$$

described by the rate coefficients k_{12} and k_{21}. The compound A_2 is assumed to carry a permanent dipole moment μ while A_1 has none. For the sake of simplicity both molecules may have the same rotational diffusion coefficient D_r. An axis of the molecules is defined by the direction of the dipole in A_2. There will be a distribution of such axes with regard to the direction of an applied field $\boldsymbol{E}$. That concentration of molecules A_i which have axes pointing into the solid angle $d\Omega = 2\pi \sin \Theta \, d\Theta$ (Θ = angle between the direction of $\boldsymbol{E}$ and the molecular axis) can be expressed as

$$dc_i = \gamma_i(\Theta, t)\, d\Omega \tag{69}$$

The distribution functions γ_i are changed not only by rotational diffusion but also by the chemical reactions. Hence we have

$$\frac{\partial \gamma_i}{\partial t} = \left(\frac{\partial \gamma_i}{\partial t}\right)_{\text{chem}} + D_r \nabla^2 \gamma_i \tag{70}$$

(∇^2 = the Laplacian operator). For small values of $\mu E/kT$ these differential equations can be solved without serious difficulties. After evaluation of the distribution of the polar molecules, the corresponding complex dielectric increment follows as [48]

$$\Delta\varepsilon^* = g\frac{N\mu^2}{3\varepsilon kT} c_2 \left(\frac{\beta_r}{1 + j\omega\tau_r} + \frac{\beta_2}{1 + j\omega\tau_2}\right) \tag{71}$$

(c_2 = concentration of A_2). We have in principle two relaxation processes, with the amplitude factors

$$\beta_r = \frac{K}{1 + K}, \quad \beta_2 = \frac{1}{1 + K} \tag{72a and b}$$

where $K = k_{12}/k_{21}$ is the equilibrium constant of the chemical process. One of the two relaxation times equals the rotational relaxation time as introduced before ($1/\tau_r = 2D_r$); the other one is related to the chemical relaxation time τ_{ch} of (68) according to

$$\frac{1}{\tau_2} = \frac{1}{\tau_r} + \frac{1}{\tau_{ch}} \tag{73}$$

where

$$\frac{1}{\tau_{ch}} = k_{12} + k_{21} \tag{74}$$

This result has been confirmed by a correlation-function treatment.[49] For sufficiently fast rotational relaxation (*i.e.* $\tau_{ch} \gg \tau_r$) we apparently obtain $\tau_2 \approx \tau_r$, so that dielectric dispersion is controlled by τ_r only. Equation (71) then becomes the classical expression of (31) and (32). A measurable dielectric increment controlled by chemical relaxation may, however, be detected if $\tau_{ch} \lesssim \tau_r$. Its absolute magnitude will be proportional to

$$\beta_2 c_2 = \frac{K}{(1 + K)^2} c_0 \qquad (75)$$

($c_0 = c_1 + c_2$ is the total weighed-in concentration of the reacting system). Accordingly there will be a maximum of the effect at $K = 1$ where equal amounts of A_1 and A_2 are present.

The chemically induced relaxation process can be expected to be clearly separable for $\tau_{ch} \ll \tau_r$. Then a distinct second dielectric dispersion range must occur at higher frequencies than those characteristic of the rotational mechanism (*cf.* Figure 2a). Its relaxation time τ_2 becomes practically equal to τ_{ch}. Therefore the phenomenon may also be exploited in order to determine the kinetics of the underlying reaction process.

Let us now briefly review the essential features of the effect under consideration. Without an applied external field we have a random distribution of molecular axes as well as a certain state of chemical equilibrium regarding the mutual interconversion of A_1 and A_2. Application of an external field determines some preferential orientation of the molecular dipoles of A_2. The system tends to establish this as rapidly as possible. Usually, this is achieved by means of rotational diffusion. In principle, however, the chemical process provides an alternative means of orienting dipoles without actually rotating them. The equilibrium constant $K(\Theta)$ of those molecules which have an angle Θ with regard to the applied field will be changed according to

$$\frac{\partial \ln K(\Theta)}{\partial E_d} = \frac{\mu \cos \Theta}{kT} \qquad (76)$$

(E_d = directing field strength) which follows from (63) owing to the fact that the corresponding molar change of M is $\Delta M = N\mu \cos \Theta$. Therefore, some unfavourably oriented dipoles (at $\Theta > 90°$) simply vanish *via* the reaction $A_2 \rightarrow A_1$ which is initiated by the field-induced decrease of $K(\Theta)$. On the other hand, some additional, more favourably oriented dipoles (at $\Theta < 90°$) are formed by the reverse reaction $A_1 \rightarrow A_2$ which is enhanced owing to the respective increase of $K(\Theta)$. The rotational and the chemical mechanism of orientation of course compete with each other. If the chemical relaxation is faster than rotation, it will be reflected in the dielectric dispersion behaviour displaying a maximum for equal amounts of the reacting species. There is, however, a certain part of the total polarization which is always controlled by rotational diffusion as can be

[49] G. Williams, *Adv. Mol. Relaxation Processes*, 1970, **1**, 409.

seen by inspection of equation (71). It should also be noted that the field does not change the total concentrations of the A_i, *i.e.*

$$\delta c_i = \delta \int_0^{\pi} \gamma_i \, d\Omega = 0 \tag{77}$$

A displacement of the overall equilibrium $A_1 \rightleftharpoons A_2$ resulting in additional dipoles can only be caused by very high field intensities.

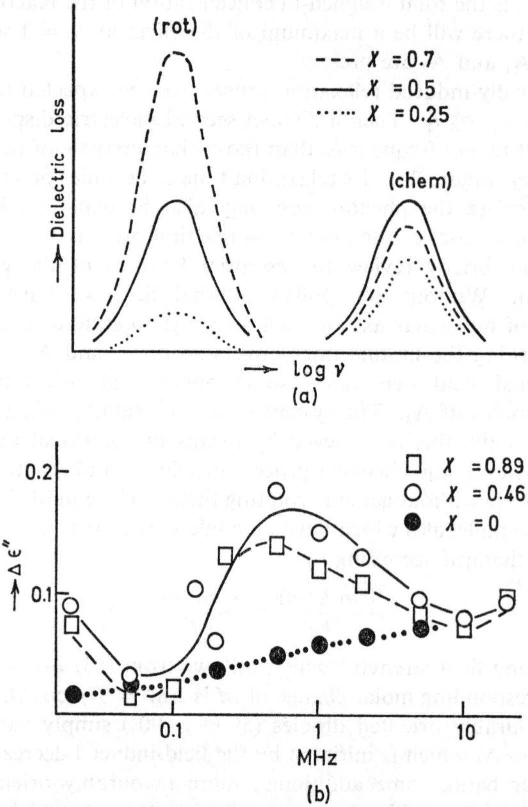

Figure 2 *Dielectric loss as a function of frequency for systems permitting chemically induced dielectric relaxation* ($\tau_{ch} \ll \tau_r$). (a) *Schematic curves reflecting the rotational as well as chemical contributions for the model process* (68). *Variations of the mole fraction x of dipolar states lead to a maximum of the chemical effect at x* = 0.5. (b) *Experimental plots of the dielectric loss increments in the range of the dispersion due to the helix–coil transition of the* PBLG-*system as described in the text* (*x* = *fraction of helix structure*). *The phenomenon is indeed most pronounced around the midpoint of the transition* (*corresponding to x* = 0.5). *The rotational effect occurs at frequencies below* 100 kHz

The essential aspects of the above model can be applied to any type of chemical reaction process which brings about changes in the dipole moments of the reactive partners. A corresponding chemically induced dielectric-dispersion effect must always occur if the chemical relaxation proceeds faster than the rotational movement of at least one of the molecular dipole species involved. This can in principle be expected for fast reactions of (bio)polymers.

Let us consider another example which is more frequently encountered in systems of practical interest, namely

$$A + B \underset{k_D}{\overset{k_R}{\rightleftharpoons}} AB \tag{78}$$

where A, B, and AB have different permanent dipole moments. If these molecules are small and not restrained regarding free rotation, no chemical effect of dielectric relaxation can be expected. If we assume, however, that B is a reaction site rigidly fixed on a macromolecule, the situation becomes quite different. Provided the macromolecule is large enough, we have for a sufficiently fast reaction process

$$\tau_A \ll \tau_{ch} \ll \tau_M \tag{79}$$

i.e. rotational relaxation of the (small) A molecules is fast compared with chemical relaxation; rotational relaxation of B and AB which is determined by τ_M will be much slower. In this case we can expect three different dispersion ranges, one at very high frequencies due to rotation of the A-dipoles, one at fairly low frequencies due to rotation of the macromolecule, and a third one in-between which is caused by the relaxation of the chemical process (78). For a detailed analysis we must discuss the reaction

$$A + B(\Theta) \rightleftharpoons AB(\Theta) \tag{80}$$

which refers to a recombination of an arbitrarily oriented A with a B-group oriented in the Θ-direction. For the corresponding chemical-relaxation time we find

$$\frac{1}{\tau_{ch}} = k_R\, c_A + k_D \tag{81}$$

This applies to a reaction process of type (78) where A is buffered.[50] In fact, our process (80) actually leads to changes of B(Θ) and AB(Θ) whereas A remains constant because the field does not produce changes of overall concentrations. The complex dielectric increment due to the chemical process turns out to be

$$\Delta\varepsilon^*_{ch} = \frac{\Delta\varepsilon^0_{ch}}{1 + j\omega\tau_{ch}} \tag{82}$$

with

$$\Delta\varepsilon^0_{ch} = g\frac{N\mu_D^2}{3\varepsilon kT}x(1 - x)\, c_0 \tag{83}$$

[50] B. H. Havsteen, in 'Physical Principles and Techniques of Protein Chemistry, Part A', ed. S. J. Leach, Academic Press, New York, 1969, p. 245.

($\mu_D = |\vec{\mu}_{AB} - \vec{\mu}_B|$, $c_0 = c_A + c_{AB}$ is the total concentration of reaction sites, and $x = c_{AB}/c_0$ = mole fraction of occupied reaction sites). Taking into account the usual experimental values of the parameters involved, it can be shown that such effects may be quite measurable. Some more definite examples which are experimentally accessible are to be discussed in the following sections.

Dielectric Relaxation Due to Fast Conformational Transitions.—Most biopolymer molecules carry an appreciable permanent dipole moment which is changed when a conformational transformation occurs. Rotational diffusion of such molecules may be much slower than the rate of the structural rearrangement. Then we have a system which is essentially equivalent to the above discussed intramolecular coversion with $\tau_{ch} \ll \tau_r$. An especially appropriate example is given by the helix–coil transition of a polypeptide in a non-aqueous solution. As was mentioned before, a very large permanent dipole moment disappears in such systems when the helix is converted to a random coil. The intermediate states of an individual chain molecule consist of extended helical fragments which are separated by coil regions. The lengths of these fragments are determined by the degree of co-operation in the transition process. The average length at the midpoint of the transition is called the co-operative length and is measured by the corresponding number of peptide bonds, N_0.[51] It amounts to about 100 (which corresponds to a length of 150 Å) in the system which has been most studied, *i.e.* poly(γ-benzyl-L-glutamate) (PBLG) in a mixture of ethylene dichloride (EDC) and dichloroacetic acid (DCA).[52] The rotational relaxation time for such a fragment can be estimated to be *ca.* 10^{-4} s. The chemical relaxation of the transition can be described by a relaxation time [51, 53]

$$\tau_{ch} = \frac{N_0^2}{k_F} x(1 - x) \qquad (84)$$

(x = helix content, *i.e.* the mole fraction of peptide bonds which are in the helical form, k_F = elementary rate coefficient of helix growth by hydrogen-bond formation at the end of helix fragments). At the mid-point of the transition (corresponding to x = 0.5) a diffusion-controlled value of $k_F \approx 10^{10}$ s^{-1} yields a τ_{ch} value of approximately 10^{-6} s. Therefore a chemically induced dielectric-dispersion effect must be expected in the region of several hundred kHz. It turns out to be necessary, however, to use fairly high concentrations of the polymer. Under these circumstances the effect was indeed clearly demonstrated with the PBLG–EDC–DCA system.[54] Its maximum was observed at x = 0.5 as predicted by the theory (*cf.* Figure 2b). Recently the phenomenon has also been detected

51 G. Schwarz, *Biopolymers*, 1968, **6**, 873.
52 Th. Ackermann and E. Neumann, *Biopolymers*, 1967, **5**, 649.
53 (*a*) G. Schwarz, *J. Mol. Biol.*, 1965, **11**, 64; (*b*) G. Schwarz, *J. Theor. Biol.*, in the press.
54 G. Schwarz and J. Seelig, *Biopolymers*, 1968, **6**, 1263.

for the analogous system of poly(β-benzyl-L-aspartate) in *m*-cresol which also undergoes a helix–coil transition.[55]

As we have pointed out before, the whole effect is based on a field-induced displacement of the chemical equilibrium involved. In the special cases under consideration this refers to a conformational change caused by direct action of the applied field. This field effect is of course extremely small for the low field intensities which occur in conventional dielectric measurements. The field strength which is required to produce an almost complete conformational change can, however, be calculated from the chemically induced dielectric increment as measured for the helix–coil transition. It turns out to be inversely proportional to the dipole moment of a helix fragment having just the co-operative length. For the PBLG system one obtains a value in the order of magnitude of several hundred kV cm^{-1}.[54] This is comparable to the field strengths encountered in biological cells as produced for instance by nerve pulses on membranes. We thus conclude that such fields may in principle be able to control biological processes by switching conformational states of biopolymers or membranes.

The Dielectric Effect of Proton Exchange with Zwitterions.—Besides structural transitions, binding or transfer reactions with small molecules or ions represent another very important class of elementary functions of biopolymers. Appreciable changes of dipole moments in such processes could also lead to chemically induced dielectric relaxation. A pertinent model binding-process was discussed earlier. An analogous typical transfer-reaction scheme can be generally written as

$$A_1 + B_1 \underset{}{\overset{\text{fast}}{\rightleftharpoons}} A_2 + B_2 \qquad (85)$$

where the macromolecular states A_1 and A_2 have different dipole moments, and the small molecules B_1 and B_2 have no dipole moments whose dielectric relaxation can interfere with that of A_1 and A_2. Proton exchange between zwitterions (*e.g.* those formed by amino-acids or proteins) and an appropriate proton donor–acceptor system represents an example of some practical significance.

Let us consider a molecule which carries an acidic and also a basic group so that it may exist in three different states:

$$HZH^+ \text{ (at low pH)}, \quad {}^-ZH^+ \text{ (at medium pH)}, \quad {}^-Z \text{ (at high pH)}$$

This applies especially to amino-acids in aqueous solution. The first dissociation step may have a pK value of pK_1. Then we find at pH $\approx$ pK_1 appreciable proton exchange according to

$$HZH^+ + A^- \underset{k_{21}}{\overset{k_{12}}{\rightleftharpoons}} {}^-ZH^+ + HA \qquad (86)$$

[55] A. Wada, *Chem. Phys. Letters*, 1971, **8**, 211.

provided the pK of HA, pK_A, does not deviate too much from pK_1. The analogous process

$$^{-}Z + HA \underset{k_{12}'}{\overset{k_{21}'}{\rightleftharpoons}} {}^{-}ZH^{+} + A^{-} \qquad (87)$$

would occur for pH $\approx$ pK_2 $\approx$ pK_A.

Both processes (86) and (87) apparently involve formation or destruction of the zwitterion dipole moment μ_z. Therefore, application of an electric field must displace the respective chemical equilibrium to some extent depending on the angle ψ between the directions of the zwitterion dipole and the field $\boldsymbol{E}$. For $\psi < 90°$ an increase of the number of zwitterions is favoured whereas for $\psi > 90°$ this number will be decreased. Therefore, the whole system tends to develop preferential orientation of dipoles parallel to the field. Proton transfer of the kind involved here has generally been proved to be practically diffusion controlled so that the reaction rates could be extremely high.[56] The chemical mechanism of dielectric polarization may thus be fast enough in comparison with the rotational diffusion of the zwitterion, especially if the latter is a macromolecule (*e.g.* a protein).

A detailed theoretical discussion of the problem can be found elsewhere.[57] The complex dielectric increment due to the orientation of permanent dipoles (except those eventually present in HA and A^-) turns out to be

$$\Delta\varepsilon^* = \frac{\Delta\varepsilon_r^0}{1 + j\omega\tau_r} + \frac{\Delta\varepsilon_2^0}{1 + j\omega\tau_2} \qquad (88)$$

The first term again refers to the part controlled by rotational diffusion. For the chemical contribution we obtain

$$\Delta\varepsilon_2^0 = g\frac{Nc_0}{3\varepsilon kT}x(1 - x)\,\mu_z^2 \qquad (89)$$

(x = mole fraction of zwitterion states). The relaxation time τ_2 is related to τ_{ch} in the way expressed by (73). The chemical relaxation time is of course somewhat different here; we have

$$\frac{1}{\tau_{ch}} = k_{12}\,c_A + k_{21}\,c_{AH} \qquad (90)$$

for the process (86). In the case of process (87) the concentration terms c_A and c_{HA} must be exchanged. As is clearly shown by equation (89), a maximum of the effect should be displayed at x = 0.5, *i.e.* for equal numbers of zwitterion and non-zwitterion states. This will be the case for pH = pK_1 (or = pK_2, respectively). Since the rate coefficients k_{12} and k_{21} can usually be anticipated to exceed values of 10^9 l mol^{-1} s^{-1}, sufficiently high concentrations of the acid–base system HA/A^- may well result in extremely small values of τ_{ch} which could even approach the small τ_r values of comparatively small zwitterions (with $\tau_r \approx 10^{-10}$ s).

[56] (*a*) M. Eigen, *Angew. Chem. Internat. Edn.*, 1964, **3**, 1; (*b*) M. Eigen, *Discuss. Faraday Soc.*, 1965, **39**, 7.

[57] G. Schwarz, *J. Phys. Chem.*, 1970, **74**, 654.

To some extent, the effect now under consideration may be considered as a generalization of the Kirkwood–Shumaker mechanism discussed before. A sufficiently small chemical lifetime of protonated states is ensured here by using high concentrations of the proton-carrying agents.

It should be noted that the effect does not require the addition of an extra proton donor–acceptor system. In fact, the proton exchange may take place between a zwitterion and a non-zwitterion state of different orientation as indicated by the reaction scheme

$$HZH^{+}(\psi) + {}^{-}ZH^{+}(\psi') \underset{k_{21}}{\overset{k_{12}}{\rightleftharpoons}} {}^{-}ZH^{+}(\psi) + HZH^{+}(\psi') \quad (91)$$

(note that $k_{12} = k_{21}$ at $\boldsymbol{E} = 0$!). This process leads to an exchange of the zwitterion dipole without any material change of the reaction system. For the chemical relaxation time one simply obtains

$$\frac{1}{\tau_{ch}} = k_{12}\, c_0 \quad (92)$$

(c_0 = total weighed-in concentration of the zwitterion molecules). Preliminary experimental evidence of this effect was found with a 1M aqueous solution of ε-aminocaproic acid ($pK_1 \approx 4$).[58] A neutral pH (corresponding to $x = 1$) dielectric relaxation in the GHz range could be attributed completely to rotational relaxation of water dipoles (at about 15 GHz) and of zwitterions (at about 0.7 GHz). After lowering the pH to a value of 4.14 (corresponding to $x \approx 0.6$) a distinct broadening of this latter part of the dispersion curve occurred. It can be readily interpreted by the appearance of the proton transfer according to (92). Numerical evaluation yields a $k_{12} \approx 3 \times 10^9\ l\ mol^{-1}\ s^{-1}$ which is a very reasonable value.[57]

[58] D. Adolph, Diplom Thesis, University of Göttingen, 1970.

7
General Molecular Theory and Electric Field Effects in Isotropic Dielectrics

BY S. KIELICH

This chapter is divided into six sections whose principal contents are as follows:

1 Introduction. Books, monographs, and review articles available on dielectrics and their molecular aspects: scope of present review: the variety of non-linear field effects.

2 Electric and magnetic properties of microsystems. Definition of multipoles: electrostatics of permanent multipoles: interaction energies for two multipoles: induced molecular multipoles: interaction energies of induced multipoles. Tables of point groups: tensor elements of multipoles: vector elements of multipoles: tensor elements of polarizabilities.

3 Statistical and electromagnetic properties of dielectrics: averaging processes and expressions: molecular correlation functions: the virial function: static and alternating fields: fluctuations: Maxwell's equations: models of dielectrics.

4 Electro-optical properties: molar refraction and molecular correlation: the Kerr effect: its component terms and molecular theory: molecular angular correlation: non-linear polarizability: generalized Kerr effect: translational and molecular field fluctuations: electric saturation.

5 Permittivity variations in a strong electric field: non-linear and saturation behaviour: progress of experimental methods: molecular behaviour and various theoretical models: molecular correlations: complete dielectric saturation: electrostriction and electrocaloric effect.

6 Concluding remarks and recapitulation. Related electric field-induced phenomena: non-linear polarizabilities of atoms and molecules: radial and angular correlations: fluctuations of molecular fields.

1 Introduction

This chapter is intended as a review of the work done, particularly during the last ten years, in the field of non-linear electric and magnetic properties of isotropic dielectrics. We shall deal with these achievements against the background of earlier results, the majority of which have, by now, become included for good in numerous monographs, handbooks, and review articles on special aspects of the physics of dielectrics. We shall consider all those non-linear effects, induced by strong electric, magnetic, and optical fields, which provide information on the structure of dielectrics (particularly in the gas and liquid phases) and which permit the drawing of conclusions regarding the properties of atoms, molecules, macromolecules, and colloid particles. Especially, we shall be concerned with certain simple methods of determining the numerical values of molecular electric multipoles (dipoles, quadrupoles, octupoles, and hexadecapoles) as well as electric, magnetic, and optical polarizabilities and hyperpolarizabilities. In this domain, enormous progress has been achieved in recent years with regard to theory, and even more so on the experimental side owing to the many new measuring methods of high sensitivity and accuracy. All this is largely due to the recent evolution of quantum and laser electronics, whose definite introduction into the majority of physical, physicochemical, and biochemical research laboratories in the form of ready-made, highly automatized electronic measuring equipment greatly facilitates the study of the electronic, atomic, molecular, and ionic properties of matter in the wide range of frequencies covering alternating currents, radio and radar frequencies, and the far-i.r. regions of the spectrum.

Books and Monographs.—The 1946 Faraday Society Discussion on dielectrics [1] opened a new era of study, bringing a series of books and monographs devoted to dielectric properties.[2–8] Böttcher's monograph [2] develops in particular the theoretical electrostatics and dynamics of dielectric materials, whereas that by Smyth [3] deals, in addition to the simpler theory, with experimental dielectric methods and the relationship between macroscopic properties and molecular structure. The fundamental principles of the theory of dielectrics are presented in Brown's monograph [4] in a molecular-statistical treatment, and Fröhlich [5] employs Kirkwood's semi-macroscopic approach.[7] In 1969, a comprehensive volume on 'Dielectric Properties and Molecular Behaviour' [8] became available, in which Hill discusses the theory

1 *Discuss. Faraday Soc.*, 1946, **42A**, 1—256.
2 C. J. F. Böttcher, 'Theory of Electric Polarisation', Elsevier, Amsterdam, 1952.
3 C. P. Smyth, 'Dielectric Behaviour and Structure', McGraw-Hill, New York, 1955.
4 W. F. Brown, jun., in 'Encyclopedia of Physics', ed. S. Flügge, Springer Verlag, Berlin, 1956,Vol. 17, p. 1.
5 H. Fröhlich, 'Theory of Dielectrics', 2nd Edn., Clarendon Press, Oxford, 1958.
6 A. von Hippel, 'Dielectrics and Waves', J. Wiley and Sons, New York, 1959; 'Molecular Relaxation Processes', *Chem. Soc. Special Publ.* No. 20, London, 1966; V. V. Daniel, 'Dielectric Relaxation', Academic Press, London, 1967; 'Complex Permittivity', ed. J. H. Calderwood, English Universities Press, London, 1971; D. W. Davies, 'The Theory of the Electric and Magnetic Properties of Molecules', Wiley, London, 1967; 'Progress in Dielectrics', ed. J. B. Birks, Heywood, London, 1960 *et seq.*, Vol. 1 *et seq.*; J. Barriol, 'Les Moments Dipolaires', Gauthier-Villars, Paris, 1957; L. Eyraud, 'Diélectriques Solides Anisotropes et Ferroélectricité', Gauthier-Villars, Paris, 1967; R. Coelho, 'Diélectriques et Isolants', Société Française des Electriciens, Paris, 1965.
7 J. G. Kirkwood, 'Dielectrics—Intermolecular Forces—Optical Rotation', Gordon and Breach, New York, 1965.
8 N. E. Hill, W. E. Vaughan, A. H. Price, and M. Davies, 'Dielectric Properties and Molecular Behaviour', Van Nostrand Reinhold, London, 1969.

of electric permittivity and loss, Vaughan experimental methods, Price the dielectric properties of gases and liquids, and Davies the range of dielectric dispersion and absorption in condensed phases, providing a first '*vademecum*' to contain a systematic coverage of the theory, experimental methods, and quantitative interpretation of the behaviour of molecular materials in electric fields below optical frequencies.

The fundamentals of the magnetic properties of matter are presented in the classical work of Van Vleck [9] and developed in numerous books and monographs.[10–12]

The highly interesting and now classical monograph of Born [13] contains the theory of the optical properties of matter covering atomic and molecular structural considerations. The fundamentals of molecular optics, presented by Born, have been developed by Volkenshteyn [14] in tensorial form. Various monographs [15–18] have been written on the laws of classical optics and the fundamentals of physical optics, including natural optical activity,[19] optical rotatory dispersion and circular dichroism,[20–22] i.r. spectroscopy,[23] and other related topics.

Light scattering in various media is the subject of quite a number of books.[24–32] Thus, molecular Rayleigh scattering has been dealt with by

9 J. H. van Vleck, 'The Theory of Electric and Magnetic Susceptibilities', University Press, Oxford, 1932.
10 J. van den Handel, in 'Handbuch der Physik', Springer Verlag, Berlin, 1956, Vol. 15, p. 1.
11 N. L. Mulay, 'Magnetic Susceptibility', Wiley, New York, 1963; A. Michel, 'Phénomènes Magnétiques et Structure', Masson et Cie, Paris, 1966.
12 A. H. Morrish, 'The Principal Properties of Magnetism', Wiley, New York, 1965; D. E. G. Williams, 'The Magnetic Properties of Matter', Longmans, London, 1966.
13 M. Born, 'Optik', Springer Verlag, Berlin, 1933.
14 M. V. Volkenshteyn, 'Molekularnaya Optika', Gostekhizdat, Moscow, 1951.
15 R. W. Ditchburn, 'Light', Blackie and Sons, London, 1963.
16 M. Born and E. Wolf, 'Principles of Optics', Pergamon Press, Oxford, 1964.
17 G. Bruhat, 'Optique', Masson et Cie, Paris, 1965.
18 M. Garbuny, 'Optical Physics', Academic Press, New York, 1965.
19 J. P. Mathieu, in 'Encyclopedia of Physics', ed. S. Flügge, Springer Verlag, Berlin, 1957, Vol. 28, p. 333.
20 C. Djerassi, 'Optical Rotatory Dispersion', McGraw-Hill, New York, 1960.
21 P. Crabbé, 'Optical Rotatory Dispersion and Circular Dichroism in Organic Chemistry', Holden Day, San Francisco, 1965.
22 L. Velluz, M. Legrand, and M. Grosjean, 'Optical Circular Dichroism', Academic Press, New York, 1965.
23 'Infra-red Spectroscopy and Molecular Structure', ed. M. Davies, Elsevier, Amsterdam, 1963.
24 J. Cabannes, 'La Diffusion Moléculaire de la Lumière', Presses Universitaires de France, 1929.
25 H. C. Van de Hulst, 'Light Scattering by Small Particles', Wiley, New York, 1957.
26 K. A. Stacey, 'Light Scattering in Physical Chemistry', Academic Press, New York, 1957.
27 I. M. Fabelinskii, 'Molecular Scattering of Light', Plenum Press, New York, 1968.
28 M. Kerker, 'The Scattering of Light', Academic Press, New York, 1969.
29 G. Placzek, in 'Marx Handbuch der Radiologie', Akademische Verlagsgesellschaft, Leipzig, 1934.
30 S. Bhagavantam, 'Scattering of Light and the Raman Effect', Chemical Publishing Co., New York, 1942.
31 F. Kohlrausch, 'Der Smekal-Raman Effekt', Springer, Berlin, 1938.
32 J. Brandmüller and H. Moser, 'Einführung in die Ramanspektroskopie, Dietrich Steinkopf, Darmstadt, 1962.

Cabannes,[24] Fabelinskii,[28] and others.[13, 14, 28–30] The theory of Raman scattering has been developed by Placzek[29] and extended in some special monographs.[30–32]

One surely should consult the classical work of Drude and Pockels[33] and Voigt,[34] to whom are due the fundamentals of crystal optics as well as of electro- and magneto-optics. A newer approach to electro-optics is to be found in the monograph by Szivessy,[35] and that of Schütz[36] provides a treatment of magneto-optics. To Agranovich and Ginzburg[37] is due a modern account of crystal optics in exciton theory approach. The importance of the magneto-optical properties of matter is obvious from the fact that in 1969 the Faraday Society organized a Symposium on Magneto-optical Effects.[38]

The 1960's witnessed the coming of pulsed and continuously operating lasers[39]—sources of collimated beams of monochromatic light coherent in space and time and conveying energy fluxes of enormous density. The rapid development of laser techniques and quantum electronics[40–42] in recent years has led to the experimental detection of a variety of non-linear optical processes taking place in isotropic and anisotropic materials under the influence of intense laser light. In this way, out of the foundations of classical optics, there grew very soon a non-linear optics,[43–50] the theoretical principles of which are formulated in the monographs of Akhmanov and Khokhlov,[43] Bloembergen,[44] and Butcher.[45] An elementary

[33] F. Pockels, 'Lehrbuch der Kristallooptik', Leipzig, 1906.
[34] W. Voigt, 'Magneto- und Elektro-Optik', Teubner, Leipzig, 1908.
[35] G. Szivessy, in 'Handbuch der Physik', Springer Verlag, Berlin, 1929, Vol. 21, p. 724.
[36] W. Schütz, in 'Handbuch der Experimental Physik', Akademische Verlagsgesellschaft, Leipzig, 1936, Vol. 16.
[37] V. M. Agranovich and V. L. Ginzburg, 'Spatial Dispersion in Crystal Optics and the Theory of Excitons', Wiley, New York, 1965.
[38] Magneto Optical Effects: Symposia of the Faraday Soc., No. 3, 1969.
[39] B. A. Lengyel, 'Lasers', Wiley, New York, 1962.
[40] 'Proceedings of the 3rd International Quantum Electronics Congress', ed. P. Grivet and N. Bloembergen, Dunod, Paris, and Columbia University Press, New York, 1964.
[41] G. Birnbaum, 'Optical Masers', Academic Press, New York, 1964; V. M. Fayn and J. I. Khanin, 'Quantum Electronics', Pergamon Press, Oxford, 1969.
[42] J. F. Ready, 'Effect of High Power Laser Radiation', Academic Press, New York, 1971.
[43] S. A. Akhmanov and R. V. Khokhlov, 'Problemy Nielinieynoy Optiki', Akad. Nauk U.S.S.R., Moscow, 1964.
[44] N. Bloembergen, 'Non-linear Optics', Benjamin, New York, 1965.
[45] P. N. Butcher, 'Non-linear Optical Phenomena', Ohio State Univ., 1965.
[46] R. Bernheim, 'Optical Pumping', Benjamin, New York, 1965.
[47] A. Piekara, 'Nowe Oblicze Optyki', PWN, Warsaw, 1968.
[48] J. L. Klimontowicz, 'Lasery i Optyka Nieliniowa', PWN, Warsaw, 1969.
[49] G. C. Baldwin, 'An Introduction to Non-linear Optics', Plenum Press, New York, 1969.
[50] 'Proceedings of the International School of Physics: Enrico Fermi—Quantum Optics', ed. R. J. Glauber, Academic Press, New York, 1969; 'Polarisation Matière et Rayonnement' (Volume Jubilaire en l'Honneur d'Alfred Kastler), ed. Soc. Française de Phys., Presses Universitaires de France, Paris, 1969.

treatment of the subject is to be found in the books of Piekara,[47] Klimontowicz,[48] and Baldwin.[49]

Our knowledge of the structure of matter and of its electric, magnetic, and optical properties is based on the theory of the electron,[51–53] quantum theories,[54–59] quantum [60, 61] and classical [62, 63] electrodynamics, statistical mechanics,[64–67] and the theory of molecular interactions.[68–71] The fundamentals of electron theory were first stated in the classical work of Lorentz [51] and then developed in a modern approach by Rosenfeld.[52] The quantum-mechanical theory of the electromagnetic properties of matter is presented semi-classically in the work of Born and Jordan,[54] and to Heitler [60] is due the complete quantum theory of interaction between matter and the electromagnetic field. The above-named methods have permitted the determination of the atomic and molecular structure of matter,[72] in

51 H. A. Lorentz, 'The Theory of Electrons', Teubner, Leipzig, 1909.

52 L. Rosenfeld, 'Theory of Electrons', North-Holland, Amsterdam, 1951.

53 R. Daudel, 'Structure Electronique des Molécules', Gauthier-Villars, Paris, 1962; J. Leite Lopez, 'Fondements de la Physique Atomique', Hermann, Paris, 1967.

54 M. Born and P. Jordan, 'Elementare Quantenmechanik', Springer, Berlin, 1930.

55 A. S. Davydov, 'Quantum Mechanics', Pergamon Press, Oxford, 1965.

56 P. W. Atkins, 'Molecular Quantum Mechanics', Clarendon Press, Oxford, 1970.

57 H. Eyring, J. Walter, and G. E. Kimbal, 'Quantum Chemistry', Wiley, New York, 1943; K. Gumiński, 'Elements of Theoretical Chemistry', PWN, Warsaw, 1964.

58 'Modern Quantum Chemistry', ed. O. Sinanoglu, Academic Press, New York, 1965, Part 1.

59 W. Kauzmann, 'Quantum Chemistry, An Introduction', Academic Press, New York, 1957.

60 W. Heitler, 'The Quantum Theory of Radiation', Clarendon Press, Oxford, 1954.

61 'Modern Quantum Chemistry' ed. O. Sinanoglu, Academic Press, New York, 1965, Part 3.

62 J. A. Stratton, 'Electromagnetic Theory', McGraw-Hill, New York, 1941.

63 L. D. Landau and E. M. Lifshitz, 'Electrodynamics of Continuous Media', Pergamon Press, New York, 1960.

64 R. H. Fowler, 'Statistical Mechanics', University Press, Cambridge, 1955; L. D. Landau and E. M. Lifshitz, 'Statistical Physics', Addison-Wesley, Reading, Mass., 1958.

65 H. Eyring, D. Henderson, B. J. Stover, and E. M. Eyring, 'Statistical Mechanics and Dynamics', Wiley, New York, 1964; 'Physical Chemistry', ed. H. Eyring, Academic Press, New York, 1967, Vol. 2.

66 T. L. Hill, 'Statistical Mechanics', McGraw-Hill, New York, 1956.

67 G. H. A. Cole, 'The Statistical Theory of Classical Simple Dense Fluids', Pergamon Press, Oxford, 1967.

68 'Intermolecular Forces', *Discuss. Faraday Soc.*, 1965, No. 40; 'Colloques Nationaux au Centre National de la Recherche Scientifique', Bordeaux, 1965, ed. C. N. R. S., Paris, 1966.

69 'Modern Quantum Chemistry', ed. O. Sinanoglu, Academic Press, New York, 1965, Part 2.

70 S. Chu, 'Molecular Forces' (based on the Baker Lectures of P. J. W. Debye), Interscience, New York, 1967.

71 H. Margenau and R. N. Kestner, 'Theory of Intermolecular Forces', Pergamon Press, London, 1969.

72 P. J. Wheatley, 'The Determination of Molecular Structure', Clarendon Press, Oxford, 2nd Edn., 1970; W. Finkelnburg, 'Structure of Matter', Springer Verlag, Berlin, and Academic Press, New York, 1964.

particular for the fluid state,[73–84] the modern statistical-kinetic fundamentals of which are largely due to Frenkel and Kirkwood.[73]

We are vastly indebted to the classical work of Drude, Langevin, and Debye [85] for our knowledge of the electric, magnetic, and optical properties of atoms and molecules. Nowadays we have at our disposal dielectric methods of determining the electric dipole moments of molecules [86] and their electric structures,[87–92] as well as methods for the investigation of their optical and magnetic structures.[88–90] Existing laser techniques,[39–42] and techniques providing intense pulsed magnetic fields [93] permit the determination of the non-linear electro-magnetic properties of molecules.[94]

[73] J. Frenkel, 'Kinetic Theory of Liquids', University Press, Oxford, 1946; J. G. Kirkwood, 'Theory of Liquids', ed. B. J. Adler, Gordon and Breach, New York, 1968.

[74] J. O. Hirschfelder, C. F. Curtis, and R. B. Bird, 'Molecular Theory of Gases and Liquids', Wiley, New York, and Chapman and Hall, London, 1954.

[75] 'The Structure and Properties of Liquids', *Discuss. Faraday Soc.*, 1967, No. 43.

[76] H. Eyring and N. S. Jhon, 'Significant Liquid Structures', Wiley, New York, 1969.

[77] I. Adamczewski, 'Ionization, Conductivity and Breakdown in Dielectric Liquids', Taylor and Francis, London, 1969.

[78] I. Prigogine (in collaboration with A. Bellemans and V. Mathot), 'The Molecular Theory of Solutions', North-Holland, Amsterdam, and Interscience, New York, 1957.

[79] J. L. Kavanau, 'Water and Solute–Water Interactions', Holden-Day, San Francisco, 1964.

[80] H. A. Stuart, 'Das Makromolekül in Lösungen,' Springer Verlag, Berlin, 1953.

[81] H. Morawetz, 'Macromolecules in Solutions', Wiley, New York, 1965; W. N. Cvietkov, W. J. Eskin, and S. J. Frenkel, 'Struktura Makrocząsteczek w Roztworach', WNT, Warsaw, 1968.

[82] 'Polymer Solutions', *Discuss. Faraday Soc.*, 1970, No. 49.

[83] A. Sheludko, 'Colloid Chemistry', Elsevier, Amsterdam, 1966.

[84] G. W. Gray, 'Molecular Structure and the Properties of Liquid Crystals', Academic Press, London, 1962; 'Liquid Crystals and Ordered Fluids', ed. J. F. Johnson and R. S. Porter, Plenum Press, New York, 1970.

[85] P. Debye, 'Polar Molecules', Dover Publications, New York, 1945; 'Collected Papers of P. J. W. Debye', Interscience, New York, 1954.

[86] J. W. Smith, 'Electric Dipole Moments', Butterworths, London, 1955; V. I. Minkin, O. A. Osipov, and J. A. Zhdanov, 'Dipolnye Momenty v Organicheskoy Khimii', Khimya, Leningrad, 1968; R. D. Nelson, D. R. Lide, and A. A. Maryott, 'National Standard Reference Data Series, National Bureau of Standards', Washington, D.C., 1967.

[87] S. Mizushima, 'Structure of Molecules and Internal Rotation', Academic Press, New York, 1954; V. N. Kondratiev, 'La Structure des Atomes et des Molécules', Masson et Cie, Paris, 1964.

[88] M. V. Volkenshteyn, 'Struktur und Physikalische Eigenschaften der Moleküle', Teubner, Leipzig, 1960; J. G. Dorfman, 'Diamagnetismus und Chemische Bindung', Teubner, Leipzig, 1963.

[89] M. Davies, 'Some Electrical and Optical Aspects of Molecular Behaviour', Pergamon Press, London, 1965.

[90] H. A. Stuart, 'Molekülstruktur', Springer Verlag, Berlin, 1967.

[91] J. C. Anderson, 'Dielectrics', Chapman and Hall, London, 1964.

[92] 'Physical Chemistry', ed. D. Henderson, Academic Press, New York, 1970, Vol. 4.

[93] 'High Magnetic Fields', ed. H. Kolm, B. Lax, F. Bitter, and R. Mills, Wiley, New York, 1962; H. Knoepfel, 'Pulsed High Magnetic Fields', North-Holland, Amsterdam, 1970.

[94] S. Kielich, 'Non-linear Molecular Optics', preprint University Bordeaux, 1971.

In addition to monographs dealing with the properties and structure of atomic and molecular fluids,[66, 67, 73–79] we have available ones on macromolecular and colloidal[80–84] as well as biological[95–98] media. This kind of research is essentially concerned with determining the macroscopic symmetry of bodies[99–101] and to an even higher degree with elucidating the microscopic symmetries of the molecules, macromolecules, and colloid particles.[102–104] An easy explanation of these matters is to be found in the books by Wheatley,[72] Volkenshteyn,[88, 97] Davies,[89] Stuart,[90] Nye,[100] and Jaffé and Orchin.[102] A systematic and modern treatment of various interatomic and intermolecular interactions is due to Margenau and Kestner.[71] Also, one should consult the encyclopaedic work of Partington[104] on the electric, magnetic, and optical properties of molecules, which moreover lists carefully all the papers from the early beginnings up to 1950.

Review Articles.—Dielectric measurements up to 1956 have been discussed by Nelson,[105] whereas de Brouckéré and Mandel[106] have reviewed the results on dielectric properties of dilute polymer solutions. Work on the permittivity of gases as a function of density and pressure is discussed by Cole and Smyth[107] and by Beaume *et al.*,[108] who moreover report some simple theories of the second dielectric virial coefficient. Dielectric relaxation phenomena (dispersion and absorption) are discussed in the articles by Cole and Smyth and by Davies[109] and Boudouris.[110] A recent article by Grant[111] deals with studies of biological molecules by dielectric methods.

95 B. Pullman and A. Pullman, 'Quantum Biochemistry', Wiley, New York, 1963.
96 J. T. Edsal and J. Wyman, 'Biophysical Chemistry', Academic Press, New York, 1958; 'Theoretical and Experimental Biophysics', ed. A. Cole, Marcel Dekker, New York, 1967, Vol. 1; 1969, Vol. 2.
97 M. V. Volkenshteyn, 'Biologia Molekularna', PWN, Warsaw, 1969; M. V. Volkenshteyn, 'Fizyka Enzymów, PWN, Warsaw, 1970.
98 'Biological Effects of Magnetic Fields', ed. M. F. Barnothy, Plenum Press, New York, 1964, Vol. 1; 1969, Vol. 2.
99 E. P. Wigner, 'Group Theory', Academic Press, New York, 1959; G. Ya. Lyubarskii, 'The Application of Group Theory in Physics', Pergamon Press, Oxford, 1960.
100 J. F. Nye, 'Physical Properties of Crystals', Clarendon Press, Oxford, 1957; H. Poulet and J. P. Mathieu, 'Spectres de Vibration et Symétrie des Cristaux', Gordon and Breach, Paris 1970.
101 R. R. Birss, 'Symmetry and Magnetism', North-Holland, Amsterdam, 1964; S. Bhagavantam, 'Crystal Symmetry and Physical Properties', Academic Press, London, 1966.
102 H. H. Jaffé and M. Orchin, 'Symmetry in Chemistry', Wiley, New York, 1965.
103 D. S. Schonland, 'Molecular Symmetry', Van Nostrand, London, 1965; R. M. Hochstrasser, 'Molecular Aspects of Symmetry', Benjamin, New York, 1966.
104 J. R. Partington, 'An Advanced Treatise on Physical Chemistry', Longmans, London, 1953, Vol. 4; 1954, Vol. 5.
105 S. M. Nelson, *Ann. Reports*, 1957, **54**, 83.
106 L. de Brouckère and M. Mandel, *Adv. Chem. Phys.*, 1958, **1**, 77.
107 R. H. Cole, *Ann. Rev. Phys. Chem.*, 1960, **11**, 149; C. P. Smyth, *ibid.*, 1966, **17**, 433.
108 R. Beaume, R. Nacasch, and R. Coulon, *J. Physique*, 1965, **26**, 597.
109 M. Davies, *Quart. Rev.*, 1954, **8**, 250; *Sci. Progr. Oxford*, 1968, **56**, 337.
110 G. Boudouris, *Rivista Nuovo Cim.*, 1969, **1**, 1.
111 E. H. Grant, in 'Solid State Biophysics', ed. S. J. Wyard, McGraw-Hill, New York, 1969, p. 295.

Pacault, Horau, and Marchand [112] have given a review of the experimental and theoretical work concerning the diamagnetic properties of molecules. Diamagnetic anisotropy of electron groups is reviewed by Bothner-By and Pople.[113]

Le Fèvre [114] published an extensive article on the work done regarding molecular refractivity. To Shore [115] is due a report on the scattering theory of absorption-line profiles and refractivity. Theories of hypochromism are reviewed in the articles of Rhodes and Chase [116] and of Weissbluth.[117] The papers on optical rotatory dispersion and circular dichroism have been reviewed by numerous authors [118–120] for various substances.

The theories of Rayleigh scattering of light by simple molecular substances are the subject of reports by Lax [121] and Mazur.[122] Numerous articles deal with the results obtained in the field of light scattering by solutions of macromolecules.[123–125] Recently, several articles reviewing the newer aspects of light-scattering studies have appeared. Thus, Candau [126] considers non-elastic light scattering by liquids, Yeh and Keeler [127] discuss investigations of the spectrum of scattered light as a new method of studying reaction kinetics, Damon *et al.*[128] review the interaction of light with ultrasound, and Fleury [129] provides a review of light scattering as a probe of phonons and other excitations in physical acoustics.

The Kerr effect, which, beside molecular refraction and light scattering methods, is an important source of information regarding the optical anisotropy of molecular systems, is discussed as to its origin and experimental

[112] A. Pacault, J. Horau, and A. Marchand, *Adv. Chem. Phys.*, 1961, **3**, 171.
[113] A. A. Bothner-By and J. A. Pople, *Ann. Rev. Phys. Chem.*, 1965, **16**, 34.
[114] R. J. W. Le Fèvre, *Adv. Phys. Org. Chem.*, 1965, **3**, 1.
[115] B. W. Shore, *Rev. Mod. Phys.*, 1967, **39**, 439.
[116] W. Rhodes and M. Chase, *Rev. Mod. Phys.*, 1967, **39**, 348.
[117] M. Weissbluth, *Quart. Rev. Biophysics*, 1971, **4**, 1.
[118] E. U. Condon, *Rev. Mod. Phys.*, 1937, **9**, 432; A. Moscowitz, *Adv. Chem. Phys.*, 1962, **4**, 67; I. Tinoco, jun., *ibid.*, p. 113; S. F. Mason, *Quart. Rev.*, 1963, **17**, 20; *Proc. Roy. Soc.*, 1967, **A297**, 3.
[119] W. Kohn, *Ann. Rev. Phys. Chem.*, 1958, **9**, 417; D. J. Caldwell and H. Eyring, *ibid.*, 1964, **15**, 281; *Rev. Mod. Phys.*, 1963, **35**, 577; D. W. Urry, *Ann. Rev. Phys. Chem.*, 1968, **19**, 477; H. Eyring, H. C. Liu, and D. Caldwell, *Chem. Rev.*, 1968, **68**, 525; P. M. Scopes, *Ann. Reports*, 1966, **63**, 265; *Ann. Reports (B)*, 1967, **64**, 47; 1968, **65**, 35; 1969, **66**, 34.
[120] V. Madison and J. Schellman, *Biopolymers*, 1970, **9**, 65.
[121] M. Lax, *Rev. Mod. Phys.*, 1955, **23**, 287.
[122] P. Mazur, *Adv. Chem. Phys.*, 1958, **1**, 309.
[123] G. Oster, *Progr. Biophys. and Biophysical Chem.*, 1950, **1**, 73; C. Sadron, *ibid.*, 1953, **3**, 237; A. Peterlin, *ibid.*, 1959, **9**, 175.
[124] W. Heller, *Rev. Mod. Phys.*, 1959, **31**, 1072.
[125] R. F. Schaufele, *Macromol. Rev.*, 1970, **4**, 67.
[126] S. Candau, *Ann. Phys.*, 1969, **4**, 21.
[127] Yin Yeh and R. N. Keeler, *Quart. Rev. Biophysics*, 1969, **2**, 315.
[128] R. W. Damon, W. T. Maloney, and D. H. McMahon, in 'Physical Acoustics', ed. W. P. Mason and R. N. Thurston, Academic Press, New York, 1970, Vol. 7, p. 273.
[129] P. A. Fleury, in 'Physical Acoustics', ed. W. P. Mason and R. N. Thurston, Academic Press, New York, 1970, Vol. 6, p. 2.

complications by Heller.[130] Peterlin and Stuart[131] discuss the theory of optical birefringence induced by a static electric field (Kerr effect), by alternating fields (dielectric dispersion and absorption of Kerr effect), and by the hydrodynamical field (Maxwell's effect). Le Fèvre and Le Fèvre[132] provide a review of experimental Kerr effect studies and of its applications in molecular chemistry. The Kerr effect applied to the investigation of macromolecular solutions and proteins is reviewed respectively by O'Konski[133] and Watanabe.[134] Stoylov[135] has recently published a long review on colloid electro-optics.

More recently, the Faraday effect (particularly its dispersion and absorption) has been found to provide new ways of studying the structure of molecules and, even more so, of macromolecules and biopolymers. This is because the Faraday effect is primarily due to the direct influence of the electric field on the optical polarizability of the microsystems (it thus differs essentially from natural optical activity). The theoretical and experimental results obtained in these studies are presented in the articles of Buckingham and Stephen,[136] and Schatz and McCaffery.[137] Existing laser and high-magnetic-field techniques permit studies of non-linear variations in the Faraday effect, particularly with regard to macromolecular and colloidal substances.[138, 139]

Cotton–Mouton studies are discussed extensively by Beams.[140] Measurements of the Cotton–Mouton constant in the gas phase permit determinations of the diamagnetic anisotropy of molecules; when performed in condensed phase, they yield the angular correlations between molecules.[141]

The theories and experimental studies of dielectric saturation in liquids are reviewed in an article by Kielich and Piekara;[142] measurements of the change in electric permittivity in strong magnetic fields are reviewed by Piekara and Chełkowski.[143] These two non-linear effects, hitherto investigated mostly by Piekara and co-workers, provide a sensitive method of studying weak and strong intra- and inter-molecular interactions in liquids and their solutions. Ordering in liquid crystals due to electric and magnetic fields has been reviewed by Carr.[144]

[130] W. Heller, *Rev. Mod. Phys.*, 1942, **14**, 390.
[131] A. Peterlin and H. Stuart, 'Doppelbrechung insbesondere Künstliche Doppelbrechung', Akademische Verlagsgesellschaft, Leipzig, 1943.
[132] C. G. Le Fèvre and R. J. W. Le Fèvre, *Rev. Pure Appl. Chem.*, 1955, **5**, 261.
[133] C. T. O'Konski, *Encyclopaedia Polymer Sci. Technol.*, 1968, **9**, 551.
[134] K. Yoshioka and H. Watanabe, in 'Physical Principles and Techniques of Protein Chemistry', Academic Press, New York, 1969, p. 335.
[135] S. P. Stoylov, *Adv. Colloid Interface Sci.*, 1971, **3**, 45.
[136] A. D. Buckingham and P. J. Stephens, *Ann. Rev. Phys. Chem.*, 1966, **17**, 399.
[137] P. N. Schatz and A. J. McCaffery, *Quart. Rev.*, 1968, **23**, 552.
[138] I. Tinoco and C. A. Bush, Biopolymers, Symposium No. 1, 1964, p. 235.
[139] S. Kielich, *J. Colloid Interface Sci.*, 1969, **30**, 159.
[140] J. Beams, *Rev. Mod. Phys.*, 1932, **4**, 133.
[141] A. Piekara and S. Kielich, *J. Phys. Radium*, 1957, **18**, 490; *J. Chem. Phys.*, 1958, **29**, 1297.
[142] S. Kielich and A. Piekara, *Acta Phys. Polon.*, 1959, **18**, 439.
[143] A. Piekara and A. Chełkowski, *J. Phys. Radium*, 1957, **18**, 49.
[144] E. F. Carr, *Adv. in Chem. Series*, 1967, No. 63, p. 76.

The article by Franken and Ward [145] was the earliest review of work done on the novel, non-linear optical effects of frequency doubling and tripling of laser light, many-photon absorption, and the like. These studies started in 1960 and have developed rapidly; their results have since been reviewed repeatedly.[146–150] The highly interesting results achieved in studies on self-focusing and self-trapping of laser beams have been reviewed by Akhmanov and Khokhlov;[151] also, much has been done in the field of many-photon spectroscopy.[152, 153] Bloembergen [154] is the author of an inspiring review of papers concerning Rayleigh, Raman, and Brillouin type stimulated light scattering. Also, important results have been obtained in the field of non-linear light [150, 155] and other kinds of laser light scattering, providing data on the molecular structure of the scatterers.[156–158] The coherence and statistical properties of electromagnetic fields,[159] as well as processes involving photoelectrons, have been the object of study.[160] Ultra-short pulses [161] and polarization processes caused by them in matter [162] have been analysed. The part played by molecular correlations and statistical processes in the evolution of non-linear optical processes (*e.g.* optical Kerr effect and multi-harmonic light scattering) in dense phases has been given consideration.[163, 164] Atkins [165] has recently published a review of natural and laser-beam-induced optical activity.

In the wake of earlier methods of determining electric dipole moments of polar molecules, recent developments have led to methods permitting the determination of electric moments of higher orders: quadrupoles,

[145] P. A. Franken and J. F. Ward, *Rev. Mod. Phys.*, 1963, **35**, 23.
[146] J. C. Canit, *J. Physique*, 1965, **26**, 431.
[147] R. W. Minck, R. W. Terhune, and C. C. Wang, *Appl. Optics*, 1966, **5**, 1595; R. W. Terhune and P. D. Maker, 'Non-linear Optics in Lasers', ed. A. K. Levine, Dekker, New York, 1968, Vol. 2, p. 295.
[148] J. A. Giordmaine, *Phys. Today*, 1969, **21**, 38.
[149] P. Görlich and C. Hofmann, *Feingerätetechnik*, 1970, **19**, 49, 101.
[150] S. Kielich, *Opto-electronics*, 1970, **2**, 125.
[151] S. A. Akhmanov, A. P. Sukhorukov, and R. V. Khokhlov, *Uspekhi fiz. Nauk*, 1967, **93**, 19; S. A. Akhmanov and R. V. Khokhlov, *ibid.*, 1968, **95**, 231.
[152] S. Kielich, *Proc. Phys. Soc.*, 1965, **86**, 709; *Acta Phys. Polonica*, 1966, **30**, 393.
[153] W. L. Peticolas, *Ann. Rev. Phys. Chem.*, 1967, **18**, 233.
[154] N. Bloembergen, *Amer. J. Phys.*, 1967, **35**, 989.
[155] S. Kielich, *J. Physique*, 1967, **28**, 519; *J. Colloid Interface Sci.*, 1968, **27**, 432; *Acta Phys. Polon.* 1964, **24**, 135.
[156] V. S. Starunov and I. L. Fabelinskii, *Uspekhi Fiz. Nauk*, 1969, **98**, 441.
[157] H. Z. Cummins, in 'Quantum Optics', ed. R. J. Glauber, Academic Press, New York, 1969, p. 247.
[158] D. A. Long, *Chem. in Britain*, 1971, **7**, 108; 'Essays in Structural Chemistry', ed. D. A. Long, A. J. Downs, and L. A. K. Stavley, Macmillan, London, 1971, p. 18.
[159] L. Mandel and E. Wolf, *Rev. Mod. Phys.*, 1965, **37**, 231.
[160] C. L. Mehta, in 'Progress in Optics', ed. E. Wolf, North-Holland, Amsterdam, 1970, Vol. 8, p. 375.
[161] G. L. Lamb, jun., *Rev. Mod. Phys.*, 1971, **43**, 99.
[162] I. L. Fabelinskii, *Uspekhi fiz. Nauk*, 1971, **104**, 77.
[163] S. Kielich, *IEEE J. Quantum Electronics*, 1968, **QE-4**, 744; *Fizyka Dielektryków, Soc Sci. Lettres, Poznań (Poland)*, 1972, Vol. 5, Part 2, p. 183.
[164] S. A. Akhmanov and A. S. Chirkin, *Izvest. V.U.Z., Radiofizika*, 1970, **13**, 787.
[165] P. W. Atkins, *Chem. in Britain*, 1971, **7**, 244.

octupoles, and hexadecapoles, which provide insight into the finer electric structure of molecular systems. Thus, dielectric and optical methods of measuring quadrupoles are discussed in review articles by Buckingham [166] and Krishnaji and Prakash.[167] Birnbaum [168] deals with microwave pressure broadening methods. Recently, Flygare and Benson [169] have reported electric quadrupole determinations from Zeeman effect studies and measurements of magnetic susceptibility. Stogryn and Stogryn [170] and more recently Buckingham [171] have published reviews of the methods leading to the determination of electric multipoles, including virial coefficient methods and quantum-mechanical direct calculation from known wave functions.

There exist various reviews [172] of the methods by which the electric and magnetic polarizabilities of atoms and molecules can be established. Particularly valuable are the comprehensive articles of Le Fèvre on the determination of the anisotropy of optical polarizability from molecular refraction measurements [114] and Kerr effect [132] and, recently, on the role of molecular polarizability in chemistry.[173] Bothorel [174] reviews the method of determining optical anisotropy from direct measurements of depolarized scattered laser light intensity and its particular utility in the study of n-alkanes. Stein *et al.*[175] have lately discussed the polarizability anisotropy of macromolecules.

In 1960, Buckingham [176] produced a review of those optical, electric, and magnetic properties of molecules which are in general accessible to a description in terms of tensorial quantities.[177] Modern methods of determining molecular hyperpolarizabilities have been considered by Buckingham and Orr [178] and, recently, by Kielich,[150] also with regard to the non-linear optical susceptibilities of various materials. A similar review is also due to Bergman and Kurtz.[179] Coulson [180] is the author of an excellent article of an introductory character, on the role of molecular and macroscopic symmetry in physics and chemistry.

166 A. D. Buckingham, *Quart. Rev.*, 1959, **23**, 183; *Chem. in Britain*, 1965, **1**, 54.
167 Krishnaji and V. Prakash, *Rev. Mod. Phys.*, 1966, **38**, 690.
168 G. Birnbaum, *Adv. Chem. Phys.*, 1967, **12**, 488.
169 W. H. Flygare and R. C. Benson, *Mol. Phys.*, 1971, **20**, 225.
170 D. E. Stogryn and A. P. Stogryn, *Mol. Phys.*, 1966, **11**, 371.
171 A. D. Buckingham, in 'Physical Chemistry', ed. D. Henderson, Academic Press, New York, 1970, Vol. 4, p. 349.
172 K. S. Pitzer, *Adv. Chem. Phys.*, 1959, **2**, 59; R. K. Nesbet, *Adv. Quantum Chem.*, 1967, **3**, 1; B. Bederson and E. J. Robinson, *Adv. Chem. Phys.*, 1966, **10**, 1.
173 R. J. W. Le Fèvre, *Rev. Pure Appl. Chem.*, 1970, **20**, 67.
174 P. Bothorel, *J. Colloid Interface Sci.*, 1968, **27**, 529.
175 M. Fukuda, G. L. Wilkes, and R. S. Stein, *J. Polymer Sci. Part A-2, Polymer. Phys.*, 1971, **9**, 1417.
176 A. D. Buckingham, *Ann. Reports*, 1960, **57**, 53.
177 S. Kielich, *Acta Phys. Polon.*, 1966, **30**, 851.
178 A. D. Buckingham and B. J. Orr, *Quart. Rev.*, 1967, **21**, 195.
179 J. G. Bergman and S. K. Kurtz, *Materials Sci. Eng.*, 1969/70, **5**, 235.
180 C. A. Coulson, *Chem. in Britain*, 1968, **4**, 113.

There have already appeared articles reviewing the progress achieved in the fields of intermolecular forces,[181] methods of statistical mechanics,[182–184] and the structures of gases, liquids, and their molecular mixtures,[185–187] including the critical region.[188] De Voe[189] reviews the theory of the conformations of biological macromolecules in solution, and Yomosa[190] that of charge transfer in the molecular compounds of biological systems.

Montgomery[191] should be consulted for the progress made in high magnetic fields, and Bradley[192] for a review of high-power laser-pulse generation and its applications to high-resolution spectroscopic techniques. Reviews have also been written on lasers[193] and their use in physical chemistry[194] and biophysics.[195]

An Outline of Non-linear Effects in Dielectrics.—*Constitutive Relations in Linear Media.* We shall be considering homogeneous and isotropic dielectrics, the electric, magnetic, and optical properties of which, in the absence of external fields, are described by the following three scalar quantities, characteristic of the material of which the medium consists:

ε = electric permittivity
μ = magnetic permeability
n = refractive index.

In the static case, the electric and magnetic field strength vectors being respectively $\boldsymbol{E}$ and $\boldsymbol{H}$, the linear constitutive relations are*

$$\boldsymbol{D} = \varepsilon \boldsymbol{E} \tag{1}$$

$$\boldsymbol{B} = \mu \boldsymbol{H} \tag{2}$$

where $\boldsymbol{D}$ is the electric displacement vector and $\boldsymbol{B}$ is the magnetic displacement vector.

181 E. B. Smith, *Ann. Reports*, 1966, **63**, 13; C. Schiler, *Ann. Rev. Phys. Chem.*, 1969, **20**, 191.
182 D. D. Fitts, *Ann. Rev. Phys. Chem.*, 1966, **17**, 59; I. Oppenheim, *ibid.*, 1964, **15**, 395.
183 R. Zwanzig, *Ann. Rev. Phys. Chem.*, 1965, **16**, 67.
184 J. L. Lebowitz, *Ann. Rev. Phys. Chem.*, 1968, **19**, 389.
185 D. Henderson, *Ann. Rev. Phys. Chem.*, 1964, **15**, 31; N. G. Parsonage, *Ann. Reports (A)*, 1967, **64**, 57; 1968, **65**, 33.
186 R. A. Horne, *Survey Progr. Chem.*, 1968, **4**, 2; S. N. Nagchi, *Adv. Phys.* 1970, **19**, 119.
187 J. S. Rowlinson, *Ann. Rev. Phys. Chem.*, 1958, **9**, 57; E. J. Alder, *ibid.*, 1961, **12**, 195; J. A. Barker, *ibid.*, 1963, **14**, 229; *Ann. Rev. Phys. Chem.*, 1964, **15**, 63.
188 M. Fixman *Adv. Chem. Phys.*, 1964, **6**, 175.
189 H. DeVoe, in 'Structure and Stability of Biological Macromolecules', ed. S. Timasheff, and G. Fasman, Dekker, New York, 1969, p. 1.
190 S. Yomosa, *Progr. Theor. Phys.*, 1967, **40**, 249.
191 D. B. Montgomery, *Reports Progr. Phys.*, 1963, **26**, 69.
192 D. J. Bradley, *Optica Acta*, 1968, **15**, 431; *Lab. Practice*, 1969, **18**, 538; *Physics Bull.*, 1970, **21**, 116.
193 W. J. Jones, *Quart. Rev.*, 1969, **23**, 73; M. Gronchi and A. Sona, *Rivista Nuovo Cim.*, 1970, **2**, 219.
194 A. F. Haught, *Ann. Rev. Phys. Chem.*, 1968, **19**, 343; L. Paterson, and G. Porter, *Chem. in Britain*, 1970, **6**, 246.
195 R. C. Smith, in 'Solid State Biophysics', ed. S. J. Wyard, McGraw-Hill, New York, 1969, p. 369.

* We conform to 'Quantities, Units and Symbols', edited by the Symbols Committee of the Royal Society, 1971, Cambridge.

The electric polarization induced in the medium by an electric field is given by the electric polarization vector:

$$\boldsymbol{P} = \boldsymbol{D} - \varepsilon_0 \boldsymbol{E} \tag{3}$$

likewise, the magnetic polarization induced by a magnetic field is described by the magnetization vector:

$$\boldsymbol{M} = \mu_0^{-1} \boldsymbol{B} - \boldsymbol{H} \tag{4}$$

Above, ε_0 is the electric permittivity of vacuum and, similarly, μ_0 is the magnetic permeability of vacuum.

Equations (1)—(4) lead to the relations:

$$(\varepsilon - \varepsilon_0) \boldsymbol{E} = \boldsymbol{P} \tag{5}$$

$$(\mu - \mu_0) \boldsymbol{H} = \mu_0 \boldsymbol{M} \tag{6}$$

Provided the strengths of the electric and magnetic field are not considerable and provided the fields are applied independently, the polarization vectors $\boldsymbol{P}$ and $\boldsymbol{M}$ are linear functions of the field strengths inducing them:

$$\boldsymbol{P} = \chi_e \varepsilon_0 \boldsymbol{E} \tag{7}$$

$$\boldsymbol{M} = \chi_m \boldsymbol{H} = \mu^{-1} \chi_m \boldsymbol{B} \tag{8}$$

where

$$\chi_e = \frac{\varepsilon - \varepsilon_0}{\varepsilon_0} = \varepsilon_r - 1 \tag{9}$$

$$\chi_m = \frac{\mu - \mu_0}{\mu_0} = \mu_r - 1 \tag{10}$$

denote electric and magnetic susceptibilities, where $\varepsilon_r (= \varepsilon/\varepsilon_0)$ is the relative electric permittivity, and $\mu_r (= \mu/\mu_0)$ is the relative magnetic permeability.

We measure ε and μ by means of an electric or magnetic uniform and static or slowly varying field (of oscillation frequency generally below that of the Debye dipole dispersion). In practice, the measuring field is weak and thus induces a polarization which is linear, in conformity with equations (7) and (8), so that ε and μ do not depend on the field strength but on the natural electric and magnetic structure of the medium, and on its thermodynamic state. The situation is no longer the same if the fields $\boldsymbol{E}$ and $\boldsymbol{H}$ are too strong for the linear relations (1), (2) and (7), (8) to remain valid. In the presence of sufficiently strong fields, the relation between $\boldsymbol{P}$ and $\boldsymbol{E}$ and between $\boldsymbol{M}$ and $\boldsymbol{H}$ is in general non-linear, and is no longer vectorial in nature, but tensorial. This is so because a naturally isotropic and linear dielectric, when acted on by an intense external field, becomes simultaneously non-linear and anisotropic. We shall now consider various special cases.

Non-linear Electric and Magnetic Effects. Electric saturation in an electric field. The medium is non-linearly polarized by a strong electric field $\boldsymbol{E}_p$ (polarizing field) and analysed by means of an electric field $\boldsymbol{E}_a$ (analysing or measuring field). The linear vectorial relation (1) has now to be replaced by:

$$\boldsymbol{D}(E_a, E_p) = \boldsymbol{\varepsilon}(E_p) \cdot \boldsymbol{E}_a \tag{11}$$

where $\boldsymbol{\varepsilon}(E_p)$ is the second-rank tensor of electric permittivity, in general a function of the strength of the polarizing electric field E_p.

Thus, when acted on by an electric (or indeed any) external field, naturally isotropic bodies become endowed with anisotropic properties like those possessed by naturally anisotropic ones, such as crystals. The electric permittivity tensor $\boldsymbol{\varepsilon}(E_p)$ describing this induced anisotropy can be represented in matrix form, as follows, in a Cartesian system of co-ordinate axes x, y, z:

$$\boldsymbol{\varepsilon}(E_p) = \begin{pmatrix} \varepsilon_{xx} & \varepsilon_{xy} & \varepsilon_{xz} \\ \varepsilon_{yx} & \varepsilon_{yy} & \varepsilon_{yz} \\ \varepsilon_{zx} & \varepsilon_{zy} & \varepsilon_{zz} \end{pmatrix} \tag{11a}$$

The number of its non-zero components will depend on the geometrical conditions of the experiment in which the induced anisotropy is measured, *i.e.* on the direction in which the electric permittivity is analysed (measured) with regard to that of the strong applied electric field.

In studies on the non-linear electric properties of dielectrics, one gains by introducing a tensor accounting for the variation induced in the electric permittivity by the polarizing field:

$$\Delta\boldsymbol{\varepsilon}(E_p) = \boldsymbol{\varepsilon}(E_p) - \varepsilon(0)\,\boldsymbol{U} \tag{12}$$

where $\varepsilon(0)$ is the scalar electric permittivity measured at zero polarizing field ($E_p = 0$). In equation (12), $\boldsymbol{U}$ is the unit tensor of rank 2; its components have the properties of Kronecker's symbol, *i.e.* the diagonal ones equal 1 and the non-diagonal ones are zero:

$$\boldsymbol{U} = \begin{pmatrix} U_{xx} & U_{xy} & U_{xz} \\ U_{yx} & U_{yy} & U_{yz} \\ U_{zx} & U_{zy} & U_{zz} \end{pmatrix} = \begin{pmatrix} 1 & 0 & 0 \\ 0 & 1 & 0 \\ 0 & 0 & 1 \end{pmatrix}$$

If, as is mostly the case in experiment, the analysing field E_a and polarizing field E_p are applied in the same direction, the electric permittivity variation tensor (12) possesses but one non-zero component, in the field direction, usually denoted by $\Delta\varepsilon_{\parallel}(E_p)$. This particular case is referred to as the effect of electric saturation in an electric field. In molecular liquids $\Delta\varepsilon_{\parallel}(E_p)$ is in general a quadratic function of the field strength E_p, as proved repeatedly by Piekara and his co-workers.[142, 196] Lately, Davies[197] has published a review on aspects of recent dielectric studies, particularly dielectric saturation in liquids and molecular solutions, as well as in solutions of macromolecules where complete dielectric saturation has been observed.

[196] A. Piekara, *J. Chem. Phys.*, 1962, **36**, 2145.
[197] M. Davies, *Ann. Reports* (*A*), 1970, **67**, 65.

In a measurement with the fields E_a and E_p mutually perpendicular, such a measurement of $\Delta\varepsilon_{\perp}(E_p)$ together with a measurement of $\Delta\varepsilon_{\parallel}(E_p)$ would permit the determination of the electric anisotropy[198] $\Delta\varepsilon_{\parallel}(E_p) - \Delta\varepsilon_{\perp}(E_p)$ induced in the naturally isotropic dielectric as a quantity that is in general independent of electrostriction and the electrocaloric effect.[142]

Electric saturation in a magnetic field. Immersion of the isotropic dielectric in a very intense magnetic field leads to a non-linear dependence of the vector $\boldsymbol{D}$ on the magnetic field strength H_p:

$$\boldsymbol{D}(E_a, H_p) = \boldsymbol{\varepsilon}(H_p) \cdot \boldsymbol{E}_a \tag{13}$$

where the electric permittivity tensor $\boldsymbol{\varepsilon}(H_p)$ is now a non-linear function of the high magnetic field.

Similarly to (12), we obtain with regard to equation (13) the following magnetic-field-induced variation of the electric permittivity tensor:

$$\Delta\boldsymbol{\varepsilon}(H_p) = \boldsymbol{\varepsilon}(H_p) - \varepsilon(0)\,\boldsymbol{U} \tag{14}$$

Piekara and Chełkowski[143] have shown that, in molecular liquids, the change in electric permittivity due to a magnetic field is a quadratic function of the field strength. The phenomenon is well observable in liquid crystals, as stated by numerous authors.[144]

Magnetic saturation in a magnetic field. With $\boldsymbol{H}_a$ denoting the weak analysing magnetic field, one has in the presence of a strong magnetic field H_p, in place of equation (2),

$$\boldsymbol{B}(H_a, H_p) = \boldsymbol{\mu}(H_p) \cdot \boldsymbol{H}_a \tag{15}$$

where $\boldsymbol{\mu}(H_p)$ is the second-rank tensor of magnetic permeability.

From equations (1) and (15), the change in magnetic permeability tensor due to a strong magnetic field is:

$$\Delta\boldsymbol{\mu}(H_p) = \boldsymbol{\mu}(H_p) - \mu(0)\,\boldsymbol{U} \tag{16}$$

$\mu(0)$ denoting the magnetic permeability measured at $H_p = 0$.

The effect of magnetic saturation in a magnetic field, though considered theoretically,[142] has as yet not been observed in diamagnetics.

Magnetic saturation in an electric field. The inverse of equation (14) takes place when the magnetic permeability is analysed with a magnetic field $\boldsymbol{H}_a$ in the presence of a strong electric field E_p. The variation in magnetic permeability, a tensorial quantity in general, is now of the form:

$$\Delta\boldsymbol{\mu}(E_p) = \boldsymbol{\mu}(E_p) - \mu(0)\,\boldsymbol{U} \tag{17}$$

The effect consisting in a change in magnetic permeability under the influence of an electric field still awaits detection, although considered theoretically for paramagnetics[9] by Van Vleck, as well as for diamagnetics.[142]

Non-linear Electro- and Magneto-optical Effects. Optical saturation in an electric field. When an optically isotropic dielectric is placed in a very

[198] S. Kielich, *Acta Phys. Polon.*, 1967, **32**, 405.

strong electric field, its refractive index n undergoes a variation and in general takes different values for different directions with respect to the strong static electric field E_p acting on the dielectric. Since, by Maxwell's well-known relation, the index is related to the electric permittivity and magnetic permeability, the tensors $\boldsymbol{\varepsilon}$ and $\boldsymbol{\mu}$ of equations (12) and (17) have now to be dealt with as being dependent on the oscillation frequency ω of the analysing light beam, with electric vector $\boldsymbol{E}_a^{\omega} = \boldsymbol{E}_a \exp(-i\omega t)$

$$\Delta\boldsymbol{\varepsilon}^{\omega}(E_p) = \boldsymbol{\varepsilon}^{\omega}(E_p) - n^2(0)\,\boldsymbol{U} \qquad (12a)$$

$$\Delta\boldsymbol{\mu}^{\omega}(E_p) = \boldsymbol{\mu}^{\omega}(E_p) - n^2(0)\,\boldsymbol{U} \qquad (17a)$$

where $n(0)$ is the refractive index in the absence of a static electric field ($E_p = 0$).

Let $n_{\|}(E_p) \approx [\varepsilon_{\|}(E_p)/\varepsilon_0]^{\frac{1}{2}}$ denote, as usual, the refractive index analysed at light vector oscillations $\boldsymbol{E}_a$ parallel to $\boldsymbol{E}_p$, and $n_{\perp}(E_p) \approx [\varepsilon_{\perp}(E_p)/\varepsilon_0]^{\frac{1}{2}}$ the index analysed with $\boldsymbol{E}_a$ perpendicular to $\boldsymbol{E}_p$. In experiments, one can measure the variations $\Delta n_{\|}(E_p)$ and $\Delta n_{\perp}(E_p)$ independently, as '*retard absolu*',[13, 14] or their difference

$$\Delta n_{\|}(E_p) - \Delta n_{\perp}(E_p) = \lambda K_{\lambda} E_p^2 \qquad (18)$$

as the optical birefringence induced in the naturally isotropic medium by the applied static electric field, *i.e.* the Kerr effect (λ is the light wavelength).

Since the early work of Kerr[199] in 1875, the results of almost all authors and particularly those of the Le Fèvre centre[114, 132] point to a value of K_{λ} independent of the electric field strength in gases and liquids, and characteristic of the substance under investigation (Kerr constant). O'Konski and co-workers[133] have shown that in solutions of macromolecules Kerr's quadratic law does not hold, and electric saturation of optical birefringence is generally observed.

The classical fundamentals of the theory of the Kerr effect are due to Voigt[200] in terms of electron theory of the atom as an anharmonic oscillator and to Langevin[201] in terms of statistical optical reorientation of anisotropic molecules in a static electric field. Buckingham[166] proposed a theory and method of measurement of the optical birefringence induced by the gradient of a static electric field permitting the determination of electric quadrupole moments.

In optically active substances, according to equations (12a) and (17a), the non-diagonal components of the refractive index tensor differ from zero and describe the variation in optical rotation due to a static electric field, an effect studied by Tinoco.[138]

Optical saturation in a magnetic field. A static magnetic field H_p can also cause optical changes in naturally isotropic bodies; by analogy with

[199] J. Kerr, *Phil. Mag.*, 1875, **50**, 337, 446.
[200] W. Voigt, *Ann. Phys.*, 1899, **67**, 345; 1901, **4**, 197.
[201] P. Langevin, *Le Radium*, 1910, **7**, 249.

equations (14) and (16), we express them in the form:

$$\Delta\varepsilon^{\omega}(H_P) = \varepsilon^{\omega}(H_P) - n^2(0)\,U \qquad (14a)$$

$$\Delta\mu^{\omega}(H_P) = \mu^{\omega}(H_P) - n^2(0)\,U \qquad (16a)$$

leading to the following magnetically induced optical birefringence:

$$\Delta n_{\parallel}(H_P) - \Delta n_{\perp}(H_P) = \lambda C_{\lambda} H_P^2 \qquad (19)$$

well known since 1905 as the Cotton–Mouton effect [202] in gases and molecular liquids.[140, 141] Detected in 1902 in colloidal solutions by Majorana,[203] it is referred to there as the Majorana effect.

It is essential at this point to stress the difference between the action of a static electric field (which is a polar vector) and that of a magnetic field (obviously, an axial vector). In naturally isotropic optically inactive media, a static electric field gives rise to an optical birefringence which is, at the least, a quadratic function of the electric field strength E_P in accordance with Kerr's law, equation (18). No effect linear in E_P occurs here (it occurs only in crystals without a centre of symmetry, as Pockels' linear electro-optical effect). In contrast, a static magnetic field H_P induces in a naturally isotropic medium not only a quadratic Cotton–Mouton effect [equation (19)] but, moreover, a linear magneto-optical effect, the effect discovered by Faraday [204] in 1845 and commonly referred to as magnetic optical rotation. The latter is due to the non-diagonal components of the tensors (14a) and (16a), which lead to the well-known relation:

$$n_{-}(H_P) - n_{+}(H_P) = \frac{\lambda}{\pi} V_{\lambda} H_P \qquad (20)$$

where n_{-} and n_{+} are the refractive indices for left and right circularly polarized light, and V_{λ} is the Verdet constant, specific to the medium and a function of the light wavelength λ.

Like the Cotton–Mouton effect (19), the Faraday effect (20) can in general depend on higher powers of the magnetic field strength.[139, 205]

Electric saturation in an optical field. Langevin, when writing his classical paper [201] on reorientation of molecules in a static electric field, considered the possibility of such reorientation by the electric field of a light wave. The light sources then available were, however, so weak that he was led to neglect the insignificant optical birefringence induced by them. The theory of optical birefringence in the electric field of light of high intensity was developed much later by Buckingham.[206] Piekara and Kielich,[142, 207] starting from these results, showed that a sufficiently intense light beam could cause optical saturation of electric permittivity.

[202] A. Cotton and M. Mouton, *Compt. rend.*, 1905, **141**, 317, 349; 1907, **145**, 229, 870.
[203] Q. Majorana, *Compt. rend.*, 1902, **135**, 159, 235.
[204] M. Faraday, Diary, 1845, Sept. 13; *Phil. Mag.*, 1846, **28**, 249; 1846, **29**, 163.
[205] C. Graham and R. E. Raab, *Proc. Phys. Soc.*, 1967, **90**, 417.
[206] A. D. Buckingham, *Proc. Phys. Soc. (B)*, 1956, **69**, 344.
[207] A. Piekara and S. Kielich, *Arch. Sci.* (fasc. spéc., 7th Colloque Ampère), 1958, **11**, 304; 'Third Conference on Quantum Electronics, Paris, 1963', Vol. 2, p. 1603.

In general, the change in electric permittivity is anisotropic, like equation (12); we have:

$$\Delta\boldsymbol{\varepsilon}(I_{\rm p}) = \boldsymbol{\varepsilon}(I_{\rm p}) - \varepsilon(0)\,\boldsymbol{U} \tag{21}$$

where $I_{\rm p}$ is the light intensity, which has to be large enough to cause a non-linear variation.

From equation (21), the optically induced electric anisotropy results in the form:[208]

$$\Delta\varepsilon_{\parallel}(I_{\rm p}) - \Delta\varepsilon_{\perp}(I_{\rm p}) = B_{\rm e}\,I_{\rm p} \tag{22}$$

where $B_{\rm e}$ is an electric anisotropy constant similar to the Kerr constant, K_λ.

By applying ingeniously the now available electronic and laser techniques, it is possible to choose experimental conditions in such a way that equation (22) can be treated as an inverse Kerr effect (18), since in the Kerr effect the static electric field induces an optical birefringence which we analyse by means of a weak light beam, whereas in the effect of equation (22) a strong laser beam induces an electric anisotropy analysable by means of a slowly varying alternating electric field (at radio- or microwave frequencies).

Magnetic saturation in an optical field. Intense light can magnetize isotropic bodies [209] and in general affect their magnetic properties.[142] In fact, as in the case of equation (17), there will be a change in magnetic permeability induced by intense light:

$$\Delta\boldsymbol{\mu}(I_{\rm p}) = \boldsymbol{\mu}(I_{\rm p}) - \mu(0)\,\boldsymbol{U} \tag{23}$$

leading to the optically induced magnetic anisotropy:[198]

$$\Delta\mu_{\parallel}(I_{\rm p}) - \Delta\mu_{\perp}(I_{\rm p}) = B_{\rm m}\,I_{\rm p} \tag{24}$$

where the magnetic anisotropy constant $B_{\rm m}$ behaves similarly to the Cotton–Mouton constant C_λ. Here, too, the effect consisting in an optically induced magnetic anisotropy can be treated as the inverse of the Cotton–Mouton effect of equation (19); however, its detection is much more difficult technically.

Pershan *et al.*[209] observed a new effect, referred to as the inverse Faraday effect, consisting in the magnetization of a non-absorbing medium by a circularly polarized laser beam:

$$M = \frac{n\lambda}{2\pi^2} V_\lambda (I_+ - I_-) \tag{25}$$

with I_+ and I_- denoting the intensities of right and left circularly polarized laser light.

Non-linear Optical Effects.—Usual light (thus, a weak analysing light beam), when traversing a transparent body, does not affect the latter's optical properties because the electric and magnetic polarizations induced are linear functions [equations (7) and (8)] of the vectors $\boldsymbol{E}_{\rm a}^{\omega}$ and $\boldsymbol{H}_{\rm a}^{\omega}$,

[208] S. Kielich, *Physica*, 1967, **34**, 365, 586.
[209] P. S. Pershan, J. P. Van Der Ziel, and L. D. Malmstrom, *Phys. Rev.*, 1966, **143**, 574.

oscillating at frequency ω:

$$P_a^\omega = \chi_e^\omega \varepsilon_0 E_a^\omega \quad (7a)$$

$$M_a^\omega = \chi_m^\omega H_a \quad (8a)$$

The electric and magnetic susceptibilities χ_e^ω and χ_m^ω depend on the light oscillation frequency.

Thus, conventional electromagnetic radiation propagates in a medium in accordance with the laws of classical, linear optics. This implies a lack of any interaction whatsoever between simultaneously propagating waves: they propagate independently of one another, each interacting with the medium in a manner as if the other waves were not present.

The reason for this state of affairs in linear optics resides in the very low intensity of classical radiation sources. The situation undergoes a radical change when, as is the case with giant lasers, the electric field of the beam has an intensity sufficient for causing non-linear changes in the optical properties of the medium. In non-linear optics, the refractive index and other optical properties vary with the light intensity. We shall now consider some of the more important non-linear optical effects, providing valuable information concerning the properties of atoms and molecules as well as the structure of isotropic media.

Optical saturation in an optical field. Intense light, when traversing an isotropic medium, induces in it an optical non-linearity apparent as a dependence of the refractive index on the light intensity I. The dynamical electric permittivity and magnetic permeability contributing to the non-linear change in index are, with regard to equations (21) and (23):

$$\Delta\varepsilon^\omega(I) = \varepsilon^\omega(I) - n^2(0)\,U \quad (21a)$$

$$\Delta\mu^\omega(I) = \mu^\omega(I) - n^2(0)\,U \quad (23a)$$

We hence obtain the following self-induced non-linear change in refractive index:

$$\Delta n(I) = n_1 I + n_2 I^2 + \ldots \quad (26)$$

where n_1 and n_2 are constants depending on the molecular structure of the medium and on the frequency of oscillation ω of the laser light beam.

To the non-linear change in refractive index (26) is due the self-trapping (and self-focusing) of laser beams predicted by Askarian,[210] observed by Hercher,[211] and explained by Townes *et al.*[212] as well as many others.[151, 163, 213] Before laser techniques became available, Piekara and Kielich[142, 207] published a theoretical discussion of optical saturation (26) in gases and liquids resulting from non-linear electron polarizability, optical reorientation,[206] and angular correlations of molecules, as well as electrostriction and the electrocaloric process.

[210] G. A. Askaryan, *Zhur. eksp. teor. Fiz.*, 1962, **42**, 1567.

[211] (*a*) M. Hercher, *J. Opt. Soc. Amer.*, 1964, **54**, 563; (*b*) G. N. Steinberg, *Phys. Rev.*, 1971, **A4**, 1182.

[212] (*a*) R. Y. Chiao, E. Garmire, and C. H. Townes, *Phys. Rev. Letters*, 1964, **13**, 479; (*b*) A. Piekara, *Appl. Phys. Letters*, 1968, **13**, 225; *Japan Appl. Phys.*, 1971, **10**, 266.

[213] E. L. Kerr ,*Phys. Rev.*, 1971, **A4**, 1195 and references therein.

The optical Kerr effect predicted by Buckingham [206] in 1956 was detected in liquids as late as 1964 by Mayer and Gires,[214] who resorted to the beam of a ruby laser. Like the static Kerr factor (18), the optical Kerr effect is given as:

$$\Delta n_{\parallel}^{\lambda}(I_L) - \Delta n_{\perp}(I_L) = \lambda_a B(\lambda_a, \lambda_L) I_L \tag{27}$$

where $B(\lambda_a, \lambda_L)$ is an optical birefringence constant dependent on the analysing light wavelength, λ_a, as well as on the laser wavelength, λ_L.

Maker *et al.*[215] observed self-induced rotation of the polarization ellipse of elliptically polarized laser light:

$$n_-(I_L) - n_+(I_L) = \lambda_L R(\lambda_L)(I_L^- - I_L^+) \tag{28}$$

$R(\lambda_L)$ being a constant characterizing the self-induced optical rotation.

A recent publication [216] of Vlasov and Zaitsev reports observations of the non-linear optical activity predicted by Kielich [217] from molecular-statistical theory and by Atkins [165] from quantum scattering theory.

The non-linear processes discussed above are listed in Table 1, which was established before laser techniques came into use.[142, 207]

Table 1 *Non-linear electric, magnetic, and optical effects*

Analysing (low) field	*Polarizing (high) field* Electric E_p	Magnetic H_p	Optical E_p^ω
Electric E_a	$\Delta\varepsilon(E_p)$	$\Delta\varepsilon(H_p)$	$\Delta\varepsilon(I_p)$
Magnetic H_a	$\Delta\mu(E_p)$	$\Delta\mu(H_p)$	$\Delta\mu(I_p)$
Optical E_a^ω	$\Delta n(E_p)$	$\Delta n(H_p)$	$\Delta n(I_p)$

Generation of harmonic optical frequencies. In 1961, Franken *et al.*[218] performed an experiment of high aesthetic value in which the red beam of a ruby laser ($\lambda = 694.3$ nm), focused on to the surface of a quartz plate, gave rise to an emerging beam which contained a weak component of doubled frequency with respect to the incident beam (u.v., $\lambda = 347.15$ nm). Since the detection of this earliest non-linear optical effect, second and third harmonics have been observed in various materials in both transmitted and reflected light.[145, 147, 150, 179]

The source of this generation of higher optical harmonics has to be sought in the circumstance that an intense incident beam, of oscillation frequency ω, induces electric polarizations of higher orders in the medium; thus, in the case of frequency-doubling 2ω:

$$\boldsymbol{P}_{2\omega} = \varepsilon_0 \boldsymbol{\chi}_e^{2\omega} : \boldsymbol{E}^\omega \boldsymbol{E}^\omega \tag{29}$$

[214] (*a*) G. Mayer and F. Gires, *Compt. rend.*, 1964, **258**, 2039; (*b*) M. Paillette, *Ann. Phys.*, 1969, **4**, 671; (*c*) F. B. Martin and J. R. Lalanne, *Phys. Rev.*, 1971, **A4**, 1275.

[215] P. D. Maker, R. W. Terhune, and C. M. Savage, *Phys. Rev. Letters*, 1964, **12**, 507; P. D. Maker and R. W. Terhune, *Phys. Rev.*, 1965, **A137**, 801.

[216] D. V. Vlasov and V. P. Zaitsev, *JETP Letters*, 1971, **14**, 112.

[217] S. Kielich, *Phys. Letters*, 1967, **25A**, 517; *Acta Phys. Polon.*, 1969, **35**, 861; *Opto-electronics*, 1969, **1**, 75; *Optics Comm.*, 1969, **1**, 129.

[218] P. A. Franken, A. E. Hill, C. W. Peters, and G. Weinreich, *Phys. Rev. Letters*, 1961, **7**, 118.

and in that of triple-frequency generation 3ω:

$$P_{3\omega} = \varepsilon_0 \chi_e^{3\omega} \vdots E^{\omega} E^{\omega} E^{\omega} \tag{30}$$

and so forth. Above, $\chi_e^{2\omega}$ is a tensor of rank 3 describing the non-linear electric susceptibility of second order at frequency 2ω, whereas $\chi_e^{3\omega}$ is a fourth-rank tensor describing the third-order non-linear electric susceptibility at frequency 3ω.

Since $\chi_e^{2\omega}$ is a third-rank tensor, it possesses non-zero components only in the case of molecules which have no centre of symmetry; the preceding process of second harmonic generation will not interest us any further here. It will be remembered, however, that a static electric field acts in a manner to lower the symmetry of bodies and can thus endow an isotropic body, naturally possessing a centre of symmetry, with the symmetry of an axially symmetric body without a centre of symmetry. In this way, a naturally isotropic body without centre of symmetry, during the time it is acted on by a static electric field E, has the ability to cause a doubling of the oscillation frequency of an incident laser beam, since now we have in equation (29):

$$\chi_e^{2\omega}(E) = \chi_e^{2\omega+0} \cdot E \tag{29a}$$

where $\chi^{2\omega+0}$ is a fourth-rank tensor and thus possesses non-zero components even in isotropic bodies. Such static-electric-field-induced second-harmonic generation has been observed by Terhune *et al.*[219] in calcite, by Mayer [220] in gases and molecular liquids, and lately in inert gases by Finn and Ward.[221] Kielich [222] has shown that such studies, when performed in molecular substances, provide data not only on the non-linear second- and third-order polarizabilities of the molecules but also on their quadrupole and octupole moments.

Non-linear light scattering. Rocard [223] considered theoretically the possibility of observing changes in Rayleigh scattering of light under the influence of a strong static electric or magnetic field. The variation in intensity of light scattered at frequency ω is a quadratic function of the electric field strength:

$$\Delta I_R^{\omega} = S_2^{\omega} E^2 \tag{31}$$

S_2^{ω} being a constant specific to the molecular structure of the scattering medium and a function of the light oscillation frequency ω. A similar expression holds for the influence of a static magnetic field or of strong laser light.[155] As yet, it has not proved possible to observe the variations

[219] R. W. Terhune, P. D. Maker, and C. M. Savage, *Phys. Rev. Letters*, 1962, **8**, 404.

[220] G. Mayer, *Compt. rend.*, 1968, **267**, *B*, 54; G. Hauchecorne, F. Kerhervé, and G. Mayer, *J. Physique*, 1971, **32**, 47.

[221] R. S. Finn and J. F. Ward, *Phys. Rev. Letters*, 1971, **26**, 285.

[222] S. Kielich, *Chem. Phys. Letters*, 1968, **2**, 596; *Acta Phys. Polon.*, 1970, **37A**, 205; *Opto-electronics*, 1970, **2**, 5.

[223] (*a*) Y. Rocard, *Ann. Phys.*, 1928, **10**, 472; (*b*) A. L. Andrews and A. D. Buckingham, *Mol. Phys.*, 1960, **3**, 183; (*c*) S. Kielich, *Acta Phys. Polon.*, 1963, **23**, 321, 819.

described by equation (31) in molecular substances,[224] but only in macromolecular ones, as done by Wippler and Benoit and others.[225]

Harmonic light scattering. Beside elastic scattering at the fundamental frequency ω, laser light of sufficiently high intensity I traversing a molecular medium gives rise to scattering processes of higher orders, at double frequency 2ω, triple frequency 3ω, and so forth:[155, 163]

$$I_S = S^{\omega} I + S^{2\omega} I^2 + S^{3\omega} I^3 + \ldots \qquad (32)$$

where the constant S^{ω} defines linear Rayleigh scattering at the fundamental frequency ω. The constant $S^{2\omega}$ defines non-linear scattering of light at doubled frequency, as observed by Terhune, Maker, and Savage[226] in substances consisting of molecules without a centre of symmetry and, quite recently, by Lalanne, Martin, and Kielich[227] in liquids consisting of centrosymmetrical molecules. Second-harmonic scattering is highly sensitive to molecular symmetry and permits the determination of non-linear polarizabilities and correlations of molecules.[163]

2 Electric and Magnetic Properties of Microsystems

Multipolar Properties of Molecules.—Matter consists of atoms and molecules or their ions, as well as of macromolecules and colloid particles, *i.e.* quite generally of microsystems. These are dynamical systems having an electromagnetic structure, which we are in some cases able to describe in terms of classical or, more strictly, quantum methods. Since, for our present aims, the quantum-mechanical structure of microsystems is not essential, we shall treat the latter classically, as electrostatic systems presenting a distribution of negative and positive electric charges.

Electric (Natural) Multipoles. Consider a system of electric point charges q_s, with radius vectors $\boldsymbol{r}_s$, and assume the electric properties of the system of charges as a whole to be observed at a point P, distant by R from the origin O. The vector

$$\boldsymbol{R}_s = \boldsymbol{R} - \boldsymbol{r}_s \qquad (33)$$

connects the point of the s-th electric charge of the microsystem with the point of observation P (Figure 1).

224 J. R. Lalanne and P. Bothorel, *Compt. rend.*, 1966, **263**, 693; J. R. Lalanne, *ibid.*, 1967, **265**, 1181.

225 (*a*) C. Whippler and H. Benoit, *Macromol. Chem.*, 1954, **13**, 7; (*b*) B. R. Jennings and H. G. Jerrard, *J. Chem. Phys.*, 1966, **44**, 1291; J. Schweitzer and B. R. Jennings, *J. Colloid Interface Sci.*, 1971, **37**, 443; (*c*) S. Stoylov, *Coll. Czech. Chem. Comm.*, 1966, **31**, 2866, 3052; (*d*) J. C. Ravey, *J. Chim. phys,*, 1970, **67**, 1787; J. C. Ravey and S. Premilat, *ibid.*, 1970, **67**, 147, 157; S. Premilat and P. Mazeron, *ibid.*, 1971, **68**, 954.

226 (*a*) R. W. Terhune, P. D. Maker, and C. M. Savage, *Phys. Rev. Letters*, 1965, **14**, 681; P. D. Maker, *Phys. Rev.*, 1970, **1A**, 923; (*b*) J. F. Verdieck, S. H. Peterson, C. M. Savage, and P. D. Maker, *Chem. Phys. Letters*, 1970, **7**, 219; S. H. Peterson, Ph.D. Thesis, University of Michigan, 1971.

227 (*a*) S. Kielich, J. R. Lalanne, and F. B. Martin, *Phys. Rev. Letters*, 1971, **26**, 1295; (*b*) S. Kielich, *Acta Phys. Polon.*, 1968, **33**, 89, 141.

When searching for a quantitative description of the electric properties of the microsystem, we shall proceed on the assumption that

$$|\boldsymbol{r}_s| \ll |\boldsymbol{R}_s|, \quad |\boldsymbol{r}_s| \ll |\boldsymbol{R}| \qquad (33a)$$

i.e. we shall be dealing with an electric system the linear dimensions of which are very small compared with the distance at which we study its properties. This assumption is obviously fulfilled with regard to atoms, molecules, and macromolecules.

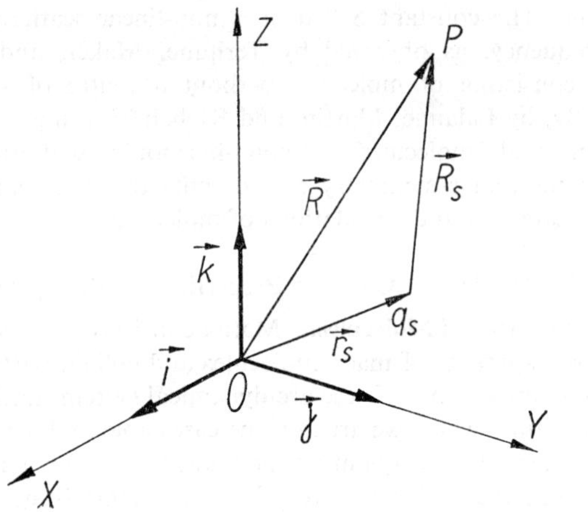

Figure 1 *System of electric charges.* $\vec{\mathbf{r}}_s$ *is the direction vector of the s-th charge* q_s. $\vec{\mathbf{R}}$ *is the observational vector* $\vec{\mathbf{R}}_s$

Within such a distribution of charges, which in general can be rather complicated, we can discern electric sub-systems of specifically defined configurations of charges. Electric sub-systems of this kind are jointly referred to as electric multipoles, and the quantities representing their properties are the multipole moments. We shall now consider these multipoles one by one.

Electric monopole (*net charge*). By definition, the total charge of the system is:

$$q = \sum_s q_s \qquad (34)$$

It can be positive or negative, or zero if the system is neutral (negative and positive charges cancelling out): $\sum_s q_s = 0$. In the last case, the charge distribution behaves as if all the charges were in one point. In an electrically neutral system the 'centres of mass' of the positive and negative charges coincide in a point which is the centre of the symmetric charge distribution.

Electric dipole. An electric system consisting of two electric, numerically equal, charges of opposite sign separated by $\boldsymbol{r}$ is termed an electric dipole.

Quantitatively, it is characterized by an electric moment, defined by the vector:

$$\boldsymbol{p} = q\boldsymbol{r} = q(\boldsymbol{r}_+ - \boldsymbol{r}_-) \tag{35}$$

$\boldsymbol{r} = \boldsymbol{r}_+ - \boldsymbol{r}_-$ being the vector connecting the points of the negative and positive charge (Figure 2).

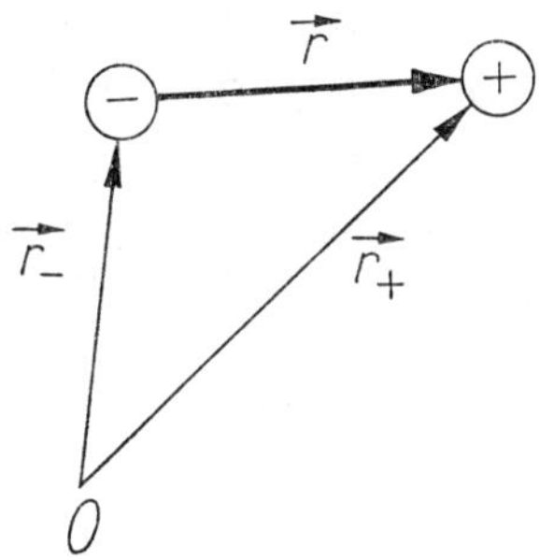

Figure 2 *An electric dipole*

In dealing with a system of many positive and negative charges, we introduce vectors of 'centres of mass of charges', *e.g.* the centre of charge of an electron shell

$$\boldsymbol{r}_- = \frac{1}{q} \sum_s q_s \boldsymbol{r}_s^- \tag{35a}$$

and of the nuclei

$$\boldsymbol{r}_+ = \frac{1}{q} \sum_s q_s \boldsymbol{r}_s^+ \tag{35b}$$

On inserting these expressions into equation (35), we obtain the general definition of the moment of the electric dipole for an arbitrary micro-system:

$$\boldsymbol{p} = \sum_s q_s \boldsymbol{r}_s \tag{36}$$

since $\boldsymbol{r}_s = \boldsymbol{r}_s^+ - \boldsymbol{r}_s^-$.

With unit vectors $\boldsymbol{i}, \boldsymbol{j}, \boldsymbol{k}$ in the directions of Cartesian axes x, y, z rigidly attached to the molecule, the permanent electric dipole moment of the latter can be expressed in terms of its components:

$$\boldsymbol{p} = p_x \boldsymbol{i} + p_y \boldsymbol{j} + p_z \boldsymbol{k} \tag{36a}$$

Electric quadrupole. Whereas the monopole and dipole characterize respectively a point and linear charge distribution, surface distributions are described by electric quadrupoles. A quadrupole is conveniently conceived of as consisting of two antiparallel dipoles, mutually displaced by some small distance (Figure 3). Just as a straight line is defined by a single vector $\boldsymbol{r}$, a plane is defined by two vectors $\boldsymbol{r}_1$ and $\boldsymbol{r}_2$; hence, the moment of an electric quadrupole is a bi-vector $q\boldsymbol{r}_1 \boldsymbol{r}_2$. In molecular physics, the tensor of an electric quadrupole moment is defined as follows:[171, 228]

$$\boldsymbol{\Theta} = \frac{1}{2}\sum_s q_s(3\boldsymbol{r}_{s1}\boldsymbol{r}_{s2} - r_s^2\,\boldsymbol{U}) \tag{37}$$

vanishing for a spherical distribution of electric charge. The second-rank tensor of the quadrupole moment is symmetric; in matrix notation:

$$\boldsymbol{\Theta} = \begin{pmatrix} \Theta_{xx} & \Theta_{xy} & \Theta_{xz} \\ \Theta_{xy} & \Theta_{yy} & \Theta_{yz} \\ \Theta_{xz} & \Theta_{yz} & \Theta_{zz} \end{pmatrix} \tag{37a}$$

it has three diagonal components Θ_{xx}, Θ_{yy}, Θ_{zz} and three non-diagonal components $\Theta_{xy} = \Theta_{yx}$, $\Theta_{yz} = \Theta_{zy}$, $\Theta_{zx} = \Theta_{xz}$, (symmetry reduces the

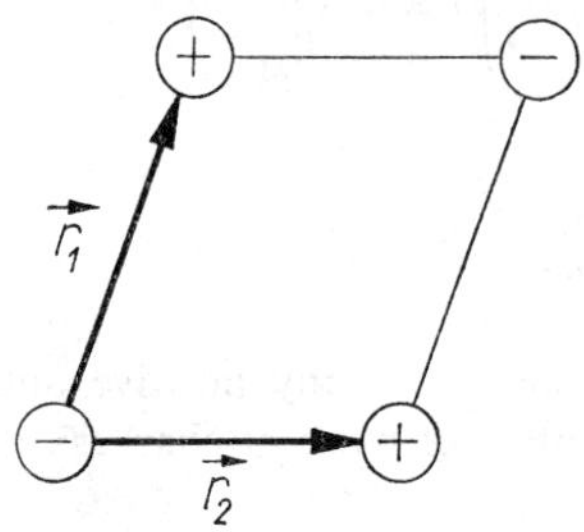

Figure 3 *An electric quadrupole*

six non-diagonal components to three mutually independent ones). Since the symmetric unit tensor is:

$$\boldsymbol{U} = \boldsymbol{i}\boldsymbol{i} + \boldsymbol{j}\boldsymbol{j} + \boldsymbol{k}\boldsymbol{k} \tag{37b}$$

we have the tensor trace:

$$\boldsymbol{\Theta} : \boldsymbol{U} = \Theta_{xx} + \Theta_{yy} + \Theta_{zz} = 0 \tag{37c}$$

leading to the relation:

$$\Theta_{zz} = -(\Theta_{xx} + \Theta_{yy}) \tag{37d}$$

which reduces the number of independent quadrupole tensor elements to five, although the non-zero elements are nine in all. A further reduction in number of independent elements of the tensor (37b) is effected by the elements of point-group symmetries listed in Table 2. Thus for the point groups C_{2v}, D_2, and D_{2h}, the tensor (37) has only three diagonal components, two of which are mutually independent:

$$\Theta_{xx},\ \Theta_{yy} \text{ and } \Theta_{zz} = -(\Theta_{xx} + \Theta_{yy})$$

a relation which we can write as follows:

$$\boldsymbol{\Theta} = \tfrac{1}{2}(\Theta_{xx} - \Theta_{yy})(\boldsymbol{i}\boldsymbol{i} - \boldsymbol{j}\boldsymbol{j}) + \tfrac{1}{2}\Theta_{zz}(3\boldsymbol{k}\boldsymbol{k} - \boldsymbol{U}) \tag{37e}$$

Now this relation in the case of axially symmetric molecules, when

[228] S. Kielich, *Physica*, 1965, **31**, 444.

$\Theta_{xx} = \Theta_{yy}$ and $\Theta = \Theta_{zz} = -2\Theta_{xx} = -2\Theta_{yy}$, reduces to the well-known result:[166]

$$\boldsymbol{\Theta} = \tfrac{1}{2}\Theta(3\boldsymbol{k}\,\boldsymbol{k} - \boldsymbol{U}) \tag{37f}$$

Electric octupole (*Figure* 4). Three-dimensional space is defined in general by three distinct vectors $\boldsymbol{r}_1, \boldsymbol{r}_2, \boldsymbol{r}_3$ and the electric octupole moment is consequently given by a tri-vector of the form $q\boldsymbol{r}_1\,\boldsymbol{r}_2\,\boldsymbol{r}_3$. The permanent

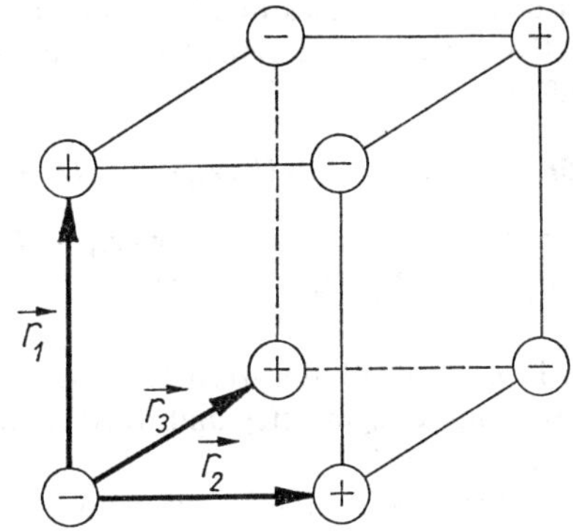

Figure 4 *An electric octupole*

moment of an octupole is a (symmetric) tensor of rank 3 defined as follows:[228]

$$\boldsymbol{\Omega} = \frac{1}{2}\sum_s q_s\{5\boldsymbol{r}_{s1}\,\boldsymbol{r}_{s2}\,\boldsymbol{r}_{s3} - r_s^2(\boldsymbol{r}_{s1}\,\boldsymbol{U}_{23} + \boldsymbol{r}_{s2}\,\boldsymbol{U}_{31} + \boldsymbol{r}_{s3}\,\boldsymbol{U}_{12})\} \tag{38}$$

Thus defined, the tensor is symmetric, and its traces vanish:

$$\boldsymbol{\Omega} : \boldsymbol{U}_{12} = \boldsymbol{\Omega} : \boldsymbol{U}_{23} = \boldsymbol{\Omega} : \boldsymbol{U}_{31} = 0 \tag{38a}$$

so that in general only seven of its 27 non-zero components are mutually independent. For the point group C_{2v} we have seven non-zero components, only two of which are independent, justifying the expression:

$$\begin{aligned}\boldsymbol{\Omega} = \tfrac{1}{2}(\Omega_{xxz} - \Omega_{yyz})\{&(\boldsymbol{i}_1\,\boldsymbol{i}_2 - \boldsymbol{j}_1\boldsymbol{j}_2)\,\boldsymbol{k}_3 \\ &+ (\boldsymbol{i}_2\,\boldsymbol{i}_3 - \boldsymbol{j}_2\boldsymbol{j}_3)\,\boldsymbol{k}_1 + (\boldsymbol{i}_3\,\boldsymbol{i}_1 - \boldsymbol{j}_3\boldsymbol{j}_1)\,\boldsymbol{k}_2\} \\ &+ \tfrac{1}{2}\Omega_{zzz}\{5\boldsymbol{k}_1\,\boldsymbol{k}_2\,\boldsymbol{k}_3 - (\boldsymbol{k}_1\,\boldsymbol{U}_{23} + \boldsymbol{k}_2\,\boldsymbol{U}_{31} + \boldsymbol{k}_3\,\boldsymbol{U}_{12})\}\end{aligned} \tag{38b}$$

For axial symmetry, $\Omega_{xxz} = \Omega_{yyz}$, $\Omega = \Omega_{zzz} = -2\Omega_{xxz} = -2\Omega_{yyz}$, and the first term of (38b) vanishes, whereas the second term becomes identical with the well-known result.[166]

In the case of tetrahedral symmetry, the dipole and quadrupole moments vanish ($\boldsymbol{p} = \boldsymbol{\Theta} = 0$); the first (lowest) non-zero electric moment is octupolar and can be written thus:[229]

$$\boldsymbol{\Omega} = \Omega_{xyz}\{\boldsymbol{i}_1(\boldsymbol{j}_2\,\boldsymbol{k}_3 + \boldsymbol{j}_3\,\boldsymbol{k}_2) + \boldsymbol{j}_1(\boldsymbol{k}_2\,\boldsymbol{i}_3 + \boldsymbol{k}_3\,\boldsymbol{i}_2) + \boldsymbol{k}_1(\boldsymbol{i}_2\boldsymbol{j}_3 + \boldsymbol{i}_3\boldsymbol{j}_2)\} \tag{38c}$$

[229] S. Kielich, *Acta Phys. Polon.*, 1963, **24**, 389.

where, by (38), only one independent component occurs, though there are six non-zero components:[171]

$$\Omega_{xyz} = \frac{5}{2}\sum_s q_s r_{sx} r_{sy} r_{sz} \tag{38d}$$

Electric hexadecapole. In microsystems of higher (*e.g.* octahedral) symmetry, as we shall see further on, all the above defined moments vanish and it is necessary to define multipoles of higher orders, beginning by a hexadecapole system the permanent moment of which is defined by a symmetric tensor [228] of rank 4:

$$\begin{aligned}\mathbf{\Phi} = \frac{1}{8}\sum_s q_s\{35\mathbf{r}_{s1}\mathbf{r}_{s2}\mathbf{r}_{s3}\mathbf{r}_{s4} - 5r_s^2(\mathbf{U}_{12}\mathbf{r}_{s3}\mathbf{r}_{s4} + \mathbf{U}_{23}\mathbf{r}_{s4}\mathbf{r}_{s1}\\ + \mathbf{U}_{34}\mathbf{r}_{s1}\mathbf{r}_{s2} + \mathbf{U}_{14}\mathbf{r}_{s2}\mathbf{r}_{s3} + \mathbf{U}_{13}\mathbf{r}_{s2}\mathbf{r}_{s4} + \mathbf{U}_{24}\mathbf{r}_{s1}\mathbf{r}_{s3})\\ + r_s^4(\mathbf{U}_{12}\mathbf{U}_{34} + \mathbf{U}_{13}\mathbf{U}_{24} + \mathbf{U}_{14}\mathbf{U}_{23})\}\end{aligned} \tag{39}$$

This tensor possesses, in general, 81 non-zero components, which reduce to 21 independent ones owing to symmetry and, further, to nine owing to relations between the traces (39).

Obviously, for spherical symmetry, the hexadecapole moment vanishes. In the tetrahedral and octahedral cases, it reduces to one component:[230]

$$\begin{aligned}\mathbf{\Phi} = \Phi_{xxzz}\{\mathbf{U}_{12}\mathbf{U}_{34} + \mathbf{U}_{13}\mathbf{U}_{24} + \mathbf{U}_{14}\mathbf{U}_{23}\\ - 5(\mathbf{i}_1\mathbf{i}_2\mathbf{i}_3\mathbf{i}_4 + \mathbf{j}_1\mathbf{j}_2\mathbf{j}_3\mathbf{j}_4 + \mathbf{k}_1\mathbf{k}_2\mathbf{k}_3\mathbf{k}_4)\}\end{aligned} \tag{39a}$$

where $\Phi_{zzzz} = -2\Phi_{xxzz} = -2\Phi_{yyzz} = -2\Phi_{xxyy} = \Phi_{xxxx} = \Phi_{yyyy}$.

Specifically, octahedrally symmetric molecules like SF_6 have, as their first electric moment, a hexadecapole, since all the lower moments vanish as a result of high symmetry ($\mathbf{p} = \mathbf{\Theta} = \mathbf{\Omega} = 0$).

In axially symmetric molecules too, the hexadecapole tensor (39) reduces to a single, independent component:[231]

$$\begin{aligned}\Phi &= \Phi_{zzzz} = \tfrac{8}{3}\Phi_{xxxx} = \tfrac{8}{3}\Phi_{yyyy}\\ &= -2\Phi_{zzxx} = -8\Phi_{xxyy} = -2\Phi_{yyzz}\end{aligned} \tag{39b}$$

Electric 2^n-poles and molecular symmetry. By extension, we are led to the concept of electric 2^n-poles with permanent moments of order n given by symmetric tensors of rank n (n-vectors):[228]

$$\mathbf{p}^{(n)} = \sum_s q_s r_s^n \mathbf{Y}^{(n)}(\mathbf{r}_s) \tag{40}$$

where

$$\mathbf{Y}^{(n)}(\mathbf{r}_s) = \frac{(-1)^n}{n!} r_s^{n+1} \nabla^n\left(\frac{1}{r_s}\right) \tag{41}$$

is the vectorial operator of degree n given in explicit form in ref. 228, and ∇ the spatial differential operator. In particular, if the system of charges presents n coinciding polar axes, the components of the operator $\mathbf{Y}^{(n)}$ are

[230] S. Kielich, *Acta Phys. Polon.*, 1964, **25**, 39.

[231] S. Kielich, *Acta Phys. Polon.*, 1962, **22**, 65; *Fizyka Dielektryków, Soc. Sci. Lettres, Poznań (Poland)*, 1964, **2**, 51.

Legendre polynomials P_n. In such cases the n-th rank tensor (40) reduces to a scalar electric moment [177] of order n, of the well-known form:

$$p^{(n)} = \sum_s q_s r_s^n P_n(\cos \theta_s) \qquad (40a)$$

An electric multipole moment of order n in the form (40) has quite generally $(2n + 1)$ independent components, and their number undergoes a further reduction for various molecular symmetries (Table 2). Buckingham,[171] by methods of group theory,[101, 232] calculated the number of independent tensor components of multipoles (40) from $n = 1$ up to $n = 6$ for 35 point-group symmetries. In Table 3, we give their numbers for the first four moments as well as the number of non-zero components.

The non-zero tensor components of multipole moments have been determined specifically for the tetrahedral [229] and octahedral [230] symmetries, beside the axial symmetry for which we have the general formula (40a). Lately, Kielich and Zawodny,[233] resorting to methods of group theory, have calculated and tabulated all non-zero and independent tensor components of electric dipole, quadrupole, octupole, and hexadecapole moments for 51 point groups (Tables 4—7).

Numerical Values of Molecular Electric Multipoles. We have at our disposal various direct and indirect methods, permitting the determination of the numerical values of molecular multipoles. Dipole moments are mostly obtained by Debye's method,[85] by measuring the temperature dependence of electric permittivities of gases and vapours, by microwave spectroscopic determinations of Stark splittings or shifts in spectral lines,[234] by resorting to molecular beams,[235] from vibrational–rotational spectral studies in the i.r. region,[236] as well as by electron paramagnetic resonance methods.[72] This has provided us with the dipole moments of some thousands of molecules.[86, 237] On the other hand, theoretical calculations are relatively few.[238, 239] Quadrupole and higher moments, also, can be calculated theoretically; however, the procedure of calculating these moments directly from known wavefunctions, though by far the most elegant, involves such

232 L. L. Boyle, *Internat. J. Quantum Chem.*, 1969, **3**, 231; 1970, **4**, 389.
233 S. Kielich and R. Zawodny, *Chem. Phys. Letters*, 1972, **12**, 20.
234 (*a*) W. Gordy, W. N. Smith, and R. F. Trambarulo, 'Microwave Spectroscopy', Wiley, New York, 1953; (*b*) C. H. Townes and A. L. Schawlow, 'Microwave Spectroscopy', McGraw-Hill, New York, 1955.
235 N. F. Ramsey, 'Molecular Beams', Clarendon Press, London, 1956.
236 (*a*) G. Herzberg, 'Molecular Spectra and Molecular Structure, II. Infrared and Raman Spectra of Polyatomic Molecules', D. Van Nostrand, New York, 1945; (*b*) E. B. Wilson, jun., J. C. Decius, and P. C. Cross, 'Molecular Vibrations, The Theory of Infrared and Raman Vibrational Spectra', McGraw-Hill, New York, 1955; (*c*) P. Barchewitz, 'Spectroscopie Infrarouge', Gauthier-Villars, Paris, 1961; (*d*) J. T. Houghton and S. D. Smith, 'Infra-red Physics', Clarendon Press, Oxford, 1966.
237 A. L. McClellan, 'Tables of Experimental Dipole Moments', Freeman, San Francisco, 1963.
238 W. Kołos and L. Wolniewicz, *J. Chem. Phys.*, 1966, **45**, 944.
239 (*a*) W. M. Huo, *J. Chem. Phys.*, 1965, **43**, 624; (*b*) P. E. Cade and W. M. Huo, *J. Chem. Phys.*, 1966, **45**, 1065; (*c*) R. L. Matcha, and R. K. Nesbet, *Phys. Rev.*, 1967, **160**, 72.

Table 2 *Point groups*

Schoenflies symbol	Plane C_s	C_6	C_5	C_4	C_3	C_2	S_n	Centre C_i	*Examples of molecules*
	Symmetry elements								
		Symmetry axes							
C_1									MeCHO, N_2H_4, C_4H_8O, CHFClBr
$C_i(S_2)$								1	$C_2H_2Cl_2F_2$, ClBrHC—CHBrCl (*trans* form)
C_2						1			H_2O_2, $H_2C{=}CCl_2$, HClC=C=CHCl
C_s	1								$MeNO_2$, NOCl, N_3H (plane)
C_{2h}						1		1	$C_6H_2Cl_2Br_2$ *trans*, ClCH=CHCl (*trans* plane)
D_2						3			C_2H_4 (partly rotated)
C_{2v}	2					1			H_2O, H_2CO, $C_6H_5NO_2$, CH_2Cl_2, $H_2C{=}CCl_2$ (plane)
D_{2h}	3					3		1	C_2H_4 (plane), C_2I_2, C_2H_2, C_2Cl_4
C_4				1					
S_4							$1S_4$		Tetraphenylmethane
C_{4h}	1			1			$1S_4(C_4)$	1	$C_4H_4Cl_4$
D_4				1		4			
C_{4v}	4			1					IF_5, BrF_5
D_{2d}	2					2 + 1	$S_4(C_2)$		$CH_2{=}C{=}CH_2$
D_{4h}	5			1		4	$S_4(C_4)$	1	C_4H_8
C_3					1				H_3CCCl_3
$C_{3i}(S_6)$							S_6	1	
D_3					1	3			C_2H_6 (partly rotated)
C_{3v}	3				1				NH_3, $CHCl_3$, CH_3Cl
D_{3d}	3				1	3	$1S_6(C_3)$	1	C_6H_{12}, C_2H_6 (staggered form)
C_6		1							
C_{3h}	1				1		$S_3(C_3)$		CN_3H_6
C_{6h}	1	1					$S_3(C_3)$ $S_6(C_6)$	1	C_6Cl_6, $C_{12}H_{18}$
D_6		1				6			
C_{6v}	6	1							

Table 2 (*cont.*)

Schoenflies symbol	*Symmetry elements* — *Plane* C_s	*Symmetry axes* C_6	C_5	C_4	C_3	C_2	S_n	*Centre* C_i	*Examples of molecules*
D_{3h}	4				1	3	$S_3(C_3)$		BCl_3, C_3H_6, $C_6H_3Cl_3$, C_2H_6 (eclipsed form)
D_{6h}	7	1				6	$S_3(C_3)$ $S_6(C_6)$	1	C_6H_6, C_6Cl_6
T					4	3			$C(CH_3)_4$ (partly rotated)
T_h	3				4	3		1	
O				3	4	6			
T_d	6				4	3	$3S_4(C_2)$		CH_4, CF_4, CCl_4, $SiCl_4$, $C(CH_3)_4$
O_h	9			3	4	6	$4S_6(C_3)$ $3S_4(C_4)$	1	SF_6, UF_6, $Fe(CN)_6$, $Cr(CO)_6$
Y			6		10	15			
Y_h	15		6		10	15	$6S_5(C_5)$ $10S_3(C_3)$	1	$B_{12}H_{12}^{2-}$
K									
K_h								1	Free atoms
$C_{\infty v}$		1	1	1	1	1			HCl, HCN, HC≡CCl, CN, COS,
$D_{\infty h}$		1	1	1	1		$S_n(C_n)$	1	H_2, CO_2, CS_2, C_2H_2, Cl_2, C_3O_2
D_{5h}	6		1			5	$S_5(C_5)$		Plane symmetrical C_5H_{10}
D_{5d}	5		1			5	$S_{10}(C_5)$	1	Ferrocene
D_{6d}	6	1				6	$S_{21}(C_6)$		
D_{4d}	4			1		4	$S_8(C_4)$		S_8
C_∞		1	1	1	1	1			

The symbol S_n (C_n) means that the element S_n (n-fold rotational–reflection axis) is rotationally coincident with C_n.

Table 3 *Number of mutually independent (I) and of non-zero (N) tensor elements of the electric dipole (**p**), quadrupole (**Θ**), octupole (**Ω**), and hexadecapole (**Φ**) moments, for all point groups*

Symbol of group		Number of independent and of non-zero elements							
		p		**Θ**		**Ω**		**Φ**	
International	*Schoenflies*	*I*	*N*	*I*	*N*	*I*	*N*	*I*	*N*
1	C_1	3	3	5	9	7	27	9	81
$\bar{1}$	$C_i(S_2)$	0	0	5	9	0	0	9	81
m	$C_s(C_{1h})$	2	2	3	5	4	14	5	41
2	C_2	1	1	3	5	3	13	5	41
2/*m*	C_{2h}	0	0	3	5	0	0	5	41
*mm*2	C_{2v}	1	1	2	3	2	7	3	21
222	$D_2(V)$	0	0	2	3	1	6	3	21
mmm	$D_{2h}(V_h)$	0	0	2	3	0	0	3	21
4	C_4	1	1	1	3	1	7	3	29
$\bar{4}$	S_4	0	0	1	3	2	12	3	29
4/*m*	C_{4h}	0	0	1	3	0	0	3	29
4*mm*	C_{4v}	1	1	1	3	1	7	2	21
$\bar{4}2m$	D_{2d}	0	0	1	3	1	6	2	21
4/*mmm*	D_{4h}	0	0	1	3	0	0	2	21
422	D_4	0	0	1	3	0	0	2	21
3	C_3	1	1	1	3	3	15	3	53
$\bar{3}$	S_6	0	0	1	3	0	0	3	53
3*m*	C_{3v}	1	1	1	3	2	11	2	37
32	D_3	0	0	1	3	1	4	2	37
$\bar{3}m$	D_{3d}	0	0	1	3	0	0	2	37

Table 3 (*cont.*)

Symbol of group		Number of independent and of non-zero elements							
		p		**Θ**		**Ω**		**Φ**	
International	*Schoenflies*	*I*	*N*	*I*	*N*	*I*	*N*	*I*	*N*
$\bar{6}$	C_{3h}	0	0	1	3	2	8	1	21
6	C_6	1	1	1	3	1	7	1	21
$6/m$	C_{6h}	0	0	1	3	0	0	1	21
$\bar{6}m2$	D_{3h}	0	0	1	3	1	4	1	21
$6mm$	C_{6v}	1	1	1	3	1	7	1	21
622	D_6	0	0	1	3	0	0	1	21
$6/mmm$	D_{6h}	0	0	1	3	0	0	1	21
23	T	0	0	0	0	1	6	1	21
$m3$	T_h	0	0	0	0	0	0	1	21
$\bar{4}3m$	T_d	0	0	0	0	1	6	1	21
432	O	0	0	0	0	0	0	1	21
$m3m$	O_h	0	0	0	0	0	0	1	21
	D_{4d}	0	0	1	3	0	0	1	21
	D_{5h}	0	0	1	3	0	0	1	21
	D_{5d}	0	0	1	3	0	0	1	21
	D_{6d}	0	0	1	3	0	0	1	21
Sphere	Y	0	0	0	0	0	0	0	0
	Y_h	0	0	0	0	0	0	0	0
	K	0	0	0	0	0	0	0	0
	K_h	0	0	0	0	0	0	0	0
∞	C_∞	1	1	1	3	1	7	1	21
∞m	$C_{\infty v}$	1	1	1	3	1	7	1	21
∞/m	$C_{\infty h}$	0	0	1	3	0	0	1	21
∞/mm	$D_{\infty h}$	0	0	1	3	0	0	1	21

Table 4 *Non-zero vector elements* **p** *of the electric dipole moment for all point groups*

System	*Symbol of group International*	*Schoenflies*	*Number of non-zero elements*	*Non-zero elements*
Triclinic	1	C_1	3	p_x, p_y, p_z
	$\bar{1}$	$C_i(S_2)$	0	
Monoclinic	m	$C_s(C_{1h})$	2	p_x, p_y
	2	C_2	1	p_z
	$2/m$	C_{2h}	0	
Orthorhombic	$mm2$	C_{2v}	1	p_z
	222	D_{2v}	0	
	mmm	$D_{2h}(V_h)$	0	
Tetragonal	4	C_4	1	p_z
	$\bar{4}$	S_4	0	
	$4/m$	C_{4h}	0	
	$4mm$	C_{4v}	1	p_z
	$\bar{4}2m$	D_{2d}	0	
	$4/mmm$	D_{4h}	0	
	422	D_4	0	
Trigonal	3	C_3	1	p_z
	$\bar{3}$	S_6	0	
	$3m$	C_{3v}	1	p_z
	32	D_3	0	
	$\bar{3}m$	D_{3d}	0	
Hexagonal	$\bar{6}$	C_{3h}	0	
	6	C_6	1	p_z
	$6/m$	C_{6h}	0	
	$\bar{6}m2$	D_{3h}	0	
	$6mm$	C_{6v}	1	p_z
	622	D_6	0	
	$6/mmm$	D_{6h}	0	
Cubic	23	T	0	
	$m3$	T_h	0	
	$\bar{4}3m$	T_d	0	
	432	O	0	
	$m3m$	O_h	0	
		D_{4d}	0	
		D_{5h}	0	
		D_{5d}	0	
		D_{6d}	0	
Isotropic	Sphere	Y	0	
		Y_h	0	
		K	0	
		K_h	0	
Axially symmetric medium	∞	C_∞	1	p_z
	∞m	$C_{\infty v}$	1	p_z
	∞/m	$C_{\infty h}$	0	
	∞/mm	$D_{\infty h}$	0	

Table 5 *Non-zero (N) and independent (I) elements of the quadrupole electric moment tensor* $\boldsymbol{\Theta}$ *for point groups*

Groups $1(C_1)$ and $\bar{1}(C_i)$: $N = 9, \quad I = 5$

$xx, \quad yy, \quad xy = yx, \quad xz = zx, \quad yz = zy, \quad zz = -(xx+yy)$

Groups $m(C_s)$, $2(C_2)$, and $2/m(C_{2h})$: $N = 5, \quad I = 3$

$xx, \quad yy, \quad xy = yx, \quad zz = -(xx+yy)$

Groups $mm2(C_{2v})$, $222[D_2(V)]$, and $mmm(D_{2h})$: $N = 3, \quad I = 2$

$xx, \quad yy, \quad zz = -(xx + yy)$

Groups $4(C_4)$, $\bar{4}(S_4)$, $4/m(C_{4h})$, $4mm(C_{4v})$, $\bar{4}2m(D_{2d})$, $422(D_4)$, $4/mmm(D_{4h})$, $3(C_3)$, $\bar{3}(S_6)$, $3m(C_{3v})$, $32(D_3)$, $\bar{3}m(D_{3d})$, $6(C_6)$, $\bar{6}(C_{3h})$, $6/m(C_{6h})$, $\bar{6}m2(D_{3h})$, $6/mmm(D_{6h})$, (D_{4d}), (D_{5h}), (D_{5d}), (D_{6d}), $\infty(C_\infty)$, $\infty m(C_{\infty v})$, $\infty/m(C_{\infty h})$, and $\infty/mm(D_{\infty h})$: $N = 3, \quad I = 1$

$xx = yy, \quad zz = -(xx + yy)$

In the groups $23(T)$, $m3(T_h)$, $\bar{4}3m(T_d)$, $432(O)$, $m3m(O_h)$, Y, Y_h, K, and K_h all the tensor elements vanish.

Table 6 *Non-zero (N) and independent (I) elements of the octupole electric moment tensor* $\boldsymbol{\Omega}$ *for point groups*

Group $1(C_1)$: $N = 27, \quad I = 7$

$xyy = yyx = yxy, \quad xxy = yxx = xyx,$
$xyz = xzy = yxz = yzx = zxy = zyx,$
$xxz = zxx = xzx, \quad yyz = zyy = yzy,$
$xzz = zzx = zxz, \quad yzz = zzy = zyz,$
$xxx = -(xyy + xzz), \quad yyy = -(xxy + yzz), \quad zzz = -(xxz + yyz)$

Group $m(C_s)$: $N = 14, \quad I = 4$

$xyy = yyx = yxy, \quad xxy = yxx = xyx,$
$xzz = zzx = zxz, \quad yzz = zzy = zyz,$
$xxx = -(xyy + xzz), \quad yyy = -(xxy + yzz)$

Group $2(C_2)$: $N = 13, \quad I = 3$

$xxz = zxx = xzx, \quad yyz = zyy = yzy,$
$xyz = xzy = yxz = yzx = zxy = zyx,$
$zzz = -(xxz + yyz)$

Group $mm2(C_{2v})$: $N = 7, \quad I = 2$

$xxz = zxx = xzx, \quad yyz = zyy = yzy,$
$zzz = -(xxz + yyz)$

Groups $222(D_2)$, $\bar{4}2m(D_{2d})$, $23(T)$, and $\bar{4}3m(T_d)$: $N = 6, \quad I = 1$

$xyz = xzy = yxz = yzx = zxy = zyx$

Groups $4(C_4)$, $4mm(C_{4v})$, $6(C_6)$, $6mm(C_{6v})$, $\infty(C_\infty)$, and $\infty m(C_{\infty v})$: $N = 7, \quad I = 1$

$xxz = zxx = xzx = yyz = zyy = yzy,$
$zzz = -(xxz + yyz)$

Table 6 (*cont.*)

Group $\bar{4}(S_4)$: $N = 12, \quad I = 2$

$xxz = zxx = xzx = -yyz = -zyy = -yzy,$
$xyz = xzy = yxz = yzx = zxy = zyx$

Group $3(C_3)$: $N = 15, \quad I = 3$

$xxx = -xyy = -yyx = -yxy,$
$xxz = zxx = xzx = yyz = zyy = yzy,$
$yyy = -yxx = -xxy = -xyx,$
$zzz = -(xxz + yyz)$

Group $3m(C_{3v})$: $N = 11, \quad I = 2$

$xxz = zxx = xzx = yyz = zyy = yzy,$
$yyy = -yxx = -xxy = -xyx,$
$zzz = -(xxz + yyz)$

Groups $32(D_3)$ and $\bar{6}m2(D_{3h})$: $N = 4, \quad I = 1$

$xxx = -xyy = -yyx = -yxy$

Group $\bar{6}(C_{3h})$: $N = 8, \quad I = 2$

$xxx = -xyy = -yyx = -yxy,$
$yyy = -yxx = -xxy = -xyx$

In the groups $\bar{1}(C_i)$, $2/m(C_{2h})$, $mmm(D_{2h})$, $4/m(C_{4h})$, $422(D_4)$, $4/mmm(D_{4h})$, $\bar{3}(S_6)$, $\bar{3}m(D_{3d})$, $6/m(C_{6h})$, $622(D_6)$, $6/mmm(D_{6h})$, (D_{4d}), (D_{5h}), (D_{5d}), (D_{6d}), $m3(T_h)$, $432(O)$, $m3m(O_h)$, $\infty/m(C_{\infty h})$, $\infty/mm(D_{\infty h})$, Y, Y_h, K, and K_h all the tensor elements vanish.

Table 7 *Non-zero (N) and independent (I) elements of the hexadecapole electric moment tensor for point groups*

Groups $1(C_1)$ and $\bar{1}C_i(S_2)$: $N = 81, \quad I = 9$

$xxyy = xyxy = xyyx = yyxx = yxyx = yxxy,$
$xxzz = xzxz = xzzx = zzxx = zxzx = zxxz,$
$yyzz = yzyz = yzzy = zzyy = zyzy = zyyz;$

$xxxx = -(yyxx + zzxx),$
$yyyy = -(xxyy + zzyy),$
$zzzz = -(xxzz + yyzz);$

$yyyx = yyxy = yxyy = xyyy,$
$zzyx = xzzy = xzyz = xyzz = yxzz = yzxz = yzzx = zxyz$
$= zyxz = zyzx = zxzy = zzxy,$
$yyzx = yyxz = yxyz = yxzy = yzyx = yzxy = xyyz = xyzy$
$= xzyy = zyyx = zxyy = zyxy;$

$zzzx = zzxz = zxzz = xzzz,$
$xxyz = xyxz = xyzx = xxzy = xzxy = xzyx = yxxz = yxzx$
$= yzxx = zxxy = zxyx = zyxx,$
$zzzy = zzyz = zyzz = yzzz;$

$xxxy = xxyx = xyxx = yxxx = -(yyyx + zzyx),$
$xxxz = xxzx = xzxx = zxxx = -(yyzx + zzzx),$
$yyyz = yyzy = yzyy = zyyy = -(xxyz - zzzy)$

Table 7 (*cont.*)

Groups $m(C_s)$, $2(C_2)$, and $2/m(C_{2h})$: $N = 41$, $I = 5$

$$xxyy = xyxy = xyyx = yyxx = yxyx = yxxy,$$
$$xxzz = xzxz = xzzx = zzxx = zxzx = zxxz,$$
$$yyzz = yzyz = yzzy = zzyy = zyzy = zyyz;$$

$$xxxx = -(yyxx + zzxx),$$
$$yyyy = -(xxyy + zzyy),$$
$$zzzz = -(xxzz + yyzz);$$

$$yyyx = yyxy = yxyy = xyyy,$$
$$zzyx = xzzy = xzyz = xyzz = yxzz = yzzx = zxyz = yzxz$$
$$= zyxz = zxzy = zzxy = zyzx,$$
$$xxxy = xxyx = xyxx = yxxx = -(yyyx + zzyx)$$

Groups $mm2(C_{2v})$, $222[D_2(V)]$, and $mmm[D_{2h}(V_h)]$: $N = 21$, $I = 3$

$$xxyy = xyxy = xyyx = yyxx = yxyx = yxxy,$$
$$xxzz = xzxz = xzzx = zzxx = zxzx = zxxz,$$
$$yyzz = yzyz = yzzy = zzyy = zyzy = zyyz;$$

$$xxxx = -(yyxx + zzxx),$$
$$yyyy = -(xxyy + zzyy),$$
$$zzzz = -(xxzz + yyzz)$$

Groups $4(C_4)$, $\bar{4}(S_4)$, and $4/m(C_{4h})$: $N = 29$, $I = 3$

$$xxyy = xyxy = xyyx = yyxx = yxyx = yxxy,$$
$$xxzz = xzxz = xzzx = zzxx = zxzx = zxxz = yyzz$$
$$= yzyz = yzzy = zzyy = zyzy = zyyz,$$
$$xxxy = xxyx = xyxx = yxxx = -yyyx = -yyxy = -yxyy$$
$$= -xyyy;$$

$$xxxx = yyyy = -(xxyy + zzyy),$$
$$zzzz = -(xxzz + yyzz)$$

Groups $422(D_4)$, $4mm(C_{4v})$, $4/mmm(D_{4h})$, and $\bar{4}2m(D_{2d})$: $N = 21$, $I = 2$

$$xxyy = xyxy = xyyx = yyxx = yxyx = yxxy,$$
$$xxzz = xzxz = xzzx = zzxx = zxzx = zxxz = yyzz$$
$$= yzyz = yzzy = zzyy = zyzy = zyyz;$$

$$xxxx = yyyy = -(xxyy + zzyy),$$
$$zzzz = -(xxzz + yyzz)$$

Groups $3(C_3)$ and $\bar{3}(S_6)$: $N = 53$, $I = 3$

$$xxyy = xyxy = xyyx = yyxx = yxyx = yxxy$$
$$= -\tfrac{1}{4}xxzz = -\tfrac{1}{4}xzxz = -\tfrac{1}{4}xzzx = -\tfrac{1}{4}zzxx$$
$$= -\tfrac{1}{4}zxzx = -\tfrac{1}{4}zxxz = -\tfrac{1}{4}yyzz = -\tfrac{1}{4}yzyz$$
$$= -\tfrac{1}{4}yzzy = -\tfrac{1}{4}zzyy = -\tfrac{1}{4}zyzy = -\tfrac{1}{4}zyyz;$$

$$xxxz = xxzx = xzxx = zxxx = -yxyz = -yyzx = -yzyx$$
$$= -yyxz = -yxzy = -yzxy = -xyyz = -xyzy$$
$$= -xzyy = -zxyy = -zyxy = -zyyx;$$

$$yyyz = yyzy = yzyy = zyyy = -xyxz = -xxzy = -xzxy$$
$$= -xxyz = -xyzx = -xzyx = -yxxz = -yxzx = -yzxx$$
$$= -zyxx = -zxyx = -zxxy;$$

$$xxxx = yyyy = 3xxyy = -\tfrac{3}{4}xxzz,$$
$$zzzz = -2xxzz = 8xxyy$$

Table 7 (*cont.*)

Groups $32(D_3)$, $3m(C_{3v})$, and $\bar{3}m(D_{3d})$: $N = 37$, $I = 2$

$$xxyy = xyxy = xyyx = yyxx = yxyx = yxxy = -\tfrac{1}{4}xxzz$$
$$= -\tfrac{1}{4}xzxz = -\tfrac{1}{4}xzzx = -\tfrac{1}{4}zzxx = -\tfrac{1}{4}zxzx$$
$$= -\tfrac{1}{4}zxxz = -\tfrac{1}{4}yyzz = -\tfrac{1}{4}yzyz = -\tfrac{1}{4}yzzy$$
$$= -\tfrac{1}{4}zzyy = -\tfrac{1}{4}zyzy = -\tfrac{1}{4}zyyz;$$

$$yyyz = yyzy = yzyy = zyyy = -xyxz = -xxzy = -xzxy$$
$$= -xxyz = -xyzx = -xzyx = -yxxz = -yxzx = -yzxx$$
$$= -zyxx = -zxyx = -zxxy;$$
$$xxxx = yyyy = 3xxyy = -\tfrac{3}{4}xxzz,$$
$$zzzz = -2xxzz = 8xxyy$$

Groups $6(C_6)$, $\bar{6}(C_{3h})$, $6/m(C_{6h})$, $622(D_6)$, $6mm(C_{6v})$, $\bar{6}m2(D_{3h})$, $\infty(C_\infty)$, $\infty/m(C_{\infty h})$, $\infty m(C_{\infty v})$, $\infty mm(D_{\infty h})$, $6/mmm(D_{6h})$, (D_{4d}), (D_{5h}), (D_{5d}), and (D_{6d}): $N = 21$, $I = 1$

$$xxyy = xyxy = xyyx = yyxx = yxyx = yxxy = -\tfrac{1}{4}xxzz$$
$$= -\tfrac{1}{4}xzxz = -\tfrac{1}{4}xzzx = -\tfrac{1}{4}zzxx = -\tfrac{1}{4}zxzx$$
$$= -\tfrac{1}{4}zxxz = -\tfrac{1}{4}yyzz = -\tfrac{1}{4}yzyz = -\tfrac{1}{4}yzzy$$
$$= -\tfrac{1}{4}zzyy = -\tfrac{1}{4}zyzy = -\tfrac{1}{4}zyyz;$$
$$xxxx = yyyy = 3xxyy = -\tfrac{3}{4}xxzz,$$
$$zzzz = -2xxzz = 8xxyy$$

Groups $23(T)$, $m3(T_h)$, $432(O)$, $\bar{4}3m(T_d)$, and $m3m(O_h)$: $N = 21$, $I = 1$

$$xxyy = xyxy = xyyx = yyxx = yxyx = yxxy = xxzz$$
$$= xzxz = xzzx = zzxx = zxzx = zxxz = yyzz$$
$$= zyzy = zyyz = zzyy = yzyz = yzzy;$$
$$xxxx = yyyy = zzzz = -2xxyy$$

In the groups Y, Y_h, K, and K_h all the tensor elements vanish.

computational difficulties as to be restricted to diatomic molecules like the hydrogen molecule, for which Kołos and Roothaan [240] succeeded in carrying out an exact calculation of the wavefunction and the quadrupole moment.[241, 242]

With regard to molecular quadrupoles we have, in addition to direct calculations from known wavefunctions,[243–245] direct experimental methods resorting to the optical birefringence induced by the gradient of an electric field,[166, 235] the anisotropy of magnetic susceptibility,[171] and high-field Zeeman linear and quadratic effects on microwave spectra.[169] Bogaard and Orr [246] have proposed a method of measuring quadrupole moments of excited states of molecules, but no experimental attempts have been

[240] W. Kołos and C. C. J. Roothan, *Rev. Mod. Phys.*, 1960, **32**, 219.
[241] (*a*) W. Kołos and L. Wolniewicz, *J. Chem. Phys.*, 1964, **41**, 3674; (*b*) L. Wolniewicz, *ibid.*, 1966, **45**, 515.
[242] G. Karl and J. D. Poll, *J. Chem. Phys.*, 1967, **46**, 2944.
[243] O. J. Sovers, M. Karplus, and C. W. Kern, *J. Chem. Phys.*, 1966, **45**, 3895.
[244] P. E. Cade, *J. Chem. Phys.*, 1967, **47**, 2390; P. E. Cade, R. F. W. Bader, and J. Pelletier, *ibid.*, 1971, **54**, 3517.
[245] A. Schweig, *Mol. Phys.*, 1968, **14**, 533.
[246] M. P. Bogaard and B. J. Orr, *Mol. Phys.*, 1968, **14**, 557.

published. Buckingham and Longuet-Higgins[247] discuss quadrupole moments of dipolar molecules.

No direct experimental methods are as yet available for the measurement of molecular octupoles, owing to the quite negligible and not easily detectable reorientation of octupoles by the gradient of a gradient of an electric field, although of course the effect is theoretically sound.[248] Theoretical calculations of dipoles, quadrupoles, and octupoles have been carried out for the molecules of water,[249] formaldehyde and carbon monoxide,[250] lithium hydride,[251] and recently for nitrogen dioxide and ozone.[252] Also, octupole and hexadecapole moments have been calculated for tetrahedral and other molecules.[170]

Indirect methods of determining quadrupoles, octupoles, and hexadecapoles consist in studying the effect of molecular multipolar correlations on various quantities measured in experiment, namely: the virial coefficients of the equation of state of gases[166, 228] and viscosities,[253] dielectric virial coefficients[229, 230, 254] (see Section 3), spectral line broadening,[167] pressure-induced absorption,[168, 255, 256] the induced i.r. spectrum,[257, 258] Raman line broadening,[259, 260] and i.r. absorption.[261, 262] These methods, as well as others based on non-linear phenomena,[94, 163] will be discussed in detail further on. Here, we shall restrict ourself to listing the values of multipoles obtained by these methods (Table 8).

[247] A. D. Buckingham and H. C. Longuet-Higgins, *Mol. Phys.*, 1968, **14**, 63.

[248] S. Kielich, *Acta Phys. Polon.*, 1965, **27**, 457.

[249] (*a*) R. M. Glaeser and C. A. Coulson, *Trans. Faraday Soc.*, 1965, **61**, 389; (*b*) D. B. Neuman and J. W. Moskowitz, *J. Chem. Phys.*, 1968, **49**, 2056.

[250] D. B. Neuman and J. W. Moskowitz, *J. Chem. Phys.*, 1969, **50**, 2216.

[251] G. P. Arrighini, J. Tomasi, and C. Guidotti, *Theor. Chim. Acta*, 1970, **18**, 329.

[252] S. Rothenberg and H. F. Schaefer, *Mol. Phys.*, 1971, **21**, 317.

[253] (*a*) T. H. Spurling and E. A. Mason, *J. Chem. Phys.*, 1967, **46**, 322; (*b*) K. K. Datta and Y. Singh, *J. Chem. Phys.*, 1971, **55**, 3541.

[254] (*a*) T. K. Bose and R. H. Cole, *J. Chem. Phys.*, 1970, **52**, 140; (*b*) T. K. Bose and R. H. Cole, *J. Chem. Phys.*, 1971, **54**, 3829; (*c*) R. D. Nelson, jun., and R. H. Cole, *J. Chem. Phys.*, 1971, **54**, 4033.

[255] (*a*) J. S. Murphy and J. E. Boggs, *J. Chem. Phys.*, 1967, **47**, 691; 1968, **49**, 3333; 1969, **50**, 3320; (*b*) D. Robert, M. Giraud, and L. Galatry, *J. Chem. Phys.*, 1969, **51**, 2192; 1970, **53**, 352.

[256] (*a*) B. T. Berendts and A. Dymanus, *J. Chem. Phys.*, 1968, **48**, 1361; 1969, **49**, 2632; (*b*) J. S. Murphy and J. E. Boggs, *ibid.*, 1969, **51**, 3891; 1971, **54**, 2443.

[257] J. D. Poll and J. Van Kranendonk, *Canad. J. Phys.*, 1962, **40**, 163; H. P. Gush and J. Van Kranendonk, *ibid.*, p. 1461.

[258] (*a*) P. Varnasi and G. D. T. Tejwani, *J. Chem. Phys.*, 1970, **53**, 4404; (*b*) G. D. T. Tejwani and P. Varnasi, *ibid.*, 1971, **55**, 1075; (*c*) P. Varnasi, G. D. T. Tejwani, and C. R. Prasad, *J. Quant. Spectroscopy Radiative Transfer*, 1971, **11**, 231; (*d*) D. Robert and L. Galtry, *J. Chem. Phys.*, 1971, **55**, 2347; C. Girardet, D. Robert, and L. Galatry, *ibid.*, 1972, **57**, 144.

[259] (*a*) J. Van Kranendonk, *Canad. J. Phys.*, 1963 **41**, 433; (*b*) J. Fiutak and J. Van Kranendonk, *ibid.*, 1962, **40**, 1085; 1963, **41**, 21; (*c*) C. G. Gray and J. Van Kranendonk, *ibid.*, 1966, **44**, 2411.

[260] D. R. Bosomworth and H. P. Gush, *Canad. J. Phys.*, 1965, **43**, 751.

[261] (*a*) A. Rosenberg and G. Birnbaum, *J. Chem. Phys.*, 1968, **48**, 1396; (*b*) G. Birnbaum and A. Rosenberg, *Phys. Letters*, 1968, **27A**, 272; (*c*) I. Ozier and K. Fox, *ibid.*, 1968, **27A**, 274; (*d*) A. Rosenberg and G. Birnbaum, *J. Chem. Phys.*, 1970, **52**, 683.

[262] (*a*) W. Ho, G. Birnbaum, and A. Rosenberg, *J. Chem. Phys.*, 1971, **55**, 1028, 1039; (*b*) A. I. Baise, *Chem. Phys. Letters*, 1971, **9**, 627.

Table 8 *Natural (permanent) multipole electric moments of simple molecules*

	Electric moment			
	Dipole	*Quadrupole*	*Octupole*	*Hexadecapole*
Molecule	$p \times 10^{30}$/Cm	$\Theta \times 10^{40}$/Cm^2	$\Omega \times 10^{50}$/Cm^3	$\Phi \times 10^{60}$/Cm^4
H_2	0	2.20[a]		
		3.17[b]		
		1.74[c]		
N_2	0	− 4.54[d]		
		6.01[b]		
		6.35[c]		
		6.85[e]		
		− 5.08[f]		
		5.01[b, g]		
O_2	0	− 1.30[f]		
		6.35[b]		
		4.94[e]		
		− 1.34[h]		
CO_2	0	16.70[b]		
		15.33[c]		
		13.69[e]		
		− 14.36[h]		
C_2H_2	0	16.73[e]		
		10.02[f]		
C_2H_4	0	13.09[c]		
		12.86[e]		
		6.68[h]		
CS_2	0	6.01[f]		
C_6H_6	0	− 30.73[i]		
		± 40.08[j]		
		− 18.70[k]		
CO	0.37[l]	9.39[c]		
		8.12[e]		
		− 8.35[h]		
NO	0.53[f]	− 6.01[f]		
		7.68[e]		
		10.02[m]		
HCl	3.57[f]	12.69[f]		13.03[f]
		19.37[n]		
OCS	2.38[l]	10.35[f]		
		9.35[m]		
		3.34[o]		
NH_3	4.91[f]	− 4.34[f]		

Table 8 (*cont.*)

		Electric moment		
Molecule	*Dipole* $p \times 10^{30}$/Cm	*Quadrupole* $\Theta \times 10^{40}$/Cm^2	*Octupole* $\Omega \times 10^{50}$/Cm^3	*Hexadecapole* $\Phi \times 10^{60}$/Cm^4
N_2O	0.60[e]	9.29[e]		
		10.12[m]		
		− 11.69[h]		
LiH	19.64[f]	− 16.70[f]	− 15.56[p]	
	− 20.04[p]	− 14.83[p]		
CH_4	0	0	15.03[f]	− 20.04[f]
			19.71[q]	− 25.72[q]
			16.70[r]	
			12.36[s]	
			8.35[t]	
CF_4	0	0	14.70[f]	
			16.03[u]	
			19.04[v]	
			45.76[s]	
CCl_4	0	0	50.10[w]	
			18.37[x]	
SiH_4	0	0	22.04[f]	− 55.44[f]
NH_4^+	0	0	21.04[f]	− 21.04[f]
SF_6	0	0	0	19.37[y]
				167.0[z]
$CClF_3$	1.70[z]		3.34[z]	

(*a*) Refs. 240 and 241; (*b*) S. Kielich, *Physica*, 1962, **28**, 511, 1121; *Mol. Phys.*, 1965, **9**, 549; (*c*) R. H. Orcut, *J. Chem. Phys.*, 1963, **39**, 605; (*d*) M. Karplus and H. J. Kolker, *J. Chem. Phys.*, 1963, **38**, 1263; (*e*) Ref. 253*a*; (*f*) Ref. 170; (*g*) W. Ho, I. Kaufman, and P. Thaddens, *J. Chem. Phys.*, 1968, **49**, 3627; (*h*) Ref. 166; A. D. Buckingham, R. L. Disch, and D. A. Dunmur, *J. Amer. Chem. Soc.*, 1968, **90**, 3104; (*i*) Ref. 245; (*j*) T. H. Spurling and A. G. De Rocco, *J. Chem. Phys.*, 1968, **49**, 2867; (*k*) R. L. Shoemaker and W. H. Flygare, *J. Chem. Phys.*, 1969, **51**, 2988; (*l*) Ref. 86; (*m*) Ref. 255; (*n*) S. Weiss and R. H. Cole, *J. Chem. Phys.*, 1967, **46**, 644; (*o*) Krishnaji and S. L. Srivastava, *J. Chem. Phys.*, 1967, **47**, 1885; (*p*) Ref. 251; (*q*) A. G. Turner, A. F. Saturno, P. Hank, and R. G. Parr, *J. Chem. Phys.*, 1964, **40**, 1919; (*r*) Refs. 228 and 229; (*s*) T. H. Spurling, A. G. De Rocco, and T. S. Storvick, *J. Chem. Phys.*, 1968, **48**, 1006; (*t*) Ref. 261*b*; (*u*) Ref. 261*a*; (*v*) Ref. 261*c*; (*w*) S. Kielich, *Phys. Letters*, 1968, **27A**, 307; *IEEE J. Quantum Electronics*, 1968, **4**, 744; (*x*) S. K. Garg, J. E. Bertie, H. Kilp, and C. P. Smyth, *J. Chem. Phys.*, 1968, **49**, 2551; (*y*) Ref. 261*d*; (*z*) Ref. 254*c*.

Magnetic (*Natural*) *Multipoles. Magnetic dipole.* In 1905 Langevin,[263] when aiming at an explanation of Curie's law, *i.e.* of the temperature dependence of magnetic susceptibility in paramagnetics, attributed *a priori* and classically a permanent magnetic moment to the microsystems.

[263] P. Langevin, *J. Physique*, 1905, **4**, 678; *Ann. Chim. phys.*, 1905, **5**, 70.

Debye [264] in 1912 extended Langevin's procedure to dielectrics, by attributing to the molecules a permanent (natural) electric dipole moment defined as in equation (34). The classical permanent magnetic dipole moment of a molecule is defined as follows (as an axial vector):[9]

$$\boldsymbol{m} = \frac{1}{2}\sum_s q_s(\boldsymbol{r}_s \times \dot{\boldsymbol{r}}_s) \tag{42}$$

where $\dot{\boldsymbol{r}}_s = \mathrm{d}\boldsymbol{r}_s/\mathrm{d}t$.

Permanent magnetic dipole moments are known for a few molecules, such as O_2, NO, OH, NO_2, and some others.[9, 265]

Magnetic 2^n*-poles.* By analogy to the definition of the electric 2^n-pole moment, a classical definition of the magnetic 2^n-pole can be introduced, with the following moment of n-th order:[266]

$$\boldsymbol{m}^{(n)} = \frac{1}{n+1}\sum_s q_s\, r_s^n\, \boldsymbol{Y}^{(n)}(\boldsymbol{r}_s) \times \dot{\boldsymbol{r}}_s \tag{43}$$

Electrostatics of Permanent Multipoles.—*Electric Potential.* In electrostatics, the electric potential produced at the point of observation P by an arbitrary system of electric charges (Figure 1) is:

$$\phi(R) = \frac{1}{4\pi\varepsilon_0}\sum \frac{q_s}{R_s} = \frac{1}{4\pi\varepsilon_0}\sum_s \frac{q_s}{|\boldsymbol{R} - \boldsymbol{r}_s|} \tag{44}$$

Provided the condition (33a) is fulfilled, the right-hand side of (44) can be expanded in a series of the Taylor type in powers of r_s about $r_s = 0$, yielding:

$$\phi(R) = \phi^{(0)} + \phi^{(1)} + \phi^{(2)} + \phi^{(3)} + \ldots = \sum_{n=0}^{\infty}\phi^{(n)} \tag{44a}$$

where, quite generally,[266]

$$\phi^{(n)} = -(-1)^n \frac{2^n\, n!}{(2n)!}\, \boldsymbol{p}^{(n)}[n]\, \boldsymbol{T}^{(n)} \tag{45}$$

is the n-th order contribution to the electric potential from the electric moment of order n defined by equation (40). In equation (44), we have introduced the interaction tensor of rank n:

$$4\pi\varepsilon_0\, \boldsymbol{T}^{(n)} = -\nabla^n\left(\frac{1}{R}\right) = (-R)^{-(n+1)}\, \boldsymbol{Y}^{(n)}(R) \tag{46}$$

The n-th order operator $\boldsymbol{Y}^{(n)}$ is defined by analogy to equation (41). In equation (45), the symbol $[n]$ in Jansen's notation [267] denotes n-fold scalar contraction of the product of two n-th rank tensors $\boldsymbol{p}^{(n)}$ and $\boldsymbol{T}^{(n)}$, with

$$\boldsymbol{p}^{(n)}[n]\, \boldsymbol{T}^{(n)} = \boldsymbol{T}^{(n)}[n]\, \boldsymbol{p}^{(n)}$$

[264] P. Debye, *Phys. Z.*, 1912, **13**, 97; 'Handbuch der Radiologie', ed. E. Marx, Leipzig, 1925, Vol. 6, p. 597.
[265] P. J. Stiles, *Mol. Phys.*, 1971, **21**, 361.
[266] S. Kielich, *Physica*, 1966, **32**, 385.
[267] L. Jansen, *Phys. Rev.*, 1958, **110**, 661.

Point-charge potential (*monopole potential*). By equation (45), the zeroth approximation

$$\phi^{(0)} = \phi_q = - p^{(0)} T^{(0)} = \frac{q}{4\pi\varepsilon_0 R} \tag{45a}$$

defines the potential due to a distribution of point charges since

$$T^{(0)} = - (4\pi\varepsilon_0 R)^{-1} \tag{46a}$$

Dipole potential. Similarly, by resorting to the equations (35) and (40), we obtain from equation (45) in a first approximation:

$$\phi^{(1)} = \phi_p = \boldsymbol{p} \cdot \boldsymbol{T}^{(1)} \tag{45b}$$

i.e. the potential caused by a dipolar system. Since, in accordance with equation (46),

$$4\pi\varepsilon_0 \boldsymbol{T}^{(1)} = - \nabla\left(\frac{1}{R}\right) = R^{-3} \boldsymbol{R} \tag{46b}$$

one can write (45b) in the usual form:[166]

$$\phi_p = \frac{\boldsymbol{p} \cdot \boldsymbol{R}}{4\pi\varepsilon_0 R^3} = \frac{p \cos\theta}{4\pi\varepsilon_0 R^2}$$

where θ is the angle between the vector $\boldsymbol{p}$ of the dipole moment and the vector of observation $\boldsymbol{R}$; thus $\boldsymbol{R} \cdot \boldsymbol{p} = pR\cos\theta$.

Quadrupole potential. In a second approximation, equation (45) yields, with equations (35) and (40), the potential due to an electric quadrupole in the form:

$$\phi^{(2)} = \phi_\Theta = - \tfrac{1}{3}\boldsymbol{\Theta} : \boldsymbol{T}^{(2)} \tag{45c}$$

Since with regard to the definition (46) the interaction tensor of rank 2 is:

$$4\pi\varepsilon_0 \boldsymbol{T}^{(2)} = - \nabla\nabla\left(\frac{1}{R}\right) = - R^{-5}(3\boldsymbol{R}_1 \boldsymbol{R}_2 - R^2 U_{12}) \tag{46c}$$

we are justified in re-writing (45c) as follows:

$$\phi_\Theta = (4\pi\varepsilon_0)^{-1} (\boldsymbol{R} \cdot \boldsymbol{\Theta} \cdot \boldsymbol{R})\, R^{-5}$$

as by equation (35c) the trace of the quadrupole moment tensor vanishes.

For an axially symmetric molecule, the quadrupole moment is of the form (35f), and we are led to the well-known formula:[166]

$$\phi_\Theta = \frac{\Theta}{8\pi\varepsilon_0 R^3} (3\cos^2\theta - 1)$$

Octupole potential. Likewise, in a third approximation, equation (45) yields the potential due to an electric octupole:

$$\phi^{(3)} = \phi_\Omega = \tfrac{1}{15}\boldsymbol{\Omega} \,\vdots\, \boldsymbol{T}^{(3)} \tag{45d}$$

since, by the definition (46), we have a tensor of the form:[230]

$$4\pi\varepsilon_0 \boldsymbol{T}^{(3)} = - \nabla\nabla\nabla\left(\frac{1}{R}\right) = 3R^{-7}\{5\boldsymbol{R}_1 \boldsymbol{R}_2 \boldsymbol{R}_3 - R^2(\boldsymbol{R}_1 U_{23} + \boldsymbol{R}_2 U_{31} + \boldsymbol{R}_3 U_{12})\} \tag{46d}$$

Now, considering that by equation (38a) the traces of the octupole moment tensor vanish, we can re-write (45d) quite generally as follows:

$$\phi_\Omega = (4\pi\varepsilon_0)^{-1}(\boldsymbol{R} \cdot \boldsymbol{\Omega} : \boldsymbol{RR})\, R^{-7}$$

On taking in equation (38b) the part referring to the axial symmetry only, and on inserting it into the preceding potential, we obtain:[166]

$$\phi_\Omega = \frac{\Omega}{8\pi\varepsilon_0 R^4}(5\cos^3\theta - 3\cos\theta)$$

Hexadecapole potential. In a fourth approximation we finally have from equation (45), by the definitions (39) and (40), the following hexadecapole potential:

$$\phi^{(4)} = \phi_\Phi = -\tfrac{1}{105}\boldsymbol{\Phi} :: \boldsymbol{T}^{(4)} \tag{45e}$$

By equation (46), the fourth-rank interaction tensor is of the more highly involved form:[230]

$$\begin{aligned} 4\pi\varepsilon_0 \boldsymbol{T}^{(4)} = & - 3R^{-9}\{35\boldsymbol{R}_1\boldsymbol{R}_2\boldsymbol{R}_3\boldsymbol{R}_4 - 5R^2(\boldsymbol{R}_1\boldsymbol{R}_2\boldsymbol{U}_{34} + \boldsymbol{R}_2\boldsymbol{R}_3\boldsymbol{U}_{41} \\ & + \boldsymbol{R}_3\boldsymbol{R}_4\boldsymbol{U}_{12} + \boldsymbol{R}_4\boldsymbol{R}_1\boldsymbol{U}_{23} + \boldsymbol{R}_1\boldsymbol{R}_3\boldsymbol{U}_{24} + \boldsymbol{R}_2\boldsymbol{R}_4\boldsymbol{U}_{13}) \\ & + R^4(\boldsymbol{U}_{12}\boldsymbol{U}_{34} + \boldsymbol{U}_{13}\boldsymbol{U}_{24} + \boldsymbol{U}_{14}\boldsymbol{U}_{23})\} \end{aligned} \tag{46e}$$

We nevertheless can write the potential (45e) simply as:

$$\phi_\Phi = (4\pi\varepsilon_0)^{-1}(\boldsymbol{RR} : \boldsymbol{\Phi} : \boldsymbol{RR})\, R^{-9}$$

This formula, like (45e), is valid for all molecular symmetries. For the axial symmetry, by equations (39) and (39b), it becomes:

$$\phi_\Phi = \frac{\Phi}{32\pi\varepsilon_0 R^5}(35\cos^4\theta - 30\cos^2\theta + 3)$$

From the foregoing considerations we see that the consecutive contributions to the potential resulting for the axial symmetry are particular cases of a contribution of the general form:

$$\phi^{(n)} = \frac{p^{(n)}}{4\pi\varepsilon_0 R^{n+1}} P_n(\cos\theta)$$

which we obtain from equation (45) with equations (40a) and (46), and where P_n stands for Legendre polynomials of the n-th degree.

Electric Field of Multipoles. By electrostatics, the electric field strength at the point of observation due to a system of electric charges is:

$$\boldsymbol{F}(\boldsymbol{R}) = -\nabla\phi(\boldsymbol{R}) \tag{47}$$

The electric multipole potential being given generally by equations (44) and (45), we have for the electric field strength:[228, 267]

$$\boldsymbol{F}(\boldsymbol{R}) = \sum_{n=0}^{\infty}(-1)^n \frac{2^n n!}{(2n)!}\, {}^{(1)}\boldsymbol{T}^{(n)}[n]\,\boldsymbol{p}^{(n)} \tag{47a}$$

This, when written out to within the hexadecapole contribution, becomes:[230]

$$\begin{aligned} \boldsymbol{F}(\boldsymbol{R}) = & {}^{(1)}\boldsymbol{T}^{(0)} q - {}^{(1)}\boldsymbol{T}^{(1)} \cdot \boldsymbol{p} + \tfrac{1}{3}{}^{(1)}\boldsymbol{T}^{(2)} : \boldsymbol{\Theta} \\ & - \tfrac{1}{15}{}^{(1)}\boldsymbol{T}^{(3)} :\cdot\, \boldsymbol{\Omega} + \tfrac{1}{105}{}^{(1)}\boldsymbol{T}^{(4)} :: \boldsymbol{\Phi} - \ldots \end{aligned} \tag{48}$$

On taking into account the properties of the various electric multipoles and interaction tensors (46), the consecutive contributions in (48) are found to be these:

(*a*) An electric point charge produces the electric field

$$\boldsymbol{F}_q(R) = \frac{q}{4\pi\varepsilon_0 R^3}\boldsymbol{R} \tag{48a}$$

directed along the vector of observation $\boldsymbol{R}$.

(*b*) An electric dipole gives rise to a field of the well-known form:

$$\boldsymbol{F}_p(R) = \frac{1}{4\pi\varepsilon_0}\{3(\boldsymbol{p}\cdot\boldsymbol{R})\boldsymbol{R} - R^2\boldsymbol{p}\}R^{-5} \tag{48b}$$

consisting of a part directed along the vector of observation R and a part directed in accordance with the dipole p.

(*c*) An electric quadrupole is the source of an electric field of the general form:

$$\boldsymbol{F}_\Theta(R) = (4\pi\varepsilon_0)^{-1}\{5(\boldsymbol{R}\cdot\boldsymbol{\Theta}\cdot\boldsymbol{R})\boldsymbol{R} - 2R^2(\boldsymbol{\Theta}\cdot\boldsymbol{R})\}R^{-7}, \tag{48c}$$

which, by equation (37f), takes the following form for axially symmetric molecules:

$$\boldsymbol{F}_\Theta(R) = \frac{3\Theta}{8\pi\varepsilon_0}\{(5\cos^2\theta - 1)\boldsymbol{R} - 2R\boldsymbol{k}\cos\theta\}R^{-5}$$

We consequently have here a component in the direction of the vector of observation $\boldsymbol{R}$ and a component in that of the symmetry axis $\boldsymbol{k}$ of the molecule.

(*d*) An electric octupole, in general, gives rise to a field:

$$\boldsymbol{F}_\Omega^{(R)} = (4\pi\varepsilon_0)^{-1}\{7(\boldsymbol{R}\cdot\boldsymbol{\Omega}:\boldsymbol{RR})\boldsymbol{R} - 3R^2(\boldsymbol{\Omega}:\boldsymbol{RR})\}R^{-9} \tag{48d}$$

This expression, when particularized for the axial symmetry, becomes:

$$\boldsymbol{F}(R) = \frac{\Omega}{8\pi\varepsilon_0}\{5(7\cos^3\theta - 3\cos\theta)\boldsymbol{R} - 3R(5\cos^2\theta - 1)\boldsymbol{k}\}R^{-6}$$

For tetrahedrally symmetric molecules we have, by equation (38c):

$$\begin{aligned}\boldsymbol{F}_\Omega(R) = \frac{3\Omega_{xyz}}{2\pi\varepsilon_0}\{&7\boldsymbol{R}\cos\theta_x\cos\theta_y\cos\theta_z \\ &- R(\boldsymbol{i}\cos\theta_y\cos\theta_z + \boldsymbol{j}\cos\theta_z\cos\theta_x \\ &+ \boldsymbol{k}\cos\theta_x\cos\theta_y)\}R^{-6}\end{aligned}$$

θ_x, θ_y, and θ_z being angles between the vector $\boldsymbol{R}$ and the unit vectors $\boldsymbol{i}, \boldsymbol{j}, \boldsymbol{k}$ of the molecular frame.

(*e*) An electric hexadecapole extends the field:

$$\boldsymbol{F}_\Phi(R) = (4\pi\varepsilon_0)^{-1}\{9\boldsymbol{R}(\boldsymbol{RR}:\boldsymbol{\Phi}:\boldsymbol{RR}) - 4R^2(\boldsymbol{R}\cdot\boldsymbol{\Phi}:\boldsymbol{RR})\}R^{-11} \tag{48e}$$

This field, for axial symmetry, becomes:

$$F_{\Phi}(R) = \frac{5\Phi}{32\pi\varepsilon_0}\{3\boldsymbol{R}(21\cos^4\theta - 14\cos^2\theta + 1) - 4R(7\cos^3\theta - 3\cos\theta)\,\boldsymbol{k}\}\,R^{-7}$$

By inserting into equation (48e) the hexadecapole tensor in the form (39a), we get for the octahedral symmetry:

$$F_{\Phi}(R) = \frac{\Phi_{xxzz}}{4\pi\varepsilon_0}\{3\boldsymbol{R}[5 - 3(\cos^4\theta_x + \cos^4\theta_y + \cos^4\theta_z)] + 4(\boldsymbol{i}\cos^3\theta_x + \boldsymbol{j}\cos^3\theta_y + \boldsymbol{k}\cos^3\theta_z)\}\,R^{-7}$$

Higher Order Electric Fields. The electric field of equation (47) can be regarded as a field of the first order, $\boldsymbol{F}^{(1)}(R)$. Similarly, we can define fields of higher order; thus $\boldsymbol{F}^{(2)} = \nabla\boldsymbol{F}^{(1)}(R)$, *i.e.* the field gradient: a field of the third order $\boldsymbol{F}^{(3)} = \nabla\nabla\boldsymbol{F}^{(1)}(R)$, *i.e.* the gradient of a field gradient: and quite generally fields of the m-th order:

$$\boldsymbol{F}^{(m)}(R) = -\nabla^m\phi(R) = \nabla^{m-1}\boldsymbol{F}^{(1)}(R)$$

On inserting above the potential (45) or the field (47b), we obtain [228]

$$\boldsymbol{F}^{(m)}(R) = \sum_{n=0}^{\infty}(-1)^n\frac{2^n\,n!}{(2n)!}\,{}^{(m)}\boldsymbol{T}^{(n)}[n]\,\boldsymbol{p}^{(n)} \tag{49}$$

as the m-th order field produced by a system of electric multipoles.

With regard to equation (49), the gradient of the electric field due to electric multipoles ($m = 2$) is, in a sufficiently good approximation,[231]

$$\boldsymbol{F}^{(2)}(R) = {}^{(2)}\boldsymbol{T}^{(0)}q - {}^{(2)}\boldsymbol{T}^{(1)}\cdot\boldsymbol{p} + \tfrac{1}{3}{}^{(2)}\boldsymbol{T}^{(2)} : \boldsymbol{\Theta} - \ldots \tag{50}$$

On taking into account equation (46c), one finds that the gradient of the electric field produced by electric charges is simply:

$$\boldsymbol{F}_q^{(2)}(R) = -\frac{q}{4\pi\varepsilon_0}(3\boldsymbol{RR} - R^2\,\boldsymbol{U})\,R^{-5} \tag{50a}$$

Electric fields of higher order are defined likewise.

Potential Energy of Multipoles in External Fields. Assuming an external static electric field of the strength:

$$\boldsymbol{E} = -\nabla\phi_e \tag{51}$$

where ϕ_e is the potential of the external field, and assuming a system of electric charges to be immersed in this field, and letting the potential ϕ_e at the point $\boldsymbol{r}_s$ (the point at which the electric charge q_s is situated) be $\phi_e(\boldsymbol{r}_s)$, then the potential energy of the system of charges in the field (51) amounts to

$$u(E) = \sum_s q_s\,\phi_e(\boldsymbol{r}_s) \tag{52}$$

The potential $\phi_e(\boldsymbol{r}_s)$ can be expanded in a power series in $\boldsymbol{r}_s$ about the centre O of the system of charges (Figure 1), leading to:

$$\phi_e(r_s) = \phi_e(0) + \boldsymbol{r}_s \cdot (\nabla\phi_e)_0 + \tfrac{1}{2}\boldsymbol{r}_s\boldsymbol{r}_s : (\nabla\nabla\phi_e)_0 + \ldots$$

Now, since the potential ϕ_e obeys Laplace's equation $\nabla \cdot \nabla\phi_e = \Delta\phi_e = 0$, and on resorting to the definitions (34)—(39) of electric moments, the energy (52) can be written with satisfactory accuracy [268] as follows:

$$u(E) = q\phi_e(0) - \boldsymbol{p} \cdot \boldsymbol{E} - \tfrac{1}{3}\boldsymbol{\Theta} : \boldsymbol{E}^{(2)} - \tfrac{1}{15}\boldsymbol{\Omega} \vdots \boldsymbol{E}^{(3)} - \tfrac{1}{105}\boldsymbol{\Phi} :: \boldsymbol{E}^{(4)} - \ldots \quad (52a)$$

with the notations:

$$\boldsymbol{E}^{(2)} = \nabla\boldsymbol{E}, \quad \boldsymbol{E}^{(3)} = \nabla\nabla\boldsymbol{E}, \ldots, \quad \boldsymbol{E}^{(n)} = \nabla^{n-1}\boldsymbol{E} = -\nabla^n\phi_e \quad (53)$$

for the field gradient $\boldsymbol{E}^{(2)}$, the gradient of the field gradient $\boldsymbol{E}^{(3)}$, ... and quite generally for the field of n-th order $\boldsymbol{E}^{(n)}$ (which results by applying n times the operator ∇ to the potential ϕ_e).

In particular, if the system of charges immersed in the external field possesses axial symmetry with respect to its z-axis, equation (52a) reduces to the form:[166]

$$u(E) = q\phi_e(0) - pE_z - \tfrac{1}{2}\Theta E_{zz} - \tfrac{1}{6}\Omega E_{zzz} - \tfrac{1}{24}\Phi E_{zzzz} \quad (52b)$$

The expressions (52a) and (52b) show that a permanent electric dipole undergoes reorientation in a uniform electric field; an electric quadrupole is reoriented in the gradient of an electric field and an octupole in the gradient of an electric field gradient, whereas it needs an electric field of order 4 to reorient a hexadecapole.

With the general definition of the electric multipole (40) and n-th order field (53), the potential energies of electrostatic interaction between electric multipoles and fields take the general form:[268]

$$u(E) = -\sum_{n=0}^{\infty} \frac{2^n n!}{(2n)!} \boldsymbol{p}^{(n)}[n]\, \boldsymbol{E}^{(n)} \quad (54)$$

where we have used Jansen's notation.[267]

Similarly to (54), one can define the potential energy of interaction between magnetic multipoles (43) and an external magnetic induction

$$\boldsymbol{B} = \nabla \times \boldsymbol{A} \quad (51a)$$

We have:

$$u(B) = -\sum_{n=0}^{\infty} \frac{2^n n!}{(2n)!} \boldsymbol{m}^{(n)}[n]\, \boldsymbol{B}^{(n)} \quad (54a)$$

with $\boldsymbol{A}$ denoting the magnetic vector potential.

Energy of Electrostatic Interaction Between Two Multipole Systems.— Consider two arbitrary electric charge systems q_i and q_j, with multipoles $\boldsymbol{p}_i^{(n)}$ and $\boldsymbol{p}_j^{(m)}$, mutually distant by $\boldsymbol{r}_{ij}$ (Figure 5). The potential energy of their interaction can be determined from equation (54) on replacing therein the n-th order field $\boldsymbol{E}^{(n)}$ by the field $\boldsymbol{F}^{(n)}$ of order n produced by the

[268] S. Kielich, *Acta Phys. Polon.*, 1965, **28**, 459.

electric charges of one of the two molecular systems. Accordingly, the potential interaction energy between the multipoles $p_i^{(m)}$ and $p_j^{(n)}$ takes the form:

$$u_{ij} = -\sum_{m=0}^{\infty} \frac{2^m\, m!}{(2m)!}\, p_i^{(m)}[m]\, F_{ij}^{(m)} \tag{55}$$

where the electric field of order m existing at the centre of the system i

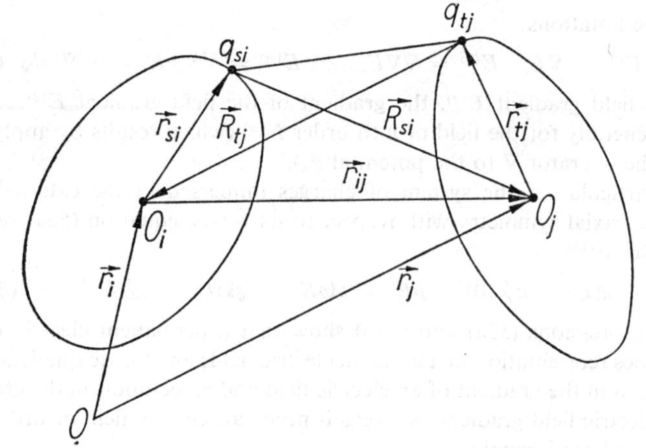

Figure 5 *Two arbitrary electric charge systems. The distance* $\vec{r}_{ij} = \vec{r}_j - \vec{r}_i$ *between their centres* O_i *and* O_j *is very large compared with the direction vectors* $\vec{r}_{si}$ *and* $\vec{r}_{tj}$ *of the charges* q_{si} *and* q_{tj} *belonging to the separate systems*

(but due to the electric charges of the system j) is, by equation (49):[268]

$$F_{ij}^{(m)} = \sum_{n=0}^{\infty} (-1)^n \frac{2^n\, n!}{(2n)!}\; {}^{(m)}T_{ij}^{(n)}[n]\, p_j^{(n)} \tag{56}$$

In the preceding formula,[228]

$${}^{(m)}T_{ij}^{(n)} = -\frac{1}{4\pi\varepsilon_0}\, \nabla^{m+n}\left(\frac{1}{r_{ij}}\right) = (-1)^{m+n}\; {}^{(n)}T_{ji}^{(m)} \tag{57}$$

is a tensor of interaction between the 2^m-pole of system i and the 2^n-pole of system j. Its properties closely resemble those of the tensor (46).

On insertion of the field (34) into equation (55) one obtains the general expression for the potential energy of electrostatic interaction between two electric multipolar systems:[228]

$$u_{ij} = \sum_{m=0}^{\infty} \sum_{n=0}^{\infty} (-1)^{n+1} \frac{2^{m+n}\, m!\, n!}{(2m)!\,(2n)!}\, p_i^{(m)}[m]\; {}^{(m)}T_{ij}^{(n)}[n]\, p_j^{(n)} \tag{58}$$

We now proceed to apply the above expressions to some of the more important particular cases most frequently occurring in molecular physics.

Ion–Ion Interaction. The potential interaction energy of two rigid ions separated by r_{ij} and bearing the charges q_i and q_j results immediately from equation (58) on putting therein $m = n = 0$:

$$u_{ij}^{q-q} = \frac{q_i q_j}{4\pi\varepsilon_0 r_{ij}} \tag{59}$$

This energy is positive in the case of charges of the same sign (repulsion) and grows as the distance r_{ij} diminishes. For ions of opposite signs the energy is negative (attraction).

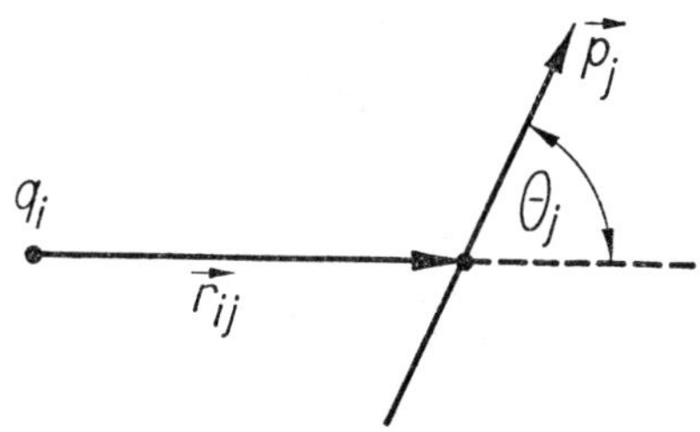

Figure 6 *Interaction energy between an ion q_i and an electric dipole $\vec{p}_j$: $\vec{p}_j \cdot \vec{r}_{ij} = p_j r_{ij} \cos\theta_j$*

Ion–Dipole Interaction. By putting in equation (58) $m = 0, n = 1$ and then $m = 1, n = 0$, one gets for the potential energy of interaction between the charges and dipoles of the systems i and j:

$$u_{ij}^{q-p} = \frac{1}{4\pi\varepsilon_0}\{q_i(\boldsymbol{p}_j \cdot \boldsymbol{r}_{ij}) - (\boldsymbol{p}_i \cdot \boldsymbol{r}_{ij}) q_j\} r_{ij}^{-3} \tag{59a}$$

Denoting by θ_i and θ_j respectively the angles between the directions of the dipoles $\boldsymbol{p}_i$ and $\boldsymbol{p}_j$ and the vector $\boldsymbol{r}_{ij}$ (Figure 6), we obtain (59a) in the well-known form:

$$u_{ij}^{q-p} = \frac{1}{4\pi\varepsilon_0}(q_i p_j \cos\theta_j - p_i q_j \cos\theta_i) r_{ij}^{-2}$$

Dipole–Dipole Interaction. On assuming the two electric systems to present rigid electric dipoles only, equation (58) with $m = n = 1$ yields their interaction energy in the general well-known form:

$$u_{ij}^{p-p} = -\frac{1}{4\pi\varepsilon_0}\{3(\boldsymbol{p}_i \cdot \boldsymbol{r}_{ij})(\boldsymbol{p}_j \cdot \boldsymbol{r}_{ij}) - r_{ij}^2(\boldsymbol{p}_i \cdot \boldsymbol{p}_j)\} r_{ij}^{-5} \tag{60}$$

Letting θ_{ij} stand for the angle between the dipoles ($\boldsymbol{p}_i \cdot \boldsymbol{p}_j = p_i p_j \cos\theta_{ij}$), equation (60) becomes:

$$u_{ij}^{p-p} = -\frac{p_i p_j}{4\pi\varepsilon_0}(3\cos\theta_i \cos\theta_j - \cos\theta_{ij}) r_{ij}^{-3} \tag{60a}$$

which is a form often better adapted to calculations, where in accordance with Figure 7 the cosine of the angle θ_{ij} is given by:

$$\cos\theta_{ij} = \cos\theta_i \cos\theta_j + \sin\theta_i \sin\theta_j \cos(\phi_j - \phi_i) \quad (60b)$$

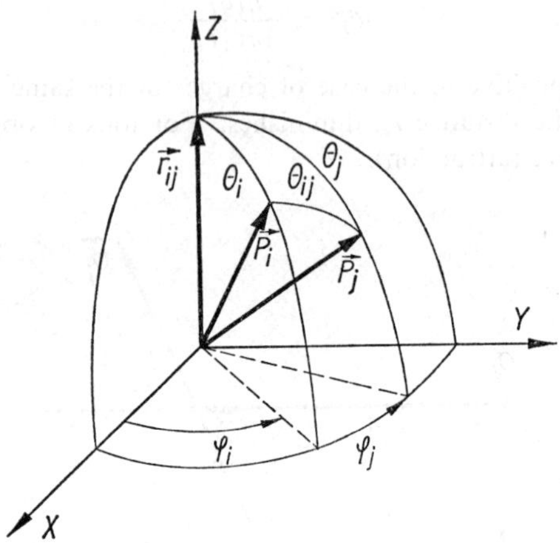

Figure 7 *Mutual orientations of two electric dipoles $\vec{p}_i$ and $\vec{p}_j$ separated by the distance r_{ij} and forming the angle θ_{ij}. ϕ_i and ϕ_j are their azimuths*

Ion–Quadrupole Interaction. From equation (58), the interaction between an ion q_i and a quadrupole Θ_j is ($m = 0, n = 2$):

$$u_{ij}^{q-\Theta} = \frac{q_i}{4\pi\varepsilon_0}(\boldsymbol{r}_{ij} \cdot \boldsymbol{\Theta}_j \cdot \boldsymbol{r}_{ij})\, r_{ij}^{-5} \quad (61)$$

For axially symmetric molecules, the quadrupole moment tensor is given by (37f), and the general expression (61) reduces to:

$$u_{ij}^{q-\Theta} = \frac{q_i\,\Theta_j}{8\pi\varepsilon_0}(3\cos^2\theta_j - 1)\, r_{ij}^{-3} \quad (61a)$$

Dipole–Quadrupole Interaction. Still resorting to equation (58), with $m = 1$, $n = 2$, the interaction of an electric dipole p_i with a quadrupole Θ_j is of the form:

$$u_{ij}^{p-\Theta} = -\frac{1}{4\pi\varepsilon_0}\{5(\boldsymbol{p}_i \cdot \boldsymbol{r}_{ij})(\boldsymbol{r}_{ij} \cdot \boldsymbol{\Theta}_j \cdot \boldsymbol{r}_{ij}) - 2r_{ij}^2(\boldsymbol{p}_i \cdot \boldsymbol{\Theta}_j \cdot \boldsymbol{r}_{ij})\}\, r_{ij}^{-7} \quad (62)$$

For axially symmetric molecules, this becomes:

$$u_{ij}^{p-\Theta} = -\frac{3p_i\Theta_j}{8\pi\varepsilon_0}(5\cos\theta_i\cos^2\theta_j - 2\cos\theta_{ij}\cos\theta_j - \cos\theta_i)\, r_{ij}^{-4} \quad (62a)$$

Quadrupole–Quadrupole Interaction. Here, equation (58) leads to the explicit expression:

$$u_{ij}^{\Theta\text{-}\Theta} = \frac{1}{12\pi\varepsilon_0}\{35(\boldsymbol{r}_{ij}\cdot\boldsymbol{\Theta}_i\cdot\boldsymbol{r}_{ij})(\boldsymbol{r}_{ij}\cdot\boldsymbol{\Theta}_j\cdot\boldsymbol{r}_{ij}) - 20r_{ij}^2(\boldsymbol{r}_{ij}\cdot\boldsymbol{\Theta}_i\cdot\boldsymbol{\Theta}_j\cdot\boldsymbol{r}_{ij}) + 2r_{ij}^4(\boldsymbol{\Theta}_i:\boldsymbol{\Theta}_j)\}\,r_{ij}^{-9} \quad (63)$$

In the case of quadrupolar, axially symmetric molecules, it reduces to:[231]

$$u_{ij}^{\Theta\text{-}\Theta} = \frac{3\Theta_i\Theta_j}{16\pi\varepsilon_0}\{1 - 5(\cos^2\theta_i + 3\cos^2\theta_i\cos^2\theta_j + \cos^2\theta_j) + 2(5\cos\theta_i\cos\theta_j - \cos\theta_{ij})^2\}\,r_{ij}^{-5} \quad (63a)$$

Dipole–Octupole Interaction. Equation (58) yields in explicit form (with $m = 1$, $n = 3$):

$$u_{ij}^{p\text{-}\Omega} = -\frac{1}{4\pi\varepsilon_0}\{7(\boldsymbol{p}_i\cdot\boldsymbol{r}_{ij})(\boldsymbol{r}_{ij}\cdot\boldsymbol{\Omega}_j:\boldsymbol{r}_{ij}\boldsymbol{r}_{ij}) - 3r_{ij}^2(\boldsymbol{p}_i\cdot\boldsymbol{\Omega}_j:\boldsymbol{r}_{ij}\boldsymbol{r}_{ij})\}\,r_{ij}^{-9} \quad (64)$$

Particularized to the case of axially symmetric molecules, the foregoing expression becomes:[231]

$$u_{ij}^{p\text{-}\Omega} = -\frac{p_i\Omega_j}{8\pi\varepsilon_0}(35\cos\theta_i\cos^3\theta_j - 15\cos\theta_{ij}\cos^2\theta_j - 15\cos\theta_i\cos\theta_j + 3\cos\theta_{ij})\,r_{ij}^{-5} \quad (64a)$$

If the dipolar molecule i is axially symmetric and the molecule j is tetrahedrally symmetric, equation (64) takes the form:

$$u_{ij}^{p\text{-}\Omega} = -\frac{3p_i\Omega_{xyz}^{(j)}}{2\pi\varepsilon_0}\{7\cos\theta_{iz}\cos\theta_{jx}\cos\theta_{jy}\cos\theta_{jz} - \cos\theta_{ij}\cos\theta_{jx}\cos\theta_{jy}\}\,r_{ij}^{-5} \quad (64b)$$

Ion–Octupole Interaction. This interaction, by equation (58) with $m = 0$ and $n = 3$, is given by:

$$u_{ij}^{q\text{-}\Omega} = \frac{q_i}{4\pi\varepsilon_0}(\boldsymbol{r}_{ij}\cdot\boldsymbol{\Omega}_j:\boldsymbol{r}_{ij}\boldsymbol{r}_{ij})\,r_{ij}^{-7} \quad (65)$$

For an octupolar molecule having the axial symmetry,[231] this becomes:

$$u_{ij}^{q\text{-}\Omega} = \frac{q_i\Omega_j}{8\pi\varepsilon_0}(5\cos^3\theta_j - 3\cos\theta_j)\,r_{ij}^{-4} \quad (65a)$$

whereas if its symmetry is tetrahedral, we have:

$$u_{ij}^{q\text{-}\Omega} = \frac{3q_i\Omega_{xyz}^{(j)}}{2\pi\varepsilon_0}(\cos\theta_{jx}\cos\theta_{jy}\cos\theta_{jz})\,r_{ij}^{-4} \quad (65b)$$

Ion–Hexadecapole Interaction. By equation (58) with $m = 0$ and $n = 4$, we obtain:

$$u_{ij}^{q\text{-}\Phi} = \frac{q_i}{4\pi\varepsilon_0}(\boldsymbol{r}_{ij}\boldsymbol{r}_{ij}:\boldsymbol{\Phi}_j:\boldsymbol{r}_{ij}\boldsymbol{r}_{ij})\,r_{ij}^{-9} \quad (66)$$

Assuming the hexadecapole as axially symmetric, we obtain:[231]

$$u_{ij}^{q-\Phi} = \frac{q_i \Phi_j}{32\pi\varepsilon_0}(35\cos^2\theta_j - 30\cos^2\theta_j + 3)\, r_{ij}^{-5} \qquad \text{(66a)}$$

whereas for the tetrahedral and octahedral symmetries, with regard to equation (39a):

$$u_{ij}^{q-\Phi} = -\frac{q_i \Phi_{xxzz}^{(j)}}{4\pi\varepsilon_0}\{5(\cos^4\theta_{jx} + \cos^4\theta_{jy} + \cos^4\theta_{jz}) - 3\}\, r_{ij}^{-5} \qquad \text{(66b)}$$

Proceeding as above, the general expression (58) can be used to calculate other particular cases involving higher multipoles,[269] on resorting to Tables 4—7, where the multipole elements are listed. Interactions between magnetic multipoles [266] are also the subject of discussion. The theory of electrostatic interactions for electric multipoles has been dealt with in various approaches by Frenkel,[270] Pople,[271] Jansen,[267] and others,[74] as well as by Gray.[272] The above presentation follows the concise uniform treatment of Kielich.[228, 266, 268]

Induced Molecular Multipoles.—We have discussed the electric and magnetic multipolar properties of microsystems in their natural state. That is to say, the charge distribution within a microsystem was assumed as rigid and unaffected either by the presence of an external electric field or by interaction with neighbouring systems of point charges. This assumption is, however, valid only as a zeroth approximation, since as a matter of fact any electric-charge distribution will undergo a modification by the action of an outer electric field: the charges will shift with respect to one another, *i.e.* the charge system polarizes.

Linear Dipolar Electric Polarization. Even atoms and microsystems with symmetrically distributed charges are known to exhibit polarization under the influence of an external electric field or the internal (molecular) electric field of a neighbour. Such a system is said to become endowed with an induced electric dipole (or higher multipoles). In many a case it suffices to assume that the induced dipole moment $\boldsymbol{p}(E)$ is proportional to the field strength inducing it:

$$\boldsymbol{p}(E) = 4\pi\varepsilon_0\, \alpha_V\, \boldsymbol{E} = \alpha\boldsymbol{E} \qquad \text{(67)}$$

where $\alpha_V = \alpha/4\pi\varepsilon_0$ is the well-known scalar volume polarizability of the atom or isotropically polarizable microsystem (in SI units, α has the dimensions $C^2\, m^2\, J^{-1}$).

However, most microsystems present asymmetric charge distributions in their natural state. As a consequence of this asymmetry, the dipole

[269] S. Kielich, J. R. Lalanne, and F. B. Martin, *J. Physique*, 1972, **33**, C1.
[270] J. Frenkel, 'Lehrbuch der Electrodynamik', Springer Verlag, Berlin, 1926, Vol. 1.
[271] J. A. Pople, *Proc. Roy. Soc.*, 1954, **A221**, 498.
[272] C. G. Gray, *Canad. J. Phys.*, 1968, **46**, 135.

moment components induced in them vary with direction, and equation (67) has to be replaced by the more general one:

$$p(E) = \boldsymbol{\alpha} \cdot E \tag{68}$$

where $\boldsymbol{\alpha}$ is now the electric polarizability tensor of rank 2 of the molecule and can be expressed in matrix form like the tensor of the quadrupole moment (37a). Thus, explicitly, the components of (68) are:

$$\left.\begin{aligned} p_x(E) &= \alpha_{xx} E_x + \alpha_{xy} E_y + \alpha_{xz} E_z \\ p_y(E) &= \alpha_{yx} E_x + \alpha_{yy} E_y + \alpha_{yz} E_z \\ p_z(E) &= \alpha_{zx} E_x + \alpha_{zy} E_y + \alpha_{zz} E_z \end{aligned}\right\} \tag{68a}$$

where $\alpha_{xx}, \alpha_{yy}, \alpha_{zz}$ are the diagonal elements of the polarizability tensor (termed the 'principal polarizabilities'); in the symmetric case, the non-diagonal tensor elements are:

$$\alpha_{xy} = \alpha_{yx}, \quad \alpha_{yz} = \alpha_{zy} = \alpha_{zx} = \alpha_{xz}$$

Like any second-rank tensor, $\boldsymbol{\alpha}$ can be separated into an isotropic part and an anisotropic part:

$$\boldsymbol{\alpha} = \alpha \boldsymbol{U} + \boldsymbol{d} \tag{69}$$

where

$$\alpha = \tfrac{1}{3}(\alpha_{xx} + \alpha_{yy} + \alpha_{zz}) \tag{69a}$$

is the scalar (or mean) polarizability, and

$$\boldsymbol{d} = \boldsymbol{\alpha} - \alpha \boldsymbol{U} \tag{69b}$$

is the deviation tensor defining the anisotropic part of the polarizability tensor.

For those molecular symmetries for which the non-diagonal elements vanish, the deviation tensor (69b) can be written in the form:

$$\begin{aligned} \boldsymbol{d} = {} & \tfrac{1}{2}(\alpha_{xx} - \alpha_{yy})(\boldsymbol{i}\,\boldsymbol{i} - \boldsymbol{j}\boldsymbol{j}) \\ & + \tfrac{1}{6}(2\alpha_{zz} - \alpha_{xx} - \alpha_{yy})(3\boldsymbol{k}\,\boldsymbol{k} - \boldsymbol{U}) \end{aligned} \tag{69c}$$

In particular, for axially symmetric molecules $\alpha_{xx} = \alpha_{yy} \neq \alpha_{zz}$, it reduces to the simple expression:

$$\boldsymbol{d} = \alpha\kappa(3\boldsymbol{k}\,\boldsymbol{k} - \boldsymbol{U}) \tag{69d}$$

where

$$\kappa = \frac{\alpha_{zz} - \alpha_{xx}}{3\alpha} = \frac{\alpha_{zz} - \alpha_{xx}}{\alpha_{zz} + 2\alpha_{xx}} \tag{69e}$$

defines the dimensionless anisotropy of clectric polarizability of an axially symmetric molecule as introduced by Langevin.[201]

Linear electric polarizabilities of atoms and polarizability anisotropies of molecules have been rather extensively evaluated by using more or less approximate wavefunctions and various perturbation methods.[273] Note-

[273] (*a*) H. P. Kelly, *Phys. Rev.*, 1966, **152**, 62; (*b*) H. D. Cohen, *J. Chem. Phys.*, 1966, **45**, 10.

worthy is the role fulfilled by the classical electrostatic theory of molecular polarizabilities of Silberstein [274] which, in a first approximation, yields results analogous to those obtained by using a quantum-mechanical approach.[172] Kołos and Wolniewicz [275] were the first to perform a stricter quantum-mechanical calculation of the principal electric polarizabilities of the hydrogen molecule. Recent years have witnessed the publication of calculations of the electric polarizabilities of atoms and ions [276] as well as of some molecules,[277–279] and also of polymer chains [280–283] and n-alkanes.[174, 284] Calculations of frequency-dependent (optical) polarizabilities will be dealt with later, as will experimental methods of determining polarizabilities and their anisotropies.[114, 132] Numerical polarizability values for atoms and molecules are listed in Table 9.

Linear Multipolar Electric Polarizabilities. Just as equation (40) served for defining the 2^n-pole permanent electric moment, one can generalize the definition of the induced dipole (68) by introducing that of an induced 2^n-pole moment. Assuming the inducing factor to be a uniform external electric field $\boldsymbol{E}$, one has:[268]

$$\boldsymbol{p}^{(n)}(\boldsymbol{E}) = {}^{(n)}\boldsymbol{\alpha}^{(1)} \cdot \boldsymbol{E} \tag{70}$$

where ${}^{(n)}\boldsymbol{\alpha}^{(1)}$ is now a symmetric tensor of rank $(n + 1)$ defining the electric multipole polarizability of order n with respect to a linear uniform electric field E. Equation (70) yields for $n = 1$ an induced dipole, which now is:

$$\boldsymbol{p}^{(1)}(E) = \boldsymbol{p}(E) = {}^{(1)}\boldsymbol{\alpha}^{(1)} \cdot \boldsymbol{E} \tag{70a}$$

for $n = 2$ an induced quadrupole:

$$\boldsymbol{p}^{(2)}(E) = \boldsymbol{\Theta}(E) = {}^{(2)}\boldsymbol{\alpha}^{(1)} \cdot \boldsymbol{E} \tag{70b}$$

for $n = 3$ an induced octupole:

$$\boldsymbol{p}^{(3)}(E) = \boldsymbol{\Omega}(E) = {}^{(3)}\boldsymbol{\alpha}^{(1)} \cdot \boldsymbol{E} \tag{70c}$$

and so forth. Equations (70b) and (70c) contain novel polarizabilities: a quadrupole polarizability ${}^{(2)}\boldsymbol{\alpha}^{(1)}$ (tensor of rank 3) and an octupole polarizability ${}^{(3)}\boldsymbol{\alpha}^{(1)}$ (fourth-rank tensor) with respect to a linear uniform electric field.

[274] L. Silberstein, *Phil. Mag.*, 1917, **33**, 92, 215, 521.
[275] W. Kołos and L. Wolniewicz, *J. Chem. Phys.*, 1967, **46**, 1426.
[276] (*a*) Tien Tson Tsong, *J. Chem. Phys.*, 1971, **54**, 4205; (*b*) D. Fu-t Tuan and A. Davidz, *ibid.*, pp. 1286, 1294.
[277] J. F. Harrison, *J. Chem. Phys.*, 1968, **49**, 3321.
[278] (*a*) H. Navangul and H. Labhart, *Theor. Chim. Acta*, 1970, **17**, 239; (*b*) G. P. Arrighini, J. Tomasi, and C. Petrongolo, *ibid.*, 1970, **18**, 341; (*c*) A. J. Sadlej, *ibid.*, 1971, **21**, 159.
[279] (*a*) D. W. Davies, *Mol. Phys.*, 1971, **20**, 605; (*b*) R. L. Fulton, *J. Chem. Phys.*, 1971, **55**, 1386.
[280] (*a*) R. P. Smith, *J. Chem. Phys.*, 1966, **44**, 2543; (*b*) P. J. Flory and Y. Abe, *ibid.*, 1971, **54**, 1351.
[281] G. Fourche and B. Lemaire, *J. Chim. phys.*, 1971, **68**, 1170.
[282] K. Nagai, *J. Chem. Phys.*, 1967, **47**, 4690.
[283] R. L. Rowell and R. S. Stein, *J. Chem. Phys.*, 1967, **47**, 2985.
[284] J. J. C. Teixeira-Dias and J. N. Murrell, *Mol. Phys.*, 1971, **19**, 329.

Table 9 *Numerical values of linear and non-linear polarizabilities of atoms and molecules calculated theoretically and determined from experimental data. Only mean values are given:* $a = (a_{11} + a_{22} + a_{33})/_3$, $b = (b_{113} + b_{223} + b_{333})/_3 = b_{333}$ *(for axial symmetry),* $b = b_{123}$ *(for tetrahedral symmetry), and* $c = c_{3333}$. *(Reference numbers in square brackets)*

Units:	10^{-40} ($C^2\ m^2\ J^{-1}$)	10^{-50} ($C^3\ m^3\ J^{-2}$)			10^{-60} ($C^4\ m^4\ J^{-3}$)			
Atom or molecule	a^{ω}	b^0	b^{ω}	$b^{2\omega}$	c^0	c^{ω}	$c^{2\omega}$	$c^{3\omega}$
He	0.23				0.003 [294*b*]	0.003 [383*a*]	0.003 [294*b*]	0.003 [294*b*]
					0.003 [295]	0.003 [294*b*]		0.003 [532]
Ne	0.44					0.006 [383*a*]	0.3 [221]	0.005 [532]
						0.007 [178]		0.03 [531]
Ar	1.84					0.07 [383*a*]	0.33 [221]	0.36 [531]
						0.09 [178]		0.08 [532]
Kr	2.80					0.17 [383*a*]	0.80 [221]	1.11 [531]
						0.19 [178]		0.20 [532]
Xe	4.56					0.47 [383*a*]	2.22 [221]	2.82 [531]
						0.48 [178]		0.64 [532]
H_2	0.88				0.02 [290]	0.03 [383*b*]	0.04 [220]	0.23 [531]
N_2	1.95							0.31 [531]
O_2	1.78					0.88 [383*d*]	0.08 [220]	
HF	2.73	− 0.06 [293*a*]						
CO	2.15	± 0.16 [293*a*]			0.12 [290]		0.22 [220]	
		0.7 [290]						
		0.07 [526]						
NO	1.91							
LiH	3.72	− 1.27 [293*c*]						
BF	3.47	0.36 [293*a*]						
CO_2	2.94					5.6 [383*d*]		0.33 [531]
NH_3	2.51	− 0.02 [293*b*]	− 1.44 [178]					

[*Continued overleaf*]

Table 9 (*cont.*)

Units:	10^{-40} ($C^2\ m^2\ J^{-1}$)	10^{-50} ($C^3\ m^3\ J^{-2}$)			10^{-60} ($C^4\ m^4\ J^{-3}$)			
Atom or molecule	a^{ω}	b^0	b^{ω}	$b^{2\omega}$	c^0	c^{ω}	$c^{2\omega}$	$c^{3\omega}$
CH_4	2.89	0.08 [178]		0.004 [226]		0.31 [178]	0.18 [220]	
		0.07 [138]				0.18 [383*b*]	0.05	
MeF	3.94	− 0.03 [293*b*]	− 0.11 [383*b*]			0.15 [383*b*]		
CHF_3	3.12		0.15 [383*b*]	− 0.02 [220]		0.11 [383*b*]	− 0.03 [220]	
$CHCl_3$	9.14		5.33 [178]	0.12 [220]		0.58 [528]	0.53 [220]	
H_2O	1.62	0.04 [293*d*]						
CS_2	9.70					6.78 [348]		
						6.01 [528]		
						6.53 [178]		
						7.67 [384*a*]		
						7.1 [383*c*]		
C_2H_6	4.96					2.4 [383*d*]	0.58 [220]	
C_6H_6	11.46					0.8 [383*c*]		
C_6H_{12}	12.07					0.14 [528]		
C_6H_5Me	13.64					3.26 [528]		
$C_6H_5NO_2$	13.51					10.20 [528]		
MeI	8.08			0.31 [220]			1.08 [220]	
SF_6	4.96					0.14 [383*a*]	0.03 [220]	
EtI				0.41 [220]			1.63 [220]	
CCl_4	11.66			0.01 [226]		0.37 [528]	0.43 [220]	
						0.86 [384*a*]	0.03	
						0.78 [348]		

However, the inducing electric field need not be uniform within the space occupied by the microsystem. In these more general cases, the induced multipole moment (70) is enriched by terms induced by the field gradient $\boldsymbol{E}^{(2)} = \nabla \boldsymbol{E}$, the gradient of the field gradient $\boldsymbol{E}^{(3)} = \nabla\nabla \boldsymbol{E} \ldots$ as follows:[268]

$$\boldsymbol{p}^{(n)}(\nabla \boldsymbol{E}) = {}^{(n)}\boldsymbol{\alpha}^{(1)} \cdot \boldsymbol{E} + \tfrac{1}{3}{}^{(n)}\boldsymbol{\alpha}^{(2)} : \boldsymbol{E}^{(2)} + \tfrac{1}{15}{}^{(n)}\boldsymbol{\alpha}^{(3)} \vdots \boldsymbol{E}^{(3)} + \ldots \tag{71}$$

with ${}^{(n)}\boldsymbol{\alpha}^{(2)}$ denoting the multipole polarizability with respect to the field gradient, ${}^{(n)}\boldsymbol{\alpha}^{(3)}$ that with respect to the gradient of the field gradient, *etc.*

Consequently, by equation (71), a non-uniform electric field acting on a molecule induces in general not the linear dipole (70a) but rather:

$$\boldsymbol{p}^{(1)}(\nabla \boldsymbol{E}) = {}^{(1)}\boldsymbol{\alpha}^{(1)} \cdot \boldsymbol{E} + \tfrac{1}{3}{}^{(1)}\boldsymbol{\alpha}^{(2)} : \nabla \boldsymbol{E} + \tfrac{1}{15}{}^{(1)}\boldsymbol{\alpha}^{(3)} \vdots \nabla\nabla \boldsymbol{E} + \ldots \tag{71a}$$

where, beside the tensor of usual dipolar polarizability there emerges a tensor of dipole polarizability ${}^{(1)}\boldsymbol{\alpha}^{(2)}$ with respect to the field gradient, a tensor ${}^{(1)}\boldsymbol{\alpha}^{(3)}$ of dipole polarizability with respect to the gradient of the field gradient, and so on.

The induced quadrupole (70b), too, has in general to be replaced by:

$$\boldsymbol{p}^{(2)}(\nabla \boldsymbol{E}) = \boldsymbol{\Theta}(\nabla \boldsymbol{E}) = {}^{(2)}\boldsymbol{\alpha}^{(1)} \cdot \boldsymbol{E} + \tfrac{1}{3}{}^{(2)}\boldsymbol{\alpha}^{(2)} : \boldsymbol{E}^{(2)} + \ldots \tag{71b}$$

where the fourth-rank tensor ${}^{(2)}\boldsymbol{\alpha}^{(2)}$ defines a quadrupolar polarizability with respect to the field gradient. It is important to state that in the case of spherical symmetry the tensors ${}^{(2)}\boldsymbol{\alpha}^{(1)} = {}^{(1)}\boldsymbol{\alpha}^{(2)}$ vanish, whereas the tensor ${}^{(2)}\boldsymbol{\alpha}^{(2)}$ does not but presents a single (independent) tensor element and can be written in the form:[285]

$${}^{(2)}\boldsymbol{\alpha}^{(2)} = 3C_{1234} = \tfrac{1}{2}C(3U_{13}U_{24} + 3U_{14}U_{23} - 2U_{12}U_{34}) \tag{71c}$$

Obviously, the expansion (71) is a particular case of the following general expression:[266]

$$\boldsymbol{p}^{(n)}(E) = \sum_{m=1}^{\infty} \frac{2^m\, m!}{(2m)!}\, {}^{(n)}\boldsymbol{\alpha}^{(m)}[m]\, \boldsymbol{E}^{(m)} \tag{72}$$

permitting the calculation of all multipole polarizabilities ${}^{(n)}\boldsymbol{\alpha}^{(m)}$ in linear (weak) non-uniform electric fields.

Non-linear Dipole Polarizabilities. In weak electric fields, the linear dipolar polarizability α appearing in equations (67) and (68) is a quantity specific to the atom or molecule (the volume polarizability α_v corresponds to the spatial dimensions of the electron shell and is of the order and dimension of 10^{-30} m^3); in fact, it describes the linear distortion of the shell under the influence of the field strength E (in the case of atoms, the electrons are, classically, performing harmonic oscillations). In the general case, the polarizability can depend on the field strength, if the latter is sufficiently large. One thus has to re-write equation (67) as follows:

$$\boldsymbol{p}(E) = \alpha(E)\, \boldsymbol{E} \tag{73}$$

[285] A. D. Buckingham, 'Permanent and Induced Molecular Moments and Long-range Intermolecular Forces', in *Adv. Chem. Phys.*, 1967, **12**, 107.

the polarizability $\alpha(E)$ now being some well-defined function of the field strength E. The shape of the function $\alpha(E)$ does not depend solely on E but moreover on the natural reaction of the microsystem subject to the action of the electric field. We shall now consider various cases separately.

Voigt distortion. The fact that atoms are non-linearly polarizable was known to Voigt [200] who, using the electron theory of Lorentz, showed that a high static electric field destroys the linear dependence between the

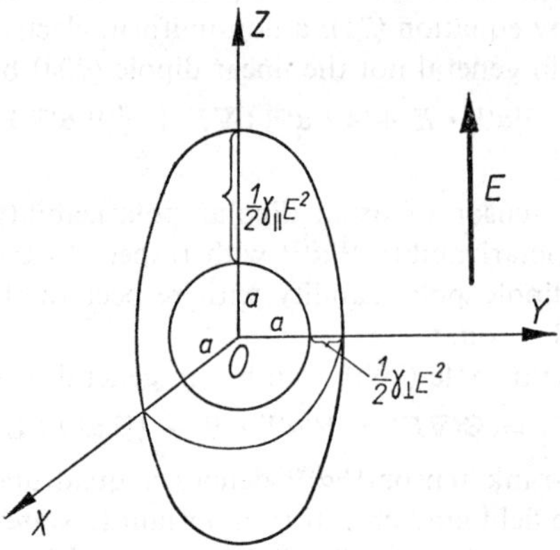

Figure 8 *The non-linear deformation of an atom due to the Voigt effect. Parallel to the applied electric field the atomic polarizability increases by a quadratic term, $\frac{1}{2}\gamma_{\parallel} E_z^2$; perpendicular to the field, it increases by $\frac{1}{2}\gamma_{\perp} E_z^2$, one third as large: $\gamma_{\perp} = \gamma_{\parallel}/3$*

elastic force and electron displacement in the atom, which now performs forced anharmonic vibrations of various frequencies with respect to the field direction. His considerations permitted the conclusion that the polarizability of the perturbed atom is a quadratic function of the electric field strength, the changes in polarizability being anisotropic, since one has, respectively parallel and perpendicular to the field, (see Figure 8):

$$\Delta\alpha_{\parallel}(E) = \tfrac{1}{2}\gamma_{\parallel} E^2$$
$$\Delta\alpha_{\perp}(E) = \tfrac{1}{2}\gamma_{\perp} E^2 \qquad (74)$$

where, according to Voigt, the non-linear distortion of the atom parallel to the field as defined by the parameter $\gamma_{\parallel}$ is three times larger than its distortion perpendicular to the field, as given by $\gamma_{\perp}$. Voigt's relation thus states that the ratio of non-linear anisotropies induced in an atom amounts to:

$$\frac{\Delta\alpha_{\parallel}(E)}{\Delta\alpha_{\perp}(E)} = \frac{\gamma_{\parallel}}{\gamma_{\perp}} = 3 \qquad (74a)$$

Quantum-mechanical calculations lead to a similar result.[54]

Born distortion. In the 1920's both the classical [264] and quantum-mechanical theories [54, 286] showed dipolar molecules to possess polarizability dependent on the field both linearly and quadratically. Consequently, equation (68) has once more to be re-written in the form:

$$p(E) = \boldsymbol{\alpha}(E) \cdot \boldsymbol{E} \tag{75}$$

where now the electric polarizability tensor $\boldsymbol{\alpha}(E)$ is a function of the field strength E and, according to Born,[13] can be rendered to within the square of the field by the expansion:

$$\boldsymbol{\alpha}(E) = \boldsymbol{\alpha}(0) + \boldsymbol{\beta} \cdot \boldsymbol{E} + \tfrac{1}{2}\boldsymbol{\gamma} : \boldsymbol{E}\boldsymbol{E} + \ldots \tag{76}$$

containing, in addition to the tensor of linear polarizability $\boldsymbol{\alpha}(0)$, the non-linear polarizabilities $\boldsymbol{\beta}$ and $\boldsymbol{\gamma}$, generally referred to as hyperpolarizabilities.[287]

The first non-linear polarizability is a tensor of rank 3 and possesses non-vanishing elements only in the case of molecules without a centre of symmetry. This Born distortion tensor was resorted to by Piekara [288] for explaining the non-linear dielectric behaviour of nitrobenzene. In the case of axially symmetric molecules, the tensor of the first non-linear distortion (hyperpolarizability [166]) can be written as follows:[289]

$$\boldsymbol{\beta} = \beta(1 - \kappa_\beta)(\boldsymbol{k}_1 \boldsymbol{U}_{23} + \boldsymbol{k}_2 \boldsymbol{U}_{31} + \boldsymbol{k}_3 \boldsymbol{U}_{12}) + 3\beta\kappa_\beta \boldsymbol{k}_1 \boldsymbol{k}_2 \boldsymbol{k}_3 \tag{76a}$$

where by analogy with the linear polarizability (69e) we have introduced:

$$\beta = \frac{\beta_{zzz} + 2\beta_{xxz}}{3}, \quad \kappa_\beta = \frac{\beta_{zzz} - \beta_{xxz}}{\beta_{zzz} + 2\beta_{xxz}} \tag{76b}$$

the mean value of the first non-linear polarizability β and its anisotropy defined by the dimensionless parameter κ_β.

The tensor $\boldsymbol{\beta}$ has non-vanishing elements in the case of tetrahedrally symmetric molecules also:[289]

$$\boldsymbol{\beta} = \beta_{xyz}\{\boldsymbol{i}_1(\boldsymbol{j}_2 \boldsymbol{k}_3 + \boldsymbol{j}_3 \boldsymbol{k}_2) + \boldsymbol{j}_1(\boldsymbol{k}_2 \boldsymbol{i}_3 + \boldsymbol{k}_3 \boldsymbol{i}_2) + \boldsymbol{k}_1(\boldsymbol{i}_2 \boldsymbol{j}_3 + \boldsymbol{i}_3 \boldsymbol{j}_2)\} \tag{76c}$$

The second non-linear polarizability $\boldsymbol{\gamma}$ is a symmetric tensor of rank 4 having non-vanishing elements even in the case of spherical symmetry, when one has:[285]

$$\boldsymbol{\gamma} = \tfrac{1}{3}\gamma(\boldsymbol{U}_{12} \boldsymbol{U}_{34} + \boldsymbol{U}_{23} \boldsymbol{U}_{41} + \boldsymbol{U}_{34} \boldsymbol{U}_{12}) \tag{76d}$$

denoting the mean polarizability:

$$\gamma = \tfrac{3}{5}(\gamma_{zzzz} + 2\gamma_{xxzz}) = \tfrac{3}{5}(\gamma_{\parallel} + 2\gamma_{\perp}) \tag{76e}$$

which, by Voigt's relation (74a), reduces to $\gamma = 3\gamma_{\perp} = \gamma_{\parallel}$.

[286] (*a*) R. de Kronig, *Z. Physik*, 1927, **45**, 458; 1928, **47**, 702; (*b*) K. F. Niessen, *Phys. Rev.*, 1929, **34**, 253.
[287] C. A. Coulson, A. McColl, and L. E. Sutton, *Trans. Faraday Soc.*, 1952, **48**, 106.
[288] A. Piekara, *Acta Phys. Polon.*, 1935, **4**, 163; 1937, **6**, 150; *Compt. rend.*, 1936, **202**, 206; *Phys. Z.*, 1937, **38**, 671.
[289] S. Kielich, *Acta Phys. Polon.*, 1964, **25**, 85.

The tensor γ can be given in explicit form for other molecular symmetries also.[155, 285, 289, 290]

The numbers of non-zero and mutually independent elements of the polarizability tensors α, β, and γ for all point groups are given in Tables 10—12.

Table 10 *Number (N) of non-zero and mutually independent (I) elements of the electric polarizability tensors* α, β, *and* γ

	Polarizability tensor elements					
	α		β		γ	
Groups	*N*	*I*	*N*	*I*	*N*	*I*
C_1	6	9	10	27	15	81
C_i	6	9	0	0	15	81
C_s	4	5	6	14	9	41
C_2	4	5	4	13	9	41
C_{2h}	4	5	0	0	9	41
C_{2v}	3	3	3	7	6	21
D_2	3	3	1	6	6	21
D_{2h}	3	3	0	0	6	21
C_4	2	3	2	7	5	29
S_4	2	3	2	12	5	29
C_{4h}	2	3	0	0	5	29
C_{4v}	2	3	2	7	4	21
D_{2d}	2	3	1	6	4	21
D_4	2	3	0	0	4	21
D_{4h}	2	3	0	0	4	21
C_3	2	3	4	15	5	53
S_6	2	3	0	0	5	53
C_{3v}	2	3	3	11	4	37
D_3	2	3	1	4	4	37
D_{3d}	2	3	0	0	4	37
C_{3h}	2	3	2	8	3	21
C_6	2	3	2	7	3	21
C_{6h}	2	3	0	0	3	21
D_{3h}	2	3	1	4	3	21
C_{6v}	2	3	2	7	3	21
D_6	2	3	0	0	3	21
D_{6h}	2	3	0	0	3	21
T	1	3	1	6	2	21
T_h	1	3	0	0	2	21
T_d	1	3	1	6	2	21
O	1	3	0	0	2	21
O_h	1	3	0	0	2	21
$C_{\infty v}$	2	3	2	7	3	21
$D_{\infty h}$	2	3	0	0	3	21
K_h	1	3	0	0	1	21
K	1	3	0	0	1	21
D_{4d}	2	3	0	0	3	21
D_{5h}	2	3	0	0	3	21
D_{5d}	2	3	0	0	3	21
D_{6d}	2	3	0	0	3	21

[290] (*a*) A. D. McLean and M. Yoshimine, *J. Chem. Phys.*, 1967, **47**, 1927; (*b*) A. D. McLean and M. Yoshimine, *ibid.*, 1967, **46**, 3682; 1967, **47**, 3256.

Table 11 *Elements of the symmetric tensor* $\boldsymbol{\beta}$ *of non-linear polarizability*

Crystallographic system	*Group*	*Number of non-zero elements*	*Number of independent elements*	*Elements of* $\boldsymbol{\beta}$ *are denoted only by their subscripts x, y, z*
Triclinic	$1(C_1)$	27	10	xxx, yyy, zzz, $xyy = yxy = yyx$, $xzz = zxz = zzx$, $xxz = xzx = zxx$, $xxy = xyx = yxx$, $yzz = zyz = zzy$, $yyz = yzy = zyy$, $xyz = xzy = yxz = yzx = zxy = zyx$
	$\bar{1}(C_i)$	0	0	
Monoclinic	$2(C_2)$	13	4	zzz, $zxx = xzx = xxz$, $zyy = yzy = yyz$, $xyz = xzy = yxz = yzx = zxy = zyx$
	$m(C_s)$	14	6	xxx, yyy, $xyy = yxy = yyx$, $yxx = xyx = xxy$, $xzz = zxz = zzx$, $yzz = zyz = zzy$
	$2/m(C_{2h})$	0	0	
Orthorhombic	$222(D_2)$	6	1	$xyz = xzy = yxz = yzx = zxy = zyx$
	$mm2(C_{2v})$	7	3	zzz, $zxx = xzx = xxz$, $zyy = yzy = yyz$
	$mmm(D_{2h})$	0	0	
Tetragonal	$4(C_4)$	7	2	zzz, $zxx = xzx = xxz = zyy = yzy = yyz$
	$\bar{4}(S_4)$	12	2	$zxx = xzx = xxz = -zyy = -yzy = -yyz$, $xyz = xzy = yxz = yzx = zxy = zyx$
	$4/m(C_{4h})$	0	0	
	$422(D_4)$	0	0	
	$4mm(C_{4v})$	7	2	zzz, $zxx = xzx = xxz = zyy = yzy = yyz$
	$\bar{4}2m(D_{2d})$	6	1	$xyz = xzy = yxz = yzx = zxy = zyx$
	$4/mmm(D_{4h})$	0	0	

[*Continued overleaf*]

Table 11 (*cont.*)

Crystallographic system	*Group*	*Number of non-zero elements*	*Number of independent elements*	*Elements of* $\boldsymbol{\beta}$ *are denoted only by their subscripts x, y, z*
Trigonal	$3(C_3)$	15	4	*zzz*, *xxx* = − *xyy* = − *yxy* = − *yyx*, *yyy* = − *yxx* = − *xyx* = − *xxy*, *zxx* = *xzx* = *xxz* = *zyy* = *yzy* = *yyz*
	$\bar{3}(C_{3i})$	0	0	
	$32(D_3)$	4	1	*xxx* = − *xyy* = − *yxy* = − *yyx*
	$3m(C_{3v})$	11	3	*zzz*, *yyy* = − *yxx* = − *xyx* = − *xxy*, *zxx* = *xzx* = *xxz* = *zyy* = *yzy* = *yyz*
	$\bar{3}m(D_{3d})$	0	0	
Hexagonal	$6(C_6)$	7	2	*zzz*, *zxx* = *xzx* = *xxz* = *zyy* = *yzy* = *yyz*
	$\bar{6}(C_{3h})$	8	2	*xxx* = − *xyy* = − *yxy* = − *yyx*, *yyy* = − *yxx* = − *xyx* = − *xxy*
	$6/m(C_{6h})$	0	0	
	$622(D_6)$	0	0	
	$6mm(C_{6v})$	7	2	*zzz*, *zxx* = *xzx* = *xxz* = *zyy* = *yzy* = *yyz*
	$\bar{6}m2(D_{3h})$	4	1	*yyy* = − *yxx* = − *xyx* = − *xxy*
	$6/mmm(D_{6h})$	0	0	
Cubic	$23(T)$	6	1	*xyz* = *xzy* = *yxz* = *yzx* = *zxy* = *zyx*
	$m3(T_h)$	0	0	
	$432(O)$	0	0	
	$\bar{4}3m(T_d)$	6	1	*xyz* = *xzy* = *yxz* = *yzx* = *zxy* = *zyx*
	$m3m(O_h)$	0	0	
Axially symmetric body	$C_{\infty v}$	7	2	*zzz*, *zxx* = *xzx* = *xxz* = *zyy* = *yzy* = *yyz*
	$D_{\infty h}$			
Isotropic body	K_h			
	K	0	0	

Table 12 *Tensor elements γ of third-order polarizability, for all crystallographic classes, on the assumption of total symmetry. The tensor elements are expressed by their subscripts only*

1. Triclinic. The two classes $1(C_1)$ and $\bar{1}(C_i)$ have 81 non-zero tensor elements, 15 of which are independent:
$xxxx$, $yyyy$, $zzzz$,
$xxyy = xyxy = xyyx = yyxx = yxyx = yxxy$,
$xxzz = xzxz = xzzx = zzxx = zxzx = zxxz$,
$yyzz = yzyz = yzzy = zzyy = zyzy = zyyz$,
$xxxy = xxyx = xyxx = yxxx$,
$yyyx = yyxy = yxyy = xyyy$,
$xzzz = zxzz = zzxz = zzzx$,
$yzzz = zyzz = zzyz = zzzy$,
$xxxz = xxzx = xzxx = zxxx$,
$yyyz = yyzy = yzyy = zyyy$,
$xxyz = xyxz = xyzx = xxzy = xzxy = xzyx$
$= yxxz = yxzx = yzxx = zxxy = zxyx = zyxx$,
$yyxz = yxyz = yxzy = yyzx = yzyx = yzxy$
$= xyyz = xyzy = xzyy = zyyx = zyxy = zxyy$,
$xzzy = xzyz = xyzz = yxzz = yzxz = yzzx$
$= zxyz = zyxz = zyzx = zxzy = zzyx = zzxy$

2. Monoclinic. All three classes $2(C_2)$, $m(C_3)$, and $2/m(C_{2h})$ have 41 non-zero tensor elements, 9 of which are independent:
$xxxx$, $yyyy$, $zzzz$,
$xxyy = xyxy = xyyx = yyxx = yxyx = yxxy$,
$xxzz = xzxz = xzzx = zzxx = zxzx = zxxz$,
$yyzz = yzyz = yzzy = zzyy = zyzy = zyyz$,
$xxxy = xxyx = xyxx = yxxx$,
$yyyx = yyxy = yxyy = xyyy$,
$xzzy = xzyz = xyzz = yxzz = yzxz = yzzx$
$= zxyz = zyxz = zyzx = zxzy = zzyx = zzxy$

3. Orthorhombic. The classes $222(D_2)$, $mm2(C_{2v})$, and $mmm(D_{2h})$ have 21 non-zero tensor elements, 6 of which are independent:
$xxxx$, $yyyy$, $zzzz$,
$xxyy = xyxy = xyyx = yyxx = yxyx = yxxy$,
$xxzz = xzxz = xzzx = zzxx = zxzx = zxxz$,
$yyzz = yzyz = yzzy = zzyy = zyzy = zyyz$

4. Tetragonal. The classes $4(C_4)$, $\bar{4}(S_4)$, and $4/m(C_{4h})$ have 29 non-zero tensor elements, 5 of which are independent:
$zzzz$, $xxxx = yyyy$,
$xxyy = xyxy = xyyx = yyxx = yxyx = yxxy$,
$xxzz = xzxz = xzzx = zzxx = zxzx = zxxz$
$= yyzz = yzyz = yzzy = zzyy = zyzy = zyyz$,
$xxxy = xxyx = xyxx = yxxx = -\ yyyx = -\ yyxy = -\ yxyy = -\ xyyy$

The classes $422(D_4)$, $4mm(C_{4v})$, $\bar{4}mm(D_{4h})$ and, $\bar{4}2m(D_{2d})$ have 21 non-zero tensor elements, 4 of which are independent:
$zzzz$, $xxxx = yyyy$,
$xxyy = xyxy = xyyx = yyxx = yxyx = yxxy$,
$xxzz = xzxz = xzzx = zzxx = zxzx = zxxz$
$= yyzz = yzyz = yzzy = zzyy = zyzy = zyyz$

Table 12 (*cont.*)

5. Trigonal. The classes $3(C_3)$ and $\bar{3}(C_{3i})$ have 53 non-zero tensor elements, 5 of which are independent:

$zzzz,$
$xxyy = xyxy = xyyx = yyxx = yxyx = yxxy,$
$xxzz = xzxz = xzzx = zzxx = zxzx = zxxz$
$= yyzz = yzyz = yzzy = zzyy = zyzy = zyyz,$
$xxxz = xxzx = xzxx = zxxx = - yxyz = - yyzx = - yzyx = - yyxz$
$= - yxzy = - yzxy = - xyyz = - xyzy = - xzyy = - zxyy$
$= - zyxy = - zyyx,$
$yyyz = yyzy = yzyy = zyyy = - xyxz = - xxzy$
$= - xzxy = - xxyz = - xyzx = - xzyx = - yxxz = - yxzx$
$= - yzxx = - zyxx = - zxyx = - zxxy,$
$(xxxx = yyyy = xxyy + xyxy + xyyx)$

The classes $3m(C_{3v})$, $\bar{3}m(D_{3d})$, and $32(D_3)$ have 37 non-zero tensor elements, 4 of which are independent:

$zzzz,$
$xxyy = xyxy = xyyx = yyxx = yxyx = yxxy,$
$xxzz = xzxz = xzzx = zzxx = zxzx = zxxz$
$= yyzz = yzyz = yzzy = zzyy = zyzy = zyyz,$
$yyyz = yyzy = yzyy = zyyy = - xyxz = - xxzy$
$= - xzxy = - xxyz = - xyzx = - xzyx = - yxxz = - yxzx$
$= - yzxx = - zyxx = - zxyx = - zxxy,$
$(xxxx = yyyy = xxyy + xyxy + xyyx)$

6. Hexagonal. The classes $6(C_6)$, $\bar{6}(C_{3h})$, and $6/m(C_{6h})$, $622(D_6)$, $6mm(C_{6v})$, $\bar{6}m2(D_{3h})$, $6/mmm(D_{6h})$ have 19 non-zero tensor elements, only 3 of which are independent:

$zzzz,$
$xxyy = xyxy = xyyx = yyxx = yxyx = yxxy,$
$xxzz = xzxz = xzzx = zzxx = zxzx = zxxz$
$= yyzz = yzyz = yzzy = zzyy = zyzy = zyyz,$
$(xxxx = yyyy = xxyy + xyxy + xyyx)$

7. Cubic. All classes of this system $23(T)$, $m3(T_h)$, $432(O)$, $\bar{4}3m(T_d)$, and $m3m(O_h)$ have 21 non-zero tensor elements, only 2 of which are independent:

$xxxx = yyyy = zzzz,$
$xxyy = xyxy = xyyx = yyxx = yxyx = yxxy$
$= zzxx = zxzx = zxxz = xxzz = xzxz = xzzx$
$= yyzz = yzyz = yzzy = zzyy = zyzy = zyyz$

8. Isotropic medium. Groups Y, Y_h, K, and K_h. Non-zero elements as in the cubic system, but only 1 independent element owing to the relation:

$xxxx = xxyy + xyxy + xyyx$

9. Axially symmetric medium.
Groups C_∞, $C_{\infty v}$, $C_{\infty h}$, and $D_{\infty h}$
Non-zero and independent tensor elements as in the classes of the hexagonal system.

Determination of non-linear polarizabilities. The expansion (76), containing the non-linear polarizabilities, was at first applied to the theory of the Kerr effect by Born [13] and Volkenshteyn,[14] as well as Buckingham and Pople,[291] and then by Piekara and Kielich [142] to the theory of other non-linear effects. The number of mutually independent elements of the tensor $\boldsymbol{\alpha}$, $\boldsymbol{\beta}$, and $\boldsymbol{\gamma}$ for all point groups has been given by Buckingham [285] and Boyle,[232] whereas Oźgo [292] determined the number of non-zero elements for $\boldsymbol{\gamma}$. The hyperpolarizability elements $\boldsymbol{\beta}$ have been calculated numerically for several dipolar molecules [290, 293] and those of the tensor $\boldsymbol{\gamma}$ for inert gases,[294] atoms and ions,[295] linear molecules,[290, 295] and planar molecules.[296] Hyperpolarizability calculations for helical polymer molecules [297] have also been proposed. For $\boldsymbol{\beta}$ and $\boldsymbol{\gamma}$, they can be determined directly from Kerr effect measurements, as has been done for many years by Buckingham and his associates,[178] and from non-linear optical effects,[150] in particular second-harmonic light scattering.[155, 226, 227] These methods of measuring the non-linear polarizabilities of atoms and molecules will be considered later. Table 9 contains numerical values of these hyperpolarizabilities.

Non-linear Multipolar Polarizabilities. Just as the polarizability tensor $\boldsymbol{\alpha}$ in equation (68) defined a linear dipole moment, or moment of order 1 (as due to the first power of E), so the tensors $\boldsymbol{\beta}$ and $\boldsymbol{\gamma}$ appearing in Born's expansions (76) can be taken to define respectively a dipole moment of order 2 (induced by the square of E):

$$p(E^2) = \tfrac{1}{2}\boldsymbol{\beta} : \boldsymbol{E}\boldsymbol{E} \qquad (77)$$

and a dipole moment of order 3 (due to the third power of E):

$$p(E^3) = \tfrac{1}{6}\boldsymbol{\gamma} \vdots \boldsymbol{E}\boldsymbol{E}\boldsymbol{E} \qquad (78)$$

The non-linear dipole moments defined by the preceding equations are relevant only if the field acting on the microsystem is a uniform field (*i.e.* one presenting no gradients). In the presence of field inhomogeneities, *i.e.* gradients, additional components related to induced quadrupoles and octupoles appear.[166] This can be taken account of by writing [by analogy with the multipolar moment of order 1, equation (72)] the following

[291] A. D. Buckingham and J. A. Pople, *Proc. Phys. Soc.* (*A*), 1955, **68**, 905.
[292] Z. Ożgo, *Acta Phys. Polon.*, 1968, **34**, 1087; Z. Ożgo and R. Zawodny, *Bull. Soc. Amis. Sci. Lettres, Poznań*, 1971, **B22**,
[293] (*a*) J. M. O'Hare and R. P. Hurst, *J. Chem. Phys.*, 1967, **46**, 2356; (*b*) G. P. Arrighini, M. Maestro, and R. Moccia, Symposia of the Faraday Soc., 1968, No. 2, p. 48; (*c*) G. P. Arrighini, J. Tomasi, and C. Petrongolo, *Theor. Chim. Acta*, 1970, **18**, 341; (*d*) S. P. Liebmann and J. W. Moskowitz, *J. Chem. Phys.*, 1971, **54**, 3622.
[294] (*a*) L. L. Boyle, A. D. Buckingham, R. L. Disch, and D. A. Dunmur, *J. Chem. Phys.*, 1966, **45**, 1318; (*b*) P. Sitz and R. Yaris, *ibid.*, 1968, **49**, 3546; (*c*) R. Locqueneux and P. Smet, *Compt. rend.*, 1968, **267**, *B*, 353, 433.
[295] (*a*) L. L. Boyle and C. A. Coulson, *Mol. Phys.*, 1966, **11**, 165; (*b*) P. W. Langhoff, J. D. Lyons, and R. P. Hurst, *Phys. Rev.*, 1966, **148**, 18; (*c*) G. W. F. Drake and M. Cohen, *J. Chem. Phys.*, 1968, **48**, 1168; (*d*) M. N. Grasso, K. T. Chung, and R. P. Hurst, *Phys. Rev.*, 1968, **167**, 1.
[296] A. Schweig, *Chem. Phys. Letters*, 1967, **1**, 195.
[297] B. Fanconi and W. L. Peticolas, *J. Chem. Phys.*, 1969, **50**, 2244.

expression for a multipolar moment of the second order:[268]

$$p^{(n)}(E^2) = \frac{1}{2} \sum_{n_1=1}^{\infty} \sum_{n_2=1}^{\infty} \frac{2^{n_1+n_2}\, n_1!\, n_2!}{(2n_1)!\,(2n_2)!}\, {}^{(n)}\beta^{(n_1+n_2)}[n_1 + n_2]\, E^{(n_1)}\, E^{(n_2)} \tag{79}$$

where ${}^{(n)}\beta^{(n_1+n_2)}$ is a tensor of rank $n + n_1 + n_2$ defining second-order multipole polarizability.

Obviously, in practice it is hardly necessary to proceed beyond the first several terms of equation (79). We thus obtain, in place of equation (77), the following dipole moment of order 2:

$$p^{(1)}(E^2) = \tfrac{1}{2}{}^{(1)}\beta^{(2)} : EE + \tfrac{1}{3}{}^{(1)}\beta^{(3)} \vdots E\nabla E + \ldots \tag{79a}$$

and quadrupole moment of order 2:

$$p^{(2)}(E^2) = \Theta(E^2) = \tfrac{1}{2}{}^{(2)}\beta^{(2)} \cdot EE + \ldots \tag{79b}$$

Hitherto, theoretical calculations have been carried out for a few multipole polarizabilities only, *i.e.* for field-gradient-induced quadrupole polarizabilities of atoms and ions[298] [defined by the tensor ${}^{(2)}\alpha^{(2)}$ of equation (69b)], multipole polarizabilities of higher orders,[299] and quadrupole hyperpolarizabilities[300] [the tensor ${}^{(2)}\beta^{(2)}$ of equation (79)].

Molecular Magnetic Susceptibilities. Our preceding considerations concerning electric polarizabilities can be carried over into the domain of molecular magnetic susceptibilities on the basis of the definitions (42) and (43). Similarly to equation (68), the induced magnetic moment of a molecule can be written as:

$$m(B) = \chi \cdot B \tag{80}$$

where χ is the tensor of linear molecular magnetic susceptibility separable, like equation (69), into an isotropic and an anisotropic part.

The tensor elements of χ can be determined from measurements of macroscopic magnetic susceptibility[112] or evaluated from molecular orbital methods[252, 279] and approximate variation–perturbation calculations.[301] Recently, calculations of the magnetic quadrupole polarizability of closed-shell atoms,[302] and magneto-electric susceptibilities of atoms,[303] have been made. These matters, which relate to the behaviour of microsystems under the simultaneous action of an electric and a magnetic field, will be dealt with in detail in subsequent sections.

298 (*a*) R. M. Sternheimer, *Phys. Rev.*, 1954, **96**, 951; 1957, **107**, 1565; 1970, **1**, 321; (*b*) A. Dalgarno and J. T. Lewis, *Proc. Roy. Soc.*, 1957, **A240**, 284; (*c*) A. Dalgarno, *Adv. Phys.*, 1962, **11**, 281.

299 (*a*) G. Burns, *Phys. Rev.*, 1959, **115**, 357; (*b*) A. L. Stewart, *Proc. Phys. Soc.*, 1961, **77**, 447; (*c*) H. P. Kelly and H. S. Taylor, *J. Chem. Phys.*, 1964, **40**, 1478; (*d*) P. W. Langhoff and R. P. Hurst, *Phys. Rev.*, 1965, **139**, 1415; (*e*) J. T. Broussard and N. Kestner, *J. Chem. Phys.*, 1970, **53**, 1507.

300 A. D. Buckingham, C. A. Coulson, and J. T. Lewis, *Proc. Phys. Soc. (A)*, 1956, **69**, 639.

301 (*a*) L. Caralp and J. Horau, *J. Chim. phys.*, 1971, **68**, 63; (*b*) W. R. Bley, *Mol. Phys.*, 1971, **20**, 491; (*c*) A. J. Sadlej, *ibid.*, p. 593.

302 H. S. Radt and R. P. Hurst, *Phys. Rev. (A)*, 1970, **2**, 696.

303 (*a*) A. D. Buckingham and J. A. Pople, *Proc. Cambridge Phil. Soc.*, 1957, **53**, 262; (*b*) J. D. Lyons and R. P. Hurst, *Phys. Rev.*, 1967, **162**, 38.

Energy of Induced Multipole Interaction.—*Induced Interaction between a Multipolar System and External Fields.* We calculated above the potential energy of electrostatic interaction between a multipole system and external fields, or the first-order energy due to the first power of the field. Besides that energy, which took account only of the reorientation of permanent multipoles, we have to take into consideration contributions due to the circumstance that an external electric field induces higher-order multipole moments given by the expressions (72) and (79).

Energy of the dipole type. Restricting ourselves at first to the dipole approximation of equations (62), (77), and (78) we write the energies of the first, second, third, fourth, ... order as follows:

$$u(E) = -\,\boldsymbol{p}(0)\cdot\boldsymbol{E} = -\,\boldsymbol{E}\cdot\boldsymbol{p}(0) \tag{81a}$$

$$u(E^2) = -\,\tfrac{1}{2}\boldsymbol{E}\cdot\boldsymbol{p}(E) = -\,\tfrac{1}{2}\boldsymbol{E}\cdot\boldsymbol{\alpha}\cdot\boldsymbol{E} \tag{81b}$$

$$u(E^3) = -\,\tfrac{1}{3}\boldsymbol{E}\cdot\boldsymbol{p}(E^2) = -\,\tfrac{1}{6}\boldsymbol{E}\cdot\boldsymbol{\beta}:\boldsymbol{E}\boldsymbol{E} \tag{81c}$$

$$u(E^4) = -\,\tfrac{1}{4}\boldsymbol{E}\cdot\boldsymbol{p}(E^3) = -\,\tfrac{1}{24}\boldsymbol{E}\cdot\boldsymbol{\gamma}\,\vdots\,\boldsymbol{E}\boldsymbol{E}\boldsymbol{E} \tag{81d}$$

The energy (81a) of order 1 is simply that of electrostatic interaction of the permanent dipole moment $\boldsymbol{p}(0)$ and a uniform electric field. The energies of order 2 and upward are energies of the inductional kind, that is to say, energies of interactions between successively induced dipole moments and a uniform electric field. We thus can quite generally write, for the m-th order energy of dipole–uniform field interaction:

$$u(E^m) = -\,\frac{1}{m}\boldsymbol{E}\cdot\boldsymbol{p}(E^{m-1}) \tag{80}$$

A similar result is obtained in the magnetic case [the magnetic dipole moment of order 1 is defined by equation (80) and higher-order moments are defined similarly]:

$$u(B^m) = -\,\frac{1}{m}\boldsymbol{B}\cdot\boldsymbol{m}(B^{m-1}) \tag{82}$$

Hence, with (42) and (80), we have for the energies of order 1, 2, 3 ...:

$$u(B) = -\,\boldsymbol{B}\cdot\boldsymbol{m}(0) = -\,\boldsymbol{m}(0)\cdot\boldsymbol{B} \tag{82a}$$

$$u(B^2) = -\,\tfrac{1}{2}\boldsymbol{B}\cdot\boldsymbol{m}(B) = -\,\tfrac{1}{2}\boldsymbol{B}\cdot\boldsymbol{\chi}\cdot\boldsymbol{B} \tag{82b}$$

$$u(B^3) = -\,\tfrac{1}{3}\boldsymbol{B}\cdot\boldsymbol{m}(B^2) = -\,\tfrac{1}{6}\boldsymbol{B}\cdot\boldsymbol{\psi}:\boldsymbol{B}\boldsymbol{B} \tag{82c}$$

where the third rank tensor $\boldsymbol{\psi}$ is that of non-linear magnetic polarizability of order 2.

Energies of multipole type. In cases when the external field is not uniform, one has to consider, in conformity with the expansion (52a), interactions between the quadrupole moment and the field gradient, the octupole and gradient of the field gradient, and so forth. The potential energy of multipole interaction of order 1 being given by equation (54), we can by extension write the potential energy of order $(m + 1)$ for a molecule as follows:[268]

$$u(E^{m+1}) = -\frac{1}{m+1}\sum_{n=1}^{\infty}\frac{2^n n!}{(2n)!}E^{(n)}[n]\,p^{(n)}(E^m) \tag{83}$$

where $p^{(n)}(E^m)$ stands for the 2^n-pole electric moment of the m-th order. The permanent electric multipoles $(m = 0)$ are defined by equation (40), the electric multipoles of first order $m = 1$ by equation (72), and those of the second order $m = 2$ by equation (79). Similarly, magnetic potential energy of order $m = 1$ can be defined, on replacing in equation (83) $\boldsymbol{E}$ by $\boldsymbol{B}$ and $\boldsymbol{p}$ by $\boldsymbol{m}$. From the general expression [equation (83)] one immediately and quite easily derives all the energies dealt with in the theory of non-linear molecular processes.[177]

Induced Interaction between Two Multipole Systems. Equation (58) defines in general form the classical electrostatic interaction of two electric systems having permanent multipoles $\boldsymbol{p}_i^{(m)}$ and $\boldsymbol{p}_j^{(n)}$, in conformity with the classical theory of Keesom.[304] In the classical approach also, as shown by Debye and Falkenhagen,[305] one has to take into consideration energies due to interactions between the permanent multipoles of the one system and electric multipoles induced in the other, and *vice versa*. Restricting the problem in a first approximation to the energy arising from the mutual interaction of dipoles, we can write:

$$V_{ij}^{\text{ind}} = -\tfrac{1}{2}(\boldsymbol{F}_{ij}\cdot\boldsymbol{\alpha}_i\cdot\boldsymbol{F}_{ij} + \boldsymbol{F}_{ji}\cdot\boldsymbol{\alpha}_j\cdot\boldsymbol{F}_{ji}) \tag{84}$$

where, in general, the electric fields $\boldsymbol{F}_{ij}$ and $\boldsymbol{F}_{ji}$ are defined by equation (56) for $m = 1$. We shall now apply equation (84) to some particular cases, discussing for the sake of simplicity only the first term—the one between permanent multipoles $\boldsymbol{p}_j^n$ and an induced dipole $\boldsymbol{\alpha}_i$.

(*a*) Induced-dipole–ion interaction, by equation (84), is given by the expression:

$$V_{ij}^{\alpha-q} = -\frac{q_j^2}{8\pi\varepsilon_0}(\boldsymbol{r}_{ij}\cdot\boldsymbol{\alpha}_{ij}\cdot\boldsymbol{r}_{ij})\,r_{vi}^{-6} \tag{85}$$

where $\boldsymbol{\alpha}_{vi}$ is the volume polarizability. For axially symmetric molecules having polarizability of the form resulting from equations (69) and (69d):

$$\boldsymbol{\alpha}_i = \alpha_i(1 - \kappa_i)\,\boldsymbol{U} + 3\kappa_i\,\boldsymbol{k}_i\,\boldsymbol{k}_i \tag{86}$$

the expression (85) reduces to:[231]

$$V_{ij}^{\alpha-q} = -\frac{\alpha_{vi}\,q_j^2}{8\pi\varepsilon_0}\{1 + \kappa_i(3\cos^2\theta_i - 1)\}\,r_{ij}^{-4} \tag{85a}$$

(*b*) Induced-dipole–permanent-dipole interaction is, by equation (84):

$$V_{ij}^{\alpha-p} = -\frac{1}{8\pi\varepsilon_0}\{9(\boldsymbol{r}_{ij}\cdot\boldsymbol{\alpha}_{vi}\cdot\boldsymbol{r}_{ij})(\boldsymbol{p}_j\cdot\boldsymbol{r}_{ij})^2$$
$$- 6r_{ij}^2(\boldsymbol{r}_{ij}\cdot\boldsymbol{\alpha}_{vi}\cdot\boldsymbol{p}_j)(\boldsymbol{p}_j\cdot\boldsymbol{r}_{ij}) + r_{ij}^4(\boldsymbol{p}_j\cdot\boldsymbol{\alpha}_{vi}\cdot\boldsymbol{p}_j)\}\,r_{ij}^{-10} \tag{87}$$

[304] W. H. Keesom, *Physik. Z.*, 1921, **22**, 129, 643; 1922, **23**, 225.
[305] (*a*) P. Debye, *Physik. Z.*, 1920, **21**, 178; 1921, **22**, 302; (*b*) H. Falkenhagen, *ibid.*, 1922, **23**, 87.

For axially symmetric molecules one has [231] by equation (86):

$$V_{ij}^{\alpha-p} = -\frac{\alpha_{vi} p_j^2}{8\pi\varepsilon_0}\{(1-\kappa_i)(3\cos^2\theta_j+1) + 3\kappa_i(3\cos\theta_i\cos\theta_j - \cos\theta_{ij})^2\} r_{ij}^{-6} \tag{87a}$$

(*c*) Induced-dipole–permanent-dipole and -quadrupole interaction, from equation (84), is, in vector notation:

$$\begin{aligned} V_{ij}^{\alpha-p\Theta} = -\frac{1}{4\pi\varepsilon_0}\{5[3(\boldsymbol{r}_{ij}\cdot\boldsymbol{\alpha}_{vi}\cdot\boldsymbol{r}_{ij})(\boldsymbol{p}_j\cdot\boldsymbol{r}_{ij}) \\ - r_{ij}^2(\boldsymbol{p}_j\cdot\boldsymbol{\alpha}_{vi}\cdot\boldsymbol{r}_{ij})](\boldsymbol{r}_{ij}\cdot\boldsymbol{\Theta}_j\cdot\boldsymbol{r}_{ij}) \\ - 2r_{ij}^2[3(\boldsymbol{r}_{ij}\cdot\boldsymbol{\alpha}_{vi}\cdot\boldsymbol{\Theta}_j\cdot\boldsymbol{r}_{ij})(\boldsymbol{p}_j\cdot\boldsymbol{r}_{ij}) \\ - r_{ij}^2(\boldsymbol{p}_j\cdot\boldsymbol{\alpha}_{vi}\cdot\boldsymbol{\Theta}_j\cdot\boldsymbol{r}_{ij})]\} r_{ij}^{-12} \end{aligned} \tag{88}$$

In the case of axially symmetric molecules, this reduces to the form:[231]

$$\begin{aligned} V_{ij}^{\alpha-p\Theta} = -\frac{3\alpha_{vi} p_j \Theta_j}{8\pi\varepsilon_0}\{4(1-\kappa_i)\cos^3\theta_j \\ + 3\kappa_i(5\cos\theta_i\cos^2\theta_j - 2\cos\theta_{ij}\cos\theta_j - \cos\theta_i) \\ \times (3\cos\theta_i\cos\theta_j - \cos\theta_{ij})\} r_{ij}^{-7} \end{aligned} \tag{88a}$$

(*d*) Induced-dipole–permanent-quadrupole interaction, defined generally by equation (84), is:

$$\begin{aligned} V_{ij}^{\alpha-\Theta} = -\frac{1}{8\pi\varepsilon_0}\{25(\boldsymbol{r}_{ij}\cdot\boldsymbol{\alpha}_i\cdot\boldsymbol{r}_{ij})(\boldsymbol{r}_{ij}\cdot\boldsymbol{\Theta}_j\cdot\boldsymbol{r}_{ij})^2 \\ - 20r_{ij}^2(\boldsymbol{r}_{ij}\cdot\boldsymbol{\alpha}_{vi}\cdot\boldsymbol{\Theta}_j\cdot\boldsymbol{r}_{ij})(\boldsymbol{r}_{ij}\cdot\boldsymbol{\Theta}_j\cdot\boldsymbol{r}_{ij}) \\ + 4r_{ij}^4(\boldsymbol{r}_{ij}\cdot\boldsymbol{\Theta}_j\cdot\boldsymbol{\alpha}_{vi}\cdot\boldsymbol{\Theta}_j\cdot\boldsymbol{r}_{ij})\} r_{ij}^{-14} \end{aligned} \tag{89}$$

For axially symmetric molecules we have:[231]

$$\begin{aligned} V_{ij}^{\alpha-\Theta} = -\frac{9\alpha_{vi}\Theta_j^2}{32\pi\varepsilon_0}\{(1-\kappa_i)(1-2\cos^2\theta_j+5\cos^4\theta_j) \\ + 3\kappa_i(5\cos\theta_i\cos^2\theta_j - 2\cos\theta_{ij}\cos\theta_j - \cos\theta_i)^2\} r_{ij}^{-8} \end{aligned} \tag{89a}$$

On neglecting, in equations (87a), (88a), and (89a), the anisotropies of polarizabilities (*i.e.* putting $\kappa_i = 0$) one obtains the Debye–Falkenhagen formulae.[305] Similarly, it is possible to calculate induced energies of interaction between higher-order multipole moments.[268]

Dispersional Interaction between Molecules. We still wish to consider briefly energies due to interaction between fluctuating induced electric charge distributions of atoms and molecules. In constrast to electrostatic and induced interactions, these are present even when the molecules do not possess permanent electric moments. These dispersional interactions cannot be dealt with on a classical electrostatics level; owing to their relation to London's quantum dispersion theory, they have been termed London dispersional interactions.

The first approximation of quantum-mechanical perturbation calculus yields the energy of electrostatic interaction (58) between permanent multipoles of the two microsystems. The diagonal part of the second quantum-mechanical approximation provides the energy of induced interaction between the permanent multipoles of the one microsystem and multipoles induced in the other microsystem, *i.e.* energies of the form (84). The remaining part of the second approximation (that related to excited states) yields additionally the energy of dispersional interaction. According to London,[306] it is of a typically quantum nature; in a 'semi-classical picture', it can be regarded as due to interaction between induced dipoles (and multipoles) of the two microsystems.

For the case of anisotropic dipolar oscillators, the dispersional forces have been discussed by London[306] and De Boer.[307] On neglecting the anisotropy of electron oscillation frequency, the dispersional energy of interaction between two anisotropic molecules becomes, in a dipole–dipole approximation:[228, 231]

$$V_{ij}^{\text{disp}} = -\frac{1}{2}\left(\frac{h\nu_i\, h\nu_j}{h\nu_i + h\nu_j}\right)\{9(\boldsymbol{r}_{ij}\cdot\boldsymbol{\alpha}_{vi}\cdot\boldsymbol{r}_{ij})(\boldsymbol{r}_{ij}\cdot\boldsymbol{\alpha}_{vj}\cdot\boldsymbol{r}_{ij}) - 6r_{ij}^2(\boldsymbol{r}_{ij}\cdot\boldsymbol{\alpha}_{vi}\cdot\boldsymbol{\alpha}_{vj}\cdot\boldsymbol{r}_{ij}) + r_{ij}^4(\boldsymbol{\alpha}_{vi}:\boldsymbol{\alpha}_{vj})\}\, r_{ij}^{-10} \tag{90}$$

where $h\nu_i$ and $h\nu_j$ are quantal characteristic energies of the two interacting molecules i and j.

For axially symmetric molecules, the polarizability tensor is of the form (86), and equation (90) leads to the following result:[231]

$$V_{ij}^{\text{disp}} = -\frac{3}{4}\left(\frac{h\nu_i\, h\nu_j}{h\nu_i + h\nu_j}\right)\alpha_{vi}\,\alpha_{vj}\{2 - \kappa_i - \kappa_j + 3\kappa_i(1-\kappa_j)\cos^2\theta_j + 3\kappa_j(1-\kappa_i)\cos^2\theta_i + 3\kappa_i\kappa_j(3\cos\theta_i\cos\theta_j - \cos\theta_{ij})^2\}\, r_{ij}^{-6} \tag{90a}$$

which is obtained from London's formula[306] on neglecting the difference between the oscillation frequencies of electrons parallel and perpendicular to the symmetry axis of the molecule.

Multipole dispersional energy calculations have been performed for atoms,[308] molecules,[309] and atom–molecule interactions.[310] Casimir and

[306] F. London, *Z. Phys.*, 1930, **63**, 245; *Trans. Faraday Soc.*, 1937, **33**, 8; *J. Phys. Chem.*, 1942, **46**, 305.

[307] J. H. De Boer and G. Heller, *Physica*, 1937, **4**, 1045; J. H. De Boer, *ibid.*, 1942, **9**, 363.

[308] (*a*) P. R. Fontana, *Phys. Rev.*, 1961, **123**, 1865, 1871; (*b*) L. L. Henry and T. W. Marshall, *Nuovo Cim.*, 1966, **41**, 592; (*c*) E. C. Darwall, M. A. D. Fluendy, and K. P. Lawley, *Mol. Phys.*, 1970, **19**, 673.

[309] (*a*) L. Salem, *Mol. Phys.*, 1960, **3**, 441; (*b*) C. Mavroyannis and M. J. Stephen, *ibid.*, 1962, **5**, 629; (*c*) B. Linder, *J. Chem. Phys.*, 1962, **37**, 963; (*d*) A. Van Der Avoird and F. Hofelich, *ibid.*, 1966, **45**, 4664.

[310] (*a*) R. M. Herman, *J. Chem. Phys.*, 1966, **44**, 1346; (*b*) A. D. Buckingham, *ibid.*, 1968, **48**, 3827; (*c*) C. G. Gray, *ibid.*, 1969, **50**, 549; (*d*) D. Robert, C. Girardet, and L. Galatry, *Chem. Phys. Letters*, 1969, **3**, 102; (*e*) J. Bonamy and D. Robert, *ibid.*, 1970, **6**, 591; (*f*) C. Girardet and D. Robert, *J. Mol. Structure*, 1971, **7**, 31.

Polder[311] have shown that, at intermolecular separations of the order of the wavelength of optical energy transitions of the constituent molecules, London's r_{ij}^{-6} dispersional energy dependence has to be replaced by an r_{ij}^{-7} dependence. Recently, non-additive three-body dispersion forces have been discussed.[312]

3 Statistical and Electromagnetic Properties of Dielectrics

Statistical methods are required for dealing with macroscopic bodies which can otherwise be characterized as thermodynamic systems consisting of a very large number of microsystems (atoms, molecules, ions, macromolecules). The simplest example of such a system is provided by the perfect gas, the atoms (molecules) of which are in a state of unceasing, random motion. Between one elastic collision and another, the microsystems move rectilinearly, exerting no action on one another (obviously, omitting the well-known forces with which they interact at collision, during an extremely short portion of the time). For such gases, Maxwell and Boltzmann proposed statistical methods which provide the basis of the kinetic theory of gases. For dense substances, in a classical approach, Gibbs' statistics are applicable, if the state is that of thermodynamic equilibrium.

Statistical methods are used for explaining the macroscopic properties of bodies from the point of view of their microscopic structure. The macroscopic physical quantities are then found to be mean statistical values of parameters characterizing the microscopic processes occurring in the system as a whole. The procedure of statistical physics consists in assuming some microscopic (classical or quantal) model of the body from which, by methods of statistics, we derive its equation of state and describe all its macroscopic, *e.g.* in the present context, electromagnetic, properties.

Statistical Perturbation Calculus.—*Statistical Mean Value.* Let Q denote some physical quantity (electric polarization, magnetic polarization, energy), in general a function of the continuously varying canonic variables Γ (generalized momenta, generalized co-ordinates), and the time t, thus $Q(\Gamma, t)$. In classical statistical mechanics, the mean value of Q is defined as:

$$\langle Q\rangle = \int_{\Gamma} Q(\Gamma, t) f(\Gamma, t)\, d\Gamma \tag{91}$$

311 (*a*) H. B. G. Casimir and D. Polder, *Phys. Rev.*, 1948, **73**, 360; (*b*) A. D. McLachlan, *Proc. Roy. Soc.*, 1963, **A274**, 80; (*c*) L. Gomberoff and E. A. Power, *ibid.*, 1966, **A295**, 476; (*d*) W. J. Meath and J. O. Hirschfelder, *J. Chem. Phys.*, 1966, **44**, 3210; (*e*) L. Gomberoff, R. R. McLone, and E. A. Power, *ibid.*, p. 4148; (*f*) E. A. Power, 'Very Long Range (Retardation Effect) Intermolecular Forces' in *Adv. Chem. Phys.*, 1967, **12**, 167.

312 (*a*) B. Linder, *J. Chem. Phys.*, 1964, **40**, 2003; (*b*) B. Linder and D. Hoernschemeyer, *ibid.*, p. 622; (*c*) I. R. McDonald and L. V. Woodcock, *J. Phys.* (*C*), 1970, **3**, 722; (*d*) D. E. Stogryn, *Phys. Rev. Letters*, 1970, **24**, 971; (*e*) W. Kołos, *Bull. Acad. Polon. Sci.* 1971, **19**, 159.

where $f(\Gamma, t)$ is the phase density of probability, or statistical distribution function, and integration extends over all phase space:

$$\Gamma = \int_{\Gamma} \mathrm{d}\Gamma$$

In the case of systems at thermodynamic equilibrium (isothermal system), the phase density of probability should not depend on time explicitly, so that in the statistics of equilibrium states we have:

$$\frac{\partial f(\Gamma, t)}{\partial t} = 0 \tag{92}$$

in this case, as derived by Gibbs, the canonical distribution is:

$$f(\Gamma, 0) = \frac{\exp\left[-\beta H(\Gamma)\right]}{\int_{\Gamma} \exp\left[-\beta H(\Gamma)\right] \mathrm{d}\Gamma} \tag{93}$$

where $\beta = 1/kT$, with k Boltzmann's constant, and T thermodynamic temperature. Hamilton's function $H(\Gamma)$ does not depend explicitly on time, and defines the total energy of the system: the kinetic energy E_k and potential energy U.

In numerous concrete cases with which we shall be dealing, the physical quantity Q does not depend explicitly on the generalized momenta but only on the generalized co-ordinates. It is then more convenient to introduce 'configurational (real) space', to each point of which corresponds a set of configurational variables $\tau = (\boldsymbol{r}, \Omega)$, with $\boldsymbol{r}$ the position and Ω the orientation of a microsystem. Instead of $\mathrm{d}\Gamma$ we now have the element of configurational space:

$$\mathrm{d}\Gamma = \mathrm{d}\boldsymbol{r}\, \mathrm{d}\Omega \tag{93a}$$

In configurational space we thus have:

$$Q = Q(\tau), \quad H = E_k + U(\tau) \tag{94}$$

and in place of the definition (91) we can apply the following statistical averaging:

$$\langle Q \rangle = \int_{\tau} Q(\tau) f(\tau)\, \mathrm{d}\tau \tag{95}$$

with the statistical distribution function:

$$f(\tau) = \frac{\exp\left[-\beta U(\tau)\right]}{\int_{\tau} \exp\left[-\beta U(\tau)\right] \mathrm{d}\tau} \tag{96}$$

Statistical Distribution for Small Perturbations. If the thermodynamical system is acted on by an external field of force F constant in time, the total potential energy of the system (on the assumption of additivity) is:

$$U(\tau, F) = U_N + U_F \tag{97}$$

where U_N is the potential energy of internal mutual interaction between the N microsystems, and U_F the potential energy of interaction between the system and the external field F. The mean statistical value of the perturbation now is:

$$\langle Q \rangle_F = \int_\tau Q(\tau, F) f(\tau, F) \, \mathrm{d}\tau \qquad (98)$$

In some cases the external potential energy U_F is small compared with the internal energy U_N and thermal energy kT, and the (perturbed) statistical distribution function

$$f(\tau, F) = \frac{\exp[-\beta(U_N + U_F)]}{\int_\tau \exp[-\beta(U_N + U_F)] \, \mathrm{d}\tau} \qquad (99)$$

appearing in the averaging formula (98) can be expressed as the expansion:

$$f(\tau, F) = f(\tau, 0) + f_1(\tau, F) + f_2(\tau, F) + f_3(\tau, F) + \ldots \qquad (100)$$

where $f(\tau, 0)$ defines the statistical distribution function of the zeroth approximation, that is to say, of the unperturbed system. The successive terms of the expansion define perturbed distribution functions of the first, second, third order, and so forth:

$$f_1(\tau, F) = -\beta f(\tau, 0)(U_F - \langle U_F \rangle) \qquad (100a)$$

$$f_2(\tau, F) = \tfrac{1}{2}\beta^2 f(\tau, 0)\{U_F^2 - \langle U_F^2 \rangle - 2(U_F - \langle U_F \rangle)\langle U_F \rangle\} \qquad (100b)$$

$$f_3(\tau, F) = -\tfrac{1}{6}\beta^3 f(\tau, 0)\{U_F^3 - \langle U_F^3 \rangle - 3(U_F^2 \langle U_F \rangle + U_F \langle U_F^2 \rangle) + 6(U_F \langle U_F \rangle + \langle U_F^2 \rangle - \langle U_F \rangle^2)\langle U_F \rangle\} \qquad (100c)$$

The expansion of equation (100) is characteristic for classical statistical perturbation calculus.

Unweighted Averaging.—One often has to deal with systems exhibiting no interaction between the microsystems ($U_N = 0$), when the potential energy does not depend explicitly on the configurational variables τ. The statistical function (96) now reduces simply to:

$$f(\tau) = \frac{1}{\int_\tau \mathrm{d}\tau} = \frac{1}{V\Omega} \qquad (96a)$$

and similarly the statistical mean value (95) to:

$$\langle Q \rangle_0 = \frac{1}{V\Omega} \int_\tau Q(\tau) \, \mathrm{d}\tau \qquad (95a)$$

In such cases, the averaging is referred to as isotropic or unweighted, since the motion of the microsystems is entirely random and all positions and orientations are equally probable in all directions of space.

If in particular the quantity Q depends only on the orientations Ω of the microsystems, the averaging (95a) reduces directly to integration over all possible orientations of microsystems:

$$\langle Q\rangle_\Omega = \frac{1}{\Omega}\int Q(\Omega)\,\mathrm{d}\Omega \tag{101}$$

where $\Omega = \int \mathrm{d}\Omega$ (the integral extends over all angles; in spherical co-ordinates θ, ϕ it defines the full body angle, since $d\Omega = \sin\theta\,\mathrm{d}\theta\,\mathrm{d}\phi$).

Averaging of Trigonometric Functions. Commonly, in molecular physics, the quantities $Q(\Omega)$ are trigonometric functions. Thus, if $Q(\theta) = \cos^n\theta$, one has by (101):

$$\langle\cos^n\theta\rangle_\Omega = \frac{1}{4\pi}\int_0^{2\pi}\mathrm{d}\phi\int_0^{\pi}\cos^n\theta\sin\theta\,\mathrm{d}\theta$$

$$= \frac{1}{2}\int_0^{\pi}\cos^n\theta\sin\theta\,\mathrm{d}\theta$$

and, on integration:[231]

$$\langle\cos^n\theta\rangle_\Omega = \begin{cases}\dfrac{1}{2k+1}, & \text{for } n = 2k\\ 0, \quad \text{for } n = 2k+1\end{cases} \tag{102}$$

with $k = 0, 1, 2, 3, \ldots$.

Likewise, if $Q(\phi) = \cos^n\phi$, one has with regard to (101):

$$\langle\cos^n\phi\rangle = \frac{1}{2\pi}\int_0^{2\pi}\cos^n\phi\,\mathrm{d}\phi$$

yielding in general

$$\langle\cos^n\phi\rangle_\Omega = \langle\sin^n\phi\rangle_\Omega = \begin{cases}\dfrac{(2k-1)!!}{2^k\cdot k!}, & \text{for } n = 2k\\ 0, \quad \text{for } n = 2k+1\end{cases} \tag{103}$$

where $(2k-1)!! = 1\cdot3\cdot5\cdots(2k-1)$.

Since formulae (102) and (103) are highly useful in calculations, we list some values in Table 13.

Table 13

k	$2k$	$\overline{\cos^{2k}\theta}$	$\overline{\sin^{2k}\theta}$	$\overline{\cos^{2k}\phi} = \overline{\sin^{2k}\phi}$
1	2	$\frac{1}{3}$	$\frac{2}{3}$	$\frac{1}{2}$
2	4	$\frac{1}{5}$	$\frac{8}{15}$	$\frac{3}{8}$
3	6	$\frac{1}{7}$	$\frac{16}{35}$	$\frac{5}{16}$
4	8	$\frac{1}{9}$	$\frac{128}{315}$	$\frac{35}{128}$
5	10	$\frac{1}{11}$	$\frac{256}{693}$	$\frac{63}{256}$

Averaging of Unit Vectors. Consider a unit vector $\boldsymbol{a}$ whose components on the reference axes x, y, z are $\boldsymbol{a}_x, \boldsymbol{a}_y, \boldsymbol{a}_z$ and are cosines of the angles subtended by the unit vector $\boldsymbol{a}$ and the axes x, y, z respectively. Obviously, such a vector, when averaged isotropically, vanishes: $\langle\boldsymbol{a}\rangle_\Omega = 0$. A bi-vector $\boldsymbol{aa}$ has to be dealt with as a symmetric tensor of rank 2 which, averaged with equal probability, is expressed by the unit symmetric tensor U, thus:

$$\langle \boldsymbol{aa} \rangle_\Omega = \tfrac{1}{3} U_{12}$$

Similarly, in the case of two arbitrary unit vectors $\boldsymbol{a}$ and $\boldsymbol{b}$, the isotropic average of the bi-vector $\boldsymbol{ab}$ is:

$$\langle \boldsymbol{ab} \rangle_\Omega = \tfrac{1}{3}(\boldsymbol{a} \cdot \boldsymbol{b})\, U_{12} \tag{104}$$

In that of three arbitrary unit vectors $\boldsymbol{a}, \boldsymbol{b}, \boldsymbol{c}$, we obtain:

$$\langle \boldsymbol{abc} \rangle_\Omega = \tfrac{1}{6}(\boldsymbol{abc} \,\vdots\, U_{123})\, U_{123} \tag{105}$$

where we come upon the antisymmetric third-rank unit tensor U_{123} known as the Levi–Cività tensor, the elements of which equal $+ 1$ for cyclic permutation of the indices 1, 2, 3:

$$U_{123} = U_{231} = U_{312} = 1 \tag{106a}$$

but equal $- 1$ for anticyclic permutation:

$$U_{213} = U_{321} = U_{132} = -1 \tag{106b}$$

whereas all remaining elements vanish (*e.g.* $U_{112} = U_{113} = \ldots = 0$). The Levi–Cività tensor can thus be written in vector form as:

$$U_{123} = \boldsymbol{ijk} + \boldsymbol{jki} + \boldsymbol{kij} - \boldsymbol{jik} - \boldsymbol{kij} - \boldsymbol{ikj} \tag{106}$$

With four unit vectors $\boldsymbol{a}, \boldsymbol{b}, \boldsymbol{c}, \boldsymbol{d}$, the result [313] is the following:

$$\begin{aligned} \langle \boldsymbol{abcd} \rangle_\Omega = \tfrac{1}{30}\{ & [4(\boldsymbol{a} \cdot \boldsymbol{b})(\boldsymbol{c} \cdot \boldsymbol{d}) - (\boldsymbol{a} \cdot \boldsymbol{c})(\boldsymbol{b} \cdot \boldsymbol{d}) - (\boldsymbol{a} \cdot \boldsymbol{d})(\boldsymbol{b} \cdot \boldsymbol{c})]\, U_{12} U_{34} \\ & + [4(\boldsymbol{a} \cdot \boldsymbol{c})(\boldsymbol{b} \cdot \boldsymbol{d}) - (\boldsymbol{a} \cdot \boldsymbol{d})(\boldsymbol{b} \cdot \boldsymbol{c}) - (\boldsymbol{a} \cdot \boldsymbol{b})(\boldsymbol{c} \cdot \boldsymbol{d})]\, U_{13} U_{24} \\ & + [4(\boldsymbol{a} \cdot \boldsymbol{d})(\boldsymbol{b} \cdot \boldsymbol{c}) - (\boldsymbol{a} \cdot \boldsymbol{b})(\boldsymbol{c} \cdot \boldsymbol{d}) - (\boldsymbol{a} \cdot \boldsymbol{c})(\boldsymbol{b} \cdot \boldsymbol{d})]\, U_{14} U_{23} \} \end{aligned} \tag{107}$$

If the four are identical, (107) reduces to:

$$\langle \boldsymbol{aaaa} \rangle_\Omega = \tfrac{1}{15}(U_{12} U_{34} + U_{13} U_{24} + U_{14} U_{23}) \tag{107a}$$

We similarly average unit poly-vectors of higher order; thus the fifth,[314, 315] the sixth,[313] and any even order.[314]

Molecular Correlation Functions.—Let us consider once more the definition of statistical averaging (95) in the absence of perturbation. We replace τ by the notation τ_N to stand for the set of configurational variables of all the N microsystems. In explicit form, this gives:

$$\langle Q \rangle = \frac{\int_N \cdots \int Q(\tau_N) \exp[-\beta U(\tau_N)]\, d\tau_N}{\int_N \cdots \int \exp[-\beta U(\tau_N)]\, d\tau_N} \tag{108}$$

where integration extends all configurations of the N microsystems.

313 (*a*) S. Kielich, *Acta Phys. Polon.*, 1961, **20**, 433; (*b*) S. Kielich, *Physica*, 1962, **28**, 511.
314 S. Kielich, *Bull. Soc. Amis. Sci. Lettres, Poznań*, 1968/69, **21B**, 47; M. Kozierowski, *ibid.*, 1970/71, **22B**, 5.
315 L. L. Boyle, *Internat. J. Quantum Chem.*, 1971, **4**, 413; L. L. Boyle and P. S. C. Matthews, *ibid.*, 1971, **5**, 381.

Here, the internal potential energy of the thermodynamic system as a whole takes the form of the following expansion:

$$U(\tau_N) = \sum_{i=1}^{N} V_i + \frac{1}{2}\sum_{i=1}^{N}\sum_{j=1}^{N}{}' V_{ij} + \frac{1}{3}\sum_{i=1}^{N}\sum_{j=1}^{N}{}'\sum_{k=1}^{N}{}'' V_{ijk}$$

where the second term defines the contribution to the total energy from pairwise interaction of microsystems, V_{ij}, discussed in detail in Section 2. The third term accounts for the contributions from three-body interactions which, for asymmetric multipolar microsystems, have been discussed in general tensor formalism by Kielich[316] and Stogryn.[317]

Let us now select, among all N molecules of the system, some set of $n \ll N$ molecules. The definition (108) now takes the form:

$$\langle Q \rangle = \int_{(n)} \cdots \int Q(\tau_N)\, P^{(n)}(\tau_n)\, d\tau_n \tag{108a}$$

where integration extends over configurations of n molecules, and

$$P^{(n)}(\tau_n) = \frac{\int_{N-n} \cdots \int \exp[-\beta U(\tau_N)]\, d\tau_{N-n}}{\int_{N} \cdots \int \exp[-\beta U(\tau_N)]\, d\tau_N} \tag{109}$$

is the probability that the selected set of n molecules shall be at configuration τ_n irrespective of the configuration of the remaining $N - n$ ones.

The probability (109) obviously fulfils the normalization condition:

$$\int_{(n)} \cdots \int P^{(n)}(\tau_n)\, d\tau_n = 1 \tag{109a}$$

In the case of random distribution of molecules (this occurs if the distances between them $r \to \infty$, or $T \to \infty$, *i.e.* for a perfect gas), the probability (109) reduces, in the limit, to the value:

$$P^{(n)}_{\text{random}} = \lim_{r\to\infty} P^{(n)}(\tau_n) = \lim_{T\to\infty} P^{(n)}(\tau_n) = \frac{1}{V^n\, \Omega^n} \tag{109b}$$

In many a problem of statistical-molecular physics it is more convenient to resort to configurational correlation functions. These, for a set of n molecules, are defined as follows:[66]

$$g^{(n)}(\tau_n) = \frac{N!}{\rho^n (N-n)!} P^{(n)}(\tau_n) \tag{110}$$

with $\rho = N/V$ is the number density of molecules in the medium. The quantity:

$$\rho^n\, g^{(n)}(\tau_1, \tau_2, \ldots, \tau_n)\, d\tau_1\, d\tau_2 \ldots d\tau_n$$

is the probability that, if the configuration of the system of N molecules is observed, a molecule (not necessarily molecule i) will be found in the

[316] S. Kielich, *Acta Phys. Polon.*, 1965, **27**, 395; 1965, **28**, 459.
[317] D. E. Stogryn, *J. Chem. Phys.*, 1969, **50**, 4967; 1970, **52**, 3671; *Mol. Phys.*, 1971, **22**, 81; 1972, **23**, 897.

element of volume $d\tau_1$ at τ_1, a second molecule in $d\tau_2$ at τ_2 ..., and another in $d\tau_n$ at τ_n. This follows because there are N choices for the molecule in $d\tau_1$, $N - 1$ choices for $d\tau_2$, ..., and $N - n + 1$ for $d\tau_n$, or a total of

$$N(N-1)\dots(N-n+1) = \frac{N!}{(N-n)!}$$

possibilities.

If the distribution of molecules is completely random, the probability is given by (109b), whence the correlation functions (110) reduce to the form:

$$g^{(n)}(\tau_p) = \frac{N!}{N^n(N-n)!} = 1 + O\left(\frac{1}{N}\right) \qquad (110a)$$

or

$$g^{(1)}(\tau_1) = 1, \quad g^{(2)}(\tau_1, \tau_2) = 1 - \frac{1}{N},$$

$$g^{(3)}(\tau_1, \tau_2, \tau_3) = 1 - \frac{3}{N} + \frac{2}{N^2}, \quad \textit{etc.} \qquad (110b)$$

The general definition (110) yields successively for $n = 1, 2, 3, \dots$ expressions for the ordinary, binary, ternary, quaternary, *etc.*, configurational correlation function:

$$\begin{aligned}
\rho g^{(1)}(\tau_1) &= N P^{(1)}(\tau_1) \\
\rho^2 g^{(2)}(\tau_1, \tau_2) &= N(N-1) P^{(2)}(\tau_1, \tau_2) \\
\rho^3 g^{(3)}(\tau_1, \tau_2, \tau_3) &= N(N-1)(N-2) P^{(3)}(\tau_1, \tau_2, \tau_3) \\
\rho^4 g^{(4)}(\tau_1, \tau_2, \tau_3, \tau_4) &= N(N-1)(N-2)(N-3) P^{(4)}(\tau_1, \tau_2, \tau_3, \tau_4)
\end{aligned} \qquad (111)$$

Of all these functions, the one of most importance in molecular problems is the binary distribution function $g^{(2)}(\tau_1, \tau_2)$ which, for real gases and liquids, depends solely on the mutual configuration τ_{12} of two molecules (pairwise configuration): $g^{(2)}(\tau_1, \tau_2) \rightarrow g^{(2)}(\tau_{12})$. In a crystal $g^{(2)}$ obviously depends in general on both $\boldsymbol{r}_1$ and $\boldsymbol{r}_2$, even if $r_{12} = |\boldsymbol{r}_1 - \boldsymbol{r}_2|$ is very large.

As already stated, the configurational variables τ_n comprise in general the set of positional variables $\boldsymbol{r}_n$ and orientational variables Ω_n of a molecule. If the distribution of molecules in a system is independent of their mutual configuration (as, to be exact, is the case only for atoms and almost spherical molecules), then on integrating both terms of equation (110) over all possible orientations of molecules one obtains the well-known radial correlation functions:[66, 67]

$$g^{(n)}(\boldsymbol{r}_n) = \frac{N!}{\rho^n(N-n)!} P^{(n)}(\boldsymbol{r}_n) \qquad (112)$$

where:

$$g^{(n)}(\boldsymbol{r}_n) = \int_{(n)} \dots \int g^{(n)}(\boldsymbol{r}_n, \Omega_n)\, d\Omega_n \qquad (112a)$$

In particular, for liquids, the binary radial correlation function is, by (112a):

$$g(r_{12}) = \iint g^{(2)}(\mathbf{r}_1, \Omega_1; \mathbf{r}_2, \Omega_2)\, d\Omega_1\, d\Omega_2 \qquad (112b)$$

For moderate densities, the function (112b) can be expanded in a series in powers of ρ:[66]

$$g(r_{12}) = \exp[-\beta U(r_{12})]\left[1 + \rho\int f_{13} f_{23}\, d\mathbf{r}_3 + O(\rho^2)\right] \qquad (113)$$

Here, $U(r_{12})$ is the total potential energy of interaction between molecule 1 and molecule 2; the functions

$$f_{13} = \exp[-\beta U(r_{13})] - 1; \quad f_{23} = \exp[-\beta U(r_{23})] - 1 \qquad (113a)$$

contain the potential energies $U(r_{13})$ and $U(r_{23})$ with which molecule 3 interacts with molecule 1 and 2 respectively.

At very low density, equation (113) permits the following satisfactory approximation:

$$g(r_{12}) = \exp[-\beta U(r_{12})] \qquad (113b)$$

At constant temperature of the system, when $r_{12} \to 0$ (because of the van der Waals repulsion between two molecules at small intermolecular distances, $U_{12} \to \infty$), the radial distribution function $g(r_{12})$ tends to zero for impermeable atoms and simple molecules. In cases when $r_{12} \to \infty$ at $T = \text{const.}$, or when $T \to \infty$, the molecules are completely independent (highly rarefied gases and vapours, $U_{12} \to 0$), and $g(r_{12}) \to 1$. The above limiting properties of the radial distribution function can be listed thus:

$$g(r_{12}) = \begin{cases} 0 & \text{for } r_{12} \to 0, \quad T = \text{const.}, \\ 1 & \text{for } r_{12} \to \infty, \quad T = \text{const.}, \\ 1 & \text{for } T \to \infty \end{cases} \qquad (113c)$$

Zernike and Prins[318] have shown that radial distribution functions $g(r)$ can be determined experimentally from *X*-ray scattering studies in liquids and real gases. We shall have occasion to show that information concerning the function $g^{(2)}(\tau_{12})$ can be gained also from experimental studies of other effects, in particular non-linear optical ones.[163]

Non-linear effects also provide valuable information on the correlation functions of higher order: $g^{(3)}$, $g^{(4)}$, and so forth. In statistical mechanics, the ternary function $g^{(3)}$ is shown to satisfy the superposition approximation of Kirkwood[66]

$$g^{(3)}(\mathbf{r}_1, \mathbf{r}_2, \mathbf{r}_3) = g^{(2)}(\mathbf{r}_1, \mathbf{r}_2)\, g^{(2)}(\mathbf{r}_1, \mathbf{r}_3)\, g^{(2)}(\mathbf{r}_2, \mathbf{r}_3) \qquad (114)$$

Virial Coefficient Method.—One of the most valuable methods of studying intermolecular forces consists in direct measurements of the deviations of real gases from the equation of state of the perfect gas. In 1873, van der

[318] F. Zernike and J. A. Prins, *Z. Phys.*, 1927, **41**, 184.

Waals derived the following equation of state of real gases:

$$\left(P + \frac{a}{V_\mathrm{m}^2}\right)(V_\mathrm{m} - b) = RT$$

in which the pressure P, molar volume V_m, and absolute temperature T were parameters of state, R the molar gas constant, and a, b constants specific to the real gas in question. By kinetic gas theory, van der Waals showed b to amount to the fourfold volume of the gas molecules and a to be related to the forces of mutual interaction between the molecules.

The state of a real gas is most commonly described by means of the following equation:

$$PV_\mathrm{m} = A + \frac{B}{V_\mathrm{m}} + \frac{C}{V_\mathrm{m}^2} + \ldots \tag{115}$$

which is an expansion of the function PV_m in a power series of $1/V_\mathrm{m}$. The equation is due to Kammerling Onnes, who termed the expansion coefficients $A, B, C, \ldots$ the first, second, third, ... virial coefficient. These coefficients depend on temperature, and their values are determined from experiment.

The first virial coefficient A fulfils the limiting condition:

$$A = \lim_{V_\mathrm{m}\to\infty} (PV_\mathrm{m}) = RT \tag{115a}$$

It thus obviously determines the state of the perfect gas. The successive, higher virial coefficients provide a measure of the deviations of a real gas from the equation of state of the perfect gas. In the range of moderate pressures, this deviation from (115a) is essentially described by the second virial coefficient B, whereas higher coefficients play an insignificant part and become important at higher pressures only.

The real-gas equation in the form (115) is also applicable to a mixture of gases, provided the virial coefficients A, B, C are expressed as:

$$\left.\begin{aligned} A_\mathrm{M} &= \sum_a x_a A^{(a)} \\ B_\mathrm{M} &= \sum_{ab} x_a x_b B^{(ab)} \\ C_\mathrm{M} &= \sum_{abc} x_a x_b x_c C^{(abc)} \end{aligned}\right\} \tag{116}$$

x_a denoting the molar fraction of component a of the mixture, and summation extending over all the components.

In classical statistical mechanics, the physical meaning of the virial coefficients is this: A corresponds to an absence of interaction between the molecules (free molecules), B takes account of pairwise interaction (the interaction of two molecules at a time), and the higher coefficients contain multiple interactions. Thus the deviation of a real gas from perfect-gas behaviour is primarily due to intermolecular interactions, the values of which can be calculated by methods of statistical mechanics with an adequate molecular model. In the moderate pressure range, only pairwise interaction of molecules i and j in the second virial coefficient [66] need be considered:

$$B = -\frac{N_A^2}{2\beta\Omega^2}\int\int\int\{\exp[-\beta U_{ij}(r_{ij}, \Omega_i, \Omega_j)] - 1\} dr_{ij} d\Omega_i d\Omega_j \qquad (117)$$

N_A denoting the Avogadro constant.

Quite generally, the total potential energy U_{ij} consists of a term dependent only on the distance separating the molecules and a term depending on their orientations (unless they are spherical):

$$U_{ij} = U_{ij}(r_{ij}) + V_{ij}(r_{ij}, \Omega_i, \Omega_j) \qquad (118)$$

$U_{ij}(r_{ij})$ being the energy related with the existence of central forces solely, and $V_{ij}(r_{ij}, \Omega_i, \Omega_j)$ that due to the non-central (tensorial) forces discussed in Section 2.

In the particular case when only central forces act between the molecules, as is true for spherical ones, equation (117) reduces to the form:

$$B_{\text{cent}} = -\frac{2\pi}{\beta}N_A^2\int_0^\infty\{\exp[-\beta U_{ij}(r_{ij})] - 1\} r_{ij}^2 dr_{ij} \qquad (117a)$$

Concrete calculations of B_{cent} are feasible only if the law according to which the forces of attraction and repulsion depend on intermolecular distance is known. For spherical molecules, this law is approximated by the Lennard-Jones formula:[74]

$$U_{ij}(r_{ij}) = -\frac{\lambda_{ij}^n}{r_{ij}^n} + \frac{\lambda_{ij}^m}{r_{ij}^m}, \quad m > n \qquad (119)$$

where the first term is the energy of attraction and the second term that of repulsion. The constants λ_{ij}^n and λ_{ij}^m as well as the exponents m and n should have such values as to ensure good conformity between the second virial coefficient calculated from (117a) and the experimental results for BT. It turns out that, for many gases, good agreement is achieved by putting $n = 6$, $m = 12$. In these cases the Lennard-Jones potential (119) reduces to the simple and convenient form:[74]

$$U_{ij}(r_{ij}) = 4\varepsilon_{ij}\left\{\left(\frac{\sigma_{ij}}{r_{ij}}\right)^{12} - \left(\frac{\sigma_{ij}}{r_{ij}}\right)^6\right\} \qquad (119a)$$

The parameters ε_{ij} and σ_{ij} have the dimensions of an energy and a length, and are specific to the interacting molecules. The constant σ_{ij} denotes that distance between the molecules at which $U_{ij}(r_{ij}) = 0$, *i.e.* at which the forces of attraction and repulsion are numerically equal (see Figure 9). The constant ε_{ij} has the meaning of the minimal value of energy, resulting for an intermolecular distance of $\sigma_{ij}\sqrt{2}$.

Calculations of the second virial coefficient (117) taking into account non-central intermolecular interactions have been performed by Stockmayer,[319] Rowlinson,[320] Pople,[321] Kielich,[228, 313] Saxena and Joshi,[322] and others.[324]

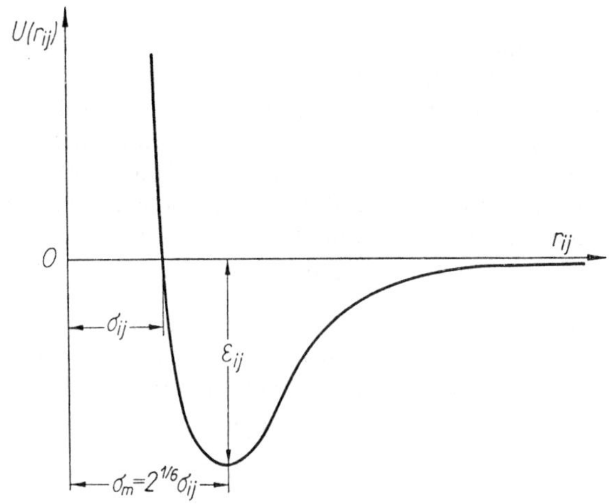

Figure 9 *Potential energy* vs. *the distance* r_{ij} *between two molecules of species* i *and* j. ε_{ij} *and* σ_{ij} *are force constants of the Lennard-Jones equation* (119a)*; at* $r_{ij} > \sigma_{ij}$ *attractive forces prevail, whereas if* $r_{ij} < \sigma_{ij}$ *repulsion predominates; at* $r_{ij} = \sigma_{ij}$ *the potential energy is zero, whence* σ_{ij} *is the shortest distance to which the molecules can approach one another*

The virial coefficient method is applicable to arbitrary quantities Q describing the property of a real gas:[324]

$$Q = A_Q + \frac{B_Q}{V_m} + \frac{C_Q}{V_m^2} + \dots \tag{120}$$

where A_Q, B_Q, C_Q, ... are the first, second, third, ... virial coefficient of the quantity Q.

In recent years, Harris and Adler,[325] as well as Buckingham and Pople [324] and others, have calculated the second virial coefficient of molecular

[319] W. H. Stockmayer, *J. Chem. Phys.*, 1941, **9**, 398.
[320] J. S. Rowlinson, *Trans. Faraday Soc.*, 1949, **45**, 984.
[321] (*a*) J. A. Pople, *Proc. Roy. Soc.*, 1954, **A221**, 498, 508; (*b*) A. D. Buckingham and J. A. Pople, *Trans. Faraday Soc.*, 1955, **51**, 1173.
[322] S. C. Saxena and K. M. Joshi, *Phys. Fluids*, 1962, **5**, 1217.
[323] (*a*) R. H. Orcutt, *J. Chem. Phys.*, 1963, **39**, 605; (*b*) A. G. de Rocco, T. H. Spurling, and T. S. Storvick, *ibid.*, 1967, **46**, 599; (*c*) T. H. Spurling, A. G. De Rocco, and T. S. Storvick, *ibid.*, 1968, **48**, 1006; (*d*) T. H. Spurling and A. G. De Rocco, *ibid.*, 1968, **49**, 2867; Y. Singh and K. K. Datta, *ibid.*, 1970, **53**, 1184.
[324] (*a*) A. D. Buckingham and J. A. Pople, *Trans. Faraday Soc.*, 1955, **51**, 1029, 1179; (*b*) A. D. Buckingham and J. A. Pople, *Discuss. Faraday Soc.*, 1956, **22**, 17; (*c*) A. D. Buckingham and R. E. Raab, *Trans. Faraday Soc.*, 1958, **54**, 623.
[325] F. E. Harris and B. J. Adler, *J. Chem. Phys.*, 1953, **21**, 1351, 1031.

electric polarization. Buckingham moreover proposed its theory with respect to molecular refraction[326] and the molecular Kerr constant.[327] Kielich developed a virial coefficient theory of the Cotton–Mouton effect and light scattering[231, 328] as well as field-gradient-induced optical birefringence.[329]

Statistical Molecular Distribution Functions in a Static and in an Alternating Electric Field.—We now shall apply the statistical distribution function expansion (100) to a molecular gas (or dilute solution of polar molecules in a non-polar solvent) immersed in an external electric field $\boldsymbol{E}$. Not taking into consideration the mutual correlations of molecules but solely their interaction with the field $\boldsymbol{E}$, we are justified by equation (81) in writing the potential energy to within the square of the field as follows:

$$U_E = -\boldsymbol{p}\cdot\boldsymbol{E} - \tfrac{1}{2}\boldsymbol{E}\cdot\boldsymbol{\alpha}\cdot\boldsymbol{E} \tag{121}$$

Since $U_N = 0$, the non-perturbed statistical mean values (95) in the expansions (100a—c) reduce to isotropic mean values over the molecular orientations (101), and by (104) and (121) we have:

$$\langle U_E\rangle_\Omega = -\tfrac{1}{2}\alpha E^2$$
$$\langle U_E^2\rangle_\Omega = -\tfrac{1}{3}p^2 E^2$$

reducing the functions (100a) and (100b) to the form:

$$f_1(\Omega, E) = \beta f(\Omega, 0)\{\boldsymbol{p}\cdot\boldsymbol{E} + \tfrac{1}{2}(\boldsymbol{\alpha} - \alpha\boldsymbol{U}) : \boldsymbol{EE}\}$$
$$f_2(\Omega, E) = \tfrac{1}{6}\beta^2 f(\Omega, 0)(3\boldsymbol{pp} - p^2\boldsymbol{U}) : \boldsymbol{EE} \tag{122}$$

$p^2 = \boldsymbol{p}\cdot\boldsymbol{p} = p_x^2 + p_y^2 + p_z^2$ being the square of the dipole moment.

For axially symmetric molecules one has $\boldsymbol{p} = p\cdot\boldsymbol{k}$ and a polarizability tensor in the form (86), whence (121) reduces simply to:

$$U_E = -pE\cos\theta - \tfrac{1}{2}\alpha[1 + \kappa_\alpha(3\cos^2\theta - 1)]E^2 \tag{121a}$$

and, similarly, the perturbed distribution functions (122):

$$f_1(\Omega, E) = \beta f(\Omega, 0)\{pE\cos\theta + \tfrac{1}{2}\alpha\kappa_\alpha(3\cos^2\theta - 1)E^2\}$$
$$f_2(\Omega, E) = \tfrac{1}{6}\beta^2 f(\Omega, 0)p^2(3\cos^2\theta - 1)E^2 \tag{122a}$$

with θ the angle subtended by the symmetry axis $\boldsymbol{k}$ of the molecule and the field direction $\boldsymbol{E}$ ($\boldsymbol{k}\cdot\boldsymbol{E} = E\cos\theta$).

This procedure of calculating perturbed distribution functions is of use if the molecules of the medium are under the effect of a static electric field only and if the system is in well-defined conditions of thermodynamic equilibrium. If, however, the system is acted on by an alternating field

$$\boldsymbol{E} = \boldsymbol{E}_0\cos\omega t = \tfrac{1}{2}\boldsymbol{E}_0[\exp(i\omega t) + \exp(-i\omega t)] \tag{123}$$

326 A. D. Buckingham, *Trans. Faraday Soc.*, 1956, **52**, 747.
327 A. D. Buckingham, *Proc. Phys. Soc.* (*A*), 1955, **68**, 910; *Discuss. Faraday Soc.*, 1967, **43**, 205.
328 S. Kielich, *Acta Phys. Polon.*, 1960, **19**, 711.
329 S. Kielich, *Bull. Acad. Polon. Sci.*, 1960, **8**, 637.

alternating in time t at the angular frequency ω, the orientation of molecules is dependent on certain times of relaxation which are functions of the molecularly effective viscosity of the medium, the temperature, and the geometrical shape of the molecules. The distribution function has now to describe the dynamical properties of the system in their dependence on the relevant relaxation times, and can be determined from the kinetic equation of diffusion:[85, 131]

$$\nabla^2 f + \beta\{(\nabla f)\cdot(\nabla U) + f\nabla^2 U\} = \beta W \frac{\partial f}{\partial t} \tag{124}$$

where W is the internal friction constant.

In the present treatment, we shall assume after Debye that the molecules, although electrically dipolar (we have here point dipoles), are nevertheless geometrically spherical in shape. Assuming them moreover to have an axis of symmetry directed along the dipole moment, we find from equations (121a) and (122a) that the energy U as well as the distribution functions f_1 and f_2 depend on the angle θ only, so that the kinetic equation (124) takes the form:[131]

$$\frac{1}{\sin\theta}\frac{\partial}{\partial\theta}\left(\sin\theta\frac{\partial f}{\partial\theta}\right) + \beta\left\{\frac{\partial U}{\partial\theta}\frac{\partial f}{\partial\theta} + \frac{1}{\sin\theta}\frac{\partial}{\partial\theta}\left(\sin\theta\frac{\partial U}{\partial\theta}\right)f\right\} = \beta W\frac{\partial f}{\partial t} \tag{124a}$$

By (123), we now write the field $\boldsymbol{E}$ as follows:

$$\boldsymbol{E} = \sum_n \boldsymbol{E}_n \exp(i\omega_n t) \tag{123a}$$

the summation extending over all positive and negative frequencies ω_n, and $\boldsymbol{E}_n$ denoting the in general complex field amplitude.

With (121a) and (122a), we obtain from equation (124a) the following distribution function describing the evolution in time of the system from the moment when the perturbation (123a) is switched on, with accuracy to within the second approximation:[331]

$$\begin{aligned} f(\Omega, t) = f(\Omega, 0)\Bigg\{1 &+ \beta p\cos\theta\sum_n \frac{E_n \exp(i\omega_n t)}{1 + i\omega_n\tau_D} \\ &+ \frac{\beta}{12}(3\cos^2\theta - 1)\sum_n\sum_m \frac{E_n E_m \exp[i(\omega_n + \omega_m)t]}{1 + i(\omega_n + \omega_m)\tau_B} \\ &\times\left[6(\alpha\kappa_\alpha)_{nm} + \frac{\beta p_n p_m}{1 + i\omega_n\tau_D} + \frac{\beta p_m p_n}{1 + i\omega_m\tau_D}\right]\Bigg\} \end{aligned} \tag{125}$$

where

$$\tau_D = \tfrac{1}{2}\beta W = \frac{4\pi r_0^3\eta}{kT}$$

is the Debye dipole relaxation time, and

$$\tau_B = \tfrac{1}{6}\beta W = \tau_D/3$$

330 F. Perrin, *J. Physique*, 1934, **5**, 497; 1936, **7**, 1.

331 S. Kielich, *Acta Phys. Polon.*, 1967, **31**, 689; *Optics Comm.*, 1969, **1**, 129.

the reorientation time of the molecule, often referred to as the birefringence relaxation time,[131] the reorientation being considered to be due to a strong electric field.

The general form of the distribution function (125) can be particularized to yield the results of Debye,[85] Peterlin and Stuart,[131] and Frenkel.[73] On putting $\omega = 0$ in the expansion (125), we obtain the result:

$$f(\Omega, E) = f_0\{1 + pE\cos\theta + \tfrac{1}{6}\beta(3\alpha\kappa_\alpha + \beta p^2)(3\cos^2\theta - 1)E^2\} \tag{125a}$$

valid for the reorientation of dipolar molecules in a static electric field.

In the other limiting case when $\omega \rightarrow \infty$, equation (125) reduces simply to the form:

$$f(\Omega, E_0) = f_0\{1 + \tfrac{1}{4}\beta\alpha\kappa_\alpha(3\cos^2\theta - 1)E_0^2\} \tag{125b}$$

One notes that an alternating electric field of very high (optical) frequency causes no reorientation of permanent electric dipoles.

With regard to loss-less systems, we are not interested in time variations of the system but rather in the state of thermal equilibrium which sets in after the perturbation has been in action for a sufficiently long time. The situation which ensues justifies our employing a distribution function with perturbation potential energy averaged in time:[142, 206]

$$f(\Omega, E) = \frac{\exp[-\beta\langle U(\Omega, E)\rangle_t]}{\int \exp[-\beta\langle U(\Omega, E)\rangle_t]\,\mathrm{d}\Omega} \tag{126}$$

Above, the time-average denoted by the symbol $\langle\ \rangle_t$ is defined as:

$$\langle Q\rangle_t = \lim_{T\rightarrow\infty}\frac{1}{2T}\int_{-T}^{T} Q(t)\,\mathrm{d}t \tag{127}$$

where T is the period of some time-dependent function $Q(t)$.

On inserting (123) into the expansion (121) and keeping in mind that, in accordance with (127), the terms in rapidly varying factors $\exp(\pm i\omega t)$, $\exp(\pm 2i\omega t)$, ... vanish on time-averaging, we obtain:

$$\langle U(\Omega, E)\rangle_t = -\tfrac{1}{4}\boldsymbol{E}_0\cdot\boldsymbol{\alpha}\cdot\boldsymbol{E}_0 \tag{128}$$

signifying that the oscillating electric field of a light wave can interact only with the electron polarizability ellipsoid of molecules that are able to keep pace with its variations.[206]

Statistical Fluctuations.—Thermodynamic systems in general present unordered random deviations from equilibrium referred to as statistical fluctuations. Even slight fluctuations are apt to give rise to experimentally accessible effects. Thus, as shown by Smoluchowski,[332] local fluctuations in density in gases and liquids cause light scattering by optically transparent media. Attempts to raise the sensitivity of measuring devices are foiled by

[332] M. Smoluchowski, *Bull. Acad. Crac. Polon.*, 1907, 1057; *Ann. Physik*, 1908, **25**, 205; *Prace Matematyczno-Fizyczne* (*Poland*), 1914, **25**, 187.

various fluctuations. All fluctuational phenomena, due to thermal motion or otherwise, are accessible to an explanation by statistical theory, whether classical or quantal.

Let Q denote some physical quantity characterizing the physical system as a whole, or some part of it. A fluctuation in Q is a deviation of the momentary value of Q from its mean value $\langle Q \rangle$:

$$\Delta Q = Q - \langle Q \rangle \tag{129}$$

Any physical quantity that is the average of the action of a restricted number of microsystems is subject to statistical fluctuations.

Since the mean value of a fluctuation vanishes, $\langle \Delta Q \rangle = \langle Q \rangle - \langle Q \rangle = 0$, the measure of a fluctuation is taken as its mean square value:

$$\langle (\Delta Q)^2 \rangle = \langle (Q - \langle Q \rangle)^2 \rangle \tag{130}$$

This is in general non-zero. In fact, we have by definition:

$$\begin{aligned} \langle (\Delta Q)^2 \rangle &= \langle Q^2 - 2Q\langle Q \rangle + \langle Q \rangle^2 \rangle \\ &= \langle Q^2 \rangle - \langle Q \rangle^2 \geqslant 0 \end{aligned} \tag{130a}$$

We wish to state clearly that, by the definition of statistical averaging $\langle Q^2 \rangle \neq \langle Q \rangle^2$, *i.e.* the average of the square is not the same as the square of the average value of Q.

By insertion of the statistical distribution function expansion (100) into the perturbed statistical average (98) and with regard to the definition (129) of a fluctuation, we obtain to within the third approximation inclusively:[333]

$$\begin{aligned} \langle Q \rangle_F = \langle Q \rangle - \beta \langle \Delta Q \Delta U_F \rangle + \tfrac{1}{2}\beta^2 \langle \Delta Q (\Delta U_F)^2 \rangle \\ - \tfrac{1}{6}\beta^3 [\langle \Delta Q (\Delta U_F)^3 \rangle - 3\langle \Delta Q \Delta U_F \rangle \langle (\Delta U)^2 \rangle] + \ldots \end{aligned} \tag{131}$$

The successive approximations to the perturbed statistical average are thus given by successive powers of the fluctuation in potential energy $(\Delta U_F)^n$.

Electromagnetic Field in Dielectrics.—*Maxwell's Equations.* Encompassing the laws of Ampère, Faraday, and Gauss, Maxwell in 1864 proposed in final form the macroscopic theory of the electromagnetic field in unbounded space filled with matter. The complete set of Maxwell's equations for a continuous material medium is, when expressed in vector symbolism, of the form:[334]

$$\nabla \times \boldsymbol{E} = -\frac{\partial \boldsymbol{B}}{\partial t}, \quad \nabla \cdot \boldsymbol{B} = 0 \tag{132}$$

$$\nabla \times \boldsymbol{H} = \frac{\partial \boldsymbol{D}}{\partial t} + \boldsymbol{J}, \quad \nabla \cdot \boldsymbol{D} = \rho_e \tag{133}$$

$\boldsymbol{E}$ and $\boldsymbol{H}$ being the electric and magnetic field strength vectors, $\boldsymbol{D}$ the electric displacement vector, $\boldsymbol{B}$ the magnetic induction vector, $\boldsymbol{J}$ the electric current density, and ρ_e the electric charge density.

[333] A. Piekara and S. Kielich, *Acta Phys. Polon.*, 1958, **17**, 209.
[334] (*a*) P. Mazur and B. R. A. Nijboer, *Physica*, 1953, **19**, 971; (*b*) S. R. de Groot and J. Vlieger, *ibid.*, 1965, **31**, 125.

Maxwell's are linear differential equations, provided the relations between the induction and field strength vectors remain linear, as is the case in linear media. The principle of superposition is a direct consequence of this linearity of the electromagnetic field equations. According to the principle of superposition, any number of light waves will propagate in a medium independently of one another, *i.e.* without interacting mutually. The linear equations have to be considered as an approximation only, and Born and Infeld[335] proposed a (canonical and relativistically invariant) phenomenological non-linear theory of the Maxwell field which, for the case of weak fields, goes over into the set of Maxwell's equations. In non-linear theory, the superposition principle no longer holds: waves of different frequencies can interact with one another, causing various non-linear effects, even in vacuum. Thus, quantum electrodynamics[336] shows that polarization occurs in a vacuum by the radiation field, rendering possible the scattering of light by light, reflection of light from light, non-linear light scattering by charges in plasma,[337] and other such effects. The order of magnitude of these effects, though extremely low, lies nevertheless within the scope of future experimental possibilities owing to the rapid progress of laser techniques. Interactions between electromagnetic waves have hitherto been detected by laser technique only in material media: gases, liquids, and solids.[43–50] Lately, authors have evolved a macroscopic quantum electrodynamics[338] covering non-linear and multipolar interactions of matter and electromagnetic fields.[339, 340]

Semi-macroscopic Approach to Dielectrics.—In describing quantitatively the non-linear electromagnetic properties of dielectrics, various classical and quantum-mechanical approaches have been followed. With regard to gaseous states, the problem is a relatively simple one on the classical level (without entering into the quantal details of the structure of matter), since in this case only properties of isolated microsystems are taken into consideration. The fundamentals of these theories have been formulated by Lorentz, Voigt, Langevin, and Debye.

The form of non-linear effects in dense media is dependent not only on the electromagnetic properties of the isolated atoms and molecules but also on their interactions and on the structure of the system as a whole. At early stages in the evolution of the theory of dense bodies, the influence of molecular interactions on the character of non-linear effects was

[335] M. Born and L. Infeld, *Proc. Roy. Soc.*, 1934, **A144**, 425; 1934, **A147**, 522; 1935, **A150**, 141.
[336] I. Białynicki-Birula and Z. Białynicka-Birula, 'Elektrodynamika Kwantowa', PWN, Warszawa, 1969.
[337] J. W. Gardner, *Nuclear Energy*, 1965, **5**, 241, 254; *New Scientist*, 1969, **41**, 18.
[338] R. L. Fulton, *J. Chem. Phys.*, 1969, **50**, 3355.
[339] S. Kielich, *Acta Phys. Polon.*, 1966, **29**, 875.
[340] (*a*) J. Humblet, *Physica*, 1944, **11**, 91; (*b*) E. A. Power and S. Zinau, *Phil. Trans.* (*A*), 1959, **251**, 427; (*c*) J. Fiutak, *Canad. J. Phys.*, 1963, **41**, 12; (*d*) P. W. Atkins and R. G. Woolley, *Proc. Roy. Soc.*, 1970, **A319**, 549.

accounted for indirectly by way of a local field acting on each atom (molecule) in the medium. The local field was dealt with as isotropic, as in the models of Lorentz [51] and Onsager,[341] or more generally as anisotropic in the case of molecules having natural electric,[342] magnetic, or optical anisotropy.[343–349] Fowler [350] and Debye [351] made a direct approach by taking into account molecular interactions (hindered rotation of dipoles in liquids), which subsequently was extended by Piekara [352] to pairwise couplings as well as couplings between two-molecule assemblages in dipolar liquids. The fundamentals of the molecular-statistical theory of non-linear effects in dielectric liquids, taking into account angular correlations between the molecules, were proposed in a general form by Buckingham [327] and Piekara and Kielich.[141, 142] Those theories, however, failed to include statistical translational fluctuations which, as shown by Kirkwood [344] and Yvon,[345] play a decisive part in the theory of molecular refraction and polarization in atomic fluids. According to Kielich [346, 348] this latter process of spatial regroupings, when it relates to atoms and isotropically polarizable molecules, causes electric and magnetic anisotropy in regions of near-range ordering in liquids.

To show the influence of various microscopic and structural factors on linear and non-linear effects in dense dielectrics, it is convenient to apply first a semi-macroscopic treatment of the theory, and then to proceed to its molecular-statistical interpretation, assuming appropriate microscopic models. The semi-macroscopic method was initially applied by Kirkwood [353] and modified by Fröhlich [5] in the theory of linear dielectrics, and has been successfully used in theories of non-linear dielectrics.[348]

In his semi-macroscopic theory Kirkwood considers a dielectric sphere of macroscopic size, of volume V containing N molecules. The spherical sample of the isotropic, homogeneous dielectric of electric permittivity ε is immersed in a uniform external field $\boldsymbol{E}_0$ applied in vacuum (of permittivity ε_0). The mean macroscopic electric field $\boldsymbol{E}$ existing within the sphere is given by:

$$\boldsymbol{E} = \frac{3\varepsilon_0}{\varepsilon + 2\varepsilon_0}\boldsymbol{E}_0 = \frac{3}{\varepsilon_r + 2}\boldsymbol{E}_0 \tag{134}$$

341 L. Onsager, *J. Amer. Chem. Soc.*, 1936, **58**, 1486.
342 T. H. G. Scholte, *Physica*, 1949, **15**, 437, 450.
343 P. Weiss, *J. Physique*, 1907, **6**, 667.
344 J. G. Kirkwood, *J. Chem. Phys.*, 1935, **3**, 300; 1936, **4**, 592.
345 J. Yvon, *Acta Scient. Ind.*, 1937, No. 542, 'Fluctuations en Densité', Herman et Cie, Paris; *ibid.*, 1937, No. 543, 'La Propagation et la Diffusion de la Lumière'; *Compt. rend.*, 1936, **202**, 35.
346 S. Kielich, *Acta Phys. Polon.*, 1960, **19**, 149; *Fizyka Dielektryków* (*Poland*), 1964, **2**, 3.
347 M. J. Stephen, *Mol. Phys.*, 1958, **1**, 223.
348 S. Kielich, *Acta Phys. Polon.*, 1962, **22**, 299; *Fizyka Dielektryków* (*Poland*), 1962, **1**, 65.
349 C. V. Raman and K. S. Krishnan, *Proc. Roy. Soc.*, 1927, **A117**, 1; *Phil. Mag.*, 1928, **5**, 498.
350 R. H. Fowler, *Proc. Roy. Soc.*, 1935, **A149**, 1.
351 P. Debye, *Phys. Z.*, 1935, **36**, 100.
352 A. Piekara, *Proc. Roy. Soc.*, 1939, **A172**, 360; *Acta Phys. Polon.*, 1950, **10**, 37, 107.
353 J. G. Kirkwood, *J. Chem. Phys.*, 1939, **7**, 919.

so that the equation (5) now becomes:

$$\frac{\varepsilon - \varepsilon_0}{\varepsilon + 2\varepsilon_0} E_0 = \frac{\varepsilon_r - 1}{\varepsilon_r + 2} E_0 = \frac{1}{3\varepsilon_0} P(E_0) \quad (135)$$

in which the electric polarization vector $P(E_0)$ is a function of the field E_0 and, in simple dielectrics, vanishes if $E_0 = 0$ (we neglect the case of spontaneous polarization presented by ferroelectrics). Obviously, $P(E_0)$ is composed of the electric dipole moments of all N molecules of the sphere of volume V, and we have:

$$P(E_0) = \frac{1}{V} \sum_{i=1}^{N} p_i(E_0) \quad (136)$$

where summation extends over all molecules of the sphere having an electric dipole moment $p_i(E_0)$.

Now within a liquid dielectric all the molecules, at configurations τ within the sphere V, have to be dealt with statistically, and the macroscopic polarization is obtained as a result of statistical averaging of the microscopic processes occurring in the sphere:

$$P(E_0) = \frac{1}{V} \left\langle \sum_{i=1}^{N} p_i(\tau, E_0) \right\rangle_{E_0} \quad (137)$$

where the statistical mean value perturbed by the field E_0, namely $\langle\,\rangle_{E_0}$, is defined at thermodynamic equilibrium in classical Gibbs statistics by equation (98).

Similarly, one has by equation (6) the following fundamental formula for the magnetic permeability:

$$\frac{\mu - \mu_0}{\mu + 2\mu_0} B_0 = \frac{\mu_r - 1}{\mu_r + 2} B_0 = \frac{\mu_0}{3} M(B_0) \quad (138)$$

Here, the magnetization vector is defined as:

$$M(B_0) = \frac{1}{V} \left\langle \sum_{i=1}^{N} m_i(\tau, B_0) \right\rangle_{B_0} \quad (139)$$

where $m_i(\tau, B_0)$ is the magnetic dipole moment of molecule i at configuration τ in the presence of the magnetic induction B_0.

Thus, the semi-macroscopic theory of the linear electric properties of isotropic dielectrics is given by the fundamental relations (135) and (137), and for their linear magnetic properties by (138) and (139). These fundamental equations can be extended to the non-linear electric and magnetic properties of dielectrics. This in fact will be the subject of detailed discussions in further Sections of this Chapter.

4 Electro-optical Properties of Dielectrics

Molar Refraction of Dense Media.—Let us consider a homogeneous medium, macroscopically isotropic in the absence of external forces, with refractive index n. With regard to the molar volume containing Avogadro's number N_A of molecules, we define the molar refraction R_m (or Lorentz–Lorenz function) as follows:

$$R_m = \frac{n^2 - 1}{n^2 + 2} V_m \tag{140}$$

In conformity with Maxwell's relation:

$$n = \left(\frac{\varepsilon}{\varepsilon_0} \frac{\mu}{\mu_0}\right)^{\frac{1}{2}} = (\varepsilon_r \mu_r)^{\frac{1}{2}} \tag{141}$$

both the electric permittivity $\varepsilon(\omega)$ and magnetic permeability $\mu(\omega)$, at a given frequency ω of light incident on the medium, contribute to the refractive index of the latter. However, in a first approximation, for the usual optically inactive dielectric bodies, it suffices to take the contribution to n from the electric wave vector $\boldsymbol{E}(\boldsymbol{r}, t)$ only, neglecting the insignificant part due to the magnetic vector $\boldsymbol{B}(\boldsymbol{r}, t)$. Assuming the molar volume V_m as being in vacuum, where the electric field strength of the light wave is $\boldsymbol{E}_0(\boldsymbol{r}, t)$, one can write by equation (135) the fundamental equation of refractive indices in the form:

$$\frac{n^2 - 1}{n^2 + 2} \boldsymbol{E}_0(\boldsymbol{r}, t) = \frac{1}{3\varepsilon_0} \boldsymbol{P}(\boldsymbol{r}, t) \tag{142}$$

Here, by equation (137), the polarization at the space-time point $(\boldsymbol{r}, t)$ is:

$$\boldsymbol{P}(\boldsymbol{r}, t) = \frac{1}{V_m} \left\langle \sum_{i=1}^{N_A} \boldsymbol{p}_i(\boldsymbol{r}, t) \right\rangle_{E_0} \tag{143}$$

$\boldsymbol{p}_i(\boldsymbol{r}, t)$ denoting the dipole moment induced in molecule i by the field $\boldsymbol{E}_0(\boldsymbol{r}, t)$ of the incident wave.

We shall not concern ourselves here with frequency dispersion and absorption related to refractive index changes, detailed discussions of which are generally accessible; neither shall we consider spatial dispersion, as we assume for simplicity the light wavelength as large with respect to the molecular size and intermolecular distances. Assuming moreover that the light intensity is small, we obtain by equations (140)—(143):

$$R_m \boldsymbol{E}_0(\boldsymbol{r}, t) = \frac{1}{3\varepsilon_0} \left\langle \sum_{i=1}^{N_A} \boldsymbol{p}_i(\boldsymbol{r}, t) \right\rangle \tag{144}$$

In the absence of interactive molecular fields in the medium, we are justified in writing the following first approximation for the induced dipole moment:

$$\boldsymbol{p}_i(\boldsymbol{r}, t) = \boldsymbol{a}_i \cdot \boldsymbol{E}_0(\boldsymbol{r}, t) \tag{145}$$

$\boldsymbol{a}_i = \boldsymbol{\alpha}_i(\omega)$ being the optical polarizability tensor of the isolated molecule at the frequency ω of the light oscillations.

On inserting (145) into the fundamental formula (144), we immediately come to the Lorentz–Lorenz result:

$$R_m^{(1)} = \frac{N_A}{3\varepsilon_0} a = \frac{4\pi}{3} N_A a_v \tag{144a}$$

where $a = \boldsymbol{a} : \boldsymbol{U}/3$ is the mean optical polarizability, a constant of the molecule at given light frequency; $a_V = a/4\pi\varepsilon_0$ is the volume optical polarizability.

Equation (144a) states that molar refraction is a constant, independent of structure and of the thermodynamic state of the medium. Numerous studies have shown this to be closely correct, at certain special conditions, in a first approximation, since in general R_m is a function of temperature and density, or pressure in the case of real gases. The non-constancy of R_m has been explained theoretically by Yvon [345] as well as by Mazur and Mandel.[354] We shall deal here only with deviations from constancy in R_m due to Kirkwood–Yvon translational statistical fluctuations, omitting for simplicity the fact that molecular interactions perturb the intrinsic polarizabilities in accordance with the Jansen–Mazur [355] model. For a molecule immersed in a statistically inhomogeneous medium, this in place of (145) leads to:

$$\boldsymbol{p}_i(\boldsymbol{r}, t) = \boldsymbol{a}_i \cdot \{\boldsymbol{E}_0(\boldsymbol{r}, t) + \boldsymbol{F}_i(\boldsymbol{r}, t)\} \tag{146}$$

On the preceding assumptions, we can adapt the solution to the molecular field $\boldsymbol{F}_i(\boldsymbol{r}, t)$, permitting us in a sufficiently good approximation to re-write equation (144) as follows:[356]

$$R_m = \frac{1}{3\varepsilon_0}\left\{\left\langle \sum_{i=1}^{N_A} \boldsymbol{e} \cdot \boldsymbol{a}_i \cdot \boldsymbol{e} \right\rangle - \left\langle \sum_{i=1}^{N_A}\sum_{j=1}^{N_A} \boldsymbol{e} \cdot \boldsymbol{a}_i \cdot \boldsymbol{T}_{ij} \cdot \boldsymbol{a}_j \cdot \boldsymbol{e} \right\rangle \right.$$
$$\left. + \left\langle \sum_{i=1}^{N_A}\sum_{j=1}^{N_A}\sum_{k=1}^{N_A} \boldsymbol{e} \cdot \boldsymbol{a}_i \cdot \boldsymbol{T}_{ij} \cdot \boldsymbol{a}_j \cdot \boldsymbol{T}_{jk} \cdot \boldsymbol{a}_k \cdot \boldsymbol{e} \right\rangle - \ldots \right\} \tag{147}$$

with $\boldsymbol{e}$ as the unit vector in the direction of $\boldsymbol{E}_0(\boldsymbol{r}, t)$.

Using the molecular correlation functions (111) we write (147) as an expansion in successive orders of interactions:[357]

$$R_m = R_m^{(1)} + R_m^{(2)} + R_m^{(3)} + \ldots \tag{148}$$

where, one by one, the contributions result by ordinary, binary, ternary, ... correlations. The first term is, from the formal point of view:

$$R_m^{(1)} = \frac{\rho}{3\varepsilon_0}\int (\boldsymbol{e} \cdot \boldsymbol{a}_i \cdot \boldsymbol{e})\, g^{(1)}(\tau_i)\, d\tau_i \tag{148a}$$

For gases and liquids, it immediately goes over into the form (144a) of the Lorentz–Lorenz formula. It thus corresponds to a lack of correlations between molecules in the medium.

The second term of the expansion (148) is of the form:

$$R_m^{(2)} = \frac{\rho^2}{3\varepsilon_0}\iint\{ - (\boldsymbol{e} \cdot \boldsymbol{a}_i \cdot \boldsymbol{T}_{ij} \cdot \boldsymbol{a}_j \cdot \boldsymbol{e})$$
$$+ (\boldsymbol{e} \cdot \boldsymbol{a}_i \cdot \boldsymbol{T}_{ij} \cdot \boldsymbol{a}_j \cdot \boldsymbol{T}_{ji} \cdot \boldsymbol{a}_i \cdot \boldsymbol{e})\}\, g^{(2)}(\tau_i, \tau_j)\, d\tau_i\, d\tau_j \tag{149}$$

354 P. Mazur and M. Mandel, *Physica*, 1956, **22**, 289, 299.
355 J. Jansen and P. Mazur, *Physica*, 1955, **21**, 193; P. Mazur and J. Jansen, *ibid.*, p. 208.
356 S. Kielich, *Acta Phys. Polon.*, 1965, **27**, 305.
357 (*a*) S. Kielich, *Physica*, 1962, **28**, 1116; (*b*) S. Kielich, *Acta Phys. Polon.*, 1962, **22**, 477.

It defines a first deviation from the Lorentz–Lorenz law (144a) due to the presence, in gases and liquids, of binary molecular correlations.

The third term of (148) is:

$$R_{\mathrm{m}}^{(3)} = \frac{\rho^3}{3\varepsilon_0}\iiint(\boldsymbol{e}\cdot\boldsymbol{a}_i\cdot\boldsymbol{T}_{ij}\cdot\boldsymbol{a}_j\cdot\boldsymbol{T}_{jk}\cdot\boldsymbol{a}_k\cdot\boldsymbol{e})\,g^{(3)}(\tau_i,\tau_j,\tau_k)\,\mathrm{d}\tau_i\,\mathrm{d}\tau_j\,\mathrm{d}\tau_k \tag{150}$$

It accounts for a further deviation from the above law due to ternary correlations.

If, in particular, the correlation functions $g^{(2)}$ and $g^{(3)}$ are independent of molecular orientations, we obtain from (149) and (150), by averaging over all possible orientations of the molecules:

$$R_{\mathrm{m}}^{(2)} = \tfrac{2}{3}R_{\mathrm{m}}^{(1)}(\boldsymbol{a}_{\mathrm{V}}:\boldsymbol{a}_{\mathrm{V}})\langle r_{ij}^{-6}\rangle + \ldots \tag{149a}$$

$$R_{\mathrm{m}}^{(3)} = 2R_{\mathrm{m}}^{(1)}\,a_{\mathrm{V}}^2\langle r_{ij}^{-3}r_{jk}^{-3}\rangle + \ldots \tag{150a}$$

where $R_{\mathrm{m}}^{(1)}$ is given by (144a) and we have introduced the following correlation parameters:[357]

$$\langle r_{ij}^{-6}\rangle = \frac{\rho}{V}\iint r_{ij}^{-6}\,g^{(2)}(\boldsymbol{r}_i,\boldsymbol{r}_j)\,\mathrm{d}\boldsymbol{r}_i\,\mathrm{d}\boldsymbol{r}_j \tag{151}$$

$$\langle r_{ij}^{-3}r_{jk}^{-3}\rangle = \frac{\rho^2}{2V}\iiint\{3(\boldsymbol{r}_{ij}\cdot\boldsymbol{r}_{jk})^2 - r_{ij}^2r_{jk}^2\} \times r_{ij}^{-5}r_{jk}^{-5}\,g^{(3)}(\boldsymbol{r}_i,\boldsymbol{r}_j,\boldsymbol{r}_k)\,\mathrm{d}\boldsymbol{r}_i\,\mathrm{d}\boldsymbol{r}_j\,\mathrm{d}\boldsymbol{r}_k \tag{152}$$

defining respectively binary and ternary molecular correlations.

Simple Liquids. The correlation parameters (151) and (152) are accessible to numerical calculation for simple models of radial interaction between the molecules. In a number of cases, evaluations for simple liquids can be usefully made [358] using the Kirkwood [344] model of hard spheres of diameter d and volume $v = \pi d^3/6$, when

$$g(r_{ij}) = \begin{cases}0 & \text{for } r_{ij} < d\\ 1 & \text{for } r_{ij} > d\end{cases} \tag{153}$$

For a binary function like this we obtain [359] for $n \geqslant 4$:

$$\langle r_{ij}^{-n}\rangle = 4\pi\rho\int_0^\infty r_{ij}^{-n+2}\,g(r_{ij})\,\mathrm{d}r_{ij} = 4\pi\rho\int_d^\infty r_{ij}^{-n+2}\,\mathrm{d}r_{ij} = \frac{4\pi\rho}{n-3}\left(\frac{\pi}{6v}\right)^{(n-3)/3} \tag{154}$$

or, as a function of the number density of molecules:

$$\langle r_{ij}^{-n}\rangle = \frac{4\pi\rho}{n-3}\left(\frac{\pi\rho}{6u}\right)^{(n-3)/3} \tag{154a}$$

where we have resorted to the circumstance that in liquids $\rho v = u$, with $u = 0.6$—0.74, as proved by experiment.[90]

358 P. Mazur and B. J. Postma, *Physica*, 1959, **25**, 251.

359 (*a*) S. Kielich, *Chem. Phys. Letters*, 1968, **2**, 112; (*b*) S. Kielich, *J. Physique*, 1968, **29**, 619.

Thus, by (154) for $n = 6$ we obtain the parameter (151) in the form:

$$\langle r_{ij}^{-6} \rangle = \frac{2\pi^2}{9v}\rho = \frac{2\pi^2}{9u}\rho^2 \tag{151a}$$

which is well adapted to numerical calculations. Significantly, in this model, the second contribution to the molar refraction (149a) is proportional to the square of the number density of molecules:

$$R_{\mathrm{m}}^{(2)} = \frac{4\pi^2}{27u} R_{\mathrm{m}}^{(1)} (\boldsymbol{a}_{\mathrm{v}} : \boldsymbol{a}_{\mathrm{v}})\, \rho^2 \tag{149b}$$

In the case of axially symmetric molecules the optical polarizability tensor $\boldsymbol{a}$ has the form (86) (similar to electric polarizability), and:

$$\boldsymbol{a}_{\mathrm{v}} : \boldsymbol{a}_{\mathrm{v}} = 3a_{\mathrm{v}}^2(1 + 2\kappa_a^2) \tag{155}$$

whence (149b) becomes:

$$R_{\mathrm{m}}^{(2)} = \frac{4\pi^2}{9u} R_{\mathrm{m}}^{(1)}\, a_{\mathrm{v}}^2(1 + 2\kappa_a^2)\, \rho^2 \tag{149c}$$

By insertion of (155) into (149a) we come to the result of Mazur and Postma.[358]

Calculation of the triple correlation parameter (152) is not so simple a matter, though such calculations can be performed on making certain assumptions.[360] This parameter, however, is highly interesting owing to the fact that it takes positive or negative values according to the model of three-body collision. It is apparent that (152) is positive if the angle subtended by the vectors $\boldsymbol{r}_{ij}$ and $\boldsymbol{r}_{jk}$ lies between 0 and 54° or 125 and 180°, whereas for angles 54—125° it is negative (Figure 10). Consequently, triple correlations raise or lower R_{m} according to the manner of collision of the three molecules.

Real Gases. In the case of moderately compressed gases we have, by (113):

$$g(r_{ij}) = \exp\left[-\beta u(r_{ij})\right]\left[1 + O(\rho)\right] \tag{156}$$

where $u(r_{ij})$ is the radial interaction energy between molecules i and j.

If in particular $u(r_{ij})$ is of the form of the Lennard-Jones potential (119a), we have instead of (154) the result:[321]

$$\langle r_{ij}^{-n} \rangle = 4\pi\rho \int r_{ij}^{2-n} \exp\left\{y_{ij}^2\left[\left(\frac{\sigma_{ij}}{r_{ij}}\right)^6 - \left(\frac{\sigma_{ij}}{r_{ij}}\right)^{12}\right]\right\} \mathrm{d}r_{ij} = \frac{\pi\rho\sigma_{ij}^{3-n}}{3y_{ij}^4} H_n(y_{ij}) \tag{157}$$

where

$$H_n(y_{ij}) = y_{ij}^{(27-n)/6} \sum_{m=0}^{\infty} \frac{y_{ij}^m}{m!} \Gamma\left(\frac{6m + n - 3}{12}\right) \tag{157a}$$

are functions first introduced by Pople and tabulated numerically by Buckingham and Pople;[321] above, $y_{ij}^2 = 4\beta\varepsilon_{ij}$.

[360] (*a*) R. W. Hellwarth, *Phys. Rev.*, 1966, **152**, 156; 1967, **163**, 205; (*b*) R. W. Hellwarth, 'Quantum Optics', ed. R. J. Glauber, Academic Press, New York, 1969, p. 563.

Similarly, the triple correlation parameter (152) can be calculated for the hard-sphere model of Lennard-Jones; this procedure yields a negative value for argon.[361]

Applying to R_m the virial expansion method (120), the relation (149a) with regard to (156) and (157) can be represented for compressed gases in

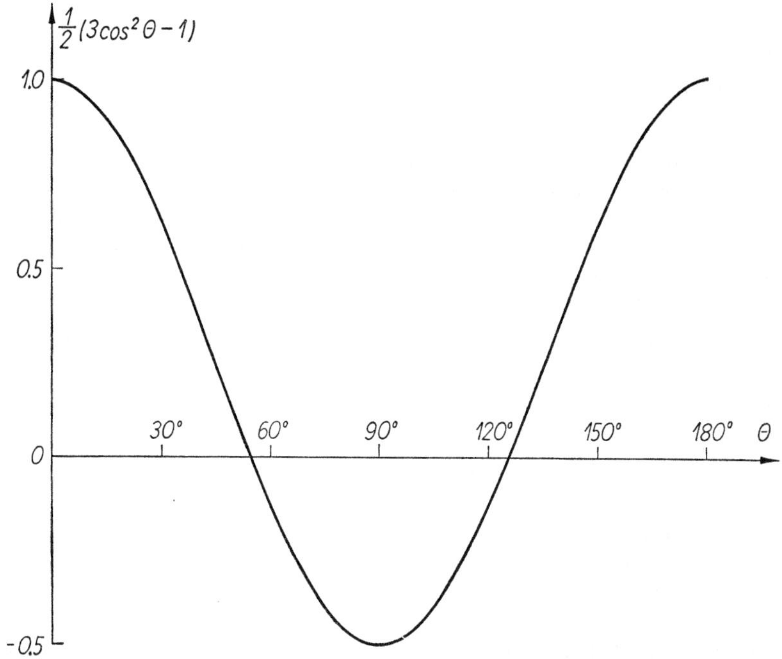

Figure 10 *Graph of the function* (3 cos² θ − 1)/2 *versus the angle* θ.

the form:

$$R_m^{(2)} = \frac{1}{V_m}\{B_R + O(\rho)\} \tag{149d}$$

where the second refractivity coefficient is given by Buckingham's formula:[326]

$$B_R = \frac{8\pi^2 a_v^3 N_A}{9\sigma^3 y^4}(1 + 2\kappa_a^2) H_6(y) \tag{149e}$$

In the general case, in addition to the contribution (149a) or (149e) to R_m resulting from linear dipole polarizability, other contributions due to quadrupole polarizability[356] and non-linear dipolar polarizabilities[326, 357] have to be taken into account. Also, contributions to $R_m^{(2)}$ have been calculated which result if the part dependent on molecular orientation (multipole interactions[357]) is included in the correlation function

[361] C. G. Gray and H. I. Ralph, *Phys. Letters* (*A*), 1970, **33a**, 165.

$g^{(2)}(\tau_i, \tau_j)$. Numerous authors have discussed certain special aspects of refractive index theory in simple dense fluids,[362] macromolecular fluids[363] and in fluids near the critical point.[364] Also, various experimental studies of R_m have been carried out for simple fluids[365] and water.[366] Frequency-dependent polarizabilities of atoms and simple molecules have been calculated.[367]

Dense Mixtures. If, quite generally, the molar volume V_m contains N_1 molecules of the first species, N_2 of the second, ... and N_a of the species a, we have:

$$N = \sum_a N_a \quad \text{and} \quad \sum_a x_a = 1 \tag{158}$$

where summation extends over all the components of the system and $x_a = N_a/N$ is the mole fraction of substance a in the mixture.

R_m is a strictly additive quantity only in mixtures of perfect gases. In imperfect gases and liquids as a result of correlations between the molecules of a single component and between those of various components, the molar refraction R_m does not satisfy the rule of additivity, and we have:[357]

$$R_m = \sum_a x_a R_m^{(a)} + \sum_{ab} x_a x_b R_m^{(ab)} + \sum_{abc} x_a x_b x_c R_m^{(abc)} + \ldots \tag{159}$$

The first term provides the well-known Lorentz result expressing additivity of the molar refraction

$$R_m^{(a)} = \frac{4\pi}{3} N_A a_V^{(a)} \tag{159a}$$

for the individual components, whose molecules possess the scalar and constant-volume polarizability $a_V^{(a)}$. The second, third, and higher terms in (159) define deviations from the rule of additivity of R_m resulting from binary, ternary, and higher molecular correlations both between like molecules and between those of the different components in a condensed system. For atoms and isotropically polarizable molecules we have:[357]

$$R_m^{(ab)} = \frac{4\pi}{3} N_A a_V^{(a)} a_V^{(b)} (a_V^{(a)} + a_V^{(b)}) \langle r_{ij}^{-6} \rangle_{ab} + \ldots \tag{159b}$$

$$R_m^{(abc)} = \frac{8\pi}{3} N_A a_V^{(a)} a_V^{(b)} a_V^{(c)} \langle r_{ij}^{-3} r_{jk}^{-3} \rangle_{abc} + \ldots \tag{159c}$$

362 (*a*) M. S. Macrakis, *J. Appl. Phys.*, 1967, **38**, 2278; (*b*) D. J. Vezzetti and J. B. Keller, *J. Math. Phys.*, 1967, **8**, 1861; (*c*) W. R. Johnson and F. D. Feoick, *Phys. Rev.*, 1968, **168**, 22; (*d*) V. G. Kaveeshwar, K. T. Chung, and R. P. Hurst, *ibid.*, 1968, **172**, 35.

363 (*a*) R. K. Bullough, *Phil. Trans.*, 1962, **254**, 397; (*b*) H. DeVoe, *J. Chem. Phys.*, 1964, **41**, 393.

364 S. Y. Larsen, R. D. Mountain, and R. Zwanzig, *J. Chem. Phys.*, 1965, **42**, 2187.

365 (*c*) C. M. Knobler, C. P. Abbiss, and C. J. Pings, *J. Chem. Phys.*, 1964, **41**, 2200; (*b*) C. P. Abbiss, C. M. Knobler, R. K. Teague, and C. J. Pings, *ibid.*, 1965, **42**, 4145. (*c*) J. W. Stewart, *ibid.*, 1964, **40**, 3297; (*d*) D. E. Diller, *ibid.*, 1968, **49**, 3097.

366 (*a*) E. Reisler and H. Eisenberg, *J. Chem. Phys.*, 1965, **43**, 3875; (*b*) H. Eisenberg. *ibid.*, p. 3887.

367 (*a*) G. A. Victor, J. C. Browne, and A. Dalgarno, *Proc. Phys. Soc.*, 1967, **92**, 42; (*c*) H. P. Kelly, *Phys. Rev.*, 1969, **182**, 84; 1970, **1**, 274; (*c*) D. B. DuPré and J. P, McTague, *J. Chem. Phys.*, 1969, **50**, 2024.

The foregoing theory of molar refraction proves that investigation of R_m in one- and multi-component systems can provide interesting information concerning the forces (radial and orientational) acting between like and unlike molecules.[326, 357]

Kerr Effect in Gases and Liquids.—The earliest investigation of the changes undergone by the optical properties of isotropic bodies in a strong static electric field was performed in the years 1875—1879 by Kerr. The electro-optical birefringence induced in such bodies with optical symmetry axis parallel to the field direction has been named the Kerr effect. Since then, electro-optical birefringence has never ceased to be the object of study, both experimental and theoretical. This is due to the fact that it provides a relatively simple method of obtaining valuable data concerning the structure as well as the optical and electrical properties of isolated atoms and molecules, and gives us information on the forces with which they interact in dense media.

We shall first deal in brief with Kerr effect theory for gases. One of the earliest theories explaining the mechanism of electro-optical birefringence in gases was due to Voigt[200] who, resorting to the electron theory of Lorentz,[51] considered the atom as a naturally isotropic oscillator which becomes anisotropic when placed in an external electric field. This is to say that its oscillation frequency in the field direction is no longer the same as in the direction perpendicular to the field. The change in the oscillator frequency thus caused by the field is proportional to the square of the field strength. A gas, consisting of a large number of such oscillators, becomes anisotropic; its optical birefringence, as shown by experiment, depends quadratically on the field.

Voigt's theory was, however, quite unable to account for the experimentally found strong temperature-dependence of the electrically induced optical birefringence. It was Langevin[201] who, by assuming the molecules of various substances as optically and electrically anisotropic in their natural state, proposed the fundamental mechanism of the Kerr effect: the anisotropic molecules of a gas interacting with the external field are regarded as tending to orient themselves in a way to make their potential energy minimal, this tendency being counteracted continually by random thermal motion, which prevents their complete alignment with axes of maximum polarizability parallel to the field. This competition between the two processes leads to a compromise in the form of statistical equilibrium when the molecules in accordance with the Maxwell–Boltzmann law achieve a degree of angular ordering along the field vector. This mechanism was able to account for the experimental facts, which stated that the birefringence of a gas grows with increasing electric field strength but decreases with increasing temperature. Born[368] generalized Langevin's theory to gases consisting of anisotropic molecules having a permanent dipole moment and non-linear polarizability.[13, 54]

[368] M. Born, *Ann. Phys.*, 1918, **55**, 177.

Definition of Kerr Constants. In the presence of an external electric field E, the electric permittivity $\varepsilon(E)$ and magnetic permeability $\mu(E)$ become functions of the field strength. As a consequence the refractive index n also depends on E:

$$\varepsilon(E)\,\mu(E) = n_E^2 = n^2 + \Delta n_E^2 \tag{160}$$

where n is the refractive index of an isotropic body in the absence of the field and Δn_E^2 defines the change in index due to the static electric field. The variation Δn_E^2 differs from one case to another, depending on the mutual configuration of the oscillation direction of the analysing light wave with respect to the applied static field vector. The optical birefringence electrically induced in the isotropic medium is defined by the difference $(n_\parallel - n_\perp)$ between the indices for light oscillations respectively parallel and perpendicular to E, with the propagation direction of the analysing light wave obviously perpendicular to E. The quantity $(n_\parallel - n_\perp)$ provides a measure of the induced optical birefringence and is a well-defined function of the field strength E. In isotropic bodies consisting of molecules, one can consider, in a good approximation, that the birefringence is proportional to the square of the field, as expressed by the following representation of Kerr's results:

$$n_\parallel - n_\perp = \lambda\, B_\lambda\, E^2 \tag{161}$$

It is costumary to refer to B_λ as the experimentally measured 'Kerr constant',[369] in spite of the fact that B_λ is in general not constant but depends on the wavelength λ of the analysing light, the structure and thermodynamic state of the medium, and (as is the case in macromolecular substances) the applied field strength.[133, 134]

In conformity with (160), the molar refraction too is a function of the static field:

$$R_m^E = R_m + \Delta R_m^E \tag{162}$$

R_m denoting the molar refraction in the absence of a field defined by (140), and the field-induced change in refraction being in explicit form:

$$\Delta R_m^E = \frac{n_E^2 - 1}{n_E^2 + 2} V_m^E - \frac{n^2 - 1}{n^2 + 2} V_m \tag{162a}$$

Here, we have duly taken into consideration the circumstance that an external electric field, if sufficiently strong, will generally affect the molar volume (by way of a change in density). We thus write:

$$V_m^E = V_m + \Delta V_m^E \tag{163}$$

whence (162a) takes the form of the sum of two components:

$$\Delta R_m^E = \frac{3\Delta n_E^2 V_m}{(n_E^2 + 2)(n^2 + 2)} + \frac{n_E^2 - 1}{n_E^2 + 2} \Delta V_m^E \tag{162b}$$

369 (*a*) R. Gans, *Z. Physik*, 1923, **17**, 353; (*b*) H. A. Stuart and W. Buchheim, *ibid.*, 1938, **111**, 36.

The first of these terms is directly related to changes in refractive index Δn_E, whereas the other term differs from zero only if the field E causes a change in volume $\Delta V_{\rm m}^E$ (*i.e.* electrostriction).[63]

Omitting terms of the type of $(\Delta n_E^2)^2$, $\Delta n_E^2 \, \Delta V_{\rm m}^E$ ..., we obtain from (162) and (162b) the following difference of molar refractions determined for light oscillations parallel and perpendicular to the static electric field:

$$\Delta R_{\rm m}^{\parallel} - \Delta R_{\rm m}^{\perp} = (K_{\rm m} + K_{\rm m}^{\rm ES})\, E_0^2 \tag{164}$$

where

$$K_{\rm m} = \left(\frac{\Delta R_{\rm m}^{\parallel} - \Delta R_{\rm m}^{\perp}}{E_0^2}\right)_{V_{\rm m}} \simeq \frac{3V_{\rm m}}{(n^2+2)^2}\left(\frac{\Delta n_{\parallel}^2 - \Delta n_{\perp}^2}{E_0^2}\right) \tag{164a}$$

is that part of the Kerr constant which arises from a difference of molar refractions due to a change in refractive index at constant volume $V_{\rm m}$. It is referred to simply as the molar Kerr constant. The other part of (164) is related to changes in molar volume. It thus defines a molar Kerr constant $K_{\rm m}^{\rm ES}$ accounting for electrostriction which can be expressed approximately as:

$$K_{\rm m}^{\rm ES} = R_{\rm m}\left(\frac{\Delta V_{\rm m}^{\parallel} - \Delta V_{\rm m}^{\perp}}{V_{\rm m}\, E_0^2}\right) \tag{164b}$$

In general, the change in volume caused by the field E consists of an isotropic contribution $\Delta V_{\rm m}^{\rm is}$ and an anisotropic one $\Delta V_{\rm m}^{\rm anis}$. The latter is chiefly due to shape deformation of the dielectric sample. In fact, it should be stated that an isotropic body can exhibit anisotropic electrostriction, according to the method adopted by the experimenter; this eventuality should be eliminated as far as possible in measurements. The constant $K_{\rm m}^{\rm ES}$ given by (164b) is non-zero only if the electrostriction apparent in the macroscopic medium is anisotropic. Anisotropy of electrostriction can to some degree affect the measurement of the Kerr constant; but if the necessary precautions are taken, the part due to $K_{\rm m}^{\rm ES}$ is eliminated completely or at least made quite insignificant.

Neglecting anisotropy of electrostriction and keeping in mind that the mean macroscopic field E depends on the field E_0 applied in vacuum by the relation:

$$\boldsymbol{E} = \frac{3\varepsilon_0}{\varepsilon + 2\varepsilon_0}\boldsymbol{E}_0 = \frac{3}{\varepsilon_r + 2}\boldsymbol{E}_0 \tag{164c}$$

and assuming with sufficient accuracy that $(\Delta n_{\parallel}^2 - \Delta n_{\perp}^2) = 2n(n_{\parallel} - n_{\perp})$, we obtain with (161) and (164a) the following formula:

$$B_\lambda = \frac{3}{2n\lambda}\left(\frac{n^2+2}{3}\right)^2\left(\frac{\varepsilon_r+2}{3}\right)^2\frac{K_{\rm m}}{V_{\rm m}} \tag{165}$$

which provides the basis for comparisons between the experimentally measured Kerr constant B and the theoretically calculated molar Kerr constant, defined as

$$K_{\rm m} = \left(\frac{\Delta R_{\rm m}^{\parallel} - \Delta R_{\rm m}^{\perp}}{E_0^2}\right)_{V_{\rm m}} \tag{166}$$

in the absence of electrostriction anisotropy.

Molecular Theory for a Molecular Dilute State. In presenting the theory of the Kerr effect, we shall begin with the simple case of a molar volume V_m free of molecular fields. This is equivalent to assuming the relation (145) for the dipole moment induced in molecule i by the electric field of the light wave $\boldsymbol{E}_0(r, t)$. Equations (142)—(145) now yield the molar refraction in the presence of an electric field $E(0)$ in the form:

$$R_m^E \, \boldsymbol{U} = \frac{1}{3\varepsilon_0} \left\langle \sum_{i=1}^{N_A} \boldsymbol{a}_i \right\rangle_E \tag{167}$$

With regard to equations (140) and (144a), the change in molar refraction

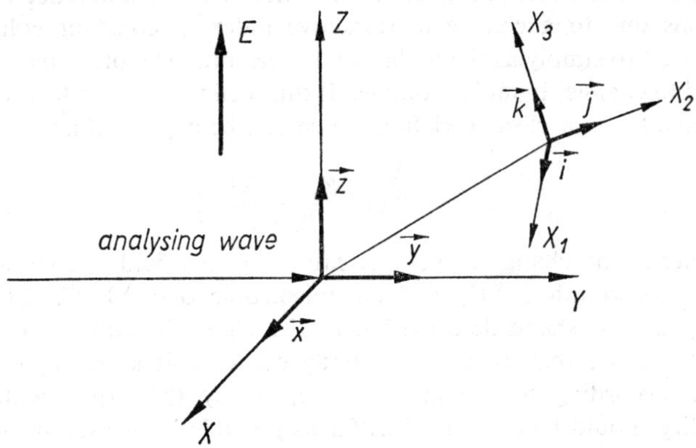

Figure 11 *To the medium under investigation is rigidly attached the laboratory Cartesian reference system x, y, z, whereas to the molecule under consideration is attached the Cartesian system of axes x_1, x_2, x_3. The static electric field* **E**, *which polarizes the medium non-linearly, acts along the z-axis. The light wave which serves for analysing the induced non-linearity propagates along y*

due to the external field is:

$$\Delta R_m^E \, \boldsymbol{U} = \frac{N_A}{3\varepsilon_0} \langle \boldsymbol{d}^\omega \rangle_E \tag{167a}$$

where

$$\boldsymbol{d}^\omega = \boldsymbol{a} - a\boldsymbol{U} \tag{168}$$

defines the deviation of the polarizability tensor of the molecule as in the electric case (69b).

Henceforth we shall be denoting by $\boldsymbol{x}, \boldsymbol{y}, \boldsymbol{z}$ unit vectors in the direction of the co-ordinate axes x, y, z of the Cartesian frame rigidly attached to the body under investigation in the laboratory, and by $\boldsymbol{i}, \boldsymbol{j}, \boldsymbol{k}$ unit vectors along the axes x_1, x_2, x_3 of the mobile frame attached to a given molecule of the medium (Figure 11). It will be assumed throughout that the analysing light wave propagates along the y-axis of the laboratory reference system and that the applied uniform static electric field $\boldsymbol{E}(0)$

acts along the z-axis (vector $\boldsymbol{z}$). On the above convention, the changes in molar refraction for light oscillations respectively parallel (directed along $\boldsymbol{z}$) and perpendicular (directed along $\boldsymbol{x}$) to the vector $\boldsymbol{E}(0)$ are, by (167a):

$$\Delta R_{\mathrm{m}}^{\parallel} = \frac{N_{\mathrm{A}}}{3\varepsilon_0}\langle \boldsymbol{z}\cdot \boldsymbol{d}^{\omega}\cdot \boldsymbol{z}\rangle_E, \quad \Delta R_{\mathrm{m}}^{\perp} = \frac{N_{\mathrm{A}}}{3\varepsilon_0}\langle \boldsymbol{x}\cdot \boldsymbol{d}^{\omega}\cdot \boldsymbol{x}\rangle_E \qquad (167\mathrm{b})$$

Inserted into equation (166), they yield the following molar Kerr constant:

$$K_{\mathrm{m}} = \frac{N_{\mathrm{A}}}{3\varepsilon_0 E_0^2}\langle(\boldsymbol{z}\cdot \boldsymbol{d}^{\omega}\cdot \boldsymbol{z}) - (\boldsymbol{x}\cdot \boldsymbol{d}^{\omega}\cdot \boldsymbol{x})\rangle_E \qquad (166\mathrm{a})$$

In order to perform the averaging denoted above by the symbol $\langle\,\rangle_E$ and defined by equation (98), we have to know the perturbed distribution function which, with regard to equation (99) in the absence of statistical correlation ($U_N = 0$), is of the form:

$$f(\Omega, E) = \frac{\exp[-\beta U(\Omega, E)]}{\int \exp[-\beta U(\Omega, E)]\,\mathrm{d}\Omega} \qquad (169)$$

In the case of molecules having a permanent electric dipole moment $\boldsymbol{p}$ and electric polarizability given by the tensor $\boldsymbol{\alpha}$, the distribution function (169) becomes, in an approximation quadratic in $\boldsymbol{E}_0 = \boldsymbol{z}E_z$ [see equation (122)]:

$$f(\Omega, E) = f(\Omega, 0)\{1 + \beta E_z(\boldsymbol{p}\cdot \boldsymbol{z}) + \tfrac{1}{2}\beta E_z^2[\boldsymbol{d}^e + \beta(\boldsymbol{pp} - \tfrac{1}{3}p^2\,\boldsymbol{U})] : \boldsymbol{zz}\} \qquad (169\mathrm{a})$$

where $\boldsymbol{d}^e$ is the deviation of the polarizability tensor of the molecule.

For a function of the form (169a), the perturbed statistical averaging of (166a) reduces to an isotropic averaging of products of unit vectors as given by formulae (104)—(107). Finally, we obtain the following molar Kerr constant:

$$K_{\mathrm{m}}^{\mathrm{S}} = \frac{N_{\mathrm{A}}\beta}{30\varepsilon_0}(\boldsymbol{d}^{\omega} : \boldsymbol{d}^e + \beta \boldsymbol{d}^{\omega} : \boldsymbol{pp}) = \frac{2\pi}{15}N_{\mathrm{A}}\beta(\boldsymbol{d}_{\mathrm{V}}^{\omega} : \boldsymbol{d}^e + \beta \boldsymbol{d}_{\mathrm{V}}^{\omega} : \boldsymbol{pp}) \qquad (170)$$

valid for dipolar molecules of arbitrary symmetry undergoing statistical electric reorientation.

If the polarizability tensors $\boldsymbol{a}$ and $\boldsymbol{\alpha}$ have elements only in the directions of the three principal axes 1, 2, 3, equation (170) yields for the anisotropic part of the Kerr constant:

$$K_{\mathrm{m}}^{\mathrm{SA}} = \frac{2\pi N_{\mathrm{A}}}{45kT}\{(a_{\mathrm{V}1} - a_{\mathrm{V}2})(\alpha_1 - \alpha_2) + (a_{\mathrm{V}2} - a_{\mathrm{V}3})(\alpha_2 - \alpha_3) + (a_{\mathrm{V}3} - a_{\mathrm{V}1})(\alpha_3 - \alpha_1)\} \qquad (170\mathrm{a})$$

and for the dipolar part:

$$K_{\mathrm{m}}^{\mathrm{SD}} = \frac{2\pi N_{\mathrm{A}}}{45k^2T^2}\{(a_{\mathrm{V}1} - a_{\mathrm{V}2})(p_1^2 - p_2^2) + (a_{\mathrm{V}2} - a_{\mathrm{V}3})(p_2^2 - p_3^2) + (a_{\mathrm{V}3} - a_{\mathrm{V}1})(p_3^2 - p_1^2)\} \qquad (170\mathrm{b})$$

We have here the well-known formulae of Born[368] whose theory leads, in addition to the Langevin anisotropic term (170a), to a supplementary dipole term of the form (170b) more strongly dependent on temperature ($\beta^2 = 1/k^2 T^2$).

Kerr Effect in an Alternating Electric Field. If the dipolar molecule under consideration is placed in a weak alternating electric field $\boldsymbol{E}(\omega)$, one of the following two processes can take place in the limit, according to the value of ω: (*i*) if ω is much smaller than the characteristic frequency, ω_0, of oscillations of the molecule, the field will have a reorienting effect on it causing some dipolar polarization of the medium; (*ii*) if ω is much larger than ω_0, no orientation of the electric dipoles can set in, so the field causes no polarization of the dipole kind, but solely electron distortional polarization; this was the situation in the case of the polarization linear in the alternating field strength, considered by Debye.[85] In a strong field the situation is somewhat different, as has been shown by Peterlin and Stuart[131] and by Frenkel.[73] However, the problem of non-linear molecular relaxation is rather complicated; it has been solved for axially symmetric molecules [*cf.* the distribution function (125)]. We consequently have to adapt the Kerr constant (167c) for axially symmetric molecules, for which the deviation tensor of optical polarizability is similar to (69d):

$$\boldsymbol{d}^{\omega} = \delta^{\omega}(\boldsymbol{k}\,\boldsymbol{k} - \tfrac{1}{3}\boldsymbol{U}) \tag{168a}$$

Insertion into (166a) now yields:

$$K_{\mathrm{m}} = \frac{4\pi}{3} N_{\mathrm{A}}\, \delta_{\mathrm{V}}^{\omega} \frac{1}{E_0^2} \langle (\boldsymbol{k}\cdot\boldsymbol{z})^2 - (\boldsymbol{k}\cdot\boldsymbol{x})^2 \rangle_E \tag{166b}$$

where $\delta_{\mathrm{V}}^{\omega} = (4\pi\varepsilon_0)^{-1}(a_3 - a_1)$ is the volume optical anisotropy.

Since the projections of the symmetry axis of the molecule $\boldsymbol{k}$ onto the laboratory unit vectors $\boldsymbol{x}, \boldsymbol{y}, \boldsymbol{z}$ are (see Figure 12): $\boldsymbol{k}\cdot\boldsymbol{x} = \sin\theta\cos\phi$, $\boldsymbol{k}\cdot\boldsymbol{y} = \sin\theta\sin\phi$, $\boldsymbol{k}\cdot\boldsymbol{z} = \cos\theta$, we can re-write equation (166b) as follows:

$$K_{\mathrm{m}} = \frac{4\pi}{3} N_{\mathrm{A}}\, \delta_{\mathrm{V}}^{\omega} \frac{\Phi(E)}{E_0^2} \tag{171}$$

Above, we have introduced the Langevin–O'Konski function:[133]

$$\Phi(E) = \langle \cos^2\theta - \sin^2\theta\sin^2\phi \rangle_E = \tfrac{1}{2}\langle 3\cos^2\theta - 1 \rangle_E \tag{172}$$

describing the reorientation of anisotropic molecules in a strong electric field.

When the molecules are acted on by an alternating electric field, the function (172) takes the form:

$$\Phi(E, t) = 2\pi \int_0^{\pi} (3\cos^2\theta - 1) f(\theta, E, t) \sin\theta \, \mathrm{d}\theta \tag{172a}$$

where the distribution function $f(\theta, E, t)$ is given in general by equation (125). We thus obtain, on reverting to the isotropic averages (44), the following result:[331]

$$\Phi(E, t) = \frac{\beta}{30} \sum_{mn} \frac{E_m E_n \exp[i(\omega_m + \omega_n) t]}{1 + i(\omega_m + \omega_n) \tau_B}$$
$$\times \left\{ 2\delta^e_{mn} + \frac{\beta p_m p_n}{1 + i\omega_m \tau_D} + \frac{\beta p_n p_m}{1 + i\omega_n \tau_D} \right\} \quad (173)$$

where $\delta^e = \alpha_3 - \alpha_1$ is the electric anisotropy of the molecule.

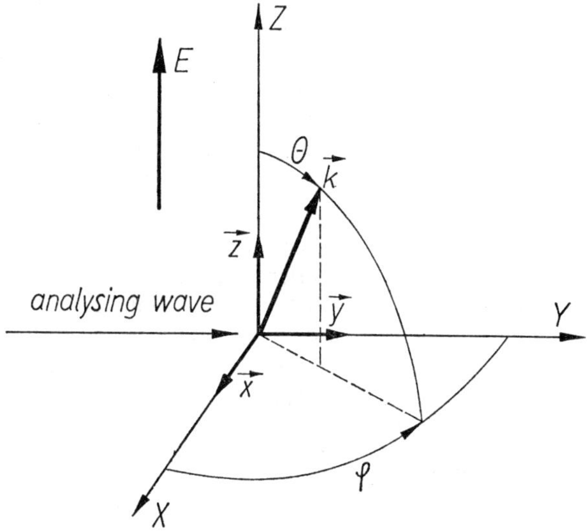

Figure 12 *Orientation of the symmetry vector* **k** *of the molecule with respect to the unit (basis) vectors* ***x***, ***y***, ***z*** *of the laboratory reference system; θ—polar angle, φ—azimuthal angle*

Consequently, in the alternating field case, the reorientation function is in general complex and can be represented in a general manner as:

$$\Phi(E, t) = \mathrm{Re}\,(\Phi) \pm i\,\mathrm{Im}\,(\Phi) \quad (173a)$$

the real part defining non-linear dispersion, and the imaginary part non-linear dielectric absorption.

In particular, if the alternating field frequency tends to zero ($\omega \to 0$, $t \to \infty$), equation (173) goes over simply into:

$$\Phi(E, \infty) = \frac{\beta}{15} (\delta^e + \beta p^2) E_0^2 \quad (173b)$$

accounting for reorientation of axially symmetric molecules in a static field. On inserting (173b) into (171) one obtains the molar Kerr constant in the form resulting from the Langevin–Born expressions (170) on the assumption of axial symmetry for the molecule.

The other limiting case arises when the oscillation frequency of the alternating electric field tends to infinity ($\omega \to \infty$, $t \to 0$). Equation (173) now reduces to:

$$\Phi(E, 0) = \frac{\beta}{15}\,\delta^{\infty}\langle E^2\rangle_t \tag{173c}$$

an expression describing reorientation of anisotropic molecules in the electric field of intense light. In this case, the medium exhibits optically induced optical birefringence, *i.e.* the optical Kerr effect considered by Buckingham [206] and detected by Mayer.[214]

From equations (171) and (173), on assuming a single oscillation frequency, one can obtain the results of Peterlin and Stuart [131] for dispersion and absorption of the dielectric Kerr constant. Also, the electron dispersion of the Kerr effect has been studied both by quantum-mechanical methods [370] and quantum field theory,[371] as well as experimentally.[372]

Direct Molecular Angular Correlations. Hitherto the molecules were assumed to be statistically independent of one another in the medium and to be able to present all possible orientations in space. In dense media, however, mutual interactions have to be taken into consideration as a result of which the mutual orientations of the molecules become statistically dependent. With regard to formula (167), one now obtains in place of (166a) the following starting-point expression for the Kerr constant:

$$K_{\mathrm{m}}^{\mathrm{SA}} = \frac{4\pi}{3E_0^2}\left\langle \sum_{i=1}^{N} \{(\boldsymbol{z}\cdot\boldsymbol{d}_{\mathrm{V}i}^{\omega}\cdot\boldsymbol{z}) - (\boldsymbol{x}\cdot\boldsymbol{d}_{\mathrm{V}i}^{\omega}\cdot\boldsymbol{x})\}\right\rangle_E \tag{174}$$

where the deviation tensor $\boldsymbol{d}_{\mathrm{V}i}$ defined by (168) relates to the volume polarizability of molecule i.

By the general definition (99) of statistical averaging, in dealing with dense systems we have to include the perturbed potential energy for the system as a whole, *i.e.* to perform a summation of the energies (121) over all the molecules in the molar volume V_{m}:

$$U_E = -\sum_{j=1}^{N_{\mathrm{A}}}(\boldsymbol{p}_j\cdot\boldsymbol{E}_0 + \tfrac{1}{2}\boldsymbol{E}_0\cdot\boldsymbol{\alpha}_j\cdot\boldsymbol{E}_0 + \ldots) \tag{175}$$

Above, for the sake of simplicity, we have provisionally omitted molecular fields.

[370] (*a*) A. D. Buckingham, *Proc. Roy. Soc.*, 1962, **A267**, 271; (*b*) L. L. Boyle and C. A. Coulson, *Proc. Phys. Soc.*, 1966, **89**, 499; (*c*) S. H. Lin, C. Y. Lin and H. Eyring, *J. Phys. Chem.*, 1966, **70**, 1756; (*d*) Sheng Hsien Lin and R. Bersohn, *J. Chem. Phys.*, 1966, **44**, 3768; (*e*) M. P. Bogaard, A. D. Buckingham, and B. J. Orr, *Mol. Phys.*, 1967, **13**, 533.

[371] (*a*) P. W. Atkins and L. D. Barron, *Proc. Roy. Soc.*, 1968, **A306**, 119; (*b*) W. C. Mackrodt, *Mol. Phys.*, 1971, **20**, 251.

[372] (*a*) R. J. W. Le Fèvre and S. C. Solomons, *Austral. J. Chem.* 1968, **21**, 1703; (*b*) H. G. Kuball and W. Galler, *Z. Naturforsch.*, 1968, **23a**, 295; (*c*) H. G. Kuball, W. Galler, R. Göb, and D. Singer, *ibid.*, 1969, **24a**, 1391; (*d*) H. G. Kuball, W. Euing, and T. Karstens, *Ber. Bunsengesellschaft phys. Chem.*, 1970, **74**, 316.

On inserting the energy (175) into the expansion of the perturbed statistical distribution function (100), we obtain to within the square of the electric field $\boldsymbol{E}_0 = \boldsymbol{z}E_0$, in place of (169a),

$$f(\tau, E) = f(\tau, 0)\Big\{1 + E_0\sum_{j=1}^{N_A}\boldsymbol{p}_j\cdot\boldsymbol{z} + \tfrac{1}{2}\beta E_0^2\sum_{j=1}^{N_A}(\boldsymbol{z}\cdot\boldsymbol{d}_j^e\cdot\boldsymbol{z})$$
$$+ \tfrac{1}{2}\beta^2 E_0^2\Big[\sum_{j=1}^{N_A}\sum_{k=1}^{N_A}(\boldsymbol{p}_j\cdot\boldsymbol{z})(\boldsymbol{p}_k\cdot\boldsymbol{z})$$
$$-\Big\langle\sum_{j=1}^{N_A}\sum_{k=1}^{N_A}(\boldsymbol{p}_j\cdot\boldsymbol{z})(\boldsymbol{p}_k\cdot\boldsymbol{z})\Big\rangle\Big] + \dots\Big\} \qquad (176)$$

Statistical averaging of the right-hand term of (174) with the distribution function (176) leads to the following molar Kerr constant, accounting for statistical-angular correlations:

$$K_m^{SA} = \frac{2\pi\beta}{15}\Big\langle\sum_{i=1}^{N_A}\sum_{j=1}^{N_A}\boldsymbol{d}_{Vi}^{\omega} : \boldsymbol{d}_j^e\Big\rangle \qquad (177)$$

$$K_m^{SD} = \frac{2\pi\beta^2}{15}\Big\langle\sum_{i=1}^{N_A}\sum_{j=1}^{N_A}\sum_{k=1}^{N_A}\boldsymbol{d}_{Vi}^{\omega} : \boldsymbol{p}_j\boldsymbol{p}_k\Big\rangle \qquad (178)$$

where the statistical averages in the absence of an external field are defined by formulae (95) and (96).

Equations (177) and (178) yield directly the results obtained by Buckingham,[327] Piekara and Kielich,[141, 142] as well as Mazur and Postma [358] for various particular cases. We shall restrict ourselves here to considering the case of axially symmetric molecules, for which the results are of a simple form providing satisfactory insight into the nature of the mechanism of direct angular correlations:[142]

$$K_m^{SA} = \frac{2\pi\beta}{45}\,\delta_V^{\omega}\,\delta^e\Big\langle\sum_{i=1}^{N_A}\sum_{j=1}^{N_A}(3\cos^2\theta_{ij} - 1)\Big\rangle \qquad (177a)$$

$$K_m^{SD} = \frac{2\pi\beta^2}{45}\,\delta_V^{\omega}p^2\Big\langle\sum_{i=1}^{N_A}\sum_{j=1}^{N_A}\sum_{k=1}^{N_A}(3\cos\theta_{ij}\cos\theta_{ik} - \cos\theta_{jk})\Big\rangle \qquad (178a)$$

Here, θ_{ij} is the angle between the symmetry axis $\boldsymbol{k}_i$ of molecule i and the symmetry axis $\boldsymbol{k}_j$ of molecule j: $(\boldsymbol{k}_i\cdot\boldsymbol{k}_j = \cos\theta_{ij})$.

Equations (177a) and (178a) are conveniently re-written as follows:[373]

$$K_m^{SA} = \frac{4\pi N_A}{45kT}\,\delta_V^{\omega}\,\delta^e\,(1 + J_A^{(2)}) \qquad (177b)$$

$$K_m^{SD} = \frac{4\pi N_A}{45k^2T^2}\,\delta_V^{\omega}p^2(1 + 2J_A^{(1)} + J_A^{(2)} + J_A^{(3)}) \qquad (178b)$$

where we have introduced the molecular angular correlation parameters

[373] S. Kielich, *Mol. Phys.*, 1963, **6**, 49; *Fizyka Dielektryków* (*Poznań*), 1964, **11**, 87.

(describing local ordering in liquids):

$$J_{\mathrm{A}}^{(1)} = \left\langle \sum_{j \neq i}^{N_{\mathrm{A}}} \cos \theta_{ij} \right\rangle \tag{179}$$

$$J_{\mathrm{A}}^{(2)} = \frac{1}{2}\left\langle \sum_{j \neq i}^{N_{\mathrm{A}}} (3\cos^2\theta_{ij} - 1) \right\rangle \tag{180}$$

$$J_{\mathrm{A}}^{(3)} = \frac{1}{2}\left\langle \sum_{j \neq i}^{N_{\mathrm{A}}} \sum_{k \neq j}^{N_{\mathrm{A}}} (3\cos\theta_{ij}\cos\theta_{ik} - \cos\theta_{jk}) \right\rangle \tag{181}$$

Kirkwood, in his statistical theory of dipolar dielectrics, derived the following angular correlation parameter:

$$g = 1 + z\langle\cos\theta\rangle = 1 + J_{\mathrm{A}}^{(1)} \tag{182}$$

involving a parameter, identical with (179), written in the form:

$$J_{\mathrm{A}}^{(1)} = z\langle\cos\theta\rangle \tag{179a}$$

where $z = N - 1$ is the number of nearest neighbours subtending the angle θ with the dipolar molecule under consideration.

The parameter (179) can also be written in the form:

$$J_{\mathrm{A}}^{(1)} = \frac{\rho}{V_m}\int\int \cos\theta_{ij}\, g^{(2)}(\tau_i, \tau_j)\, \mathrm{d}\tau_i\, \mathrm{d}\tau_j \tag{179b}$$

in which it has been calculated for various models of correlation between dipolar molecules.[352, 374]

Similarly to (179a) and (179b), the parameter (180) can be put in the particular form:[141]

$$J_{\mathrm{A}}^{(2)} = \tfrac{1}{2}z\langle 3\cos^2\theta - 1\rangle \tag{180a}$$

containing the 'number of neighbours', and in more general explicit form:[348]

$$J_{\mathrm{A}}^{(2)} = \frac{\rho}{2V_{\mathrm{m}}}\int\int (3\cos^2\theta_{ij} - 1)\, g^{(2)}(\tau_i, \tau_j)\, \mathrm{d}\tau_i\, \mathrm{d}\tau_j \tag{180b}$$

$J_{\mathrm{A}}^{(2)}$ can be determined numerically for non-dipolar molecules from Kerr constant measurements,[375, 376] or calculated theoretically for arbitrary weak molecular correlations [377] as well as for strong dipolar [352] and multipolar ones.[269]

In the two limiting cases of liquid structure discussed by Prins and Prins [378] when the molecules correlate (*i*) into parallel pairs (paratropism, $J_{\mathrm{A}}^{(2)} = z$), (*ii*) into pairs of mutually perpendicular partners (diatropism,

[374] A. I. Anselm, *Zhur. eksp. teor. Fiz.*, 1942, **12**, 274; 1943, **13**, 432; 1944, **14**, 364.

[375] (*a*) A. I. Anselm, *Zhur. eksp. teor. Fiz.*, 1947, **17**, 489; (*b*) G. Deželić, *J. Chem. Phys.*, 1966, **45**, 185; *Pure Appl. Chem.*, 1970, **23**, 327.

[376] (*a*) S. Kielich, *Bull. Acad. polon. Sci.*, *Sér. Sci. math. astr. et phys.*, 1958, **6**, 215; (*b*) B. Kasprowicz and S. Kielich, *Acta Phys. Polon.*, 1968, **33**, 495; (*c*) J. R. Lalanne, *J. Physique*, 1969, **30**, 643.

[377] (*a*) S. Kielich, *J. Chem. Phys.*, 1967, **46**, 4090; (*b*) S. Kielich, *Acta Phys. Polon.* 1968, **33**, 65.

[378] J. A. Prins and W. Prins, *Physica*, 1956, **22**, 576; 1957, **23**, 253.

$J_A^{(2)} = -z/2$), one can gain from (180a) information regarding the number z of nearest neighbours (*cf.* Figure 10).

The parameter (181) can be expressed as follows:

$$J_A^{(3)} = \tfrac{1}{2}z(z-1)\langle 3\cos\theta_{12}\cos\theta_{13} - \cos\theta_{23}\rangle \qquad (181a)$$

or in the form:[373]

$$J_A^{(3)} = \frac{\rho^2}{2V_m}\iiint(3\cos\theta_{ij}\cos\theta_{ik} - \cos\theta_{jk})\,g^{(3)}(\tau_i,\tau_j,\tau_k)\,d\tau_i\,d\tau_j\,d\tau_k \qquad (181b)$$

It thus is seen to express the contribution to the dipolar part of the Kerr constant from ternary correlations. Like the parameter (152) occurring in molar refraction, it is positive or negative according to the model of orientational near-range ordering.

All this shows that the Kerr effect in liquids depends largely on various angular correlations. Numerous experimental studies,[352, 379] especially of K_m as a function of temperature and pressure,[380] also those performed recently using new measuring techniques,[381] confirm this conclusion. The role of the angular correlation parameters (179)—(181) is, moreover, apparent in studies of the quadratic n.m.r. electric-field effect in liquids.[382]

Non-linear Polarizability of Molecules. Hitherto, we assumed the molecules as linearly polarizable and the optical polarizability tensor $\boldsymbol{a}_i$ as constant for a given molecule i in the ground state. However, as mentioned earlier (p. 248), according to the Voigt–Born hypothesis molecules in a strong field $\boldsymbol{E}$ polarize non-linearly, and the polarizability tensor now becomes a function of the externally applied field E:

$$\boldsymbol{a}_i(E) = \boldsymbol{a}_i(0) + \boldsymbol{b}_i^e \cdot \boldsymbol{E}_0 + \tfrac{1}{2}\boldsymbol{c}_i^e : \boldsymbol{E}_0\boldsymbol{E}_0 + \ldots \qquad (183)$$

where the tensor $\boldsymbol{b}_i^e$ of rank 3 defines a linear variation of the polarizability tensor $\boldsymbol{a}_i(E)$ and the tensor $\boldsymbol{c}_i^e$ of rank 4 a quadratic variation of $\boldsymbol{a}_i(E)$ due to the static electric field E_0.

With regard to the expansion (183), the deviation tensor $\boldsymbol{d}_i^\omega$ of the Kerr constant equation (174) has now to be replaced by a tensor $\Delta\boldsymbol{a}_i(E)$ expressing the change in polarizability as a function of the field:

$$\Delta\boldsymbol{a}_i(E) = \boldsymbol{b}_i^e \cdot \boldsymbol{E}_0 + \tfrac{1}{2}\boldsymbol{c}_i^e : \boldsymbol{E}_0\boldsymbol{E}_0, \quad \ldots \qquad (183a)$$

379 (*a*) W. H. Orttung and J. A. Meyers, *J. Phys. Chem.*, 1963, **67**, 1905; (*b*) H. G. Kuball and R. Göb, *Z. Naturforsch.*, 1967, **22**, 737; *Z. phys. Chem. (Frankfurt)*, 1968, **62**, 237; (*c*) J. V. Champion, G. H. Meeten, and C. D. Whittle, *Trans. Faraday Soc.*, 1970, **66**, 2671.

380 (*a*) H. Grodde, *Physik. Z.*, 1938, **39**, 772; (*b*) E. Kuss and H. A. Stuart, *ibid.*, 1941, **42**, 95.

381 (*a*) A. Piekara and R. Konopka, *Brit. J. Appl. Phys.*, 1961, **12**, 50; (*b*) N. George, R. W. Hellwarth, and C. R. Cooke, *Electron Technology (Warsaw)*, 1969, **2**, 229; (*c*) J. M. Brown, A. D. Buckingham, and D. A. Ramsay, *Canad. J. Phys.*, 1970, **49**, 914; (*d*) E. A. Volkova, V. A. Zamkov, and L. V. Nalbandov, *Optika i Spektroskopiya*, 1971, **30**, 556.

382 (*a*) J. M. Deutch and J. S. Waugh, *J. Chem. Phys.*, 1965, **43**, 2568; (*b*) C. W. Hilbers, C. Maclean, and M. Mandel, *Physica*, 1971, **51**, 246.

and one has to perform an averaging with the distribution function (176) in linear approximation. This yields, beside (177) and (178), an additional Kerr constant resulting from the non-linear polarizability of the molecules:

$$K_{\mathrm{m}}^{\mathrm{NL}} = \frac{4\pi}{9}\left\{\left\langle \sum_{i=1}^{N_{\mathrm{A}}} c_{\mathrm{V}i}^{e} \right\rangle + \frac{6}{5}\beta\left\langle \sum_{i=1}^{N_{\mathrm{A}}}\sum_{j=1}^{N_{\mathrm{A}}} \boldsymbol{b}_{\mathrm{V}i}^{e}\cdot\boldsymbol{p}_j \right\rangle\right\} \tag{184}$$

where for the sake of brevity we have introduced the scalar quantity:

$$c_{\mathrm{V}i}^{e} = \frac{1}{40\varepsilon_0}(3\boldsymbol{U}_{13}:\boldsymbol{c}_i^{e}:\boldsymbol{U}_{24} - \boldsymbol{U}_{12}:\boldsymbol{c}_i^{e}:\boldsymbol{U}_{34}) \tag{185}$$

and vectorial quantity:

$$\boldsymbol{b}_i^{e} = \frac{1}{24\varepsilon_0}(3\boldsymbol{U}_{13}:\boldsymbol{b}_i^{e} - \boldsymbol{U}_{12}:\boldsymbol{b}_i^{e}) \tag{186}$$

describing the non-linear electro-optical properties of the molecules.

Spherical symmetry. For spherical symmetry, equation (184) reduces to the form:

$$K_{\mathrm{m}}^{\mathrm{NL}} = \frac{4\pi N_{\mathrm{A}}}{9}c_{\mathrm{V}}^{e} = \frac{N_{\mathrm{A}}}{6\varepsilon_0}(c_{\parallel}^{e} - c_{\perp}^{e}) \tag{184a}$$

where, on taking into consideration that by Voigt's hypothesis in the absence of absorption $c_{\parallel}^{e} = 3c_{\perp}^{e}$, the quantity

$$c_{\mathrm{V}}^{e} = \frac{3}{8\pi\varepsilon_0}(c_{3333}^{e} - c_{1133}^{e}) = \frac{3}{8\pi\varepsilon_0}(c_{\parallel}^{e} - c_{\perp}^{e}) \tag{185a}$$

has the meaning of a mean non-linear polarizability.[291]

Equation (184a) provides, in a molecular approach, the essence of Voigt's deformational theory. A gas consisting of atoms or isotropic molecules becomes optically birefringent owing to the non-linear anisotropy $(c_{\parallel}^{e} - c_{\perp}^{e})$ induced in its molecules (atoms) by the deforming action of a strong electric field. The effect of non-linear electron distortion is, by (184a), independent of temperature, and provides direct information concerning the changes undergone by an atom or isotropic molecule in a strong static electric field. Boyle *et al.*[294a] were the first to measure *via* the Kerr effect the mean non-linear electro-optical polarizability c^e of the helium atom; Buckingham *et al.*[383] succeeded in this with regard to other atoms and isotropic molecules.

Axial symmetry. In the case of molecules with axial symmetry along the k-axis, one has:

$$\boldsymbol{p}_j = p_j\boldsymbol{k}_j, \quad \boldsymbol{b}_i^{e} = b_i^{e}\boldsymbol{k}_i$$

and equation (184) reduces to:

$$K_{\mathrm{m}}^{\mathrm{NL}} = \frac{4\pi N_{\mathrm{A}}}{9}\{c_{\mathrm{V}}^{e} + \tfrac{6}{5}\beta b_{\mathrm{V}}^{e}p(1 + J_{\mathrm{A}}^{(1)})\} \tag{184b}$$

[383] (*a*) A. D. Buckingham and D. A. Dunmur, *Trans. Faraday Soc.*, 1968, **64**, 1776; (*b*) A. D. Buckingham and B. J. Orr, *Proc. Roy. Soc.*, 1968, **A305**, 259; *Trans. Faraday Soc.* 1969, **65**, 673; (*c*) M. P. Bogaard, A. D. Buckingham, and G. L. D. Ritchie, *Mol. Phys.*, 1970, **18**, 575; (*d*) A. D. Buckingham, M. P. Bogaard, D. A. Dunmur, C. P. Hobbs, and B. J. Orr, *Trans. Faraday Soc.*, 1970, **66**, 1548.

where

$$c_V^e = \frac{1}{60\pi\varepsilon_0}(3c_{3333}^e + 12c_{3311}^e + 8c_{1111}^e) \tag{185b}$$

$$b_V^e = \frac{1}{12\pi\varepsilon_0}(b_{333}^e + 2b_{113}^e) \tag{186a}$$

$J_A^{(1)}$ being the angular correlation parameter defined by (179).

In the expressions (184) and (184b) the second, temperature-dependent term defines the Born effect [13] due to superposition of the two non-linear processes of second-order distortion and reorientation of permanent dipole moments in the electric field. Buckingham *et al.*[383] determined non-linear polarizabilities b^e and c^e for numerous molecules by Kerr effect measurements in gases as a function of temperature and pressure. It is here convenient to use the virial expansion of the molar Kerr constant, when the first and second virial coefficients A_K and B_K result immediately from equations (177), (178), and (184). Meeten *et al.*[384] determined non-linear molecular polarizabilities by measuring K_m in liquids as a function of temperature.

General Theory of the Kerr Effect.—Closer investigation of the Kerr effect in liquids has shown that a part, by no means smaller than that of the direct angular correlations (179)—(181), is played by the molecular fields, which affect the electric properties specific to the isolated molecule. Many authors have attempted to solve this problem by resorting to anisotropy of the local Lorentz [349] and Onsager [346, 385] field. The method of mean local fields is, however, not suitable for simple liquids consisting of atoms or isotropic molecules, in which the molecular field fluctuates with the thermal motion of the microsystems and produces a fluctuating induced dipole moment. According to the Yvon–Kirkwood translational fluctuational theory, the induced dipole thus created possesses anisotropic electric properties even in simple fluids. We have accordingly to formulate the Kerr effect theory in a sufficiently general way to include, beside molecular reorientation and angular correlation, the various statistical–fluctuational processes. The semi-macroscopic method of Kirkwood [353] permits the formulation of such a general theory [348] of the Kerr effect applicable to various cases of anisotropic molecules [327, 358, 373] and intrinsically non-anisotropic molecules.[360, 386, 387]

Fundamental Formulae. We have seen that a strong electric field makes an isotropic body optically anisotropic. By this we mean that its frequency-dependent electric permittivity and magnetic permeability take the form

[384] (*a*) G. H. Meeten, *Trans. Faraday Soc.*, 1968, **64**, 2267; (*b*) J. V. Champion, G. H. Meeten, and Ch. D. Whittle, *J. Chim. phys.*, 1970, **67**, 1864.

[385] (*a*) Von Reinhard Küster, *Z. Naturforsch.*, 1961, **16a**, 366; (*b*) G. Klages, *ibid.*, 1954, **9a**, 602; (*c*) A. D. Buckingham and R. E. Raab, *J. Chem. Phys.*, 1957, **449**, 2341; (*d*) D.A.A.S.N. Rao, *Trans. Faraday Soc.*, 1958, **54**, 954.

[386] J. A. Brace and J. V. Champion, *J. Phys.* (*C*), 1969, **2**, 2408.

[387] (*a*) S. Kielich, *Acta Phys. Polon.*, 1966, **30**, 683; (*b*) S. Kielich, *ibid.*, 1968, **34**, 1093.

of tensorial quantities, ε_E^ω and μ_E^ω. Restricting, as previously, our considerations to the contribution of the electric light wave vector to the refractive index, we have:

$$(\boldsymbol{\varepsilon}_E^\omega - \varepsilon_0^\omega \boldsymbol{U}) \cdot \boldsymbol{E}(r, t) = \boldsymbol{P}_E(r, t) \tag{187}$$

Since, in the Kerr effect measurements, light, acting solely as an analysing agent, is of low intensity and thus gives rise to linear polarization only, we write:

$$\boldsymbol{P}_E(r, t) = \frac{1}{V_m}\langle \boldsymbol{A}^\omega \rangle_E \cdot \boldsymbol{E}_0(r, t) \tag{187a}$$

where $\boldsymbol{A}^\omega$ is the second-rank tensor of optical polarizability of the macroscopic sample of volume V_m. The tensor $\boldsymbol{A}^\omega$ is in general a function of the static field strength, but with regard to simplicity we shall omit the non-linear polarizabilities.

The total potential energy of the perturbation is, to the accuracy of terms in the square of the field:[388]

$$U_E = -\boldsymbol{M}_0^e \cdot \boldsymbol{E}_0 - \tfrac{1}{2}\boldsymbol{E}_0 \cdot \boldsymbol{A}^e \cdot \boldsymbol{E}_0 - \ldots \tag{188}$$

where $\boldsymbol{M}_0^e$ is the total electric moment of the sample of volume V_m in the absence of an external field, and $\boldsymbol{A}^e$ is the electric polarizability tensor of the sample in the field $\boldsymbol{E}_0$.

On performing in (187a) an averaging with the statistical function (100) for the perturbation (188), we obtain by equation (187) and the general expression for the mean macroscopic field $\boldsymbol{E}_i$ within a dielectric sphere of permittivity ε_i placed in a strong external field E_e, *viz.*

$$(\boldsymbol{\varepsilon}_i + 2\varepsilon_e \boldsymbol{U}) \cdot \boldsymbol{E}_i = 3\varepsilon_e \boldsymbol{E}_e \tag{188a}$$

where $\boldsymbol{U}$ is the unit symmetric tensor, the following general fundamental expression for the change in frequency-dependent permittivity tensor caused by a strong electric field:

$$\Delta\boldsymbol{\varepsilon}_E^\omega = \frac{1}{3V_m}\{Q_m E_0^2 \boldsymbol{U} + K_m(3\boldsymbol{E}_0\boldsymbol{E}_0 - E_0^2 \boldsymbol{U})\} \cdot (\boldsymbol{\varepsilon}_E^\omega + 2\boldsymbol{U}) \tag{189}$$

Here, we have introduced the following constants:

$$Q_m = Q_m^S + Q_m^{NL}, \quad K_m = K_m^S + K_m^{NL} \tag{189a}$$

defining respectively the isotropic and anisotropic parts of the change in permittivity tensor. The constant:

$$Q_m^S = \frac{\beta}{2\varepsilon_0}\{\langle \Delta A^\omega \Delta A^e \rangle + \frac{\beta}{3}\langle \Delta A^\omega \Delta(\boldsymbol{M}_0^e \cdot \boldsymbol{M}_0^e)\rangle\} \tag{190}$$

accounts for isotropic variations due to statistical processes.

One sees that in arbitrarily dense media the isotropic variations $\Delta\varepsilon_E^\omega$ are due to the presence of statistical fluctuations of the mean optical

[388] S. Kielich, *Acta Phys. Polon.* 1958, **17**, 239.

polarizability $\Delta A^{\omega} = A^{\omega} - \langle A^{\omega}\rangle$ and mean electrical polarizability $\Delta A^{e} = A^{e} - \langle A^{e}\rangle$ as well as to fluctuations of the squared dipole moment $\Delta(\boldsymbol{M}_0^e \cdot \boldsymbol{M}_0^e) = \boldsymbol{M}_0^e \cdot \boldsymbol{M}_0^e - \langle \boldsymbol{M}_0^e \cdot \boldsymbol{M}_0^e\rangle$. Clearly, the mean polarizabilities of the medium are defined as follows:

$$A = \tfrac{1}{3}\boldsymbol{A} : \boldsymbol{U} = \tfrac{1}{3}(A_{xx} + A_{yy} + A_{zz})$$

In a first approximation, equation (190) is reducible to fluctuations in the number of molecules $\langle(\Delta N_A)^2\rangle$ which are the source of light scattering.[332]

The anisotropic changes in the permittivity tensor $\Delta\boldsymbol{\varepsilon}_E^{\omega}$ due to statistical processes define the molar Kerr constant:

$$K_{\mathrm{m}}^{\mathrm{S}} = \frac{\beta}{30\varepsilon_0}(\langle \boldsymbol{D}^{\omega} : \boldsymbol{D}^{e}\rangle + \beta\langle \boldsymbol{D}^{\omega} : \boldsymbol{M}_0^e \boldsymbol{M}_0^e\rangle) \tag{191}$$

involving the deviation tensors of, respectively, optical and electric polarizability:

$$\boldsymbol{D}^{\omega} = \boldsymbol{A}^{\omega} - A^{\omega}\boldsymbol{U}, \quad \boldsymbol{D}^{e} = \boldsymbol{A}^{e} - A^{e}\boldsymbol{U} \tag{192}$$

From the fundamental relation (189), we derive for changes in optical permittivity in two mutually perpendicular directions:

$$\Delta\varepsilon_{xx}^{\omega} = \frac{1}{3V_{\mathrm{m}}}(Q_{\mathrm{m}} - K_{\mathrm{m}})(\varepsilon_{xx} + 2)$$

$$\Delta\varepsilon_{zz}^{\omega} = \frac{1}{3V_{\mathrm{m}}}(Q_{\mathrm{m}} + 2K_{\mathrm{m}})(\varepsilon_{zz} + 2) \tag{193}$$

yielding the ratio:[387]

$$\frac{\Delta\varepsilon_{zz}^{\omega}}{\Delta\varepsilon_{xx}^{\omega}} = \left(\frac{Q_{\mathrm{m}} + 2K_{\mathrm{m}}}{Q_{\mathrm{m}} - K_{\mathrm{m}}}\right)\left(\frac{\varepsilon_{zz} + 2}{\varepsilon_{xx} + 2}\right) \tag{194}$$

which can be considered as representing a generalized Havelock relation.[389]

By equation (193), the Kerr constant K_{m} raises the value of $\Delta\varepsilon_{zz}^{\omega}$ but lowers $\Delta\varepsilon_{xx}^{\omega}$. The general relation (194) states that in order to have the value -2 resulting from Langevin's reorientation theory one has to neglect the quantity Q_{m} due to statistical fluctuations and non-linear polarizabilities. On neglecting $Q_{\mathrm{m}}^{\mathrm{S}}$ and $K_{\mathrm{m}}^{\mathrm{S}}$ and taking into account that[331] $2Q_{\mathrm{m}}^{\mathrm{NL}} = 5K_{\mathrm{m}}^{\mathrm{NL}}$, equation (194) yields $+3$, in conformity with Voigt's non-linear deformation theory.

It would be highly desirable to measure $\Delta\varepsilon_{zz}^{\omega} \simeq 2n\Delta n_{\parallel}$ and $\Delta\varepsilon_{xx}^{\omega} \simeq 2n\Delta n_{\perp}$ independently of one another and to use the result in conjunction with (161) and (194) for deciding which of the mechanisms predominates in the liquid under consideration: isotropic statistical fluctuations ($Q_{\mathrm{m}}^{\mathrm{S}}$), non-linear polarizability ($Q_{\mathrm{m}}^{\mathrm{NL}}, K_{\mathrm{m}}^{\mathrm{NL}}$), or anisotropic statistical processes ($K_{\mathrm{m}}^{\mathrm{S}}$). The last two can be separated, with some degree of accuracy, also by measurements as a function of temperature and pressure or of analysing-light frequency.

[389] T. H. Havelock, *Proc. Roy. Soc.*, 1906, **77**, 170; 1908, **80**, 28; 1911, **84**, 492; *Phys. Rev.*, 1909, **28**, 136; *Phil. Mag.*, 1923, **45**, 560.

Translational Fluctuations. We shall now discuss the general expression (191) for the Kerr constant, starting from the simple case of atomic liquids and ones composed of isotropic molecules. In substances like these, the above derived results (170) and (177), as well as (178), reduce to zero since in the case of isolated atoms the deviation tensors $\boldsymbol{d}^{\omega}$ and $\boldsymbol{d}^{e}$ vanish. Formula (191) now contains deviation tensors relating to the medium as a whole, which presents in general intense molecular fields. Taking into account Kirkwood–Yvon translational statistical fluctuations, one obtains in a first approximation of interaction between induced dipoles:

$$\boldsymbol{D}^{\omega} = \sum_{i=1}^{N_A} \boldsymbol{d}_i^{\omega} - \sum_{i=1}^{N_A} \sum_{k \neq i}^{N_A} \{\boldsymbol{a}_i^{\omega} \cdot \boldsymbol{T}_{ik} \cdot \boldsymbol{a}_k^{\omega} - \tfrac{1}{3}\boldsymbol{U}(\boldsymbol{a}_i^{\omega} \cdot \boldsymbol{T}_{ik} \cdot \boldsymbol{a}_k^{\omega}) : \boldsymbol{U}\} + \ldots \tag{195}$$

For molecules without intrinsic anisotropy $\boldsymbol{a}_i = a_i \boldsymbol{U}$, and (195) reduces simply to:

$$\boldsymbol{D}^{\omega} = - \sum_{i=1}^{N_A} \sum_{k \neq i}^{N_A} a_i^{\omega} a_k^{\omega} \boldsymbol{T}_{ik} \tag{195a}$$

Since the electric polarizability deviation tensor is of similar form, one has by (191), on taking the first anisotropic term only:[346]

$$K_{\mathrm{m}}^{\mathrm{SA}} = \frac{2\pi\beta}{15} a_{\mathrm{V}}\, a\alpha^2 \left\langle \sum_{i=1}^{N_A} \sum_{j=1}^{N_A} \sum_{k \neq i}^{N_A} \sum_{l \neq j}^{N_A} \boldsymbol{T}_{ik} : \boldsymbol{T}_{jl} \right\rangle \tag{196}$$

This result is accurate in general to within the square of the dipole interaction tensor $\boldsymbol{T}$.

Binary correlations. One notes immediately that (196) vanishes in the absence of correlations, *i.e.* if $i = j = k = l$. In order to have a non-vanishing result, it is sufficient to assume binary correlations, namely (*a*) $i = j$ and $k = l$, or (*b*) $i = l$ and $j = k$. These two contributions are mutually equivalent (because molecule i polarizes molecule j in the same way as molecule j polarizes molecule i) and are additive. Hence, equation (196) in the approximation of binary correlations becomes:[348]

$$K_{\mathrm{m}}^{\mathrm{SA}} = \frac{4\pi\beta}{15} a_{\mathrm{V}}\, a\alpha^2 \left\langle \sum_{i=1}^{N_A} \sum_{j \neq i}^{N_A} \boldsymbol{T}_{ij} : \boldsymbol{T}_{ij} \right\rangle \tag{196a}$$

With regard to the definition of multipole interaction tensors (57), we have for dipole–dipole interaction:

$$\boldsymbol{T}_{ij} = - \frac{1}{4\pi\varepsilon_0} (3\boldsymbol{r}_{ij}\boldsymbol{r}_{ij} - r_{ij}^2 \boldsymbol{U}) r_{ij}^{-5} \tag{197}$$

or

$$\boldsymbol{T}_{ij} : \boldsymbol{T}_{ij} = \frac{6 r_{ij}^{-6}}{(4\pi\varepsilon_0)^2} \tag{197a}$$

and (196a) leads to the following result:

$$K_{\mathrm{m}}^{\mathrm{SA}} = \frac{8\pi N_A}{5kT} a_{\mathrm{V}}^2\, \alpha_{\mathrm{V}}\, \alpha \left\langle \sum_{j \neq i}^{N_A} r_{ij}^{-6} \right\rangle \tag{196b}$$

This result (derived by Kielich [348]) states that in dense substances composed of atoms and isotropic molecules the presence of translational fluctuations is able to cause optical birefringence without the participation of the non-linear Voigt effect present in the gaseous state.

Equation (196b) reduces in particular to the form:

$$K_{\mathrm{m}}^{\mathrm{SA}} = \frac{8\pi N_{\mathrm{A}}}{5kT} a_{\mathrm{V}}^{2} \alpha_{\mathrm{V}} \alpha \langle r_{ij}^{-6} \rangle \tag{196c}$$

where $\langle r_{ij}^{-6} \rangle$ is defined statistically by (151). In this approximation, a result corresponding to (196c) for the perfect gas [*i.e.* when $\langle r_{ij}^{-6} \rangle$ is given by (157) at $n = 6$] has been derived by Buckingham and Dunmur [313] with Silberstein's model [274] of two mutually polarizing atoms in a molecule.

The significance of (196c) lies in the conclusion that already binary correlations suffice for inducing local anisotropy in simple liquids. Pairs of atoms (or isotropic molecules), thus correlated, behave like anisotropic assemblages subject to reorientation by an external electric field, and cause an observable Kerr effect in carbon tetrachloride [372] and similar liquids with almost symmetric molecules.[381b]

Ternary correlations. Formula (196) contains, in addition to binary correlations, also ternary and quaternary ones. For ternary correlations, we have pairs of possibilities leading to identical non-vanishing results, namely: (*a*) $i = j \neq k \neq l$ and $i = l \neq k \neq j$, and (*b*) $k = j \neq i \neq l$ and $k = l \neq i \neq j$; equation (196) now yields the additional contribution:[346]

$$K_{\mathrm{m}}^{\mathrm{SA}} = \frac{4\pi\beta N_{\mathrm{A}}}{15} a_{\mathrm{V}}\, a\alpha^{2} \left\langle \sum_{j \neq i}^{N_{\mathrm{A}}} \sum_{k \neq j}^{N_{\mathrm{A}}} \boldsymbol{T}_{ij} : \boldsymbol{T}_{jk} + \sum_{j \neq i}^{N_{\mathrm{A}}} \sum_{k \neq i}^{N_{\mathrm{A}}} \boldsymbol{T}_{ij} : \boldsymbol{T}_{ki} \right\rangle \tag{198}$$

With regard to (197), we finally have:

$$K_{\mathrm{m}}^{\mathrm{SA}} = \frac{8\pi N_{\mathrm{A}}}{5kT} a_{\mathrm{V}}^{2} \alpha_{\mathrm{V}} \alpha (\langle r_{ij}^{-3} r_{jk}^{-3} \rangle + \langle r_{ij}^{-3} r_{ki}^{-3} \rangle) \tag{198a}$$

involving the statistical averages defined by (152).

Although the Silberstein model [274] has been extended to tri-atomic triangular and linear [390] as well as more complicated [24, 90] molecules, some amount of caution would seem necessary when applying it directly to liquids, where generally the situation in the range of near ordering is entirely different. Silberstein's model, though excellent, is hardly applicable to certain situations, in particular to collisions. In fact, by the general formula (196), when dealing with dense media we have not only the problem of electrostatic interaction between induced molecular dipoles given by the tensor $\boldsymbol{T}_{ij}$, but also the complicated problem of statistical averaging with an appropriate correlation function (in the presence of statistical fluctuations, $\langle r^{-6} \rangle \neq \langle r^{-3} \rangle^{2}$).

[390] K. Ramanathan, *Proc. Roy. Soc.*, 1925, **A107**, 684; 1926, **A110**, 123; K. L. Wolf, G. Briegleb, and H. A. Stuart, *Z. Phys. Chem.*, 1929, **6B**, 163.

Since from one model to another the correlation parameters (152) take positive or negative values, the local anisotropy induced by fluctuations in bimolecular regrouping (196c) can undergo a lowering or an enhancement under the influence of multi-molecular regrouping in the near order region (198a).

Fluctuations of the Molecular Fields. The second right-hand term of Kerr's constant (191), in the case of dipolar molecules, leads directly to the result (178). We shall now show that this part of the Kerr constant is non-zero even in liquids composed of non-dipolar molecules. This is due to the circumstance that in dense media, even if no external field is applied, intense molecular fields fluctuating in time and space have to be considered locally. The molecular fields $\boldsymbol{F}_{\mathrm{m}}$ induce electric dipoles in the molecules in such regions, giving rise in the medium to the non-zero total dipole moment $\boldsymbol{M}_0$ occurring in the second part of the constant (191). In fact, we have in a linear approximation:

$$\boldsymbol{M}_0 = \sum_{j=1}^{N_A} \boldsymbol{\alpha}_j \cdot \boldsymbol{F}_j = \sum_{j=1}^{N_A} (\alpha_j \boldsymbol{F}_j + \boldsymbol{d}_j^e \cdot \boldsymbol{F}_j) \tag{199}$$

Quadrupolar molecules. In a first approximation, we can assume the molecular electric fields to polarize the molecules isotropically, enabling us to write (199) for molecules of a single species, and on neglecting anisotropy ($\boldsymbol{d}^e = 0$):

$$\boldsymbol{M}_0 = \sum_{j=1}^{N} \alpha_j \boldsymbol{F}_j = \alpha \sum_{j=1}^{N_A} \boldsymbol{F}_j \tag{199a}$$

Inserting now into (191) the moment (199a) and the deviation tensor (195) without the translational fluctuation term, we have:

$$K_{\mathrm{m}}^{\mathrm{SD}} = \frac{2\pi\beta^2\alpha^2}{15} \left\langle \sum_{i=1}^{N_A} \sum_{j=1}^{N_A} \sum_{k=1}^{N_A} \boldsymbol{d}_{\mathrm{vi}}^{\omega} : \boldsymbol{F}_j \boldsymbol{F}_k \right\rangle \tag{200}$$

One notes that in general, if a non-dipolar medium contains molecular fields, its Kerr constant (200) is not zero. As an example, we take quadrupolar axially symmetric molecules, for which the molecular field is of the form:

$$\boldsymbol{F}_j = \frac{3}{8\pi\varepsilon_0} \sum_{l \neq j}^{N_A} \Theta_l \{[5(\boldsymbol{k}_l \cdot \boldsymbol{r}_{jl})^2 - r_{jl}^2]\, \boldsymbol{r}_{jl} - 2r_{jl}^2\, \boldsymbol{k}_l(\boldsymbol{k}_l \cdot \boldsymbol{r}_{jl})\}\, r_{jl}^{-7} \tag{201}$$

and the constant (200) is, in the approximation of binary correlations:[391]

$$K_{\mathrm{m}}^{\mathrm{SD}} = \frac{4\pi N_A}{105 k^2 T^2}\, \delta_{\mathrm{V}}^{\omega}\, \alpha_{\mathrm{V}}^2\, \Theta^2 \left\langle \sum_{j \neq i}^{N_A} G_{ij}\, r_{ij}^{-8} \right\rangle \tag{200a}$$

where we have introduced the following function describing the mutual

[391] (*a*) S. Kielich, *Chem. Phys. Letters*, 1971, **10**, 516; (*b*) S. Kielich, *Compt. rend.*, 1972, **274**, *B*, 121; (*c*) S. Kielich, *Acta Phys. Polon.* (*A*), 1972, **41**, 653.

orientation of correlating molecules:

$$G_{ij} = \tfrac{21}{16}\{75(\cos^2\theta_i - \cos^2\theta_j)(\cos^4\theta_i - \cos^4\theta_j) - 70(\cos^2\theta_i - \cos^2\theta_j)^2 - 32\cos\theta_i\cos\theta_j\cos\theta_{ij} + 4(\cos^2\theta_i + \cos^2\theta_j)(3\cos^2\theta_{ij} + 1)\} \qquad (202)$$

with θ_i and θ_j denoting the angles between the symmetry axes $\boldsymbol{k}_i$ and $\boldsymbol{k}_j$ of molecules i and j and the vector $\boldsymbol{r}_{ij}$ connecting their centres.

The orientation function (202) is highly sensitive with regard to the two extreme models of a liquid discussed by Prins and Prins,[378] namely diatropism, where G_{ij} is very large, and paratropism, where G_{ij} tends to zero. For not too strong angular correlations, equation (202) can be averaged in a first approximation over all orientations, leading to:

$$G_{ij} = 1 + O(T^{-1}) \qquad (202a)$$

whence equation (200a) immediately takes the form:

$$K_{\mathrm{m}}^{\mathrm{SD}} = \frac{4\pi N_{\mathrm{A}}}{105k^2T^2}\,\delta_{\mathrm{V}}^{\omega}\,\alpha_{\mathrm{V}}^2\,\Theta^2\langle r_{ij}^{-8}\rangle \qquad (200b)$$

The radial correlation parameter $\langle r_{ij}^{-8}\rangle$ occurring in (200b) can be computed numerically for Kirkwood's model from equation (154) or determined by direct experimental study of double-photon scattering in liquids with centrosymmetric molecules[227] and static-electric-field-induced second-harmonic laser light generation in liquids.[222] For real gases, equation (200b) goes over into the result of Buckingham.[327]

The constant (200b) is highly characteristic in that its sign depends on the sign of the optical anisotropy $(a_3 - a_1)$ of the molecule, which is negative in planar molecules of the type C_6H_6 but positive in linear ones, *e.g.* CS_2. The ratio of the constants (200b) and (177b) is, in the present approximation,

$$\frac{K_{\mathrm{m}}^{\mathrm{SD}}}{K_{\mathrm{m}}^{\mathrm{SA}}} = \frac{3\alpha_{\mathrm{V}}^2\,\Theta^2}{7kT\delta^e}\,\frac{\langle r^{-8}\rangle}{1 + J_{\mathrm{A}}^{(2)}} \qquad (203)$$

Calculations from extant data yield, for C_6H_6, a negative ratio (203) amounting to 20 or 40% at 20 °C, according to the value taken for the quadrupole moment; for CS_2 the ratio (203) is positive and amounts to 25 or 30%, according to the model. For both these liquids, on taking into consideration the new term (200b), one obtains a temperature-dependence of the molar Kerr constant K_{m} which is in agreement with the latest measurements.[384] In the analysis of the experimental results, the term (200b) eliminates the question of the negative non-linear polarizability of benzene,[384] which measurements in the gaseous state[383c] show to be positive. In this way, the positive contribution from K^{NL} does not exceed 6% of the total Kerr constant. The semi-diatropic structure of liquid benzene assumed in the preceding evaluations and the suggestion of a larger role of the fields of its molecular electric quadrupoles than of London dispersional forces are in agreement with the latest calculations by Kihara and Yamazaki [392] of the enhanced solubilities of quadrupolar molecules in benzene.

Isotropic molecules. For spherical symmetry, the contribution (200) vanishes and we have to revert to the general formula (191), inserting therein the deviation tensor (195a) accounting for the presence of translational fluctuations. By (199a), this yields:[208]

$$K_{\mathrm{m}}^{\mathrm{SD}} = -\frac{2\pi\beta^2 a_{\mathrm{V}} a\alpha^2}{15}\left\langle \sum_{i=1}^{N_A}\sum_{j=1}^{N_A}\sum_{k\neq i}^{N_A}\sum_{l=1}^{N_A} \boldsymbol{T}_{ik} : \boldsymbol{F}_j \boldsymbol{F}_l \right\rangle \tag{204}$$

The preceding quantity is also in general non-zero.

As examples of molecules having, in a linear approximation, isotropic polarizability, we adduce ones of tetrahedral symmetry such as CH_4, CCl_4, and the like. Here, neither a permanent dipole nor quadrupole is available, and the lowest non-zero moment is an octupole having the non-zero components Ω_{123}. The electric field of such octupoles, given by equation (48e), induces a dipole in another molecule, and we have by (204) in the approximation of pairwise correlations:[208]

$$K_{\mathrm{m}}^{\mathrm{SD}} = \frac{64\pi N_{\mathrm{A}}}{35k^2T^2} a_{\mathrm{V}}^2 \alpha_{\mathrm{V}}^2 \Omega_{123}^2 \left\langle \sum_{j\neq i}^{N_A} r_{ij}^{-13} \right\rangle \tag{204a}$$

This contribution is rather insignificant compared to the previously calculated one (196b) resulting from translational fluctuations alone, and can be of importance only in the case of tetrahedral molecules with a large octupole moment.

Other Contributions to the Kerr Constant. In the case of naturally anisotropic molecules, the deviation tensor $\boldsymbol{d}_i^{\omega}$ in the expansion (195) does not vanish, causing a cross-effect as well as higher-order fluctuational effects to appear, in addition to the previous angular correlation effect (177a). The cross-effect appears owing to co-operation between molecular angular correlations—the first term of equation (195) and the translational fluctuations described by the second term of (195). We shall consider this effect in the next subsection.

In some cases of highly condensed media, the fact that the molecular electric fields $\boldsymbol{F}$ largely exceed the externally applied field E and can induce not only linear electric dipoles (199) but also non-linear ones has to be taken into consideration. We thus write quite generally, for dipolar systems:

$$\boldsymbol{M}_0 = \sum_{j=1}^{N_A}\{\boldsymbol{p}_j + \boldsymbol{\alpha}_j \cdot \boldsymbol{F}_j + \tfrac{1}{2}\boldsymbol{\beta}_j : \boldsymbol{F}_j\boldsymbol{F}_j \ldots\} \tag{205}$$

where in addition to the permanent dipole moment $\boldsymbol{p}_j$ and linear polarizability tensor $\boldsymbol{\alpha}_j$ we have tensors $\boldsymbol{\beta}_j$ of non-linear polarizabilities of order 2 and so forth, induced by the molecular field.

[392] T. Kihara and K. Yamazaki, *Chem. Phys. Letters*, 1971, **11**, 62.

Expansions similar to (205) hold for the tensors of optical and electric polarizability of the medium:

$$A^{\omega} = \sum_{i=1}^{N_A} \{\boldsymbol{a}_i + \boldsymbol{b}_i \cdot \boldsymbol{F}_i + \tfrac{1}{2}\boldsymbol{c}_i : \boldsymbol{F}_i \boldsymbol{F}_i + \ldots\} \tag{206}$$

$$A^{e} = \sum_{j=1}^{N_A} \{\boldsymbol{\alpha}_j + \boldsymbol{\beta}_j \cdot \boldsymbol{F}_j + \tfrac{1}{2}\boldsymbol{\gamma}_j : \boldsymbol{F}_j \boldsymbol{F}_j + \ldots\} \tag{207}$$

The expansions (205)—(207) in conjunction with formulae (191) and (192) define the complete molecular-statistical theory of the Kerr effect in very dense media. They can be particularized to various cases.[359]

Comparison with Experiment. The parameters $\langle r_{ij}^{-n}\rangle$ and $\langle r_{ij}^{-n} r_{jk}^{-m}\rangle$ of pairwise and triple radial correlations, as well as those of angular correlations (179)—(181), can be calculated numerically on making assumptions as to the model. Since the optical and electric parameters of atoms and molecules are known, the various contributions to the molecular Kerr constant can be calculated numerically from the above-derived formulae. The results of these calculations are compared with the experimental data in Table 14.

Table 14 *Theoretically calculated, and experimentally measured Kerr constants of some non-dipolar liquids (at room temperature)*

Liquids	*Argon* (85 K)	CCl_4	C_6H_6	CS_2
α_1 (× $10^{40}/C^2\,m^2\,J^{-1}$)	1.814	11.69	7.05	6.15
α_2 (× $10^{40}/C^2\,m^2\,J^{-1}$)	1.814	11.69	13.66	6.15
α_3 (× $10^{40}/C^2\,m^2\,J^{-1}$)	1.814	11.69	13.66	16.81
$\gamma \times 10^{60}/C^4\,m^4\,J^{-3}$ [a]	0.073	0.433	0.8	6.2
$\Theta \times 10^{40}/C\,m^2$	—	—	40.0	30.0
$V_m \times 10^6/m^3\,mol^{-1}$	28.3	96.9	89.4	60.2
n (λ = 5461 Å)	1.224	1.463	1.505	1.636
ε_r	1.516	2.236	2.282	2.641
K_m^{NL} (× $10^{25}/C^2\,m^5\,J^{-2}\,mol^{-1}$)	0.01	0.03	0.06	0.47
K_m^{SA} (× $10^{25}/C^2\,m^5\,J^{-2}\,mol^{-1}$)	0.11	1.56	0.73	1.83
K_m^{SD} (× $10^{25}/C^2\,m^5\,J^{-2}\,mol^{-1}$)	—	0.08	− 0.18	0.33
K_m^{Total} (× $10^{25}/C^2\,m^5\,J^{-2}\,mol^{-1}$)	0.12	1.67	0.61	2.63
$B_\lambda^{theor} \times 10^{16}/C^2\,m\,J^{-2}$	8.59	121.25 8.78[b]	52.57	410.09
$B_\lambda^{exp} \times 10^{16}/C^2\,m\,J^{-2}$	—	9.51[c]	47.3[c]	396.1[c]

[a] Hyperpolarizability values are taken from ref. 150 and Table 9.
[b] Value calculated with correlation parameters determined from the depolarization ratio of light scattering.
[c] G. H. Meeten, *Trans. Faraday Soc.*, 1968, **64**, 2267.

The Kerr Effect in Multicomponent Systems.—In recent years, especial interest has come to bear on the study of the Kerr effect in solutions of liquids over the entire concentration range. Such studies permit conclusions regarding interactions between molecules of different species, thus between non-dipolar molecules and dipolar or quadrupolar ones, and so forth.

On the Additivity of the Kerr Constant. The principle of additivity states that the molecular Kerr constant K_m of a system is the sum of the molecular Kerr constants of its components. For a two-component system, this principle yields:

$$K_m = x_1 K_m^{(1)} + x_2 K_m^{(2)} \tag{208}$$

x_1 and x_2 being the molar fractions of components 1 and 2, and $K_m^{(1)}$ standing for the molecular constant of the solvent and $K_m^{(2)}$ for that of the solute.

In order that the constant K_m of the solution be a strictly additive quantity, it is necessary that the constants $K_m^{(1)}$ and $K_m^{(2)}$ of the components shall not depend on the concentration. In reality, however, earlier [393] and newer [394] experimental results yield beyond doubt deviations from the additivity rule (208) even in dilute solutions. Particularly large deviations from (208) in the entire range of concentrations are observed in solutions of dipolar liquids in non-dipolar solvents, whereas in systems containing only non-dipolar liquids these deviations are much smaller.

The Langevin–Born theory of molecular orientation, discussed above, permits the expression of $K_m^{(1)}$ and $K_m^{(2)}$ for the case of a mixture of a non-dipolar and a dipolar gas as follows:

$$\begin{aligned} {}_{\mathrm{gas}}K_m^{(1)} = \frac{2\pi N_A}{45kT}\{&(a_{v1}^{(1)} - a_{v2}^{(1)})(\alpha_1^{(1)} - \alpha_2^{(1)}) \\ &+ (a_{v2}^{(1)} - a_{v3}^{(1)})(\alpha_2^{(1)} - \alpha_3^{(1)}) + (a_{v3}^{(1)} - a_{v1}^{(1)})(\alpha_3^{(1)} - \alpha_1^{(1)})\} \end{aligned} \tag{208a}$$

$$\begin{aligned} {}_{\mathrm{gas}}K_m^{(2)} = \frac{2\pi N_A}{45kT}\{&(a_{v1}^{(2)} - a_{v2}^{(2)})(\alpha_1^{(2)} - \alpha_2^{(2)}) \\ &+ (a_{v2}^{(2)} - a_{v3}^{(2)})(\alpha_2^{(2)} - \alpha_3^{(2)}) + (a_{v3}^{(2)} - a_{v1}^{(2)})(\alpha_3^{(2)} - \alpha_1^{(2)}) \\ &+ \frac{1}{kT}[(a_{v1}^{(2)} - a_{v2}^{(2)})(p_1^{(2)2} - p_2^{(2)2}) \\ &+ (a_{v2}^{(2)} - a_{v3}^{(2)})(p_2^{(2)2} - p_3^{(2)2}) + (a_{v3}^{(2)} - a_{v1}^{(2)})(p_3^{(2)2} - p_1^{(2)2})]\} \end{aligned} \tag{208b}$$

where $a_{v1}^{(1)}$, $a_{v2}^{(1)}$, $a_{v3}^{(1)}$ and $\alpha_1^{(1)}$, $\alpha_2^{(1)}$, $\alpha_3^{(1)}$ are respectively components of the volume optical and electric polarizabilities of the molecule of species 1 in the three principal directions, and $p_1^{(2)}$, $p_1^{(2)}$, $p_3^{(2)}$ and $a_{v1}^{(2)}$, $a_{v2}^{(2)}$, $a_{v3}^{(2)}$ are the components of the dipole moment and volume optical polarizability of the molecule of species 2 in these directions. One notes that by equations (208a) and (208b) the constants $K_m^{(1)}$ and $K_m^{(2)}$ depend neither on the density nor on the concentration of the solution, so that the additivity principle (208) is strictly fulfilled.

393 (*a*) G. Briegleb, *Z. Phys. Chem.* (*B*), 1931, **14**, 97; (*b*) H. A. Stuart and H. Volkman, *Z. Physik*, 1933, **83**, 444; (*c*) G. Otterbein, *Physik. Z.*, 1934, **35**, 249; (*d*) H. Friedrich, *ibid.*, 1937, **38**, 318; (*e*) L. Kozłowski, *Prace Matematyczno Fizyczne* (*Poland*), 1938, **46**, 31.

394 (*a*) R. Konopka, *Bull. Soc. des Amis des Sci. et des Lettres de Poznań* (*B*), 1960–61, **16**, 115; R. Konopka and S. Surma, *Acta Phys. Polon.* 1967, **31**, 1075; (*b*) J. Gibalewicz, *Fizyka Dielektryków i Radiospektroskopia* (*Poznań*), 1966, **3**, 131; 1968, **5/1**, 19; (*c*) R. Hellwarth and N. George, *Opto-electronics*, 1969, **1**, 213.

Kerr-effect measurements in gaseous systems[178] are by no means easy and, in many a case, can be performed more conveniently on highly dilute solutions. Le Fèvre *et al.*[132] made a study of a large variety of dilute solutions and proposed a handy method of extrapolating $K_m^{(2)}$ to infinite dilution, yielding a value $K_m^{(2)}$ which can be applied for determining the anisotropy of polarizability in molecules of the solute from equation (208b) and from formulae for molecular refraction and the depolarization ratio of scattered light[395] as well as depolarized intensity scattering.[396]

Kerr-effect studies in the entire concentration range in liquids provide valuable information regarding the intermolecular forces acting between molecules of the solute and between those of the solvent and solute. The earliest attempts in this direction are due to Friedrich,[393d] Piekara,[352] and Peterlin and Stuart,[397] who considered the case of solvents with spherical molecules and solutes with axially symmetric ones having a dipole moment sufficiently large to justify the omission, in equation (208b), of the first anisotropic–anisotropic term as compared with the second anisotropic–dipolar term. On these assumptions, $K_m^{(1)} = 0$, and from (208) and (208b) the following formula is obtained:

$$\frac{K_m}{x_2} = K_m^{(2)} = \frac{4\pi N_A}{45k^2 T^2} \delta_V^{\omega(2)} p_3^{(2)^2} R_K^{(2)} = {}_{\text{gas}}K_m^{(2)} R_K^{(2)} \qquad (208c)$$

where $R_K^{(2)}$ is a factor accounting for the mutual interaction between the dipolar molecules in solution. Friedrich calculated the factor $R_K^{(2)}$ for Kerr's constant on the basis of the Debye model[351] applied in the theory of the permittivity of dipolar liquids. In that model, a dipolar molecule of the liquid is acted on by the electric field of a large assemblage of neighbouring molecules, arrayed almost regularly, as in a crystal. The molecule is thus

[395] (*a*) R. J. W. Le Fèvre and B. P. Rao, *J. Chem. Soc.* (*B*), 1957, 3644; 1958, 1465; (*b*) C. G. Le Fèvre and R. J. W. Le Fèvre, *ibid.*, 1959, 2670; (*c*) R. J. W. Le Fèvre, D. V. Radford, G. L. D. Ritchie, and P. J. Stiles, *ibid.*, 1968, 148; (*d*) R. J. W. Le Fèvre, L. Radom, and G. L. D. Ritchie, *ibid.*, 1967, 595; 1969, 912; (*e*) M. J. Aroney, R. J. W. Le Fèvre, W. Luttke, G. L. D. Ritchie, and P. J. Stiles, *Austral. J. Chem.*, 1968, **21**, 2551; (*f*) M. J. Aroney, G. M. Hoskins, and R. J. W. Le Fèvre, *J. Chem. Soc.* (*B*), 1969, 979; (*g*) K. E. Calderbank, R. J. W. Le Fèvre, and R. K. Pierens, *ibid.*, 1970, 1115, 1463, 1608; (*h*) M. J. Aroney, G. Cleaver, R. J. W. Le Fèvre, and R. K. Pierens, *ibid.*, 1971, 82; P. A. Hopkins, R. J. W. Le Fèvre, L. Radom, and G. L. D. Ritchie, *ibid.*, 1971, 574; M. J. Aroney, R. J. W. Le Fèvre, R. K. Pierens, and M. G. N. The, *ibid.*, 1971, 1132; C. L. Cheng, R. J. W. Le Fèvre, G. L. D. Ritchie, P. A. Goodman, and P. H. Gore, *ibid.*, 1971, 1198.

[396] (*a*) A. Pacault and A. Rousset, *Compt. rend.*, 1954, **238**, 1705; (*b*) P. Bothorel, *Ann. Chim.* (*Paris*), 1959, 669; (*c*) C. Clément and P. Bothorel, *Compt. rend.*, 1960, **251**, 2323; *J. Chim. phys.*, 1964, 878, 1262, 1282; (*d*) A. Unanue, C. Clément, and P. Bothorel, *Compt. rend.*, 1964, **258**, 4757; R. Lapouyade and P. Bothorel, *ibid.*, 1967, **265**, 707; H. R. Craig and P. Botherel, *ibid.*, p. 1384; (*e*) A. Caristan and P. Bothorel, *J. Chim. phys.*, 1969, **66**, 740; A. Caristan, P. Bothorel, and H. Bodot, *ibid.*, p. 1009; P. Clement and C. Malval, *ibid.*, p. 1161; C. Clément, A. Caristan, and P. Marval, *ibid.*, 1970, **67**, 542; C. Clément and R. Seurin, *ibid.*, 1971, **68**, 22; (*f*) A. Caristan, H. Bodot, and P. Bothorel, *Bull. Soc. chim. France*, 1969, 2589; P. Foulani and C. Clément, *ibid.*, p. 3462; 1970, 450, 4308; J. P. Rioux and C. Clément, *ibid.*, p. 2139.

[397] A. Peterlin and H. A. Stuart, *Z. Physik*, 1939, **113**, 663.

not free to rotate but performs more or less extensive oscillations about an axis determined by the molecular field. A similar model had been proposed simultaneously by Fowler.[350]

With the aim of accounting for the experimental values, Piekara[352] showed that in some cases it was necessary (in addition to Debye–Fowler coupling) to consider particularly the coupling between a given dipole and its nearest neighbour or next nearest neighbours of closest vicinity, leading to a factor $R_K^{(2)}$ in the form of a function of two or three dipole-coupling parameters. In both Friedrich's and Piekara's theories, the dependence of the coupling parameters $R_K^{(2)}$ on the concentration had to be determined experimentally from a set of other effects (electric polarization, Cotton–Mouton effect, non-linearity in response to the electric field). Consequently, since it involved the concomitant measurement of a whole series of phenomena, that theoretical treatment failed to find wide application.

To Buckingham *et al.*[398] is due the theory of the Kerr effect in dilute solutions, whereas Kielich[373] developed a statistical theory of the effect for multi-component systems of an arbitrary degree of concentration and showed the Kerr constant of real solutions to be a non-additive quantity.

Divergences from Additivity of the Kerr Constant due to Angular Correlations. To a sufficiently good approximation, the non-linear part of Kerr's constant K_m^{NL} can be assumed as additive, and we shall not discuss it further. In very highly concentrated solutions, the deviation from additivity in K_m is principally due to the statistical fluctuational part of the Kerr constant given generally by formula (191). For a multi-component system, by (208), we can write, in the absence of molecular fields:

$$\mathbf{D}^{\omega} = \sum_a \sum_{i=1}^{N_a} \mathbf{d}_{ia}^{\omega}, \quad \mathbf{D}^{e} = \sum_b \sum_{j=1}^{N_b} \mathbf{d}_{jb}^{e}, \quad \mathbf{M}_0^{e} = \sum_c \sum_{k=1}^{N_c} \mathbf{p}_{kc}, \tag{209}$$

where $\mathbf{d}_{ia}^{\omega}$ is the deviation tensor of optical polarizability of molecule i of species a, $\mathbf{d}_{jb}^{e}$ is the deviation tensor of electric polarizability of molecule j of species b, and $\mathbf{p}_{kc}$ the permanent electric dipole moment of molecule k of species c.

By insertion of (209) into (191), we obtain for an arbitrary multi-component system:

$$K_m^{SA} = \frac{\beta}{30\varepsilon_0}\left\{\sum_{ab}\left\langle \sum_{i=1}^{N_a}\sum_{j=1}^{N_b} \mathbf{d}_{ia}^{\omega} : \mathbf{d}_{jb}^{e}\right\rangle + \beta \sum_{abc}\left\langle \sum_{i=1}^{N_a}\sum_{j=1}^{N_b}\sum_{k=1}^{N_c} \mathbf{d}_{ia}^{\omega} : \mathbf{p}_{jb}\mathbf{p}_{kc}\right\rangle\right\} \tag{210}$$

On resorting to methods of statistical mechanics we can write the Kerr constant of a multi-component system in the following form:[373]

$$K_m = \sum_a x_a K_m^{(a)} + \sum_{ab} x_a x_b K_m^{(ab)} + \sum_{abc} x_a x_b x_c K_m^{(abc)} + \ldots \tag{211}$$

[398] (*a*) A. D. Buckingham, *Trans. Faraday Soc.*, 1956, **52**, 611; (*b*) A. D. Buckingham, P. J. Stiles, and G. L. D. Ritchie, *ibid.*, 1971, **67**, 577.

where the first term describes Kerr constant additivity for ideal components

$$K_{\mathrm{m}}^{(a)} = \frac{\beta N_{\mathrm{A}}}{30\varepsilon_0}(\boldsymbol{d}_a^{\omega} : \boldsymbol{d}_a^{e} + \beta \boldsymbol{d}_a^{\omega} : \boldsymbol{p}_a \boldsymbol{p}_a) \tag{212}$$

not involving molecular interactions. Obviously, equation (212) leads to the Langevin–Born expressions (208a) and (208b).

The second term of the expansion (211) describes a first deviation from the additivity rule for the Kerr constant. Here, for axially symmetric molecules [373] we have in particular by (210):

$$K_{\mathrm{m}}^{(ab)} = \frac{2\pi\beta}{45} N_{\mathrm{A}}\{2\beta(\delta_{\mathrm{v}a}^{\omega} + \delta_{\mathrm{v}b}^{\omega}) p_a p_b J_{ab}^{(1)} + [\delta_{\mathrm{v}a}^{\omega}(\delta_b^{e} + \beta p_b^2) + \delta_{\mathrm{v}b}^{\omega}(\delta_a^{e} + \beta p_a^2)] J_{ab}^{(2)}\} \tag{213}$$

where we have introduced the following binary angular correlation factors:

$$J_{ab}^{(1)} = \frac{\rho}{V_{\mathrm{m}}} \iint \cos \theta_{ij}^{(ab)} g_{ab}^{(2)}(\tau_i, \tau_j) \, \mathrm{d}\tau_{ia} \, \mathrm{d}\tau_{jb} \tag{214}$$

$$J_{ab}^{(2)} = \frac{\rho}{2V_{\mathrm{m}}} \iint (3 \cos^2 \theta_{ij}^{(ab)} - 1) g_{ab}^{(2)}(\tau_i, \tau_j) \, \mathrm{d}\tau_{ia} \, \mathrm{d}\tau_{jb} \tag{215}$$

$\theta_{ij}^{(ab)}$ being the angle between the symmetry axes of molecules i and j of species a and b at configurations τ_{ia} and τ_{jb} statistically interdependent in accordance with the binary function $g_{ab}^{(2)}(\tau_i, \tau_j)$.

The third term in the expansion (211) describes a second-order deviation from Kerr-constant additivity due to ternary correlations, which are described for axially symmetric molecules as follows:[373]

$$K_{\mathrm{m}}^{(abc)} = \frac{4\pi}{135} \beta^2 N_{\mathrm{A}}(\delta_{\mathrm{v}a}^{\omega} p_b p_c + \delta_{\mathrm{v}b}^{\omega} p_c p_a + \delta_{\mathrm{v}c}^{\omega} p_a p_b) J_{abc}^{(3)} \tag{216}$$

with ternary angular correlation factor of the form:

$$J_{abc}^{(3)} = \frac{\rho^2}{2V_{\mathrm{m}}} \iiint (3 \cos \theta_{ij}^{(ab)} \cos \theta_{ik}^{(ac)} - \cos \theta_{jk}^{(bc)}) g_{abc}^{(3)}(\tau_i, \tau_j, \tau_k) \times d\tau_{ia} \, \mathrm{d}\tau_{jb} \, \mathrm{d}\tau_{kc} \tag{217}$$

$g_{abc}^{(3)}$ denoting the ternary correlation function for three molecules i, j, k of species a, b, c respectively, at configurations $\tau_{ia}, \tau_{jb}, \tau_{ik}$.

In particular, if the system presents no interaction between the molecules, the latter assume all possible configurations in volume V_{m} with equal probability, *i.e.* the correlation function then fulfils the condition:

$$\Omega^1 g_a^{(1)}(\tau_i) = \Omega^2 g_{ab}^{(2)}(\tau_i, \tau_j) = \Omega^3 g_{abc}^{(3)}(\tau_i, \tau_j, \tau_k) = \ldots = 1$$

The angular correlation factors (214), (215), and (217) now vanish, as do the constants (213) and (216) responsible for deviations from additivity in the Kerr constant. As a result of this, equation (211) reduces to the form:

$$K_{\mathrm{m}} = \sum_a x_a K_{\mathrm{m}}^{(a)} \tag{211a}$$

Hence, K_m is a strictly additive quantity only in mixtures of perfect gases, that is to say, in cases when the configuration of a molecule does not statistically depend on the configuration of the remaining molecules of the system. In dense multi-component systems, as a result of correlations between molecules of a given component or between those of different components of the system, the quantities $K_m^{(ab)}$ and $K_m^{(abc)}$ are non-zero, and K_m does not obey the additivity rule.

The Role of Fluctuation Processes. Deviations from additivity of the Kerr constant are caused not exclusively by direct angular correlations, but in many cases primarily by various fluctuational processes occurring in solutions.

Translational–orientational fluctuations. Re-writing the expansion (195) of the optical polarizability deviation tensor in a manner adapted to multi-component systems (as done previously for electric polarizability), one obtains from the general formula an additional contribution from binary correlations of axially symmetric molecules, of the form:

$$K_m^{(ab)} = \frac{8\pi N_A}{135kT}\{[\delta_{va}^{\omega}(3a_b - \delta_b)(3a_a + \delta_a - \delta_b) + \delta_{vb}^{\omega}(3a_a - \delta_a)(3a_b + \delta_b - \delta_a)]J_{ab}^{RA} + 2\delta_{va}^{\omega}\,\delta_b(3a_a + 3a_b + \delta_a + \delta_b)K_{ab}^{RA}\} \tag{218}$$

involving the following parameters describing pairwise radial–angular correlations:[391]

$$J_{ab}^{RA} = \frac{\rho}{4V_m}\iint(3\cos^2\theta_{ia} + 3\cos^2\theta_{jb} - 2)\,r_{ij,ab}^{-3}\,g_{ab}^{(2)}(\tau_i, \tau_j)\,d\tau_{ia}\,d\tau_{jb} \tag{219}$$

$$K_{ab}^{RA} = \frac{\rho}{2V_m}\iint(3\cos\theta_{ia}\cos\theta_{jb}\cos\theta_{ij}^{(ab)} - \cos^2\theta_{ij}^{(ab)}) \times r_{ij,ab}^{-3}\,g_{ab}^{(2)}(\tau_i, \tau_j)\,d\tau_{ia}\,d\tau_{jb} \tag{220}$$

θ_{ia} denoting the angle subtended by the symmetry axis of molecule i of species a and the vector $\mathbf{r}_{ij,ab}$ connecting its centre with the centre of molecule j of species b; the angle θ_{jb} is defined similarly.

The contribution (218) resulting from the cross-process of translational–orientational fluctuations is highly important, as it can be positive or negative, depending on the sign of the optical anisotropy of the molecule as well as on the near-order structure, the latter influencing the sign of the correlation parameters (219) and (220). This cross-process leads to a contribution due to ternary correlations, but we shall refrain from considering it here as it is a matter of some complexity.[391]

Translational fluctuations. In addition to (218), certain other contributions can be derived from equation (191), among which the most important role in the approximation of binary correlations belongs (on neglecting optical anisotropy) to the one given by the following expression:[208]

$$K_{\mathrm{m}}^{(ab)} = \frac{4\pi N_{\mathrm{A}}}{5kT}(a_{\mathrm{v}a}^{2}\,\alpha_{\mathrm{v}b}\,\alpha_{b} + a_{\mathrm{v}b}^{2}\,\alpha_{\mathrm{v}a}\,\alpha_{a})\langle r_{ij}^{-6}\rangle_{ab} \tag{221}$$

Translational fluctuations related to ternary molecular correlations yield (on neglecting anisotropy of polarizability):[399]

$$K_{\mathrm{m}}^{(abc)} = \frac{8\pi N_{\mathrm{A}}}{5kT}\,a_{\mathrm{v}a}\,\alpha_{\mathrm{v}b}(a_{\mathrm{v}b}\,\alpha_{c}\langle r_{ij}^{-3}\,r_{jk}^{-3}\rangle_{abc} + \alpha_{a}\,a_{\mathrm{v}c}\langle r_{ij}^{-3}\,r_{ki}^{-3}\rangle_{abc}) \tag{222}$$

The binary and ternary radial correlation parameters occurring in equations (221) and (222) are defined by (151) and (152), respectively.

Fluctuations of molecular fields. As in the case of pure liquids, fluctuations of molecular fields in solutions contribute to the effects under consideration. In particular, they lead to non-zero values of the constant $K_{\mathrm{m}}^{\mathrm{SD}}$ in non-dipolar substances. For instance, if the molecules of the various components are axially symmetric and if they possess permanent quadrupole moments, we obtain for the binary correlations, in a first approximation:

$$K_{\mathrm{m}}^{(ab)} = \frac{2\pi N_{\mathrm{A}}}{105k^{2}T^{2}}(\delta_{\mathrm{v}a}^{\omega}\,\alpha_{\mathrm{v}b}^{2}\,\Theta_{a}^{2} + \delta_{\mathrm{v}b}\,\alpha_{\mathrm{v}a}^{2}\,\Theta_{b}^{2})\langle r_{ij}^{-8}\rangle_{ab} \tag{223}$$

Significantly, this contribution is positive for molecules having positive anisotropy but negative for ones having negative anisotropy.

Likewise, one can calculate contributions due to fluctuations of dispersional fields, as well as fluctuations of molecular fields of higher electric multipoles.

Application to a Two-component System. In the case of a two-component system, equation (211) becomes:

$$\begin{aligned} K_{\mathrm{m}} = {} & x_1 K_{\mathrm{m}}^{(1)} + x_2 K_{\mathrm{m}}^{(2)} + x_1^2 K_{\mathrm{m}}^{(11)} + 2x_1 x_2 K_{\mathrm{m}}^{(12)} + x_2^2 K_{\mathrm{m}}^{(22)} \\ & + x_1^3 K_{\mathrm{m}}^{(111)} + 3x_1^2 x_2 K_{\mathrm{m}}^{(112)} + 3x_1 x_2^2 K_{\mathrm{m}}^{(122)} + x_2^3 K_{\mathrm{m}}^{(222)} + \ldots \end{aligned} \tag{224}$$

On the other hand, for a mixture of two components, K_{m} can be expanded in a series in powers of x (= x_2, the molar fraction of the solute):[373]

$$K_{\mathrm{m}} = A_0 + A_1 x + A_2 x^2 + A_3 x^3 + \ldots = \sum_{n=0}^{\infty} A_n x^n \tag{225}$$

The coefficients

$$A_n = \frac{1}{n!}\left(\frac{\partial^n K_{\mathrm{m}}}{\partial x^n}\right)_{x=0} \tag{226}$$

of the expansion can be determined directly from the experimental data. The coefficient A_0 accounts for the properties of the solvent only, whereas the others, A_1, A_2, A_3, ... characterize the properties of the solution.

[399] (*a*) S. Kielich, *Proc. Phys. Soc.*, 1967, **90**, 847; (*b*) S. Kielich, *J. Physique*, 1968, **29**, 619.

Using equation (224), we express the phenomenological coefficients (226) as follows:

$$A_0 = (f_0)_{x=0}$$

$$A_1 = \left(\frac{\partial f_0}{\partial x} + f_1\right)_{x=0}$$

$$A_2 = \left(\frac{1}{2}\frac{\partial^2 f_0}{\partial x^2} + \frac{\partial f_1}{\partial x} + f_2\right)_{x=0}$$

$$A_3 = \left(\frac{1}{6}\frac{\partial^3 f_0}{\partial x^3} + \frac{1}{2}\frac{\partial^2 f_1}{\partial x^2} + \frac{\partial f_2}{\partial x} + f_3\right)_{x=0} \tag{226}$$

whereas the quantities

$$f_0 = K_{\rm m}^{(1)} + K_{\rm m}^{(11)} + K_{\rm m}^{(111)} + \ldots$$

$$f_1 = -K_{\rm m}^{(1)} + K_{\rm m}^{(2)} + 2(K_{\rm m}^{(12)} - K_{\rm m}^{(11)}) + 3(K_{\rm m}^{(112)} - K_{\rm m}^{(111)}) + \ldots$$

$$f_2 = K_{\rm m}^{(11)} - 2K_{\rm m}^{(12)} + K_{\rm m}^{(22)} + 3(K_{\rm m}^{(111)} - 2K_{\rm m}^{(112)} + K_{\rm m}^{(122)}) + \ldots$$

$$f_3 = -K_{\rm m}^{(111)} + 3K_{\rm m}^{(112)} - 3K_{\rm m}^{(122)} + K_{\rm m}^{(222)} - \ldots \tag{226b}$$

are in general functions of the molar fraction x.

Let us now consider the case of dilute solutions, for which we have by equation (225):

$$\Delta K_{\rm m\infty} = \left\{\frac{\partial}{\partial x}(K_{\rm m} - A_0)\right\}_{x=0} = K_{\rm m}^{(2)} - K_{\rm m}^{(1)} + 2(K_{\rm m}^{(12)} - K_{\rm m}^{(11)}) + 3(K_{\rm m}^{(112)} - K_{\rm m}^{(111)}) + \ldots \tag{227}$$

Solution of a dipolar liquid in a quadrupolar solvent. Let us assume the solvent (component 1) to be non-dipolar, and the solute (component 2) as consisting of dipolar molecules. Moreover, for the sake of simplicity, let the molecules of both components be axially symmetric. Equation (227) now yields, on neglecting fluctuations and non-linear polarizability:

$$K_{\rm m\infty} = \frac{4\pi N_{\rm A}}{45kT}\left\{\delta^{\omega}_{\rm v2}\left(\delta^{e}_{2} + \frac{p_2^2}{kT}\right) - \delta^{\omega}_{\rm v1}\,\delta^{e}_{1}(1 + 2J_{11}^{(2)})\right\} + \frac{4\pi N_{\rm A}}{45kT}\left\{\delta^{\omega}_{\rm v1}\,\delta^{e}_{2} + \delta^{\omega}_{\rm v2}\,\delta^{e}_{1} + \frac{\delta^{\omega}_{\rm v1}\,p_2^2}{kT}\right\}J_{12}^{(2)} \tag{228}$$

Hence, once the correlation parameter $J_{11}^{(2)}$ of the pure solvent is available, equation (228) permits the determination of the parameter of cross-angular correlations $J_{12}^{(2)}$. Recently, Stiles [400] determined the value of $J_{12}^{(2)}$ for several dipolar liquids by measuring the Kerr constant in dilute solutions in benzene. However, the $J_{12}^{(2)}$ values thus obtained, *e.g.* for the benzene–nitrobenzene correlation, $J_{\rm BN}^{(2)}$, differ in sign from the value

[400] P. J. Stiles, Ph.D. Thesis, University of Sydney, 1969.

determined from light depolarization studies.[401] This discrepancy can be explained by the above-calculated new constant (223), which stems from fluctuations of quadrupolar fields and contributes the following amount to (228):

$$K_{m\infty}^{quadr} = \frac{2\pi N_A}{105k^2T^2}\{\delta_{v1}^{\omega}\,\Theta_1^2(\alpha_{v2}^2\langle r_{ij}^{-8}\rangle_{12} - 2\alpha_{v1}^2\langle r_{ij}^{-8}\rangle_{11}) + \delta_{v2}^{\omega}\,\Theta_2^2\,\alpha_{v1}^2\langle r_{ij}^{-8}\rangle_{12}\} \quad (228a)$$

Other corrections will obviously result from the contributions due to translational–orientational fluctuations.

Solution of a polar liquid in a solvent with isotropic molecules. Assume the molecules of the solvent to be isotropically polarizable, and those of the solute to be dipolar. Equation (227) now yields:

$$\Delta K_{m\infty} = K_{m\infty}^{TF} + \frac{4\pi N_A}{45kT}\,\delta_{v2}^{\omega}\left(\delta_2^e + \frac{p_2^2}{kT}\right) \quad (229)$$

where

$$K_{m\infty}^{TF} = \frac{8\pi N_A}{45kT}\{[\delta_{v2}^{\omega}\,\alpha_{v1}(3\alpha_2 + \delta_2^e) + \delta_{v2}^e\,a_{v1}(3a_2 + \delta_2^{\omega})]\,J_{12}^{RA} - 9a_{v1}^2\,\alpha_{v1}\,\alpha_1\langle r_{ij}^{-6}\rangle_{11}\} \quad (230)$$

is a translational–fluctuational contribution hitherto not taken into account in the analysis of experimental measurements. It defines the Kerr-constant solvent effect studied recently by Buckingham *et al.*[398b] and it can be positive or negative, according to the properties of the solution.

If the solvent with isotropically polarizable molecules contains in solution a liquid whose molecules are quadrupolar, we obtain:

$$K_{m\infty} = K_{m\infty}^{TF} + \frac{4\pi N_A}{45kT}\,\delta_{v2}^{\omega}\left\{\delta_2^e + \frac{3\alpha_{v1}^2}{3kT}\,\Theta_2^2\langle r_{ij}^{-8}\rangle_{12}\right\} \quad (231)$$

This formula contains, in addition to (230), another term stemming from fluctuations of quadrupolar fields which, as yet, has not been taken into account in analyses of experimental values.

It is to be presumed that by using formulae (227)—(231) in determining the optical anisotropy δ_2^{ω} of molecules of the solute, better agreement can be obtained between the optical anisotropy values derived by the method of Le Fèvre *et al.*[173, 395] from studies of the static Kerr effect and those derived by the method of Pacault, Bothorel, *et al.*[132, 396] from intensity-depolarized Rayleigh scattering. Also, the latest measurements of the optical Kerr effect[214] point to the possibility of achieving such agreement.

[401] (*a*) B. Fechner, *Acta Phys. Polon.*, 1969, **36**, 297; (*b*) C. Such, C. Clément, and P. Bothorel, *Compt. rend.*, 1970, **271**, *B*, 228; (*c*) C. Such, Thèse, Bordeaux, 1970; (*d*) H. C. Lucas and D. A. Jackson, *Mol. Phys.*, 1971, **20**, 801.

The Kerr Effect in Polymer Solutions and Colloids.—Strong electro-optical birefringence is observed in colloidal systems [402–404] and in macromolecular solutions.[405] Numerous theoretical papers [406] have been devoted to the quadratic optical birefringence exhibited by solutions of small macromolecules in weak electric fields. O'Konski *et al.*[407] and others [408–412] have shown that complete alignment of macromolecules in a static electric field does in fact take place in various macromolecular solutions, permitting the observation of the electric saturation of optical birefringence.[402c]

Electric Saturation of the Kerr Effect. When calculating the electric reorientation function of molecules (172) for molecular gases, it was justifiable to use the approximation (169a) for the perturbed statistical distribution function. In the general case, calculations have to be carried out with a distribution function of the form (169) where, for axially

[402] (*a*) H. Disselhorst and H. Freundlich, *Physik. Z.*, 1915, **16**, 419; C. Bergholm and Y. Björnståhl, *ibid.*, 1921, **21**, 137; (*b*) S. Procopin and M. Tibu, *Compt. rend.*, 1968, **267**, *B*, 1369; (*c*) J. Errera, J. Th. G. Overbeek and H. Sack, *J. Chim. phys.*, 1935, **32**, 681; (*d*) H. Mueller, *Phys. Rev.*, 1939, **55**, 508, 792; H. Mueller and B. W. Sackmann, *ibid.*, 1939, **56**, 615; B. W. Sackmann, *J. Opt. Soc. Amer.*, 1945, **35**, 66.

[403] A. Peterlin and H. A. Stuart, *Z. Physik*, 1939, **112**, 1, 129.

[404] (*a*) V. N. Tsvetkov, *Izvest. Akad. Nauk, S.S.S.R., Ser. fiz.*, 1941, 57; V. N. Tsvetkov, Yu. V. Mitin, V. R. Glushenkova, A. E. Grishchenko, N. N. Boitsova, and S. Ya. Lubina, *Vysokomol. Soedineniya*, 1963, **5**, 453; V. N. Tsvetkov, I. N. Shetennikova, E. I. Ryumtsev, and V. S. Skazka, *ibid.*, 1965, **7**, 1111; (*b*) N. A. Tolstoi and P. P. Feofilov, *Doklady Akad. Nauk S.S.S.R.*, 1949, **66**, 617; *Zhur. eksp. teor. Fiz.*, 1949, **19**, 421.

[405] (*a*) H. Benoit, *Compt. rend.*, 1949, **228**, 1716; 1949, **229**, 30; (*b*) H. Benoit, *Ann. Physique*, 1951, **6**, 561; *J. Chim. phys.*, 1952, **49**, 517; (*c*) G. Boeckel, J. C. Genzling, G. Weil, and H. Benoit, *J. Chim. phys.*, 1962, 999.

[406] (*a*) E. Kluk, *Acta Phys. Polon.*, 1962, **22**, 249; (*b*) S. Ikeda, *J. Chem. Phys.*, 1963, **38**, 2839; (*c*) D. A. Dows, *ibid.*, 1964, **41**, 2656; (*d*) K. Nagai and T. Ishikawa, *ibid.*, 1965, **43**, 4508; (*e*) K. Nagai, *ibid.*, 1969, **51**, 1091; (*f*) R. S. Stein, *J. Polymer Sci. Part A-2, Polymer Phys.*, 1969, **7**, 1021.

[407] (*a*) Ch. T. O'Konski, K. Yoshioka, and W. H. Orttung, *J. Chem. Phys.*, 1959, **63**, 1558; (*b*) S. Krause and Ch. T. O'Konski, *J. Amer. Chem. Soc.*, 1959, **81**, 5082; *Biopolymers*, 1963, **1**, 503; (*c*) K. Yoshioka and Ch. T. O'Konski, *Biopolymers*, 1966, **4**, 499; *J. Polymer Sci.*, 1968, **6**, 421; (*d*) Ch. T. O'Konski and S. Krause, *J. Phys. Chem.*, 1970, **74**, 3243.

[408] (*a*) I. Tinoco, jun., and K. Yamaioka, *J. Phys. Chem.*, 1959, **63**, 423; (*b*) A. E. V. Haschemeyer and I. Tinoco, jun., *Biochemistry*, 1962, **1**, 996; (*c*) D. N. Holcomb and I. Tinoco, jun., *J. Phys. Chem.*, 1963, **67**, 2691.

[409] (*a*) M. J. Shah, *J. Phys. Chem.*, 1963, **67**, 2215; (*b*) M. J. Shah and D. C. Thompson, *Soil Sci.*, 1963, **96**, 347; (*c*) M. J. Shah, D. C. Thompson, and C. M. Hart, *J. Phys. Chem.*, 1963, **67**, 1170.

[410] (*a*) W. H. Orttung and J. A. Meyers, *J. Phys. Chem.*, 1963, **67**, 1911; (*b*) W. H. Orttung, *J. Amer. Chem. Soc.*, 1965, **87**, 924; (*c*) W. H. Orttung, *J. Phys. Chem.*, 1968, **72**, 4058.

[411] (*a*) H. Watanabe and K. Yoshioka, *Biopolymers*, 1964, **2**, 91; (*b*) H. Nakayama and K. Yoshioka, *J. Polymer Sci.* (*A*), 1965, **3**, 813; (*c*) H. Matsumoto, H. Watanabe, and K. Yoshioka, *Biopolymers*, 1970, **19**, 1307; 1968, **6**, 929; (*d*) K. Nishinari and K. Yoshioka, *Kolloid-Z.*, 1969, **235**, 1189; 1970, **240**, 831.

[412] (*a*) J. C. Powers, jun., and W. L. Peticolas, *Adv. in Chem. Series*, 1967, **63**, 217; *J. Phys. Chem.*, 1967, **71**, 3191; (*b*) B. L. Brown, B. R. Jennings, and H. Plummer, *Appl. Optics*, 1969, **8**, 2019; (*c*) S. P. Stoylov, *Bulg. Acad. Sci. Comm. Dept. Chem.*, 1969, **2**, 449; (*d*) A. K. Wright and W. Keefe, *Biopolymers*, 1970, **9**, 1503; (*e*) S. Kobayasi, *ibid.*, 1971, **10**, 915.

symmetric microsystems, one has:

$$-\beta U(\Omega, E) = X_0 + X_1 \cos\theta \pm X_2 \cos^2\theta + \ldots \tag{232}$$

X_0 being a constant, and

$$X_1 = \frac{p^* E}{kT} \tag{232a}$$

a dimensionless Langevin–Debye parameter describing the orientation of permanent electric dipoles p^* in the static electric field E. Similarly,

$$X_2 = \frac{|\alpha_3^* - \alpha_1^*|}{2kT} E^2 \tag{232b}$$

is a dimensionless Langevin parameter accounting for the orientation of an induced dipole of absolute effective electric anisotropy $|\alpha_3^* - \alpha_1^*|$.

In equation (232), the sign 'plus' at X_2 is for microsystems having positive electric anisotropy ($\alpha_3 - \alpha_1 > 0$), *i.e.* for cigar-shaped ones, where both the permanent electric dipole moment and the largest principal electric polarizability are directed along the symmetry axis at an angle θ to the field E. The sign 'minus' refers to microsystems with negative electric anisotropy ($\alpha_3 - \alpha_1 < 0$), *i.e.* disc-shaped microsystems having their dipole moment in the direction of the symmetry axis but their largest polarizability lying in the plane perpendicular thereto.

On inserting (232) into the distribution function (169) one obtains from (172) the following reorientation function for microsystems in a high static electric field:[413]

$$\Phi(X_1, \pm X_2) = \tfrac{1}{2}\{3L_2(X_1, \pm X_2) - 1\} \tag{233}$$

where we have introduced the generalized Langevin function:

$$L_2(X_1, \pm X_2) = \frac{\int_0^\pi \cos^2\theta \exp(X_1 \cos\theta \pm X_2 \cos^2\theta) \sin\theta \, d\theta}{\int_0^\pi \exp(X_1 \cos\theta \pm X_2 \cos^2\theta) \sin\theta \, d\theta} \tag{234}$$

which is reducible to a form adapted to direct numerical calculations [413] when $X_2 \neq 0$:

$$L_2(X_1, \pm X_2) = \frac{X_1^2 \mp 2X_2}{4X_2^2} \pm \frac{e^{X_1} + e^{-X_1}}{4X_2^{\frac{1}{2}} I(X_1, \pm X_2)} - \frac{X_1(e^{X_1} - e^{-X_1})}{8X_2^{\frac{3}{2}} I(X_1, \pm X_2)} \tag{234a}$$

involving the integrals:

$$I(X_1, \pm X_2) = \frac{1}{2} \exp\left\{\mp \left(\frac{X_1^2}{4X_2} + X_2\right)\right\} \int_{l_1}^{l_2} \exp(\pm t^2) \, dt \tag{235}$$

with the integration limits:

$$l_1 = -X_2^{\frac{1}{2}} \pm \frac{X_1}{2X_2^{\frac{1}{2}}}, \quad l_2 = X_2^{\frac{1}{2}} \pm \frac{X_1}{2X_2^{\frac{1}{2}}} \tag{235a}$$

[413] (*a*) S. Kielich, *Acta Phys. Polon.* (*A*), 1970, **37**, 447; (*b*) S. Kielich, *J. Colloid Interface Sci.*, 1970, **33**, 142.

In particular, Davson integrals of the form $\int_0^l \exp(t^2)\,dt$ up to $l = 10$ are tabulated in the paper of Matsumoto *et al.*[414] Integrals $\int_0^l \exp(-t^2)\,dt = (\sqrt{(\pi)}/2)\,\mathrm{erf}\,l$ can be calculated numerically from the tabulated error function $\mathrm{erf}\,l$.

One notes that the optical birefringence induced by the static electric field is no longer a quadratic function of the field strength but, by (164a) and (171), is given as follows:

$$n_\parallel - n_\perp = \pm\, 2\pi\rho n\left(\frac{n^2+2}{3n}\right)^2 |\,\delta_V^\omega\,|\,\Phi(X_1, \pm X_2) \qquad (236)$$

where the Kerr-effect reorientation function $\Phi(X_1, \pm X_2)$ is given by equation (233).

The electro-optical birefringence in the form (236) with the 'plus' sign defines the result of O'Konski *et al.*[407] for cigar-shaped particles. With 'minus' sign, it defines those of Shah [409] for disc-like ones and of Yoshioka *et al.*[411, 414] for both cases.

Birefringence of Strongly Dipolar Microsystems. In dealing with strongly dipolar but weakly polarizable microsystems, *i.e.* when $X_2 \ll X_1$, the term in X_2 in the Langevin function (234) can be omitted. The reorientation function (233) now becomes:

$$\Phi(X_1) = \tfrac{1}{2}\{3L_2(X_1) - 1\} = 1 - 3\frac{L(X_1)}{X_1} \qquad (237)$$

where $L(X_1) = \coth X_1 - 1/X_1$ is the well-known Langevin function [263] of dipole microsystems. Figure 13 shows the reorientation function (237) plotted *versus* X_1.

Weak reorientation of permanent dipoles occurs by (232a) if the electric field E is weak or the temperature of the system T is high, leading to $X_1 < 1$. Equation (237) now yields:

$$\Phi(X_1) = \frac{X_1^2}{15} - \frac{2X_1^4}{315} + \frac{X_1^6}{1575} - \ldots \qquad (237a)$$

One sees that on taking the first term of the expansion, an optical birefringence (236) dependent on the square of the external field is obtained in accordance with Kerr's law.

Strong reorientation of permanent dipoles takes place if the field strength E is large or the temperature T is low, when in general $X_1 > 1$. At $X_1 \geqslant 10$, the quantity (237) can be computed from the following limiting formula:

$$\Phi(X_1) = 1 - \frac{3}{X_1} + \frac{3}{X_1^2} \qquad (237b)$$

In the case of electric saturation, *i.e.* when all the permanent electric dipoles

[414] M. Matsumoto, H. Watanabe, and K. Yoshioka, *Scientific Papers of the College of General Education, University of Tokyo*, 1967, **17**, 173.

are completely aligned in the direction of the applied electric field, the reorientation function (237b) tends to 1, and the optical birefringence (236) becomes:

$$(n_{\parallel} - n_{\perp})_{\infty} = \pm 2\pi\rho n\left(\frac{n^2+2}{3n}\right)^2 |\delta_V^{\omega}| \qquad (236a)$$

This formula permits the determination, from birefringence saturation studies, of not only the absolute value of the volume optical anisotropy $|\delta_V^{\omega}| = |a_{V3}^{\omega} - a_{V1}^{\omega}|$ but moreover of its sign, thus revealing to us the shape of the particles.

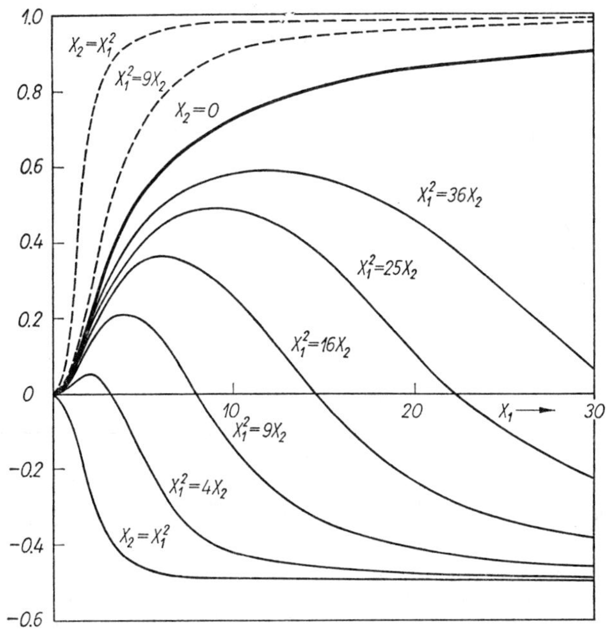

Figure 13 *Kerr-effect reorientation functions $\Phi(X_1, \pm X_2)$. Above the continuous bold-face curve $\Phi(X_1, 0)$ lie the dashed curves $\Phi(X_1, +Y_2)$ for cigar-shaped microsystems; below it lie the continuous-line curves $\Phi(X_1, -Y_2)$ for disc-like microsystems*

Birefringence of Non-dipolar Microsystems. In the case of anisotropic non-dipolar or very weakly dipolar molecules, $X_1^2 \ll X_2$, and equation (233) leads to the following reorientation function:

$$\Phi(\pm X_2) = \tfrac{1}{2}\{3L_2(\pm X_2) - 1\} = -\frac{1}{2} \mp \frac{3}{2X_2} \pm \frac{3}{2X_2^{\frac{1}{2}} I(\pm X_2)} \qquad (238)$$

where the integral

$$I(\pm X_2) = \exp(\mp X_2) \int_0^{\sqrt{X_2}} \exp(\pm t^2)\,dt \qquad (235b)$$

results from (235) at $X_1 = 0$.

Figure 14 shows the reorientation function (238) for both positive and negative anisotropy of the microsystems.

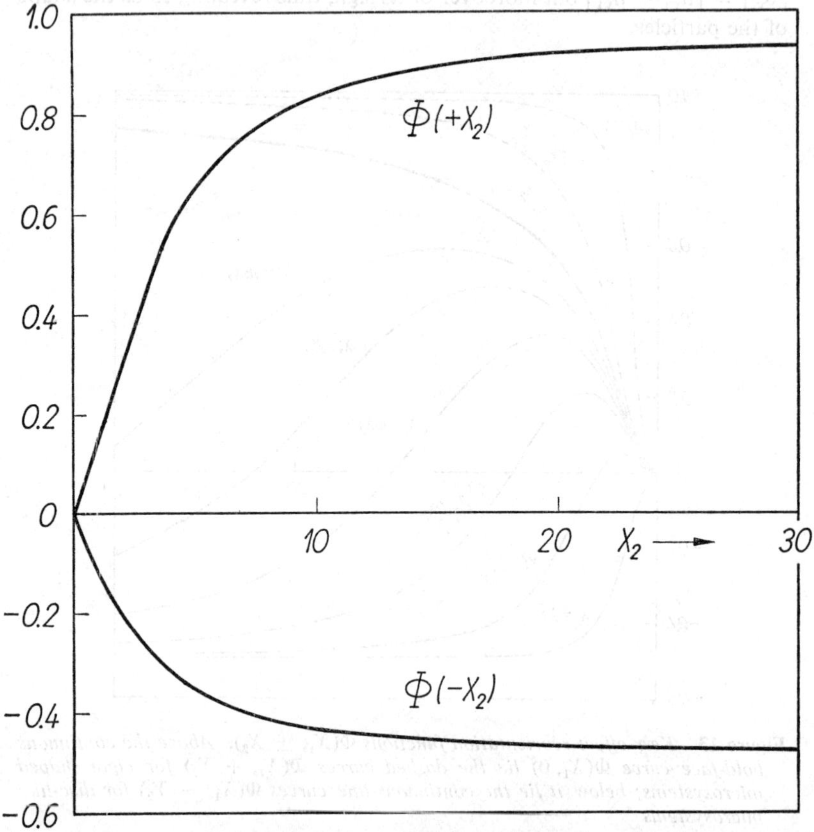

Figure 14 *Kerr-effect functions, calculated from induced dipole reorientation alone:* $\Phi(+X_2)$ *for cigar-like microsystems,* $\Phi(-X_2)$ *for disc-like ones*

In the case of weak reorientation, $X_2 < 1$, the reorientation function (238) is expressed in sufficient approximation by the series:

$$\Phi(\pm X_2) = \pm \frac{2X_2}{15} + \frac{4X_2^2}{315} \mp \frac{8X_2^3}{4725} + \ldots \qquad (238a)$$

which, in a first approximation, leads to a quadratic dependence of the birefringence (236) on E.

For strong reorientation, $X_2 \geqslant 10$, (238) leads to the following limiting formulae:

(*i*) For molecules with positive anisotropy:[413]

$$\Phi(+X_2) = 1 - \frac{3}{2X_2} + O(X_2^{-2}) \tag{238b}$$

(*ii*) For molecules with negative anisotropy:

$$\Phi(-X_2) = -\frac{1}{2} + \frac{3}{4X_2} + O(X_2^{-2}) \tag{238c}$$

For the case of electric saturation, *i.e.* when $X_2 \to \infty$, we obtain respectively:

$$\Phi(+\infty) = 1 \quad \text{and} \quad \Phi(-\infty) = -\tfrac{1}{2} \tag{238d}$$

and the birefringence (236) assumes the following two limiting values for positive and negative anisotropy:

$$(n_\parallel - n_\perp)_+ = 2\pi\rho n\left(\frac{n^2+2}{3n}\right)^2 |\,\delta_V^\omega\,|_+ \tag{236b}$$

$$(n_\parallel - n_\perp)_- = \pi\rho n\left(\frac{n^2+2}{3n}\right)^2 |\,\delta_V^\omega\,|_- \tag{236c}$$

In this case, as we see, only the absolute values of the optical anisotropy of the particle can be determined.

Birefringence of Dipolar Anisotropically Polarizable Microsystems. In the general case of microsystems which are dipolar and at the same time anisotropically polarizable in an external field E, the reorientation function for the Kerr effect is given by equation (233). Graphs of this reorientation function are shown in Figure 13 against the parameter X_1 at parametrical values of $X_2 = X_1^2/n$ for $n = 1, 4, 9, 16, 25, 36, \ldots$.

Strong reorientation. In the case of cigar-shaped microsystems (positive anisotropy), the permanent electric dipoles and induced electric dipoles tend to orient themselves in a concordant manner in the direction of the static field. As a result of this, with growing static field strength and growing parameter X_2 the value of $\Phi(X_1, +X_2)$ tends very steeply from 0 to the limiting value 1, which defines a state of complete alignment of the microsystems in the direction of the static electric field (electric saturation of orientation). This is illustrated by the dashed curves of Figure 13. For disc-like microsystems (negative anisotropy), the situation is entirely different: here, with the permanent dipole moment p_3 directed along the symmetry axis of the microsystem, the polarizability α_3 along this axis will be less than the polarizability α_1 perpendicular to it. Consequently, the torque acting on the permanent dipole will tend to orient the microsystem into alignment with the static field E^0, whereas the torque on the induced dipole will act in a manner to orient the microsystem perpendicular to E^0. Accordingly, a disc-shaped microsystem is subject to two mutually orthogonal torques. With increasing field strengths E, the induced-dipole parameter [equation (232b)] increases, thus causing the microsystem to go over from parallel orientation to perpendicular orientation with respect to the field E, and $\Phi(X_1, -X_2)$ decreases in value correspondingly, tending rapidly to 0 as the parameter X_2 of induced-dipole reorientation grows and, after exhibiting a reversal of sign at a well-defined value of X_2, tending to the value $\frac{1}{2}$ at complete saturation. This is illustrated in Figure 13 by the full curves lying below the curve for $X_2 = 0$.

Weak reorientation. In this case, the reorientation function (233) is, in a first approximation,

$$\Phi(X_1, \pm X_2) = \frac{X_1^2}{15} \pm \frac{2X_2}{15} \tag{233a}$$

and insertion into (236) yields the well-known Born–Langevin result for birefringence of axially symmetric microsystems.

As long as reorientation concerns single molecules, the approximation of equation (233a) suffices, since the parameters (232a) and (232b) are numerically rather small, being of the order $X_1 = 10^{-8} E$ and $X_2 = 10^{-18} E^2$ at the most (for nitrobenzene, $p \approx 14 \times 10^{-30}$ C m, $|\alpha_3 - \alpha_1| \approx 6 \times 10^{-40}$ C^2 m^2 J^{-1}). In a very intense field of the order 10^7 V m^{-1} one has $X_1 = 0.1$ and $X_2 = 10^{-4}$ so that no perceptible deviation from the quadratic field dependence of the Kerr effect can take place. The situation, however, can be different in molecular substances which admit of multiple association of the molecules or of the formation of molecular assemblages or even quasi-domains.[76] In such liquids, the reorientation parameters X_1 and X_2 increase by one order of magnitude or more, so that it is possible to observe experimentally the approach to the electric saturation of the Kerr effect by applying a sufficiently intense electric field and low temperatures.

A totally different situation exists in solutions of polymers and colloids, where the macromolecules (particles) possess large electric moments and polarizabilities, by several orders of magnitude in excess of those of molecules of the nitrobenzene type. In the class of synthetic polypeptides, the most often studied are macromolecules of poly-γ-benzyl-L-glutamate[133] in the shape of long cylinders presenting large dipole moments of the order of 3×10^{-27} C m and a large positive anisotropy of electric polarizability amounting to 10^{-33} C^2 m^2 J^{-1}. For such macromolecules, the reorientation parameters amount respectively to $X_1 = 10^{-5} E$ and $X_2 = 10^{-12} E^2$, which means that reorientation of the permanent dipoles predominates over reorientation of induced dipoles, and electric saturation can set in already at $E \geqslant 10^5$ V m^{-1}. As an example of disc-shaped particles, we mention bentonite suspensions,[409] where $X_1 = 5 \times 10^{-5} E$ and $X_2 = 7 \times 10^{-8} E^2$. Here, $X_2 > X_1^2$, and with growing field strength the reorientation of induced dipoles rapidly becomes predominant. At electric saturation, which sets already in a field of $E \geqslant 10^4$ V m^{-1}, this reorientation brings the particles into alignment, with their short axes parallel to E, causing the $\Phi(X_1, -X_2)$ curves to fall steeply to zero and leading to reversal of sign. Similarly, high saturation of electric orientation can be achieved in solutions of collagen and tobacco mosaic virus [133] as well as in solutions of other biomacromolecules [134] and colloid particles.[135]

5 Electric Permittivity Variations in a Strong Electric Field

Molar Electric Polarization of Dense Media.—In Kirkwood's semi-macroscopic treatment [353] of the linear properties of isotropic dielectrics, one has the following relation between the relative electric permittivity ε_r and the polarization $P(E_0)$ induced in the medium:

$$\frac{\varepsilon_r - 1}{\varepsilon_r + 2} E_0 = \frac{1}{3\varepsilon_0} P(E_0) \tag{239}$$

or

$$P_m E_0 = \frac{V_m}{3\varepsilon_0} P(E_0) \tag{239a}$$

where we have introduced the molar electric polarization:

$$P_m = \frac{\varepsilon_r - 1}{\varepsilon_r + 2} V_m = P_m^D + P_m^O \tag{240}$$

consisting of a distortional part P_m^D and a part P_m^O related to statistical processes of molecular orientation under the influence of the electric field. The distortional part P_m^D of the molar polarization is due to the polarizability of the atoms and molecules in the field; for rarefied gases, it is of the form:

$$P_m^D = \frac{N_A}{3\varepsilon_0} \alpha = \frac{4\pi}{3} N_A \alpha_v \tag{240a}$$

where α is the mean electric polarizability of the atom (molecule), and $\alpha_v = \alpha/4\pi\varepsilon_0$ is the volume electric polarizability.[415] In dense substances, P_m^D depends indirectly on molecular correlation, similarly to the molecular refraction R_m, widely discussed for various models [344, 354–356] and studied experimentally as a function of pressure in simple gases [416] (see Section 4).

The other part of the molar polarization (240) is, in Kirkwood's theory, of the following form:[353]

$$P_m^O = \frac{\langle M^2 \rangle}{9\varepsilon_0 kT} \tag{241}$$

[415] (*a*) K. H. Illinger and C. P. Smyth, *J. Chem. Phys.*, 1960, **32**, 787; 1961, **35**, 392; K. H. Illinger, *ibid.*, 1961, **35**, 409; 1962, **36**, 1084; (*b*) J. W. Stewart, *ibid.*, 1964, **40**, 3297.

[416] (*a*) W. F. Brown, jun., *J. Chem. Phys.*, 1950, **18**, 1193; (*b*) J. De Boer, F. Van Der Maesen, and C. A. Ten Seldam, *Physica*, 1953, **19**, 265; (*c*) D. R. Johnston, G. J. Oudemans, and R. H. Cole, *J. Chem. Phys.*, 1960, **33**, 1310; (*d*) R. H. Orcutt and R. H. Cole, *Physica*, 1965, **31**, 1779; (e) R. H. Orcutt and R. H. Cole, *J. Chem. Phys.*, 1967, **46**, 697.

where M is the total electric dipole moment of a spherical sample of the dielectric of volume V.

If the dielectric consists of N_A molecules, having mutually non-interacting electric dipole moments p, (241) reduces to the well-known Langevin–Debye formula:

$$P_m^O = \frac{N_A p^2}{9\varepsilon_0 kT} \tag{241a}$$

providing the basis for determination of p from measurements of the molar polarization $P_m(T)$ as a function of temperature. In dense dielectrics, interactions of the electric dipoles become appreciable, and formula (241a) has to be modified in accordance with the discussions of Onsager,[341] Kirkwood,[353] and others.[417]

In non-dipolar dielectrics sufficiently dense for molecular interaction, the temperature-dependent polarization (241) is generally non-zero. Such interaction will lead to an effect consisting in the induction, in any given molecule immersed in the dense medium, of a dipole moment M by the fluctuating electric field of the permanent quadrupoles,[325, 418] octupoles,[229, 254, 419] hexadecapoles,[230] and in general multipoles[419] of its neighbours.

For axially symmetric molecules having a permanent electric quadrupole moment Θ, polarizability α, and electric anisotropy κ, separated by the distance r_{ij}, we have Jansen's formula[418] (in the absence of angular correlations):

$$P_m^O = \frac{N_A \alpha_v^2}{15\varepsilon_0 kT}(5 + 4\kappa^2)\,\Theta^2\langle r_{ij}^{-8}\rangle + \ldots \tag{241b}$$

where the statistical average $\langle r_{ij}^{-8}\rangle$ is defined by equation (154) for $n = 8$.

For tetrahedral molecules the linear polarizability is isotropic, and the first non-zero permanent electric moment is the octupole $\Omega = \Omega_{123}$. The electric fields of these octupoles induce a dipole moment in any given molecule, whence $\langle M^2\rangle \neq 0$ and we have by equation (241) to a sufficiently good approximation:[229]

$$P_m^O = \frac{16 N_A \alpha_v^2}{15\varepsilon_0 kT}\,\Omega^2\langle r_{ij}^{-10}\rangle \tag{241c}$$

Molecules of still higher, *e.g.* octahedral, symmetry possess a (lowest) hexadecapole moment; the electric fields of these hexadecapoles now induce a dipole moment in neighbouring molecules. Here, too, $\langle M^2\rangle \neq 0$, and

417 (*a*) W. F. Brown, jun., *J. Chem. Phys.*, 1953, **21**, 1327; (*b*) F. E. Harris, *ibid.*, 1955, **23**, 1663; (*c*) A. D. Buckingham, *Proc. Roy. Soc.*, 1956, **A238**, 235; (*d*) B. K. P. Scaife, *Proc. Phys. Soc.*, 1957, **22B**, 314; (*e*) M. Mandel and P. Mazur, *Physica*, 1958, **24**, 116; (*f*) H. Sutter and R. H. Cole, *J. Chem. Phys.*, 1970, **52**, 132; 1971, **54**, 4988; (*g*) A. D. Buckingham, R. H. Cole, and H. Sutter, *ibid.*, p. 5960.

418 (*a*) R. W. Zwanzig, *J. Chem. Phys.*, 1956, **25**, 211; (*b*) L. Jansen, *Phys. Rev.*, 1958, **112**, 434; L. Jansen and A. D. Solem, *ibid.*, 1956, **104**, 1291.

419 S. Kielich, *Mol. Phys.*, 1965, **9**, 549; *Acta Phys. Polon.*, 1965, **28**, 95.

we obtain the formula:[230, 419]

$$P_{\mathrm{m}}^{\mathrm{O}} = \frac{80 N_{\mathrm{A}} \alpha_{\mathrm{V}}^2}{21 \varepsilon_0 kT} \Phi^2 \langle r_{ij}^{-12} \rangle \tag{241d}$$

In many a case, when calculating $P_{\mathrm{m}}^{\mathrm{O}}$ for multipolar molecules, one is justified in neglecting the anisotropy of linear electric polarizability, leading in general (Kielich [419]) to the formula:

$$P_{\mathrm{m}}^{\mathrm{O}} = \frac{N_{\mathrm{A}} \alpha_{\mathrm{V}}^2}{9 \varepsilon_0 kT} \sum_{n=2}^{\infty} \frac{2^n (n+1)!\, n!}{(2n)!} \{\boldsymbol{p}^{(n)} [n]\, \boldsymbol{p}^{(n)}\} \langle r_{ij}^{-2(n+2)} \rangle \tag{242}$$

where $\boldsymbol{p}^{(n)}$ denotes the 2^n-pole tensorial electric moment defined by equation (40).

In particular, for axially symmetric molecules the general expression (242) reduces simply to:[419]

$$P_{\mathrm{m}}^{\mathrm{O}} = \frac{N_{\mathrm{A}} \alpha_{\mathrm{V}}^2}{9 \varepsilon_0 kT} \sum_{n=2}^{\infty} (n+1) \{p^{(n)}\}^2 \langle r_{ij}^{-2(n+2)} \rangle \tag{242a}$$

$p^{(n)}$ standing for the 2^n-pole scalar electric moment of an axially symmetric molecule defined by equation (40a). Obviously, at $n = 2$, this yields equation (241b) on neglecting the anisotropy and putting $p^{(2)} = \Theta$.

The preceding considerations lead to the conclusion that, whereas numerical determinations of electric dipoles require, in accordance with equation (241a), the investigation of substances in the absence of molecular correlations, the quadrupole, octupole, and hexadecapole moments are accessible to determination only if at least pairwise radial correlations are present in the non-dipolar medium. Numerical values of electric moments of higher order are determined by the method of measuring second dielectric virial coefficients worked out and successfully applied by Cole *et al.*[254] It has yielded quadrupole, octupole, and hexadecapole moments in good numerical agreement with the values obtained by other methods (see Section 2).

High Electric Field Effects.—*Determination of Variations in Electric Permittivity.* The electric permittivity ε of an isotropic dielectric is usually measured by means of a slowly varying electric field, of a field strength E_{0a} so small as to cause only linear polarization of the medium:

$$P(E_{\mathrm{a}}) = \frac{3\varepsilon_0}{V_{\mathrm{m}}} P_{\mathrm{m}} E_{0\mathrm{a}} \tag{243}$$

where P_{m} is the linear molar polarizability. In this case, in accordance with equations (240) and (241), the permittivity does not depend on the field strength E_{a} and it is characteristic of intrinsic electric properties of the medium, its structure, and thermodynamic state. If, however, the dielectric sample is simultaneously acted on by another (polarizing) electric field of strength E_{p} sufficiently high to induce non-linear polarization, the permittivity measured becomes a function of this applied polarizing field: $\varepsilon = \varepsilon(E_{\mathrm{p}})$. Obviously, the polarization as such is now a function of the two fields, *i.e.* of the analysing field E_a and polarizing field E_p, and one has: $\boldsymbol{P} = \boldsymbol{P}(E_a, E_p)$. As we have said, the weak analysing field causes a linear polarization only. This justifies us in writing the expansion:

$$\boldsymbol{P}(E_a, E_p) = \left\{\frac{\partial \boldsymbol{P}(E_a, E_p)}{\partial E_{0a}}\right\}_{E_{0a}=0} \cdot \boldsymbol{E}_{0a} \tag{244}$$

where the expansion coefficient in braces is a function of the polarizing field E_p only, but does not depend on E_a.

In the absence of a strong electric field (at $E_p = 0$), the measured relative electric permittivity of an isotropic dielectric is a scalar quantity, defined as follows:

$$(\varepsilon_r - 1)\,\boldsymbol{E}_a = \frac{1}{\varepsilon_0}\boldsymbol{P}(E_a) \tag{243a}$$

where in accordance with (164c) the macroscopic electric field of a spherical dielectric in vacuum is:

$$\boldsymbol{E}_a = \frac{3}{\varepsilon_r + 2}\boldsymbol{E}_{0a} \tag{243b}$$

Under the influence of a strong field, an isotropic dielectric becomes not only non-linear but also anisotropic, with the following electric permittivity tensor:

$$\{\boldsymbol{\varepsilon}_r(E_p) - \boldsymbol{U}\} \cdot \boldsymbol{E}_a = \frac{1}{\varepsilon_0}\boldsymbol{P}(E_a, E_p) \tag{244a}$$

The non-linearity, induced in a naturally isotropic dielectric by an electric field, is defined in terms of the variation in the permittivity tensor, *cf.* equation (12):

$$\Delta\boldsymbol{\varepsilon}_r(E_p) = \boldsymbol{\varepsilon}_r(E_p) - \varepsilon_r(0)\,\boldsymbol{U} \tag{245}$$

where $\boldsymbol{\varepsilon}_r(E_p)$ is the permittivity tensor measured in the presence of the strong field, and $\varepsilon_r = \varepsilon_r(0)$ the permittivity in the absence of the latter.

In a high static electric field, isotropic dielectrics are endowed with properties similar to those of an axial crystal; consequently, in the absence of absorption and dispersion, only the three diagonal elements of the permittivity variation tensor (245) are non-zero, and two of them are mutually independent. Thus, assuming the strong field E_p to be applied along the z-axis, we have by (245):

$$\Delta\varepsilon_{xx}(E_p) = \Delta\varepsilon_{yy}(E_p) \equiv \Delta\varepsilon_r^{\perp}(E_p) = \varepsilon_r^{\perp}(E_p) - \varepsilon_r(0) \tag{245a}$$

$$\Delta\varepsilon_{zz}(E_p) = \Delta\varepsilon_r^{\parallel}(E_p) = \varepsilon_r^{\parallel}(E_p) - \varepsilon_r(0) \tag{245b}$$

that is to say, we are dealing with changes in permittivity measured perpendicular ($\boldsymbol{E}_a \perp \boldsymbol{E}_p$) and parallel ($\boldsymbol{E}_a \parallel \boldsymbol{E}_p$) to the direction of the strong field along which lies the axis of electric anisotropy, induced in the naturally isotropic dielectric.

The usually convenient experimental set-up is that with the measuring field $\boldsymbol{E}_a$ and strong polarizing field $\boldsymbol{E}_p$ acting in the same direction (say, parallel to the z-axis). As a consequence, we measure permittivity variations (245b), which we can calculate from the equation:

$$\frac{\varepsilon_r^{\parallel} - 1}{\varepsilon_r^{\parallel} + 2} V_m^{\parallel} = P_m + \Delta P_m^{\parallel}(E_p) \qquad (244b)$$

resulting from equations (244), (243b), and (244a). Equation (244b) contains, in addition to the linear molar polarization P_m, a non-linear change in molar polarization that is induced by the strong field and is a function of its strength E_p:

$$\Delta P_m^{\parallel}(E_p) = \frac{V_m}{3\varepsilon_0} \left\{ \frac{\partial P_{\parallel}(E_a, E_p)}{\partial E_{0a}} \right\}_{E_a=0} - P_m \qquad (246)$$

Taking into account that a strong electric field generally modifies the molar volume in a manner expressed by equation (163), we obtain [equations (240) and (244b)] the experimentally measured change in electric permittivity due to a strong electric field in the form:

$$\Delta\varepsilon_r^{\parallel}(E_p) = \frac{(\varepsilon_r + 2)^2}{3V_m} \Delta P_m^{\parallel}(E_p) - (\varepsilon_r - 1)\left(\frac{\varepsilon_r + 2}{3}\right) \frac{\Delta V_m^{\parallel}}{V_m} \qquad (247)$$

We see that the variation $\Delta\varepsilon_r^{\parallel}$ is in general due to a change in molar polarization, $\Delta P_m^{\parallel}$, and a change in molar volume, $\Delta V_m^{\parallel}$, of the dielectric under the effect of the strong field. In an isotropic dielectric, ΔV_m is to a sufficiently good approximation a quadratic function of the applied field strength. We shall consider ΔV_m in more detail later. As to $\Delta P_m(E_p)$, its form depends not only on the value of E_p but largely on the microscopic structure of the dielectric and its thermodynamic state. In numerous liquid dielectrics, having molecular structure, ΔP_m depends quadratically on the field strength,[142, 196] whereas in macromolecular substances $\Delta P_m(E_p)$ is not a quadratic function but is of a more general non-linear nature.[197] It is customary to refer to variations in electric permittivity under the effect of a strong electric field as dielectric saturation,[196] notwithstanding the fact that in naturally isotropic dielectrics these variations are mostly quadratic in the field strength. In order to describe these quadratic variations which represent the initial departures from low-field conditions, they are best referred to as the non-linear effects: their upper limit, *i.e.* when a further rise in field strength no longer produces a change in ΔP_m, is then adequately differentiated from the initial effects by its designation as dielectric saturation, a term which only correctly applies at that limit.[197]

Methods for the Measurement of Non-linear Electric Field Effects. The measurement of changes in electric permittivity $\Delta\varepsilon$ as a function of electric field strength is particularly difficult in molecular substances, where the induced electric non-linearities are relatively insignificant. To be sure, these non-linearities increase as one applies stronger and stronger fields, but this cannot be done indefinitely as a medium will in general undergo electric breakdown or some other form of dielectric deterioration, seriously perturbing the process of measurement. Hence equipment for the measurement of electric field effects in molecular liquids has to have very high sensitivity and stability. The requirements with regard to sensitivity are defined by the variations in relative permittivity $\Delta\varepsilon/\varepsilon$ which, in molecular liquids, usually lie between 10^{-3} and 10^{-6} with the currently employed fields of the order of 10^7 V m^{-1}.

Experimental methods of dielectric study have been extensively discussed by Vaughan.[8] Measurements of dielectric permittivity ε and its variations $\Delta\varepsilon$ are performed by the bridge, resonance, and beat methods. The highest sensitivity is obtained with the beat method, in a steady-state version if the inducing field is constant in time, or in a pulsed version if the field is applied in short pulses of millisecond duration. With the equipment now currently in use for such measurements, the variation in dielectric permittivity of the substance $\Delta\varepsilon$ causes a change in frequency $\Delta\nu$ in the measuring generator. According to the technique adopted, measurement is either direct by compensating $\Delta\nu$ with a change in capacitance ΔC of the measuring condenser, or else the change in frequency $\Delta\nu$ is converted into a voltage or current signal and ΔC is related to the observed signal by appropriate scaling of the device; the change in permittivity is then determined simply from the relation $\Delta\varepsilon = \Delta C/C_0$, where C_0 is the capacitance of the empty condenser. Block diagrams of typical set-ups are shown in Figures 15—17; for the details of the construction of the various sub-circuits and liquid condensers we refer the reader to the original papers.

To Herweg[420a] are due the earliest correct measurements of $\Delta\varepsilon$ in ethyl ether as a function of applied electric field strength. He resorted to the beat method which, with various improvements, is still in use. The changes in permittivity in ethyl ether found by Herweg were quadratic in the field and negative, $\Delta\varepsilon < 0$, in agreement with the original dipole theory of Debye.[85] Herweg's measurements were repeated by Kautzsch[420b] using the same method: he extended them to chloroform and chlorobenzene, obtaining $\Delta\varepsilon < 0$ in agreement with Debye's theory as to sign and order of magnitude.

With the aim of measuring $\Delta\varepsilon$ in water, nitrobenzene, and alcohols, Malsch[421a] proposed a pulsed-discharge method, based on a Wien bridge adapted to the measurement of changes in capacitance. By the same method, Gunderman[421b] studied the behaviour of $\Delta\varepsilon$ in some alcohols. In all cases under investigation, both Malsch and Gunderman obtained a negative effect, $\Delta\varepsilon < 0$; albeit, as to order of magnitude, the values were far smaller than would result from the first theory of Debye.[85] This discrepancy between the theory and the experimental $\Delta\varepsilon$ values induced Debye[351] to work out, in 1935, a new treatment of electric field effects in liquid dielectrics, taking into account hindered rotation of the molecular dipoles in the

[420] (*a*) J. Herweg, *Z. Physik*, 1920, **3**, 36; J. Herweg and W. Potsch, *ibid.*, 1922, **8**, 1; (*b*) F. Kautzsch, *Physik. Z.*, 1928, **29**, 105.

[421] (*a*) J. Malsch, *Physik. Z.*, 1928, **29**, 770; 1929, **30**, 837; (*b*) H. Gundermann, *Ann. Phys.*, 1930, **6**, 545.

internal field.[73] Subsequent measurements by numerous authors [422–426] did not confirm the results of Malsch and Gunderman, which were probably influenced by thermal effects and by the electric conductivity of the liquids.

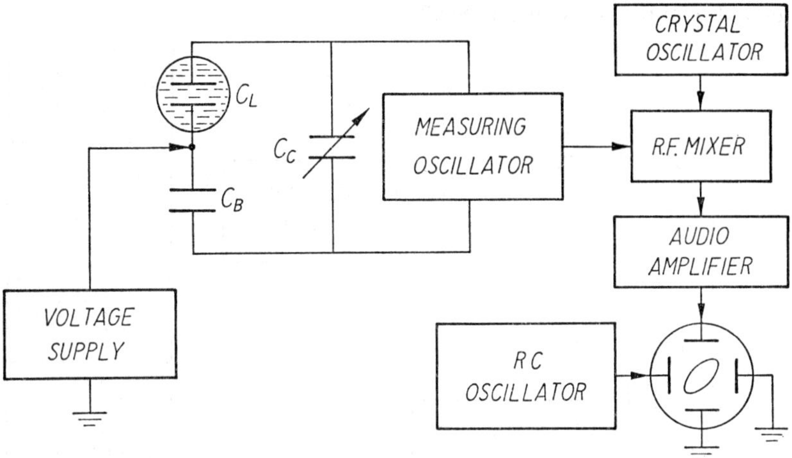

Figure 15 *Block diagram of beat device for dielectric saturation studies according to Piekara.*[423] *The liquid is in a condenser of capacitance* C_L. *Variations in* ε *due to the strong field applied from the high-voltage supply cause a change in frequency* Δf *in the measuring oscillator, whose signal is brought to beat with that of the stabilized quartz generator. The resulting acoustic difference frequency is conveyed to the cathode oscilloscope, the RC oscillator permitting the formation of Lissajous figures on the screen. The change in measuring oscillator frequency is apparent with high accuracy as a change in shape of the Lissajous curve which, in turn, is compensated for by retuning the calibrated condenser* C_c *by an amount* ΔC. *To Chełkowski* [425] *is due a bridge version of this method yielding a higher sensitivity* ($\Delta\varepsilon/\varepsilon \approx 10^{-8}$)

In 1936, Piekara and Piekara [422] studied dielectric saturation in nitrobenzene and its solutions in benzene and hexane. Their results did not confirm those of Malsch, as pure nitrobenzene was found to exhibit a positive effect, $\Delta\varepsilon > 0$. Successive dilution of the nitrobenzene with benzene caused a rapid fall of $\Delta\varepsilon$ to zero, followed by a negative effect

[422] A. Piekara and B. Piekara, *Compt. rend.*, 1936, **203**, 852, 1058.

[423] A. Piekara, *Acta Phys. Polon.*, 1959, **18**, 361; *Suppl. Nuovo Cim*, 1958, **9**, 192.

[424] (*a*) A. Piekara and A. Chełkowski, *J. Chem. Phys.*, 1956, **25**, 794; (*b*) A. Piekara, A. Chełkowski, and S. Kielich, *Z. phys. Chem.* (*Leipzig*), 1957, **206**, 375; A. Piekara, S. Kielich, and A. Chełkowski, *Arch. Sci.*, 1959, **12**, 59.

[425] (*a*) A. Chełkowski, *J. Chem. Phys.*, 1958, **28**, 1249; (*b*) A. Chełkowski, *Fizyka Dielektryków* (*Poznań*), 1962, **1**, 3; (*c*) A. Chełkowski, *Acta Phys. Polon.*, 1963, **24**, 165; L. Tomawski and J. Zioto, *ibid.*, 1971, **A40**, 91.

[426] (*a*) J. Małecki, *J. Chem. Phys.*, 1962, **36**, 2144; 1965, **43**, 1351; (*b*) J. Małecki, *Acta Phys. Polon.*, 1962, **21**, 13; 1963, **24**, 107; (*c*) J. Małecki, *Fizyka Dielektryków* (*Poznań*), 1966, **3**, 97; (*d*) J. Małecki, *Wiadomości Chemiczne* (*Poland*), 1970, **24**, 751; 1971, **25**, 23.

below 80%, decreasing to zero in pure benzene. These results were confirmed by Piekara,[423] Piekara and Chełkowski,[424] and Chełkowski,[425] using different versions of the beat method (Figure 15). Chełkowski [425] observed a similar behaviour of $\Delta\varepsilon$ as a function of concentration in solutions of *o*- and *m*-nitrotoluene in benzene.*

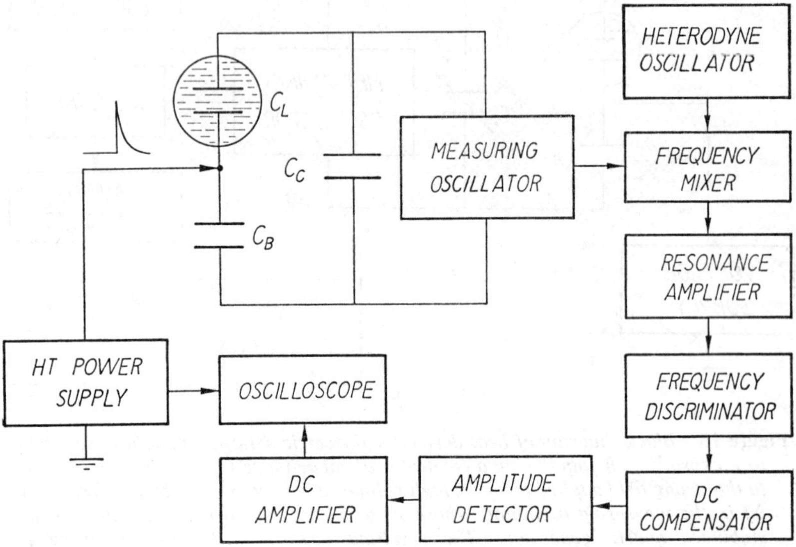

Figure 16 *Block diagram of pulsed beat device for the study of dielectric saturation in substances of some electric conductivity, according to Małecki [426] and Krupkowski,[427] consisting essentially of a superheterodyne receiver for signals with modulated frequency. The difference frequency signal on the output of the mixer is amplified by the resonance amplifier, whose last degree (a frequency discriminator) converts the modulation of the signal frequency to linear changes in voltage. On detection and amplification by the amplifier of direct current, the output voltage is supplied to the vertical deflection plates of the cathode-ray oscilloscope. Variations of this voltage are caused by variations in frequency of the measuring generator at the moment when the electric field pulse reaches the liquid condenser. The voltage signal observed on the oscilloscope screen is an exact mapping of the shape and duration of the high-voltage pulse which causes the variation ΔC in capacitance of the condenser containing the liquid. A different version of this method has been used by Szurkowski [429] and Przeniczny,[431] yielding a higher sensitivity of the measurements*

The imposition of an electric field on a liquid is always accompanied by a flow of electric current (due to conductivity) and dissipation of Joule heat. The ensuing rise in temperature of the liquid can lead to a change in permittivity much in excess of the saturation effect under investigation.

* It is to Piekara and his collaborators that we owe almost all the currently available significant data on non-linear electric field effects in condensed molecular systems, including at least three different forms of molecular response. See *Ann. Reports* (*A*), 1970, **67**, 87—96. (Note by Senior Reporter.)

In order to avoid this undesirable situation, the liquids have to be purified thoroughly when using the above methods; their specific electric conductivity σ should not be more than* 10^{-9} S m^{-1}. For the study of liquids of relatively high electric conductivity, Małecki[426] proposed a pulsed device (superheterodyne receiver with frequency modulation) where

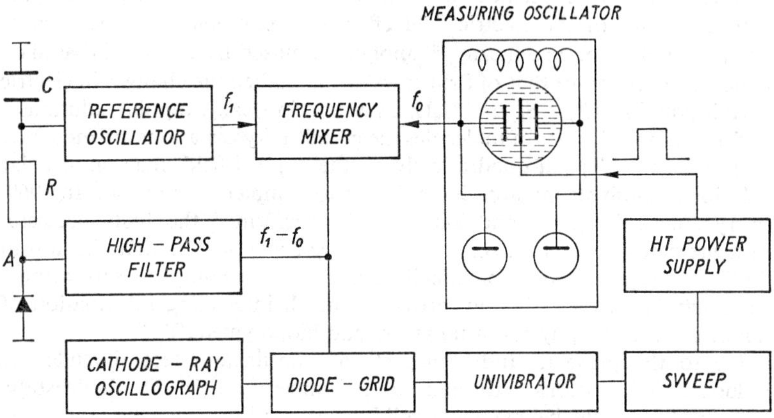

Figure 17 *Pulse-wise operating beat device of high sensitivity for* $\Delta\varepsilon$ *studies according to Aussenegg.*[430] *The high-voltage sweep-controlled generator supplies rectangular voltage pulses at intervals of about* 3 s *for the build-up of a strong electric field between the middle and outer electrodes. The frequency* f_0 *of the measuring oscillator and frequency* f_1 *of the reference oscillator* (f_1 *exceeds* f_0 *by about* 2 kHz) *are supplied to the mixer by way of a splitting amplifier* (*not shown*). *The difference frequency,* $f_1 - f_0$, *is then supplied to the high-pass filter, whose limit lies at* 2 kHz, *and is subsequently rectified. Between A and the shield, a DC voltage arises when* $\nu_1 - \nu_0 > 2$ kHz *but is zero if* $\nu_1 - \nu_0 < 2$ kHz. *The resistance R and capacitance C cause the DC voltage across C to vary but insignificantly with short-duration changes in voltage between A and the shield. The voltage across C serves for retuning the reference oscillator. The time constant of the RC element is set at* 1 s *so that the studied change in frequency, lasting about* 5 ms, *cannot trigger a retuning. The measuring signal* $f_1 - f_0$ *is supplied to the cathode-ray oscillograph by way of the univibrator and diode grid*

the duration of the field pulse was of the order of 1 ms (Figure 16). Compared to the static methods, this procedure led to a reduction in Joule heat by three orders of magnitude, thus permitting the measurement of $\Delta\varepsilon$ in a wider class of liquids, of specific conductivity $\sigma \leqslant 10^{-6}$ S m^{-1}. Małecki[126] studied alcohols and their solutions, obtaining $\Delta\varepsilon$ values one order of magnitude smaller than those obtained by Malsch and Gunderman.[421] In pure alcohols the effect was negative, $\Delta\varepsilon < 0$; it decreased to zero with dilution, changing its sign to positive as dilution proceeded, to fall to zero in the pure solvent (benzene or hexane). Newer measurements of the

* 10^{-9} Siemens metre^{-1} = 10^{-9} ohm^{-1} metre^{-1}.

non-linear behaviour of the higher alcohols and their solutions are due to Krupkowski,[427] Danielewicz-Ferchmin,[428] and Szurkowski.[429]

With the exception of Kautzsch,[420b] who first found a weak positive effect ($\Delta\varepsilon > 0$) in CS_2, none of the earlier papers reported changes in permittivity in pure non-dipolar solvents. Recently, however, Aussenegg [430] described a pulsed method of high sensitivity (Figure 17), permitting determinations of the electric field effect in non-dipolar liquids, such as CCl_4 and CS_2 also. Przeniczny,[431] applying a pulsed-beat method (based on a modified version of that of Szurkowski [429]), studied the changes in electric permittivity in CCl_4, CS_2, and C_6H_6 as well as in weakly dipolar solutions.*

Whereas in solutions of molecules the permittivity variations are moderate and in general depend quadratically on the applied field, macromolecular solutions exhibit strong non-linearities up to complete saturation. In 1969, Parry Jones, Gregson, and Davies [432, 433] performed the first successful measurements of $\Delta\varepsilon$ in a solution of poly-γ-benzyl-L-glutamate in dioxan finding a negative effect practically amounting to complete saturation.† These studies are continuing, providing much interesting information of importance to the physics of polymers and biopolymers.[434–436]

Up to the present, studied aspects of non-linear effects include the influence of various solvents and temperatures,[437–439] hydrostatic pressure,[440] and the behaviour of solid dielectrics.[441]

[427] T. Krupkowski, *Fizyka Dielektryków i Radiospektroskopia*, 1966, **3**, 125; 1972, **5/2**, 243; *Bull. Acad. Polon. Sci., Sér. Sci. chim.* (*Poland*), 1971, **19**, 461.

[428] (*a*) I. Danielewicz-Ferchmin, *Bull. Acad. Sci. Sér. Sci. math. astr et phys.* (*Poland*), 1966, **14**, 51; 1968, **16**, 965; (*b*) I. Danielewicz-Ferchmin and J. Wejner, *Acta Phys. Polon.*, 1970, **A38**, 923; (*c*) I. Danielewicz-Ferchmin and D. Jęczkowska, *Fizyka Dielektryków i Radiospektroskopia* (*Poznań*), 1972, **5/2**, 227.

[429] (*a*) B. Szurkowski, Doctor's Thesis, University of Poznań, 1970; (*b*) B. Szurkowski and I. Danielewicz-Ferchmin, *Acta Phys. Polon.*, 1971, **A39**, 227.

[430] (*a*) F. Aussenegg, *Acta Phys. Austriaca*, 1967, **26**, 43; (*b*) F. Aussenegg and T. Horwath, *Z. Naturforsch.*, 1967, **22a**, 142.

[431] (*a*) Z. Przeniczny, *Acta. Phys. Polon.* (*A*), 1970, **37**, 479; (*b*) Z. Przeniczny, *Bull. Soc. Amis Sci. Lettres de Poznań*, 1971–72, **22B**, 67.

[432] G. Parry Jones, M. Gregson, and M. Davies, *Chem. Phys. Letters*, 1969, **4**, 33.

[433] M. Gregson, G. Parry Jones, and M. Davies, *ibid.*, 1970, **6**, 215.

[434] M. Gregson, G. Parry Jones, and M. Davies, *Trans. Faraday Soc.*, 1971, **67**, 1630.

[435] B. Kasprowicz, S. Kielich, and Z. Przeniczny, *Bull. Soc. Amis Sci. Lettres de Poznań*, 1970/71, **22B**, 47.

[436] M. Davies, *Acta Phys. Pol.* (*A*), 1971, **40**, 561.

[437] (*a*) D. Surma, T. Janikowska, and A. Chełkowski, *Acta Phys. Polon.*, 1968, **34**, 181; (*b*) M. Dutkiewicz and A. Chełkowski, *ibid.*, 1969, **35**, 651.

[438] (*a*) J. Małecki and J. Nowakowa, *Bull. PTPN Pracc Komisji Mat. Przyr.*, 1968, **4**, 35; (*b*) J. Małecki, *ibid.*, p. 47.

[439] J. M. Thiébaut, A. Weisbecker, and Ch. Ginet, *Compt. rend.*, 1968, **267**, *B*, 661.

[440] (*a*) T. Hilczer, *Fizyka Dielektryków* (*Poznań*) 1962, **1**, 39; 1964, **2**, 137; (*b*) T. Hilczer and E. Malyszko, *Fizyka Dielektryków i Radiospektroskopia*, 1968, **4**, 51; (*c*) T. Hilczer and J. Tluchowski, *ibid.*, 1972, **6/1**, 115.

[441] W. Nawrocik, *Fizyka Dielektryków i Radiospektroskopia* (*Poznań*), 1968, **4**, 69.

* Przeniczny's values for $\Delta\varepsilon$ in non-dipolar media are not confirmed by other recent studies (M. Gregson; P. Deft; J. Małecki; personal communications to the Senior Reporter). This matter is being pursued further.

† See also H. Block and E. F. Hayes, *Chem. Comm.*, 1969, 76; *Trans. Faraday Soc.*, 1970, **66**, 2512.

This concise review of the experimental methods and results will now be supplemented with a closer discussion of the principal theories of non-linear electric field responses in rarefied and dense states.

Quadratic Permittivity Variations in Gases and Liquids.—*Gases and Dilute Dipolar Media.* We shall calculate $\Delta\varepsilon$ within the framework of the classical Langevin–Debye theory. The total electric moment of a dipolar molecule having the permanent electric moment $\boldsymbol{p}$ and linear electric polarizability $\boldsymbol{\alpha}$ immersed in a field $\boldsymbol{E}_0$ is:

$$\boldsymbol{p}(\boldsymbol{E}_0) = \boldsymbol{p} + \boldsymbol{\alpha} \cdot \boldsymbol{E}_0 \tag{248}$$

whereas its potential energy amounts to:

$$u(\boldsymbol{E}_0) = u(0) - \boldsymbol{p} \cdot \boldsymbol{E}_0 - \tfrac{1}{2}\boldsymbol{E}_0 \cdot \boldsymbol{\alpha} \cdot \boldsymbol{E}_0 \tag{249}$$

By the field $\boldsymbol{E}_0$ we mean, in general, the sum of two fields, namely the weak analysing field $\boldsymbol{E}_{0a}$ and the strong polarizing (non-linearity-inducing) field $\boldsymbol{E}_{0p}$:

$$\boldsymbol{E}_0 = \boldsymbol{E}_{0a} + \boldsymbol{E}_{0p} \tag{250}$$

These fields can obviously be static or slowly varying in time, but usually of a frequency lying below the Debye dipole dispersion.

In a medium so dilute as to make molecular correlations negligibly small, the electric polarization (137) of the macroscopic medium of number density $\rho = N/V$ can be expressed in the form:

$$\boldsymbol{P}(\boldsymbol{E}_0) = \rho\langle \boldsymbol{p}(\boldsymbol{E}_0)\rangle_{E_0} \tag{251}$$

where the mean statistical value of the dipole moment of a molecule in the presence of the electric field (250) is, at thermodynamic equilibrium:

$$\langle \boldsymbol{p}(\boldsymbol{E}_0)\rangle_{E_0} = \int \boldsymbol{p}(\boldsymbol{E}_0) f(\Omega, \boldsymbol{E}_0)\, \mathrm{d}\Omega \tag{251a}$$

At moderate non-linearities of the medium, the statistical distribution function can be expanded in a series [*cf.* formulae (122) and (169a)]:

$$f(\Omega, \boldsymbol{E}_0) = f(\Omega, 0)\Big\{1 + \frac{1}{kT}(\boldsymbol{p} \cdot \boldsymbol{E}_0) + \frac{1}{2kT}\Big[\boldsymbol{d}^e + \frac{1}{3kT}(3\boldsymbol{pp} - p^2\,\boldsymbol{U})\Big] : \boldsymbol{E}_0\,\boldsymbol{E}_0 + \frac{1}{6k^3T^3}(\boldsymbol{p} \cdot \boldsymbol{E}_0)(\boldsymbol{pp} - p^2\,\boldsymbol{U}) : \boldsymbol{E}_0\,\boldsymbol{E}_0 + \ldots\Big\} \tag{252}$$

where

$$\boldsymbol{d}^e = \boldsymbol{\alpha} - \alpha\boldsymbol{U} \tag{252a}$$

is the deviation tensor of linear electric polarizability of the molecule.

By inserting the expansions (248) and (252) into the definition (251a) and unweighted averaging of the unit vectors [the procedure is discussed in Section 3, formulae (104) and (107)], one has:

$$\langle p(E_0)\rangle_{E_0} = \left\{\left(\alpha + \frac{p^2}{3kT}\right) + (\Theta_1 + 2\Theta_2 - \Theta_3)\,E_0^2\right\} E_0 \quad (251b)$$

where, after Debye,[442] we have introduced the following parameters of the molecular electric field effect:

$$\Theta_1 = \frac{1}{15kT}(d^e : d^e) \quad (253)$$

$$\Theta_2 = \frac{1}{15k^2\,T^2}(p \cdot d^e \cdot p) \quad (254)$$

$$\Theta_3 = \frac{(p \cdot p)^2}{45k^3\,T^3} \quad (255)$$

For a molecule having the dipole moment components p_1, p_2, p_3 and electric polarizability components $\alpha_1, \alpha_2, \alpha_3$ in the three directions of its principal axes, the parameters (253)—(255) can be written in the form proposed by Debye:[442]

$$\Theta_1 = \frac{1}{45kT}\{(\alpha_1 - \alpha_2)^2 + (\alpha_2 - \alpha_3)^2 + (\alpha_3 - \alpha_1)^2\} \quad (253a)$$

$$\Theta_2 = \frac{1}{45k^2\,T^2}\{(\alpha_1 - \alpha_2)(p_1^2 - p_2^2) + (\alpha_2 - \alpha_3)(p_2^2 - p_3^2) + (\alpha_3 - \alpha_1)(p_3^2 - p_1^2)\} \quad (254a)$$

$$\Theta_3 = \frac{1}{45k^3\,T^3}(p_1^2 + p_2^2 + p_3^2)^2 \quad (255a)$$

If in particular the molecule is symmetric with respect to the 3-axis along which the dipole moment p lies, then $p_1 = p_2 = 0$, $p_3 \neq 0$, $\alpha_1 = \alpha_2 \neq \alpha_3$, and the parameters reduce to:

$$\Theta_1 = \frac{2}{45kT}(\alpha_3 - \alpha_1)^2 \quad (253b)$$

$$\Theta_2 = \frac{2}{45k^2\,T^2}(\alpha_3 - \alpha_1)\,p_3^2 \quad (254b)$$

$$\Theta_3 = \frac{p_3^4}{45k^3\,T^3} \quad (253b)$$

As we see, the parameter Θ_1 results from Langevin reorientation of the polarizability ellipsoid and is always positive. The second of the above parameters, Θ_2, corresponds to Born's term in the Kerr effect and can be positive or negative, depending on the electric structure of the molecule. The third, the Debye parameter Θ_3, has no counterpart in other phenomena of molecular orientation, and is specific to the non-linear dielectric behaviour of dipolar substances.

[442] P. Debye, *Marx Hdb. d. Radiologie*, 1925, 6, 633; P. Debye and H. Sack, *ibid.*, 1934, 6/2, 69.

By inserting the mean dipole moment (252a) into the polarization formula (251) and using the definitions (246) and (250), one obtains the change in molar polarization of a gas due to the square of the electric field strength:

$$\Delta P_m(E_{0p}) = \varepsilon_0^{-1} N_A(\Theta_1 + 2\Theta_2 - \Theta_3) E_{0p}^2 \qquad (256)$$

On restricting considerations to the change in electric permittivity due to changes in molar polarization only, insertion of (256) into (247) finally yields Debye's [442] formula in the form:

$$\Delta\varepsilon_r(E_p) = 3 \frac{S_m}{V_m} \left(\frac{\varepsilon_r + 2}{3}\right)^4 E_p^2 \qquad (257)$$

where we have introduced the molar constant of the non-linear effect:

$$S_m = \varepsilon_0^{-1} N_A(\Theta_1 + 2\Theta_2 - \Theta_3) \qquad (258)$$

The preceding formulae (257) and (258) state that the quadratic variations $\Delta\varepsilon$ are always positive in non-dipolar substances ($\Theta_2 = \Theta_3 = 0$), whereas in dipolar ones the sign of $\Delta\varepsilon$ will depend on the electric structure of the molecules, namely on the numerical value of the dipole moment and its polarizability, *i.e.* on the mutual numerical relationship between the constants:

$$\left.\begin{aligned} S_m^{(1)} &= \varepsilon_0^{-1} N_A \Theta_1 \\ S_m^{(2)} &= 2\varepsilon_0^{-1} N_A \Theta_2 \\ S_m^{(3)} &= -\varepsilon_0^{-1} N_A \Theta_3 \end{aligned}\right\} \qquad (259)$$

Obviously, in strongly dipolar substances, the purely dipolar constant $S_m^{(3)}$ will be the predominant one and consequently this treatment of gas theory results in a negative saturation effect. Numerical values of molar non-linear electric field constants (259) are listed in Table 15 for some of the more important non-dipolar as well as dipolar molecules.

The calculations of Table 15 refer to the gas state, for which as yet there is a lack of $\Delta\varepsilon$ measurements. Saturation measurements in gases are extremely difficult owing to the smallness of the non-linearity, since densities are low, and dielectric breakdown in gases imposes a low limit on the externally applied electric field strength. The accuracy of available measuring techniques has restricted the measurement of such small values of $\Delta\varepsilon$ as 10^{-10} in gases or 10^{-8} in vapours for strongly dipolar substances.

The Onsager Model in the Non-linear Electric Field Effect. The non-linear electric field theory of Debye disagrees with most of the measurements of $\Delta\varepsilon$ in liquids and their solutions. To Van Vleck [443] is due the attempt to reduce the discrepancies by applying Onsager's local electric field model in the treatment of the field effect. Aiming at simplicity, he considered only the influence of reorientation of rigid (not polarizable) electric dipoles p,

[443] J. H. Van Vleck, *J. Chem. Phys.*, 1937, **5**, 556.

Table 15 *$\Delta\varepsilon$ values, calculated by Debye's 'gas' theory and measured in some of the more important liquids. (Reference numbers in square brackets)*

Liquids	C_6H_6	CS_2	$CHCl_3$	C_5H_5N	C_6H_5Cl	$C_6H_5NO_2$
$p \times 10^{30}$/C m	0	0	3.83	7.32	5.17	14.06
α_1 ⎫	7.05	6.15	10.00	6.42	8.25	8.22
α_2 ⎬ $\times 10^{40}$/C^2 m^2 J^{-1}	13.66	6.15	10.00	13.19	15.38	15.75
α_3 ⎭	13.66	16.81	7.41	12.03	17.26	19.05
$S_m^{(1)}$ ⎫	3.26	8.49	0.50	2.94	5.06	2.93
$S_m^{(2)}$ ⎪	0	0	− 14.08	11.52	53.95	517.55
$S_m^{(3)}$ ⎬ $\times 10^{25}$/C^2 m^5 J^{-2} mol^{-1}	0	0	− 49.78	− 657.35	− 163.57	− 8947.33
S_m^{total} ⎭	3.26	8.49	− 63.36	− 642.89	− 104.56	− 8426.85
$V_m \times 10^{-6}$/m^3 mol^{-1}	89.4	60.2	80.2	80.45	101.68	102.3
ε_r (20 °C)	2.28	2.64	4.81	13.23	5.7	34.82
$\Delta\varepsilon^{theor}/E_p^2 \times 10^{18}$/m^2 V^{-2}	+ 0.04	+ 0.24	− 6.28	− 1592.4	− 133.89	− 560733.2
$\Delta\varepsilon^{exp}/E_p^2 \times 10^{18}$/m^2 V^{-2}	+ 0.3 [430]	+ 0.1 [420*b*]	− 4.0 [420*b*]		− 1.1 [420*b*]	− 577 [421]
		+ 0.37 [430]	− 3.5 [426]		− 73.9 [421*b*]	+ 361 [424]

obtaining the following formula for the change in electric permittivity:*

$$\Delta\varepsilon(E_p) = -\left(\frac{3\varepsilon_r^2}{2\varepsilon_r^2+1}\right)\left(\frac{3\varepsilon_r}{2\varepsilon_r+1}\right)^2 \frac{\rho p^4}{15\varepsilon_0 k^3 T^3} E_p^2 \qquad (260)$$

On taking into consideration the electric dipoles as isotropically polarizable, we get in place of Van Vleck's formula (260) the relation:[388]

$$\Delta\varepsilon(E_p) = -\left(\frac{3\varepsilon_r^2}{2\varepsilon_r^2+\varepsilon_{r\infty}^2}\right)\left(\frac{3\varepsilon_r}{2\varepsilon_r+\varepsilon_{r\infty}}\right)^2\left(\frac{\varepsilon_{r\infty}+2}{3}\right)^4 \frac{\rho p^4}{15\varepsilon_0 k^3 T^3} E_p^2 \qquad (261)$$

where $\varepsilon_{r\infty}$ is the high-frequency electric permittivity of the substance, determined from the well-known formula of Mossotti [444] and Clausius:[445]

$$\frac{\varepsilon_{r\infty}-1}{\varepsilon_{r\infty}+2} = \frac{\rho\alpha^\infty}{3\varepsilon_0} = \frac{4\pi}{3}\rho\alpha_V^\infty \qquad (262)$$

Obviously, if one neglects the effect of polarizability, $\varepsilon_{r\infty} \to 1$, equations (260) and (261) become identical.

Assembling the preceding results, one can write the quadratic change in electric permittivity as:

$$\Delta\varepsilon_r(E_p) = -\frac{3\rho}{\varepsilon_0} A(\varepsilon)\, \Theta_3\, E_p^2 \qquad (263)$$

where $A(\varepsilon)$ is a quantity resulting from the local field model applied:

$$A(\varepsilon) = \begin{cases} \left(\dfrac{\varepsilon_r+2}{3}\right)^4 & \text{Debye}^{442} \quad (263a) \\ \left(\dfrac{3\varepsilon_r^2}{2\varepsilon_r^2+1}\right)\left(\dfrac{3\varepsilon_r}{2\varepsilon_r+1}\right)^2 & \text{Van Vleck}^{443} \quad (263b) \\ \left(\dfrac{3\varepsilon_r^2}{2\varepsilon_r^2+\varepsilon_{r\infty}^2}\right)\left(\dfrac{3\varepsilon_r}{2\varepsilon_r+\varepsilon_{r\infty}}\right)^2\left(\dfrac{\varepsilon_{r\infty}+2}{3}\right)^4 & \text{Kielich}^{388} \quad (263c) \end{cases}$$

Recently, Eyring *et al.*,[76, 446] resorting to the domain theory of liquids and Onsager's model, proposed a theory of the electric permittivity of hydrogen-bonded liquids. Danielewicz-Ferchmin,[447] starting from this model of Eyring *et al.*, succeeded in modifying formula (261) for $\Delta\varepsilon$ in a manner which achieves much better agreement with available measurements on alcohols.[426–429]

[444] O. F. Mosotti, *Mem. Soc. Ital.*, 1850, **14**, 49.
[445] R. Clausius, 'Die Mechanische Wärmetheorie', Vieweg, Braunschweig, 1879, Vol. 2.
[446] M. E. Hobbs, M. S. Jhon, and H. Eyring, *Proc. Nat. Acad. Sci. U.S.A.*, 1966, **56**, 31; M. S. Jhon, J. Grosh, T. Ree, and H. Eyring, *J. Chem. Phys.*, 1966, **44**, 1465; M. S. Jhon, E. R. Van Arsdalen, J. Grosh, and H. Eyring, *ibid.*, 1967, **47**, 2231; M. S. Jhon and H. Eyring, *J. Amer. Chem. Soc.*, 1968, **90**, 3071.
[447] I. Danielewicz-Ferchmin, *Acta Phys. Polon.*, 1971, **A40**, 263; 1972, **A41**, 627.

* Van Vleck [443] and Böttcher [2] calculated a variation $\Delta\varepsilon'$, defined as the difference between the electric permittivity in the presence of a strong field $\varepsilon'(E) = D(E)/E$ and in its absence, $\varepsilon(0)$. In a quadratic approximation, we have the relation $\Delta\varepsilon = 3\Delta\varepsilon'$ between $\Delta\varepsilon(E_p)$, measured by means of a weak field E_a in the presence of the strong field E_p, and the quantity $\Delta\varepsilon' = D(E)/E - \varepsilon(0)$, of theoretical significance.

The Models of Debye and Piekara. In 1935 Debye,[351] under the impact of Malsch's measurements,[421] strove to improve his first non-linear theory by including coupling between molecular dipoles. According to him, as well as Fowler,[350] the reorientation of a given dipolar molecule by the external electric field is hindered by a field of forces F'_m due to the direct environment of its neighbours, which constitute a quasi-crystalline ordering.[90] In the absence of an external field, this field F'_m stems from short-range forces of the dipole kind, given by formula (48b). Let us denote by d the mean distance separating the dipoles; then the order of magnitude of the internal field F'_m can be evaluated from the formula:

$$F'_m \approx \frac{p}{4\pi\varepsilon_0 d^3} = \frac{p}{71.76\varepsilon_0} \tag{264}$$

d^3 having been eliminated by resorting to the formula:[90]

$$\frac{\pi}{6} d^3 \rho = 0.74 \tag{265}$$

In the case of nitrobenzene, $\rho = 5.9 \times 10^{27}$ m^{-3}, $p = 14 \times 10^{-30}$ C m, and with regard to (264) the internal field is of the order of 6×10^8 V m^{-1}, which is high compared to the external applied fields, of order 10^7 V m^{-1}; in consequence, reorientation of the dipoles is indeed subject to a hindering. Considerations of this kind led Debye to the following expression for the change in electric permittivity of a strongly dipolar liquid:[351]

$$\Delta\varepsilon_r^L(E_p) = \Delta\varepsilon_r(E_p)\, R_s(Y) \tag{266}$$

here, $\Delta\varepsilon_r(E_p)$ is the change in permittivity calculated without dipole coupling according to (263) and (263a), whereas $R_s(Y)$ is a reduction factor, always positive but less than unity. Its numerical value depends on the parameter of coupling between the dipole under consideration and the internal field (264):

$$Y = \frac{pF'_m}{kT} \approx \frac{p^2}{17.76\varepsilon_0 kT} \tag{267}$$

Debye's theory, however, was unable to account for the circumstance that in nitrobenzene, according to Piekara and Piekara,[422] the variation $\Delta\varepsilon$ was positive. It was Piekara himself[352] who supplied a theoretical explanation of the positive effect of dielectric saturation by taking into account coupling between a dipole and its nearest dipolar neighbour as well as that between it and its quasi-crystalline environment as a whole.

In Piekara's model, the dipolar molecule under consideration and its nearest neighbour momentarily form an aggregate, with the partners in nearly parallel or nearly antiparallel mutual orientation, according to the structure of the molecule. The coupling energy of such a pair of dipoles is written in the following approximate form:

$$V_{12} = V_0 - V_1 \cos\theta_{12} \tag{268}$$

with θ_{12} the angle between the two dipoles.

This model, too, leads to an expression of the form of equation (266), albeit with a different correlation factor $R_S(Y)$ which, for pairs nearly in parallel, is always positive and larger than unity, is less than unity for nearly antiparallel pairs, and undergoes a change in sign to become negative at $Y = 1.4$ (Figure 18). Thus Piekara [352] was able to explain the change in sign of $\Delta\varepsilon$ from positive to negative observed by him in solutions of nitrobenzene in benzene.[422]

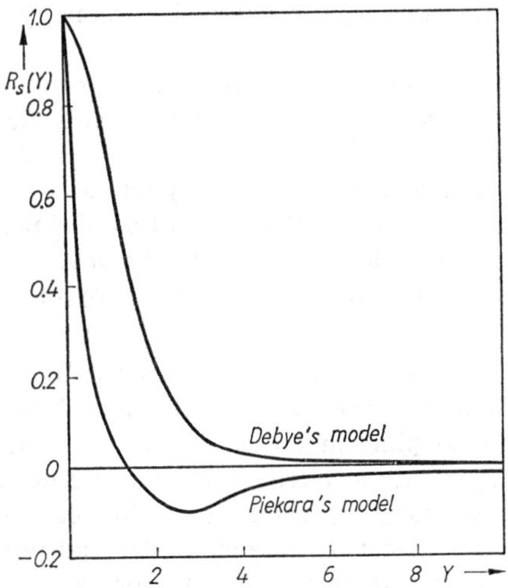

Figure 18 *Saturation correlation factor versus dipole interaction, for the models of Debye and Piekara. With Debye's model, R_S is always positive, decreasing from 1 to 0 as dipole coupling increases. With Piekara's model, in the case of nearly antiparallel pairs, R_S falls rapidly from 1 to 0, is negative for dipole coupling $Y > 1.33$, and then vanishes for association into a rigid pair of antiparallel dipoles*

Peterlin and Stuart [397] extended Debye's theory [351] to the case of polarizable dipoles, as well as to that of polarizable non-dipolar molecules. For axially symmetric molecules, the expression for the change in permittivity now takes the form:

$$\Delta\varepsilon(E_P) = \frac{3\rho}{\varepsilon_0}\left(\frac{\varepsilon_r + 2}{3}\right)^4 \{\Theta_1 R_{CM}(Y) + 2\Theta_2 R_K(Y) - \Theta_3 R_S(Y)\} E_P^2 \tag{269}$$

with Θ_1, Θ_2, and Θ_3 defined respectively by (253b), (254b), and (255b).

The correlation factor $R_{CM}(Y)$ is specific to and determinable from the Cotton–Mouton effect [448–450] as well as from anisotropic light-scattering

[448] A. Piekara, *Compt. rend.*, 1939, **208**, 990.

[449] (*a*) S. Kielich and M. Surma, *Fizyka Dielektryków (Poznań)*, 1962, **1**, 15; (*b*) M. Surma, *Acta Phys. Polon.*, 1964, **25**, 485.

[450] R. J. W. Le Fèvre, D. S. N. Murthy, and P. J. Stiles, *Austral. J. Chem.*, 1969, **22**, 1421.

studies.[401] The second factor, $R_K(Y)$, is characteristic for the Kerr effect in dipolar substances.[352] In the case of non-dipolar molecules, the only term to appear in equation (269) is $\Theta_1 R_{CM}(Y)$, where the correlation factor $R_{CM}(Y)$ was calculated by Müller [451] and Snellman [452] on the assumption that the energy of interaction of two non-dipolar molecules can be expressed approximately as:

$$V_{12} = V_0 - V_2 \cos^2 \theta_{12} \tag{270}$$

with θ_{12} now denoting the angle between the symmetry axes of the two molecules.

With regard to dipolar liquids and their solutions in non-dipolar solvents, Piekara [352] developed a complete theory of dipolar couplings calculating all three correlation factors R_{CM}, R_K, and R_S. These studies revealed that nitrobenzene molecules coalesce momentarily into antiparallel aggregates, which then tend to couple mutually into almost parallel pairs. Peterlin and Stuart [397] performed a detailed study of the influence on $\Delta\varepsilon$ not only of Debye dipole couplings but moreover of anisotropy of the local Lorentz field.

Angular Correlations between Molecules. Complete and critical analyses of the theories of molecular orientation effects in an electric field based on Debye and Müller's coupling model have been made by Frenkel [73] and Anselm.[374] With regard to those criticisms, and in order to liberate non-linear theory from too strongly specified assumptions concerning the model, Piekara and Kielich [141, 142] formulated a molecular-statistical theory leading to angular correlation factors of the following form:[333]

$$R_{CM} = \frac{1}{2N_A}\left\langle \sum_{i=1}^{N_A}\sum_{j=1}^{N_A} (3\cos^2\theta_{ij} - 1)\right\rangle \tag{271}$$

$$R_K = \frac{1}{2N_A}\left\langle \sum_{i=1}^{N_A}\sum_{j=1}^{N_A}\sum_{k=1}^{N_A} (3\cos\theta_{ij}\cos\theta_{ik} - \cos\theta_{jk})\right\rangle \tag{272}$$

$$R_S = \frac{1}{2N_A}\left\{5\left\langle \sum_{i=1}^{N_A}\sum_{j=1}^{N_A}\cos\theta_{ij}\right\rangle\left\langle \sum_{k=1}^{N_A}\sum_{l=1}^{N_A}\cos\theta_{kl}\right\rangle - 3\left\langle \sum_{i=1}^{N_A}\sum_{j=1}^{N_A}\sum_{k=1}^{N_A}\sum_{l=1}^{N_A}\cos\theta_{ij}\cos\theta_{kl}\right\rangle\right\} \tag{273}$$

θ_{ij} denoting the angle between the symmetry axes of molecules i and j.

The factors (271) and (272) reduce to the form:

$$R_{CM} = 1 + J_A^{(2)} \tag{271a}$$

$$R_K = 1 + 2J_A^{(1)} + J_A^{(2)} + J_A^{(3)} \tag{272a}$$

the angular correlation parameters $J_A^{(1)}$, $J_A^{(2)}$, and $J_A^{(3)}$ being given by the expressions (179)—(181).

[451] H. Müller, *Phys. Rev.*, 1936, **50**, 547.
[452] O. Snellman, *Phil. Mag.*, 1949, **40**, 938.

The angular correlation factor for dielectric saturation (273) can be expressed similarly:

$$R_S = (1 + z)(1 + 2z\langle\cos\theta_{ij}\rangle + \tfrac{5}{2}z^2\langle\cos\theta_{ij}\rangle^2) - 3z\{\langle\cos^2\theta_{ij}\rangle + 2(z-1)\langle\cos\theta_{ij}\cos\theta_{ik}\rangle + \tfrac{1}{2}(z-1)(z-2)\langle\cos\theta_{ij}\cos\theta_{kl}\rangle\} \quad (273a)$$

with $z = (N - 1)$ as the number of nearest neighbours around a molecule.

Restricting ourselves, for simplicity, to the binary correlations, which contribute predominantly to the correlation factors (271)—(273), we obtain:

$$R_{CM} = 1 + \tfrac{1}{2}z\langle 3\cos^2\theta_{ij} - 1\rangle \quad (271b)$$

$$R_K = R_{CM} + 2z\langle\cos\theta_{ij}\rangle \quad (272b)$$

$$R_S = (1 + z)(1 + 2z\langle\cos\theta_{ij}\rangle + \tfrac{5}{2}z^2\langle\cos\theta_{ij}\rangle^2) - 3z\langle\cos^2\theta_{ij}\rangle \quad (273b)$$

Above, one still has to compute the statistical mean values:

$$\langle\cos^n\theta_{ij}\rangle = \frac{\iint \cos^n\theta_{ij}\exp(-U_{ij}/kT)\,d\tau_i\,d\tau_j}{\iint \exp(-U_{ij}/kT)\,d\tau_i\,d\tau_j} \quad (274)$$

Now, as explained in Section 2, the potential energy U_{ij} of interaction between dipolar molecules is a mathematically involved expression and the averages (274) are not accessible to numerical computation generally unless certain simplifying assumptions are made. For some special models, computations of (274) can be carried through completely, in particular in the three cases when the molecules are seen as tending to interact mutually (*i*) in nearly parallel configuration (paratropism), (*ii*) in nearly perpendicular configuration (diatropism, *e.g.* C_6H_6), and (*iii*) in nearly antiparallel configuration (*e.g.* $C_6H_5NO_2$). In these special cases, the potential energy of interaction can be put in the form:[269]

$$U_{ij} = \sum_{n=0}^{\infty} V_n\cos^n\theta_{ij} = V_0 + V_1\cos\theta_{ij} + V_2\cos^2\theta_{ij} + \ldots \quad (275)$$

where θ_{ij} is the angle between the electric dipoles (in the case of dipolar molecules) or between the symmetry axes (in that of non-dipolar ones) of the partners. The expansion coefficients A_n depend on the electric parameters of the molecules, their polarizability and multipole moments, and the distance r_{ij} between them: the sign at the coefficients A_n varies according to the kind of interaction.

Strong Correlations of Dipolar Molecules. For strongly dipolar molecules, it suffices to take the electrostatic energy of dipolar interaction [*cf.* equation (60a)]:

$$U_{ij} = -\frac{p_i p_j}{4\pi\varepsilon_0}(3\cos\theta_i\cos\theta_j - \cos\theta_{ij})\,r_{ij}^{-3}$$

where:

$$\cos\theta_{ij} = \cos\theta_i\cos\theta_j + \sin\theta_i\sin\theta_j\cos\phi_{ij}$$

If in particular the dipoles tend towards parallel array, one may assume $\theta_i = 0$, $\theta_j = \theta_{ij}$ (the angle between the dipoles), whence:

$$U_{ij} = -V_1\cos\theta_{ij}$$

with $V_1 = 2p_ip_j/4\pi\varepsilon_0 r_{ij}^3$. On the other hand, in the case of nearly anti-parallel coupling, we write $\theta_i = \theta_j = 90°$ and $\phi_{ij} = 180° - \theta_{ij}$, whence:

$$U_{ij} = -V_1'\cos\theta_{ij}$$

Thus, between the two cases, the potential energy depends identically on θ_{ij} and has the same sign, but the constants differ by a factor of two, since $2V_1' = V_1$. Here, consequently, (274) becomes:[388]

$$\langle\cos^n\theta_{ij}\rangle = (\pm 1)^n L_n(Y_1) \qquad (274a)$$

where the sign '+' is for parallel and '−' for antiparallel configuration, and the L_n's are Langevin functions of the n-th order,[352] defined as:

$$L_n(Y_1) = \frac{\int_0^\pi \cos^n\theta_{ij}\exp(Y_1\cos\theta_{ij})\sin\theta_{ij}\,d\theta_{ij}}{\int_0^\pi \exp(Y_1\cos\theta_{ij})\sin\theta_{ij}\,d\theta_{ij}} \qquad (276)$$

whereas $Y_1 = V_1/kT$ denotes the coupling energy of the molecular dipoles in kT units.

On integrating (276) *per partes* we obtain:[388]

$$L_n(Y_1) = \sum_{k=1}^{n}\frac{n!}{(n-k)!}\left(-\frac{1}{Y_1}\right)^k\frac{e^{Y_1} - (-1)^{n-k}e^{-Y_1}}{e^{Y_1} - e^{-Y_1}} \qquad (276a)$$

i.e. especially,

$$\left.\begin{aligned} L_1(Y_1) &= \coth Y_1 - \frac{1}{Y_1} = L(Y_1) \\ L_2(Y_1) &= 1 - 2\frac{L(Y_1)}{Y_1}\ldots \end{aligned}\right\} \qquad (276b)$$

In this model of dipole couplings, the angular correlation factors (271b)—(273b) assume the following form, well adapted to direct numerical computations:

$$R_{CM} = 1 + z\left\{1 - 3\frac{L(Y_1)}{Y_1}\right\} \qquad (271c)$$

$$R_K = R_{CM} \pm 2zL(Y_1) \qquad (272c)$$

$$R_S = (1 + z)\{1 \pm 2zL(Y_1) + \tfrac{5}{2}z^2L^2(Y_1)\} - 3z\left\{1 - 2\frac{L(Y_1)}{Y_1}\right\} \qquad (273c)$$

As the dipole coupling attains its maximum value [saturation of coupling, $Y_1 \to \infty$ and $L(Y_1) \to 1$], the correlation factors take the limiting values:

$$R_{CM} = 1 + z \tag{271d}$$

$$R_K = 1 + z \pm 2z \tag{272d}$$

$$R_S = 1 - 2z + (1 + z)(\pm 2z + \tfrac{5}{2}z^2) \tag{273d}$$

On putting the number of neighbours as $z = 1$, the preceding formulae reduce to those derived by Piekara [352] for the case of binary dipole coupling. One notes that the factor R_{CM} is not sensitive to whether the partners of a pair are arrayed parallel or antiparallel, taking the value $R_{CM} = 2$ for association into pairs. The factors R_K and R_S are, however, sensitive to the distinction between these two limiting kinds of coupling: for association to parallel pairs $R_K = 4 = 2^2$ and $R_S = 8 = 2^3$, whereas for antiparallel binary aggregates $R_K = R_S = 0$.

Semi-macroscopic Theory of Dielectric Saturation.—The theories of dielectric saturation discussed above were all based on some particular molecular model, and were thus of restricted generality. Essentially, they took into consideration only mechanisms of reorientation of the molecules in the applied field, as is the case with gases, and angular correlations in dense states. Their applicability is restricted to substances composed of dipolar and non-dipolar, though anisotropically polarizable, molecules. But they are not applicable to substances consisting *e.g.* of atoms, or of molecules the linear polarizability of which is isotropic (CCl_4). Neither do they take into consideration other molecular mechanisms, such as translational fluctuations, fluctuations of density, fluctuations of the electric multipole fields of the molecules, and the non-linear molecular polarizabilities in external and molecular electric fields. The inclusion of these statistical–fluctuational processes can be performed generally within the framework of the semi-macroscopic theory of dielectrics as proposed by Kirkwood.[353] The first to attempt an extension of Kirkwood's theory to the case of strong electric fields was Anselm,[374] followed by O'Dwyer,[453] Booth,[454] and Schelman.[455] They calculated variations $\Delta\varepsilon$ due to dipole effect only, omitting anisotropy of polarizability. The general theory of quadratic variations $\Delta\varepsilon$, including the complete set of statistical molecular processes as well as non-linear polarizability, has been developed by Buckingham [456] and by Kielich.[388]

Dipolar Saturation. The vector of electric polarization of a macroscopic sample of volume V_m in an electric field E_0 is:

$$\boldsymbol{P}(E_0) = \frac{1}{V_m}\langle \boldsymbol{M} \rangle_{E_0} \tag{277}$$

[453] J. J. O'Dwyer, *Proc. Phys. Soc. (A)*, 1951, **64**, 1125.
[454] F. Booth, *J. Chem. Phys.*, 1951, **19**, 391; 1955, **23**, 453.
[455] J. A. Schellman, *J. Chem. Phys.*, 1957, **26**, 1225.
[456] A. D. Buckingham, *J. Chem. Phys.*, 1956, **25**, 428.

where the mean statistical value of the dipole moment in the presence of the external field is:

$$\langle \boldsymbol{M} \rangle_{E_0} = \int \boldsymbol{M}(\tau) f(\tau, E_0)\, d\tau \tag{278}$$

and integration extends over all configurations τ of the system.

Taking the potential energy of the system to comprise perturbation in dipole approximation only

$$U(\tau, \boldsymbol{E}_0) = U(\tau, 0) - \boldsymbol{M} \cdot \boldsymbol{E}_0 \tag{279}$$

we have the distribution function in the form:

$$f(\tau, E_0) = \frac{\exp\left\{-\dfrac{U(\tau,0) - \boldsymbol{M}\cdot\boldsymbol{E}_0}{kT}\right\}}{\int \exp\left\{-\dfrac{U(\tau,0) - \boldsymbol{M}\cdot\boldsymbol{E}_0}{kT}\right\} d\tau} \tag{280}$$

Dealing with the energy in the field E_0 as a perturbation to the zero energy $U(\tau, 0)$ we have, on expanding the Boltzmann factors in (280) in a series in powers of E_0, and on insertion into (278):

$$\langle \boldsymbol{M} \rangle_{E_0} = \frac{\sum_{m=0}^{\infty} \dfrac{1}{m!}\left(\dfrac{E_0}{kT}\right)^m \langle \boldsymbol{M}(\boldsymbol{M}\cdot\boldsymbol{e})^m \rangle}{\sum_{m=0}^{\infty} \dfrac{1}{m!}\left(\dfrac{E_0}{kT}\right)^m \langle (\boldsymbol{M}\cdot\boldsymbol{e})^m \rangle} \tag{278a}$$

where $\boldsymbol{e}$ denotes the unit vector in the field direction, $\boldsymbol{E}_0 = E_0\,\boldsymbol{e}$.

In the absence of an external field, when all orientations of $\boldsymbol{M}$ with respect to $\boldsymbol{e}$ are equally probable, one obtains:[388]

$$\langle (\boldsymbol{M}\cdot\boldsymbol{e})^m \rangle_{E=0} = \begin{cases} \dfrac{1}{2n+1} \langle (\boldsymbol{M}\cdot\boldsymbol{M})^n \rangle & \text{for } m = 2n \\ 0 \quad \text{for } m = 2n+1 \end{cases} \tag{281}$$

With this averaging, we finally get by (278a) and (277) the non-linear polarization in the form:

$$P_{\parallel}(E_0)\, V_{\mathrm{m}} = \frac{\sum_{n=0}^{\infty} \dfrac{1}{(2n+1)!\,(2n+3)} \left(\dfrac{E_0}{kT}\right)^{2n+1} \langle (\boldsymbol{M}\cdot\boldsymbol{M})^{n+1} \rangle}{\sum_{n=0}^{\infty} \dfrac{1}{(2n+1)!} \left(\dfrac{E_0}{kT}\right)^{2n} \langle (\boldsymbol{M}\cdot\boldsymbol{M})^n \rangle} \tag{282}$$

Provided that the field-induced non-linearity is not too large, the preceding formula yields, with regard to the definition (246) and equation (250), the following quadratic variation of the molar polarization:

$$\Delta P_{\mathrm{m}}(E_{\mathrm{p}}) = -\frac{1}{90\varepsilon_0 k^3 T^3} \{5\langle \boldsymbol{M}\cdot\boldsymbol{M} \rangle^2 - 3\langle (\boldsymbol{M}\cdot\boldsymbol{M})^2 \rangle\} \left(\frac{\varepsilon_{\mathrm{r}}+2}{3}\right)^2 E_{\mathrm{p}}^2 \tag{283}$$

On taking equation (247), omitting therein the term in molar volume variations, and combining it with equation (283), we obtain for the quadratic change in electric permittivity resulting within the framework of semi-macroscopic dipole dielectrics theory:

$$\Delta\varepsilon_r(E_p) = -\frac{1}{30\varepsilon_0 k^3 T^3 V_m}\{5\langle M \cdot M\rangle^2 - 3\langle(M \cdot M)^2\rangle\}\left(\frac{\varepsilon_r + 2}{3}\right)^4 E_p^2 \tag{284}$$

The preceding expression can be re-written in the form:

$$\Delta\varepsilon_r(E_p) = -\frac{\rho p^4}{15\varepsilon_0 k^3 T^3}\left(\frac{\varepsilon_r + 2}{3}\right)^4 R_S E_p^2 \tag{284a}$$

involving the correlation factor of dielectric saturation:

$$R_S = \frac{1}{2p^4 N_A}\{5\langle M \cdot M\rangle^2 - 3\langle(M \cdot M)^2\rangle\} \tag{285}$$

In the general case, the total dipole moment of the dielectric specimen in the absence of an external field takes the form

$$M = \sum_{i=1}^{N_A}(p_i + \alpha_i \cdot F_i + \ldots) \tag{286}$$

permitting the calculation of the effect on $\Delta\varepsilon$ of fluctuations of the electric molecular fields F_i. Within the approximation of non-polarizable dipoles ($F = 0$), the correlation factor (283) immediately reduces to the previously discussed form (273).

Semi-macroscopic theory in Fröhlich's approach yields for $\Delta\varepsilon$, instead of (284a), the expression:[388]

$$\Delta\varepsilon_r(E_p) = -\frac{p^4}{15\varepsilon_0 k^3 T^3}\left(\frac{3\varepsilon_r^2}{2\varepsilon_r^2 + \varepsilon_{r\infty}^2}\right)\left(\frac{3\varepsilon_r}{2\varepsilon_r + \varepsilon_{r\infty}}\right)^2\left(\frac{\varepsilon_{r\infty} + 2}{3}\right)^4 R_S E_p^2 \tag{287}$$

The semi-macroscopic theory of dielectric saturation has been applied to water by Booth [454] and to associating liquids by Schellman,[455] proving the possibility in them of a change in sign of the correlation factor (285). To Kluk [457] is due the extension of the theory to molecules with two axes of internal rotation.

General Treatment of Fluctuational Processes. The previous treatment is good only as long as we deal with strongly dipolar substances and all other polarizational effects remain negligible. In the majority of substances, besides reorientation of permanent dipoles, one has to consider reorientation of the polarizability ellipsoids as well as statistical–fluctuational processes. In calculating the electric polarization (277), one has to include the term accounting for linear distortional polarizability of the dielectric (non-linear polarizabilities are dealt with below):

$$P(E_0) = \frac{1}{V_m}\int\{M(\tau) + A(\tau) \cdot E_0 + \ldots\}f(\tau, E_0)\,d\tau \tag{288}$$

[457] E. Kluk, *Acta Phys. Polon.*, 1961, **20**, 845.

where $A(\tau)$ is the linear electric polarizability tensor of the isotropic dielectric specimen.

In the case now under consideration, we make use of the distribution functions in the form of the expansion (100) with the potential energy (188). This leads to the following quadratic change in molar polarization:

$$\Delta P_{\mathrm{m}}(E_{\mathrm{p}}) = S_{\mathrm{m}}\left(\frac{\varepsilon_{\mathrm{r}} + 2}{3}\right)^2 E_{\mathrm{p}}^2 \tag{289}$$

where, now, the molar constant of dielectric saturation is:

$$S_{\mathrm{m}} = S_{\mathrm{m}}^{\mathrm{Fl}} + S_{\mathrm{m}}^{(1)} + S_{\mathrm{m}}^{(2)} + S_{\mathrm{m}}^{(3)} \tag{290}$$

Compared to the formulae (258) and (259) derived by Debye's 'gas' theory, equation (290) contains the new constant:[458]

$$S_{\mathrm{m}}^{\mathrm{Fl}} = \frac{1}{18\varepsilon_0 kT}\left\langle\left(\Delta A + \frac{1}{kT}\Delta M^2\right)^2\right\rangle \tag{291}$$

due to the processes of statistical fluctuation of the polarizability of the medium, $\Delta A = A - \langle A\rangle$, and of the square of the electric dipole, moment, $\Delta M^2 = M^2 - \langle M^2\rangle$.

The other molar constants of dielectric saturation are of the form:[458]

$$S_{\mathrm{m}}^{(1)} = \frac{1}{15\varepsilon_0 kT}\langle \boldsymbol{D} : \boldsymbol{D}\rangle \tag{292}$$

$$S_{\mathrm{m}}^{(2)} = \frac{2}{15\varepsilon_0 k^2 T^2}\langle \boldsymbol{M}\cdot\boldsymbol{D}\cdot\boldsymbol{M}\rangle \tag{293}$$

$$S_{\mathrm{m}}^{(3)} = -\frac{1}{45\varepsilon_0 k^3 T^3}\langle(\boldsymbol{M}\cdot\boldsymbol{M})^2\rangle \tag{294}$$

where we have introduced the deviation tensor of electric polarizability:

$$\boldsymbol{D} = \boldsymbol{A} - A\boldsymbol{U}$$

The constants (292)—(294) differ in general from the previous constants (251) and (253)—(255) in that they account for all isotropic media presenting statistical molecular correlations. We now proceed to apply the molar constants (291)—(294) to some particular cases.

Atomic liquids. The essential role here belongs to the contribution (291). The latter, in binary correlation approximation, takes the form:

$$S_{\mathrm{m}}^{\mathrm{Fl}} = \frac{\alpha^2}{2\varepsilon_0 kT}\{\langle(\Delta N_{\mathrm{A}})^2\rangle + 8\alpha_{\mathrm{v}}^2 N_{\mathrm{A}}\langle r_{ij}^{-6}\rangle + \ldots\} \tag{291a}$$

where the first term relates to fluctuations in number density of the molecules which, according to Smoluchowski,[332] can be expressed in thermodynamic form:

$$\langle(\Delta N_{\mathrm{A}})^2\rangle = N_{\mathrm{A}}\,\rho kT\kappa_T \tag{295}$$

where $\kappa_T = -(1/V)(\partial V/\partial P)_T$ is the isothermal compressibility coefficient, and P the pressure.

[458] S. Kielich, *Chem. Phys. Letters*, 1970, 7, 347.

The second term of (291a) is due to translational fluctuations of the atoms. With regard to (295), we re-write equation (291a) in the following form, which is well adapted to numerical calculation:[358]

$$S_{\mathrm{m}}^{\mathrm{Fl}} = \frac{N_{\mathrm{A}}\,\alpha^2}{2\varepsilon_0\,kT}\{\rho kT\kappa_T + 8\alpha_{\mathrm{V}}^2\langle r_{ij}^{-6}\rangle + \ldots\} \tag{291b}$$

As a result of Yvon–Kirkwood translational fluctuations, the anisotropic constant of molar saturation (292) is non-zero, but finally takes the form:

$$S_{\mathrm{m}}^{(1)} = \frac{2N_{\mathrm{A}}\,\alpha^2}{5\varepsilon_0\,kT}\,\alpha_{\mathrm{V}}^2\{\langle r_{ij}^{-6}\rangle + \langle r_{ij}^{-3}\,r_{jk}^{-3}\rangle + \langle r_{ij}^{-3}\,r_{ki}^{-3}\rangle + 2\alpha_{\mathrm{V}}(\langle r_{ij}^{-9}\rangle + \ldots) + \ldots\} \tag{292a}$$

This electric anisotropy is induced in atomic liquids by the translational fluctuations, which cause the atoms to re-group momentarily into assemblages of two, three, or more atoms (in general, multiple assemblages) which can undergo orientation by the external electric field.[458]

Quadrupolar molecules. In a liquid composed of anisotropic non-polar molecules, the molar constant (291b) is still applicable in a first approximation (it defines the contribution from fluctuations in number density of the molecules, as in isotropic light scattering). The anisotropic molar constant (292) reduces to the contribution:

$$S_{\mathrm{m}}^{(1)} = \frac{2N_{\mathrm{A}}}{45\varepsilon_0\,kT}(\alpha_3 - \alpha_1)^2(1 + J_A^{(2)}) + \ldots \tag{292b}$$

arising by Langevin reorientation and by angular correlations, as well as to the contribution arising from translational molecular fluctuations which, in a sufficiently good approximation, can be written in the form (292a).

If, now, the non-dipolar molecules possess a permanent quadrupole moment Θ, fluctuations of their molecular electric fields induce dipole moments owing to which the molar anisotropic–dipolar constant (293) also becomes non-zero. In a first approximation, it takes the form:

$$S_{\mathrm{m}}^{(2)} = \frac{2N_{\mathrm{A}}\,\alpha_{\mathrm{V}}^2\,\Theta^2}{105\varepsilon_0\,k^2\,T^2}\{(\alpha_3 - \alpha_1)\langle r_{ij}^{-8}\rangle + \tfrac{168}{5}\alpha\alpha_{\mathrm{V}}\langle r_{ij}^{-11}\rangle + \ldots\} \tag{293a}$$

It is highly significant here that the first term is positive if the molecules have positive anisotropy ($\alpha_3 > \alpha_1$, as in CS_2) but is negative if they have negative electric anisotropy ($\alpha_3 < \alpha_1$, as in C_6H_6).

For quadrupolar molecules, the dipole contribution (294) is non-zero also, since

$$\langle(\boldsymbol{M}\cdot\boldsymbol{M})^2\rangle \approx \alpha^4\,\Theta^4\,N_{\mathrm{A}}\langle r_{ij}^{-16}\rangle$$

though unimportant unless temperatures are sufficiently low.

In a similar manner, the molar constants (292)—(294) can be applied to liquids composed of tetrahedral molecules, having permanent electric octupole moments.[458]

The Role of Non-linear Polarizability. We have just proved that the anisotropic molar saturation constants are non-zero even in liquids consisting of atoms and non-polar molecules. Beside the previous statistical–fluctuational processes, simple liquids reveal mechanisms that are due to the non-linear polarizabilities of atoms and such molecules. This is so because, in general, in addition to linear electric polarizability in the form (288), if a sufficiently strong electric field is applied, we have to deal with the non-linear polarizability:

$$\boldsymbol{P}^{\mathrm{NL}}(\boldsymbol{E}_0) = \frac{1}{2V_{\mathrm{m}}}\int\{\boldsymbol{B}(\tau) + \tfrac{1}{3}\boldsymbol{C}(\tau)\cdot\boldsymbol{E}_0 + \ldots\} : \boldsymbol{E}_0\,\boldsymbol{E}_0 f(\tau, \boldsymbol{E}_0)\,\mathrm{d}\tau \tag{296}$$

where the third-rank tensor $\boldsymbol{B}(\tau)$ defines the second-order non-linear polarizability of the medium due to the square of the electric field E_0^2, whereas the fourth-rank tensor $\boldsymbol{C}(\tau)$ describes the third-order non-linear electric polarizability induced by E_0^3.

The existence of a polarization (296) in an isotropic dielectric causes the molar saturation constant (290) to contain the new contribution:

$$S_{\mathrm{m}}^{\mathrm{NL}} = \frac{1}{30\varepsilon_0}\left\{\langle\boldsymbol{U}:\boldsymbol{C}:\boldsymbol{U}\rangle + \frac{4}{kT}\langle\boldsymbol{U}:\boldsymbol{B}\cdot\boldsymbol{M}\rangle\right\} \tag{297}$$

Atomic liquids. For simple atomic liquids, equation (297) reduces to the form:

$$S_{\mathrm{m}}^{\mathrm{NL}} = \frac{N_{\mathrm{A}}}{6\varepsilon_0}\gamma(1 + \tfrac{48}{5}\alpha_{\mathrm{V}}^2\langle r_{ij}^{-6}\rangle + \ldots) \tag{297a}$$

where the first term, involving the mean non-linear polarizability γ of an atom, defines the effect calculated by Buckingham.[456] We note that, even in atomic gases, and in the absence of radial correlations ($\langle r_{ij}^{-6}\rangle \approx 0$), some dielectric saturation can be induced by direct non-linear electronic distortion of the atoms, *i.e.* by Voigt's effect.

Polar liquids. Here, in addition to Voigt's effect as given by formula (297a), we have to deal in consequence of the second term of (297) with a degree of distortional–reorientational effect:

$$S_{\mathrm{m}}^{\mathrm{NL}} = \frac{2N_{\mathrm{A}}}{5\varepsilon_0 kT}\beta p(1 + J_{\mathrm{A}}^{(1)}) \tag{297b}$$

where $\beta = (\beta_{333} + 2\beta_{113})/3$ is the mean second-order non-linear electric polarizability of a molecule possessing the permanent dipole moment p. The angular correlation parameter $J_{\mathrm{A}}^{(1)}$ of dipolar molecules is defined by formula (179). Quite obviously, the effect (297b) is especially important in dipolar liquids, like nitrobenzene.[288]

Equation (297) is similarly applicable to quadrupolar and octupolar molecules, *etc.* The particularized forms of the molar constants (291)—(294) and (297) thus derived are well adapted for direct numerical calculations.

The results thus calculated are listed in Table 16, along with the available experimental results of Aussenegg[430] and Przeniczny[431] for non-dipolar liquids.

Table 16 *Calculated and measured variations, $\Delta\varepsilon$, in some non-dipolar liquids. (Reference numbers in square brackets)*

	Liquids	*Argon at* 85 K	CCl_4	C_6H_6	CS_2
$\kappa_T \times 10^{10}/m^2\,N^{-1}$		18.0	10.6	9.5	9.4
S_m^{F1}	$10^{25}/C^2\,m^5\,J^{-2}\,mol^{-1}$	1.27	17.44	18.38	21.07
S_m^{NL}	$10^{25}/C^2\,m^5\,J^{-2}\,mol^{-1}$	0.01	0.11	0.09	1.00
$S_m^{(1)}$	$10^{25}/C^2\,m^5\,J^{-2}\,mol^{-1}$	0.22	3.12	6.08	14.78
$S_m^{(2)}$	$10^{25}/C^2\,m^5\,J^{-2}\,mol^{-1}$	—	0.07	− 1.89	1.43
S_m^{total}	$10^{25}/C^2\,m^5\,J^{-2}\,mol^{-1}$	1.50	20.74	22.66	38.28
$\Delta\varepsilon^{theor}/E_p^2 \times 10^{20}/m^2\,V^{-2}$		3.0	25.5	31.5	109.3
$\Delta\varepsilon^{exp}/E_p^2 \times 10^{20}/m^2\,V^{-2}$		—	18.4 [430] 47.3 [431]	29.4 [430] 63.1 [431]	10.8 [420*b*] 36.9 [430] 122 [431]

Liquid Molecular Solutions.—The molar constants of dielectric saturation in the form (291)—(294) and (297) are of such generality as to be applicable not only to atomic, non-polar, polar, and other liquids but moreover to multi-component systems, for which we have:

$$S_m = \sum_a x_a S_m^{(a)} + \sum_{ab} x_a x_b S_m^{(ab)} + \sum_{abc} x_a x_b x_c S_m^{(abc)} + \ldots \qquad (298)$$

The first term of this expansion corresponds to the case of the ideal mixture, *i.e.* to mutually non-interacting components. This additivity principle for the molar saturation constant S_m assumes the following form in the case of a two-component mixture:

$$S_m = x_A S_m^{(A)} + x_B S_m^{(B)} \qquad (299)$$

with x_A and x_B denoting the molar fractions of the components A and B of the mixture.

Available experimental results on dielectric saturation in liquid solutions predominantly point to an influence of the solvent on the numerical value and sign of $\Delta\varepsilon$ of the solute. In other words, the additivity principle (299) is not obeyed, and any theoretical analysis of $\Delta\varepsilon$ measurements has to take into account higher terms in (298) resulting from interactions between the molecules of the various components.

Since theoretical analysis based on the expansion (298) is somewhat complicated (though in principle possible), the following procedure can be recommended in cases of two-component mixtures. Let $x = x_B$ be the molar fraction of the solute and $1 - x = x_A$ that of the solvent. The molar constant $S_m(x)$ of the mixture, a function of x, can now be expanded:

$$S_m(x) = \sum_{n=0}^{\infty} Q_n x^n = Q_0 + Q_1 x + Q_2 x^2 + Q_3 x^3 + \ldots \qquad (300)$$

The expansion coefficients:

$$Q_n = \frac{1}{n!}\left(\frac{\partial^n S_{\mathrm{m}}}{\partial x^n}\right)_{x=0} \tag{301}$$

are then determined on the one hand directly from experimental measurements and on the other from statistical theory leading to the expansion (298).

The zeroth coefficient

$$Q_0 = S_{\mathrm{m}}^{(A)} + S_{\mathrm{m}}^{(AA)} + S_{\mathrm{m}}^{(AAA)} + \ldots \tag{301a}$$

obviously defines the properties of the solvent alone, whereas the higher ones Q_1, Q_2, ... define those of the solution. The dielectric saturation of the solute is given as the variation:

$$\Delta S_{\mathrm{m}}(x) = \frac{S_{\mathrm{m}}(x) - Q_0}{x} = \sum_{n=1}^{\infty} Q_n x^{n-1}$$
$$= Q_1 + Q_2 x + Q_3 x^2 + Q_4 x^3 + \ldots \tag{302}$$

At infinite dilution ($x \to 0$), we obtain:

$$\Delta S_{\mathrm{m}\infty} = \lim_{x\to 0}\left\{\frac{S_{\mathrm{m}}(x) - Q_0}{x}\right\} = Q_1$$
$$= S_{\mathrm{m}}^{(B)} - S_{\mathrm{m}}^{(A)} + 2\{S_{\mathrm{m}}^{(AB)} - S_{\mathrm{m}}^{(AA)}\}$$
$$+ 3\{S_{\mathrm{m}}^{(AAB)} - S_{\mathrm{m}}^{(AAA)}\} + \ldots \tag{302a}$$

With regard to the expansion (302), as concentration grows, one has to take higher and higher terms beyond Q_1:

$$Q_2 = S_{\mathrm{m}}^{(AA)} - 2S_{\mathrm{m}}^{(AB)} + S_{\mathrm{m}}^{(BB)}$$
$$+ 3(S_{\mathrm{m}}^{(AAA)} - 2S_{\mathrm{m}}^{(AAB)} + S_{\mathrm{m}}^{(ABB)}) + \ldots \tag{302b}$$

$$Q_3 = -S_{\mathrm{m}}^{(AAA)} + 3S_{\mathrm{m}}^{(AAB)} - 3S_{\mathrm{m}}^{(ABB)} + S_{\mathrm{m}}^{(BBB)} + \ldots \tag{302c}$$

The constants occurring above can be determined from the semi-macroscopic theory of $\Delta\varepsilon$ developed in the preceding sections.

The expressions simplify considerably if the solution is chosen so that the solvent (*e.g.* CCl_4, C_6H_6) under the influence of the electric field becomes endowed with a weak non-linearity, negligible by comparison with the non-linearity of the solute (*e.g.* nitrobenzene). If, moreover, the solute is strongly dipolar, one is justified in considering only the effect of reorientation of the permanent electric dipoles and omitting other mechanisms contributing to dielectric saturation. The molar saturation constant of the solute (302) now becomes:

$$S_{\mathrm{m}}(x) = S_{\mathrm{m}}^{(B)} + S_{\mathrm{m}}^{(BB)} x + S_{\mathrm{m}}^{(BBB)} x^2 + S_{\mathrm{m}}^{(BBBB)} x^3 + \ldots \tag{303}$$

where $S_{\mathrm{m}}^{(B)}$ is for the system without interactions, and the successive constants $S_{\mathrm{m}}^{(BB)}$, $S_{\mathrm{m}}^{(BBB)}$, ... correspond to binary, ternary, ... molecular correlations.

We thus see that, in accordance with (302) and (303), the shape of $\Delta S_m(x)$ as a function of concentration depends on the numerical value and sign of the various constants responsible for the specific interactions. By accumulating measurements of $\Delta\varepsilon$ as a function of concentration of the solutions, it is thus possible to investigate in the light of this theory the various types of molecular correlations and to ascertain the concentrations at which binary, or ternary, or more highly multiple correlations predominate. The graphs of Figure 19 show the various possible types of

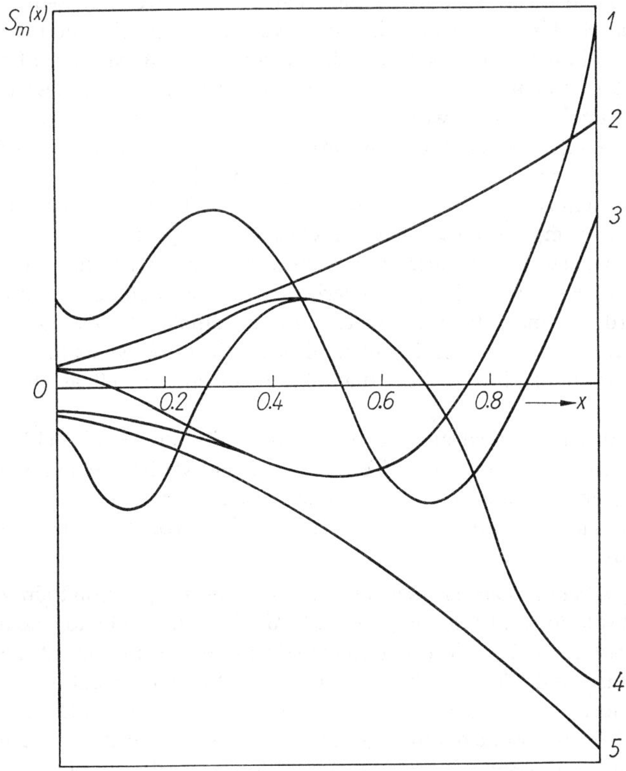

Figure 19 *Various shapes of dielectric saturation as a function of concentration.* (1) *Dipolar substances with a strongly positive effect in solution in a non-dipolar liquid presenting a weak positive effect* (e.g. *nitrobenzene–benzene* [422]) *or in a dipolar liquid presenting a weak negative effect* (e.g. 1,1-*dichloroethane–nitrobenzene* [459]). *The inversion is due to dipolar intermolecular interactions;*[352] (2) *Substances dissolved with field-stimulated molecular configurational change* [424] (e.g. *benzene*–1,2-*dibromoethane* [425])*;* (3) *Solutions of two liquids having positive effects whose molecules correlate variably;* (4) *Liquids with a strongly negative effect dissolved in solvents with a small, positive or negative, effect* (e.g. *alcohols in non-dipolar solvents* [426, 461] *where a field-stimulated proton shift in* H-*bonded dimers occurs* [196]*;* (5) *Liquids with a strongly negative effect and weak or no correlations* [424] (e.g. *diethyl ether or* 1,3-*dibromopropane in benzene* [425])

dielectric saturation in liquid solutions resulting from formulae (302) and (303) and existing experimental studies.[459] The shapes of the curves depend on whether $\Delta S_m(x)$ is linear (curves of the types 2 and 5), quadratic, cubic, or of still higher order in x. The complete theory of dielectric saturation in solutions [460] derived by semi-macroscopic treatment takes into consideration the co-existence in solutions of various microscopic mechanisms, competing or co-operating with one another at one concentration or another. By the expansion (300), the molar saturation constant is generally a polynomial of the n-th degree with respect to x and its graph is a curve of the n-th degree of the parabolic kind. Functions like these vary continuously, and the number of points of intersection as well as the existence or non-existence of extrema (with alternating maxima and minima) depends on whether n is even or odd.

In particular, curves with a single maximum followed by a minimum (or *vice versa*) can occur too, as is the case in certain alcohols,[461] in solutions of nitrobenzene in cyclohexane,[459] and of piperidine in benzene.[431]

Hitherto most authors, when studying $\Delta\varepsilon$ as a function of concentration, chose the components so that the effect in the pure solvent should be negligibly small in comparison with the very large non-linear effect of the solute. In cases when the non-linearities of the two components are of the same order of magnitude, *i.e.* when both are dipolar [459] or non-dipolar,[431] the accuracy of the measuring method has to be raised so as to permit the measurement of either effect: that of the solvent as well as that of the solute.

Available theoretical and experimental results on electric field induced non-linearities in isotropic dielectrics prove that we have here an effective method for the study of the electric structure of molecules, as well as of their mutual interactions in dense states, *i.e.* of the structure of near ordering in liquids.

Investigation of Molecular Structure. We are thus in a position to investigate internal rotation and *trans–gauche* conformations in 1,2-dihalogenoalkanes. *Rotational isomerism.* In order to explain the positive $\Delta\varepsilon$-effect measured by Piekara and Chełkowski [424, 425] in the entire concentration range of 1,2-dichloroethane and 1,2-dibromoethane in benzene and CCl_4, a thermal equilibrium is assumed (after other authors [3, 87]) to exist between three forms of the molecule: a *trans* conformation, of higher stability, and two less stable equivalent *gauche* conformations. With regard to the symmetry of the steric system of chemical bonds the dipole moment of the *trans* conformation is zero, $p_t = 0$, whereas that of the *gauche* conformation is non-zero, $p_g \neq 0$. Taking this into consideration, the rise in electric permittivity in a strong field can be attributed to field-stimulated transitions from *trans* to *gauche*. In this way, the external electric field shifts the

[459] M. Dutkiewicz, Doctor's Thesis, University of Poznań, 1971; *Acta Phys. Polon.* (*A*), 1972, **A42**, in press.
[460] S. Kielich, 1965, unpublished results.
[461] I. Korzeniowska and Z. Przeniczny, 1970, unpublished results.

equilibrium favouring *gauche* configuration, thus causing an increase in effective dipole moment of the system in the field. By resorting to this mechanism and classical Boltzmann statistics, Piekara and Kielich[424] derived the following formula for the correlation factor of dipole saturation:

$$R_S^g = \frac{2 - 3r_t}{2(1 + r_t)^2} \tag{304}$$

where $r_t = N_t/N_g$ is the ratio of the numbers of molecules of *trans* (N_t) and *gauche* (N_g) structure at zero field and is given by:

$$r_t = A_g^{-1} \exp\left\{\frac{\Delta U}{kT}\right\} \tag{305}$$

A_g being a numerical factor accounting for the statistical weight of the occurrence of the two *gauche* forms,[87] and $\Delta U = U_g - U_t$ the difference between the energy values corresponding to the *gauche*- and *trans*-forms of the molecule.

The study of variations $\Delta\varepsilon$ based on the model (304), as developed by Chełkowski,[425] in conjunction with measurements of the permittivity ε, permits determination of the dipole moment p_g of the unstable *gauche*-form as well as of ΔU. In this way, the steric and electrical structures of conformational molecules can be determined.

Internal rotation. Chełkowski's measurements can also be explained by internal rotation. In molecules presenting internal rotation (*e.g.* XH_2C—CH_2X) we have to deal with interaction between two dipole moments localized at the two radicals in rotation about the C—C bond. After Mizushima,[87] we assume the existence of but a single potential energy minimum for the *trans* configuration and an angular dependence of the potential energy given by the expression:

$$U = U_0(1 - \cos\psi) \tag{306}$$

where $\psi = \pi - \varphi$, φ being the azimuthal angle between the two moments p_1, and $2U_0$ is the potential barrier. Here, we still have equation (284a), albeit with $p = p_1\sqrt{2}$ and the saturation correlation factor:[424b]

$$R_S = \frac{1}{4} - 2\frac{I_1(Z)}{I_0(Z)} + \frac{5}{2}\left\{\frac{I_1(Z)}{I_0(Z)}\right\}^2 - \frac{3}{4}\frac{I_2(Z)}{I_0(Z)} \tag{307}$$

where $I_0(Z)$, $I_1(Z)$, and $I_2(Z)$ are Bessel functions, and $Z = U_0/kT$ is the intramolecular interaction parameter.

The graphs of the correlation factors (304) and (307) shown in Figure 20 reveal, as should be expected, a similar Z-dependence in the two approaches of rotational isomerism and internal rotation.

Intermolecular Interaction Investigation. In previous Sections, stress has often been laid on the role of molecular correlations in dielectric saturation in pure liquids. Their study is in many cases, made simpler by dissolving the liquid in an appropriately chosen solvent. It would be of particular interest to study dielectric saturation in strongly dilute solution, when the expression (302a) becomes appropriate for the theoretical analysis of $\Delta\varepsilon$. Thus, in dealing with a strongly dipolar substance at high dilution in a non-dipolar solvent, such that $S_m^{(A)} \ll S_m^{(B)}$, one would have by (302a), in satisfactory approximation:

$$\Delta S_{m\infty} = S_m^{(B)} + 2S_m^{(AB)} + 3S_m^{(AAB)} \qquad (308)$$

from which direct information can be obtained regarding the pairwise interaction between a molecule A of the solvent and a dipolar solute molecule B (the constant $S_m^{(AB)}$) as well as regarding ternary interaction between two molecules A and one molecule B. The method could serve

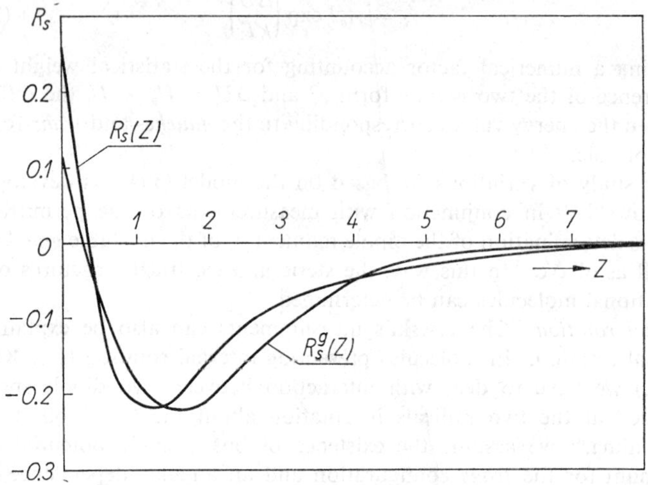

Figure 20 *Graphs of dielectric saturation correlation factors.* (i) $R_S^g(Z)$ *calculated from formula* (304) *for the rotational isomerism model with* $Z = U/kT$; (ii) $R_S(Z)$ *calculated from formula* (307) *for internal rotation, with* $Z = U_0/kT$. *In either case an inversion in sign of* $R_S(Z)$ *is found to occur very rapidly with growing Z*

for the study of weak interactions between dipolar and non-dipolar molecules largely due to fluctuational–redistributional effects.

Hitherto, $\Delta\varepsilon$ has been measured in solutions at concentrations upward of 5%, and good reproducibility of results has been achieved at intermediate and very high concentrations. No techniques are as yet available for the measurement of $\Delta\varepsilon$ in highly dilute solutions, chiefly because of the difficulties of constructing operational, stable equipment of high sensitivity for measuring small variations $\Delta\varepsilon$ in not very strong electric fields. The majority of existing $\Delta\varepsilon$ measurements in the entire concentration range (at $x > 0.1$) provide information on strong correlations of solute, mostly strongly dipolar, molecules. In cases like these it is often sufficient to take into consideration the correlation factor of dipole saturation R_S in the form (273), or the more general form (285). This is the situation with the solutions of nitrobenzene in benzene and carbon tetrachloride studied extensively by Piekara,[352] who assumed the high interaction energy to give rise, at some concentration, to at least pairs of molecules with antisymmetric alignment of their dipoles. At higher concentrations the probability for double pairs, *etc.* grows, so that the distribution of nitrobenzene molecules ceases to be random (as is the case at high dilution) but presents a well-defined structure of near-range order. An electric field, when applied externally, modifies the region of near-range order in a manner to augment its resultant dipole moment, leading to a rise in electric permittivity, in agreement with the measurements.[422] The negative $\Delta\varepsilon$ values of dilute solutions of nitrobenzene in benzene are easily explained by a gradual decay of this ordering leading to the emergence of single nitrobenzene molecules, surrounded by molecules of benzene. Such isolated electric dipoles undergo electric reorientation in agreement with Debye's theory, making negative saturation apparent. Other dipolar solutions [423–425] exhibit a similar concentration- and field-dependent change in structure.

Dielectric saturation is a source of valuable data on interactions not only of the electrostatic variety between multipoles but also those of the inductional and dispersional kinds between non-dipolar, electrically anisotropic molecules.

Dielectric studies in associate-forming liquids like water, the alcohols, phenols, amines, *etc.*, though rather difficult, provide information regarding the hydrogen bonds.[196] The difficulties of investigating dielectric saturation of such liquids are due, on the one hand, to their quite considerable electric conductivity and, on the other, to their complicated structure. They contain simultaneously a wide range of molecular associates (multimers and complexes), and their solutions are strongly dependent on the concentration of the solute, and temperature. Hitherto studies of the association process by polarization and saturation measurements have dealt essentially with aliphatic alcohols and their solutions.[426–429] The theoretical elucidation of these studies is very difficult, and the first attempts in this direction have proved to be rather complicated [462] and have hitherto failed to provide satisfactory quantitative agreement with experiment.[447]

In this context, it may be worthwhile stressing the importance of studying dielectric saturation at the critical point of solubility,[422] where a steep rise in $\Delta\varepsilon$ should occur similar to critical opalescence in light scattering.[463] Piekara and Piekara,[422] when studying nitrobenzene in hexane, observed a maximum of $\Delta\varepsilon > 0$ near the critical concentration $x = 0.5125$ at 22 °C. It would be highly interesting to trace the dependence of $\Delta\varepsilon$ on the field strength E in the neighbourhood of the critical point in order to find out whether it is quadratic or of some other form of non-linearity.

[462] (*a*) J. Małecki, *Acta Phys. Polon.*, 1965, **28**, 891; 1966, **29**, 45; (*b*) J. Małecki and Z. Dopierała, *ibid.*, 1969, **36**, 385, 401, 409.
[463] B. Fechner, Doctor's Thesis, University of Poznań, 1965.

Complete Dielectric Saturation.—Our previous considerations have dealt essentially with permittivity variations $\Delta\varepsilon(E_P)$ which were quadratic in the applied strong electric field E_P. The quadratic dependence is confirmed by $\Delta\varepsilon$ measurements performed on molecular substances, acted on by moderately strong electric fields. In macromolecular substances, $\Delta\varepsilon$ is no longer quadratic in the field strength: as is the case with the Kerr effect,[407–412] complete electric saturation sets in.

Experimentally, complete electric saturation of electric permittivity variations has been observed recently, in solutions of dipolar macromolecules where the effect is negative,[432] and in ones of non-dipolar and weakly dipolar macromolecules where it is positive in the entire range of variability of the electric field strength.[431, 433]

Booth[454] calculated the deviation of $\Delta\varepsilon$ from quadraticity by Onsager's model. Calculations of non-linear $\Delta\varepsilon$-variations of higher order have also been performed by Kielich[388] in the Kirkwood–Fröhlich semi-macroscopic approach taking into consideration statistical molecular correlations. Results such as these can be derived with the non-linear polarization (282). This treatment, however, is not directly applicable to the description of complete electric saturation, and we shall not develop it further here. It appears preferable, for simplicity, to proceed within the framework of classical Langevin–Debye theory, which yields results well adapted to numerical computations.[464]

Non-linear Electric Polarization. Consider a polymer or colloidal solution so dilute that interaction between the macromolecules or particles can be omitted. Also, assume the solvent not to affect their properties. On these assumptions, the electric polarization of the system is given by (251), with the total dipole moment of the macromolecule $\boldsymbol{p}(E_0)$ given in a linear approximation by (248). Of particular relevance is the polarization component in the field direction:

$$P_{\parallel}(E_0) = \rho\langle \boldsymbol{p}(E_0)\cdot\boldsymbol{e}\rangle_{E_0} \tag{309}$$

Restricting the problem to the important case of macromolecules having axial symmetry (with respect to the 3-axis), we have by (248):

$$\boldsymbol{p}(E_0)\cdot\boldsymbol{e} = \alpha E_0 + p\cos\theta + (\alpha_3 - \alpha_1)(\cos^2\theta - \tfrac{1}{3})\,E_0 \tag{248a}$$

where θ is the angle between the molecular symmetry axis and the electric field direction, denoted by $\boldsymbol{e}$.

Insertion of (248a) into (309) yields:

$$P_{\parallel}(E_0) = \rho\alpha E_0 + \rho\{p\langle\cos\theta\rangle_{E_0} + (\alpha_3 - \alpha_1)(\langle\cos^2\theta\rangle_{E_0} - \tfrac{1}{3})\,E_0\} \tag{309a}$$

involving the following statistical mean values in the presence of a strong electric field:

$$\langle\cos^n\theta\rangle_{E_0} = \int \cos^n\theta\, f(\theta, E_0)\,\mathrm{d}\Omega \tag{310}$$

[464] S. Kielich, *Acta Phys. Polon.*, 1969, **36**, 495; *Acta Phys. Polon.* (*A*), 1970, **37**, 447; *Optoelectronics*, 1970, **2**, 5.

In the case now under consideration, the statistical distribution function is of the form:

$$f(\theta, E_0) = \frac{\exp(X_1 \cos\theta \pm X_2 \cos^2\theta)}{2\pi \int_0^\pi \exp(X_1 \cos\theta \pm X_2 \cos^2\theta) \sin\theta \, d\theta} \tag{311}$$

resulting immediately from formulae (169) and (232); the reorientation parameter of permanent dipoles is:

$$X_1 = \frac{pE_0}{kT} \tag{312}$$

and that of induced electric dipoles:

$$X_2 = \frac{|\alpha_3 - \alpha_1|}{2kT} E_0^2 \tag{313}$$

As previously, the sign '+' in (311) is for rod-like macromolecules and the sign '−' for disc-like ones.

With regard to (311), the mean values (310) can be expressed as follows:

$$\langle \cos^n \theta \rangle_{E_0} = L_n(X_1, \pm X_2) \tag{310a}$$

where we have introduced the generalized Langevin functions:[464]

$$L_n(X_1, \pm X_2) = \frac{\int_0^\pi \cos^n\theta \exp(X_1 \cos\theta \pm X_2 \cos^2\theta) \sin\theta \, d\theta}{\int_0^\pi \exp(X_1 \cos\theta \pm X_2 \cos^2\theta) \sin\theta \, d\theta} \tag{314}$$

In particular, for $n = 1$, we obtain:

$$L_1(X_1, \pm X_2) = \mp \frac{X_1}{2X_2} \pm \frac{e^{X_1} - e^{-X_1}}{4X_2^{\frac{1}{2}} I(X_1, \pm X_2)} \tag{314a}$$

where the integrals $I(X_1, \pm X_2)$ are given by formula (235).

The Langevin function of degree 2 is given in explicit form by equation (234a).

By resorting to (310a), the non-linear polarization (309a) can be expressed as a function of the electric reorientation parameters (312) and (313) as follows:

$$P_\parallel(E_0) = \rho\alpha E_0 + \rho p L_1(X_1, \pm X_2) + \rho(\alpha_3 - \alpha_1)\{L_2(X_1, \pm X_2) - \tfrac{1}{3}\} E_0 \tag{309b}$$

In order to calculate the change in non-linear molar polarization in accordance with the definition (246), we have to keep in mind that the reorientation parameters (312) and (313) depend, by way of (250), on the analysing field strength E_{0a} *as well as* on the polarizing field strength E_{0p}. Consequently, taking the derivative of the Langevin function (314) with respect to the measuring field, we find:

$$\left\{\frac{\partial L_n(X_1, \pm X_2)}{\partial E_{0a}}\right\}_{E_{0a}=0} = \{L_{n+1}(X_1, \pm X_2)$$

$$- L_n(X_1, \pm X_2)\, L_1(X_1, \pm X_2)\} \left(\frac{\partial X_1}{\partial E_{0a}}\right)_{E_{0a}=0}$$

$$\pm \{L_{n+2}(X_1, \pm X_2) - L_n(X_1, \pm X_2)\, L_2(X_1, \pm X_2)\} \left(\frac{\partial X_2}{\partial E_{0a}}\right)_{E_{0a}=0}$$

This relation, together with definition (246) and equation (309b), leads to the following non-linear change in molar polarization:

$$\Delta P_{\mathrm{m}}(E_{\mathrm{p}}) = \frac{N_{\mathrm{A}}}{9\varepsilon_0}\left\{\pm 2\,|\,\alpha_3 - \alpha_1\,|\, R_{\alpha}(X_1, \pm X_2) - \frac{p^2}{kT} R_{\mathrm{p}}(X_1, \pm X_2)\right\} \tag{315}$$

where, for brevity, we have introduced the following reorientation function of dielectric saturation:

$$R_{\alpha}(X_1, \pm X_2) = \Phi(X_1, \pm X_2) - \tfrac{3}{2}X_1\{L_3(X_1, \pm X_2) - L_1(X_1, \pm X_2)\, L_2(X_1, \pm X_2)\} \pm 3X_2\{L_4(X_1, \pm X_2) - L_2^2(X_1, \pm X_2)\} \tag{316}$$

$$R_{\mathrm{p}}(X_1, \pm X_2) = 3L_1^2(X_1, \pm X_2) - 2\Phi(X_1, \pm X_2) \mp 6\frac{X_2}{X_1}\{L_3(X_1, \pm X_2) - L_1(X_1, \pm X_2)\, L_2(X_1, \pm X_2)\} \tag{317}$$

$\Phi(X_1, \pm X_2)$ being the reorientation function of Kerr effect defined by (233).

In order to be able to compute the functions (316) and (317) numerically, we have to know in explicit form the Langevin functions of orders 3 and 4, which are:[464]

$$L_3(X_1, \pm X_2) = \mp \frac{X_1(X_1^2 \mp 6X_2)}{8X_2^3} - \frac{X_1(e^{X_1} + e^{-X_1})}{8X_2^{\frac{3}{2}} I(X_1, \pm X_2)} \pm \frac{(X_1^2 \mp 4X_2 + 4X_2^2)(e^{X_1} - e^{-X_1})}{16X_2^{\frac{5}{2}} I(X_1, \pm X_2)} \tag{314b}$$

$$L_4(X_1, \pm X_2) = \frac{X_1^4 + 12X_2(X_2 \mp X_1^2)}{16X_2^4} \pm \frac{(X_1^2 \mp 6X_2 + 4X_2^2)(e^{X_1} + e^{-X_1})}{16X_2^{\frac{5}{2}} I(X_1, \pm X_2)} - \frac{X_1(X_1^2 \mp 10X_2 + 4X_2^2)(e^{X_1} - e^{-X_1})}{32X_2^{\frac{7}{2}} I(X_1, \pm X_2)} \tag{314c}$$

Strongly dipolar macromolecules. In the case of macromolecules which are strongly dipolar but have a low anisotropy of polarizability ($X_1^2 \gg X_2$), the permittivity variation (318) reduces to:

$$\Delta\varepsilon_r(E_p) = -\frac{\rho p^2}{3\varepsilon_0 kT}\left(\frac{\varepsilon_r + 2}{3}\right)^2 R_p(X_1) \tag{319}$$

where the reorientation function R_p of rigid dipoles in the electric field is:

$$R_p(X_1) = 3L_1^2(X_1) - 2\Phi(X_1) \tag{320}$$

The shape of this function *vs.* X_1 is shown in Figure 21.

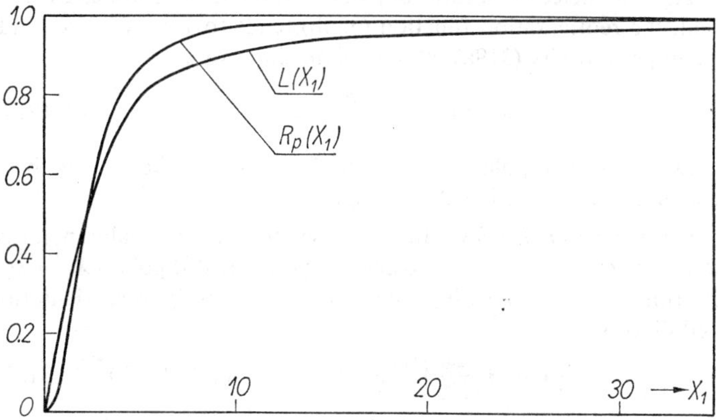

Figure 21 *Shape of the reorientation function for dipole saturation $R_p(X_1)$ as a function of X_1. For comparison, Langevin's function $L(X_1) = \coth X_1 - X_1^{-1}$ is adduced; it deviates but insignificantly from $R_p(X_1)$*

For the case of weak reorientation $X_1 < 1$ (weak field E, or small value of the dipole p, or high temperature T), the Langevin functions assume the form:

$$L_1(X_1) = \frac{X_1}{3} - \frac{X_1^3}{45} + \frac{2X_1^5}{945} - \frac{X_1^7}{4725} + \ldots$$

$$L_2(X_1) = \frac{1}{3} + \frac{2X_1^2}{45} - \frac{4X_1^4}{945} + \frac{2X_1^6}{4725} - \ldots$$

and the reorientation function (320) takes the form of the series:

$$R_p(X_1) = \frac{X_1^2}{5} - \frac{2X_1^4}{63} + \frac{X_1^6}{225} - \ldots \tag{320a}$$

On taking only the first term, formula (319) reduces immediately to Debye's expression, yielding a quadratic variation of $\Delta\varepsilon$ with the field E.

On the other hand, if dipole reorientation is very strong, $X_1 > 10$ (large values of E or p, or low temperature T), the Langevin functions become:

$$L_1(X_1) = 1 - \frac{1}{X_1}, \quad L_2(X_1) = 1 - \frac{2}{X_1} + \frac{2}{X_1^2} \tag{321}$$

so that the reorientation function (320) assumes the limiting form:

$$R_p(X_1) = 1 - \frac{3}{X_1^2} \tag{320b}$$

Thus, in dealing with a macromolecular solution exhibiting strong dipolar reorientation, we have a variation (319) of the form:

$$\Delta\varepsilon_r(E_p) = -\frac{\rho p^2}{3\varepsilon_0 kT}\left(\frac{\varepsilon_r + 2}{3}\right)^2\left\{1 - 3\left(\frac{kT}{pE_p}\right)^2\left(\frac{3}{\varepsilon_r + 2}\right)^2\right\} \tag{319a}$$

At complete dielectric saturation, when all the dipoles are aligned along the field, the reorientation function (320b) tends to 1 as $X_1 \to \infty$, and the change in permittivity (319a) attains its maximal value:

$$\Delta\varepsilon_r(\infty) = -\frac{\rho p^2}{3\varepsilon_0 kT}\left(\frac{\varepsilon_r + 2}{3}\right)^2 \tag{319b}$$

Thus, at complete dipolar saturation, the variation $\Delta\varepsilon$ is negative and independent of the electric field strength.

Non-dipolar macromolecules. In the case of anisotropically polarizable macromolecules ($X_2 \neq 0$) not possessing permanent dipoles ($X_1 = 0$), we obtain from (318) the following $\Delta\varepsilon$ variation, due solely to reorientation of induced dipoles:

$$\Delta\varepsilon_r(E_p) = \pm\frac{2\rho}{3\varepsilon_0}\left(\frac{\varepsilon_r + 2}{3}\right)^2 |\alpha_3 - \alpha_1|\, S_\alpha(\pm X_2) \tag{322}$$

where, on insertion of $X_1 = 0$ into (316), we have the saturation reorientation function for non-dipolar dielectrics:

$$S_\alpha(\pm X_2) = \Phi(\pm X_2) \pm 3X_2\{L_4(\pm X_2) - L_2^2(\pm X_2)\} \tag{323}$$

For weak reorientation, the preceding function can be written in the form of the series:

$$S_\alpha(\pm X_2) = \pm\frac{2X_2}{5} + \frac{4X_2^2}{63} \mp \frac{8X_2^3}{675} + \ldots \tag{323a}$$

The first term, inserted into (322), yields Debye's expression for the quadratic variation in $\Delta\varepsilon$ in non-dipolar substances.

At complete saturation ($X_2 \to \infty$), the Langevin functions for prolate macromolecules and likewise the reorientation function (323) tend to 1, so that the variation (322) takes the maximal value:

$$\Delta\varepsilon_r(+\infty) = \frac{2\rho}{3\varepsilon_0}\left(\frac{\varepsilon_r + 2}{3}\right)^2 |\alpha_3 - \alpha_1| \tag{322a}$$

If the macromolecules of the solution at saturation are of the oblate variety, the Langevin functions tend to 0 at $X_2 \to \infty$ and the correlation factor (323) tends to the value

$$S_\alpha(-\infty) = -\tfrac{1}{2} \tag{323b}$$

causing formula (322) to reduce to the form:

$$\Delta\varepsilon_r(-\infty) = \frac{\rho}{3\varepsilon_0}\left(\frac{\varepsilon_r + 2}{3}\right)^2 |\alpha_3 - \alpha_1| \tag{322b}$$

Consequently, in solutions of non-dipolar macromolecules, complete saturation of the electric permittivity variations is always positive. Significantly, one has the following relationship:

$$\Delta\varepsilon_r(+\infty) = 2\Delta\varepsilon_r(-\infty) \tag{322c}$$

stating that complete dielectric saturation for prolate macromolecules is twice as large as that for oblate macromolecules having negative anisotropy of polarizability.

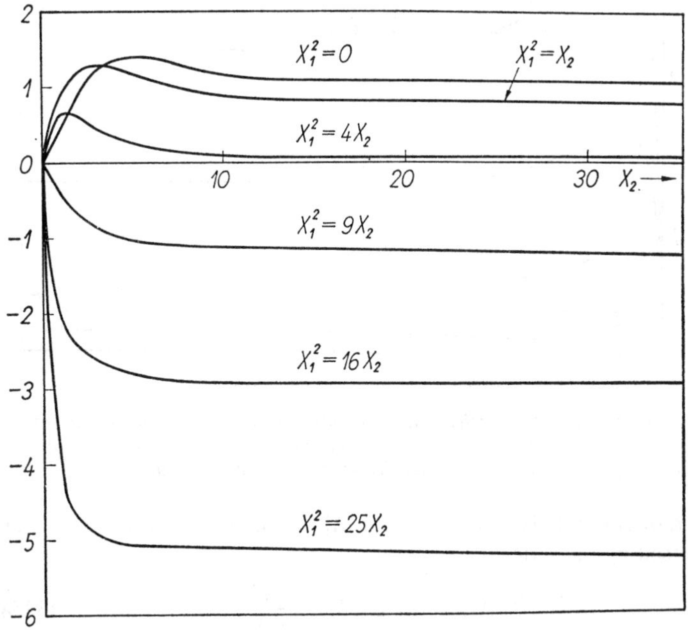

Figure 22 *Shape of the reorientation function $S_\alpha(X_1, +X_2)$ of dielectric saturation due to superposition of reorientation of permanent dipoles X_1 on reorientation of induced dipoles X_2 of macromolecules having positive anisotropy*

The function (323) is shown plotted against X_2 for arbitrary reorientation in Figure 22 for prolate macromolecules, and in Figure 23 for oblate ones (curves for $X_1 = 0$).

Dipolar, anisotropically polarizable macromolecules. If the solution contains macromolecules which are at the same time dipolar and anisotropically polarizable, one has to apply formula (318). For comprehensive analysis, the latter is more conveniently expressed in the following form:[435]

$$\Delta\varepsilon_r(E_p) = \pm \frac{2\rho}{3\varepsilon_0}\left(\frac{\varepsilon_r + 2}{3}\right)^2 |\alpha_3 - \alpha_1| S_\alpha(X_1, \pm X_2) \tag{324}$$

involving the general reorientation function of dielectric saturation:

$$S_\alpha(X_1, \pm X_2) = R_\alpha(X_1, \pm X_2) \mp \left(\frac{X_1^2}{4X_2}\right) R_p(X_1, \pm X_2) \quad (325)$$

Graphs of this function are shown in Figure 22 for the case of positive anisotropy and in Figure 23 for negative anisotropy.

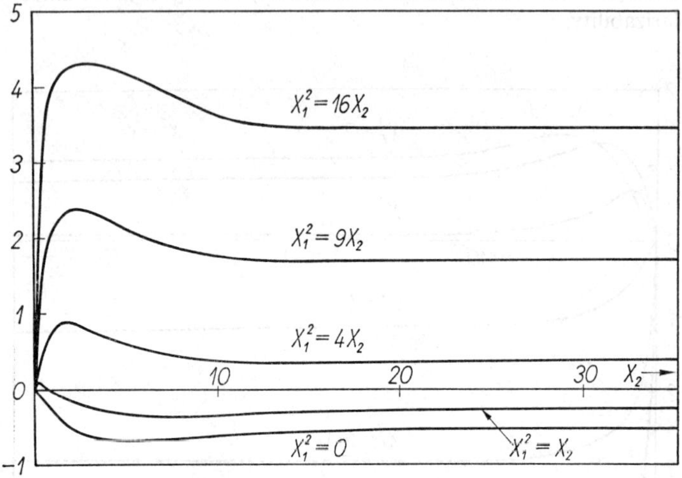

Figure 23 *Shape of the reorientation function $S_\alpha(X_1, -X_2)$ of dielectric saturation due to superposition of reorientation of permanent dipoles X_1 on reorientation of induced dipoles X_2 of macromolecules with negative anisotropy*

If reorientation is weak ($X_1 \ll 1, X_2 \ll 1$), the function (325) can be expressed by the series:

$$S_\alpha(X_1, \pm X_2) = \frac{1}{5}\left(\pm 2X_2 - \frac{X_1^4}{4X_2} + X_1^2\right) + \ldots \quad (325a)$$

and the variation (324) reduces to the Debye–Langevin formula (257) of the quadratic approximation.

Discussion and Applications. Figure 24 permits a comparison of the graphs of the function $R_\alpha(\pm X_2)$ describing electric saturation of electric permittivity variations for non-dipolar macromolecules on the one hand and of functions $\Phi(\pm X_2)$ describing electric saturation of the Kerr effect on the other. One easily sees that, in the region of intermediate orientation (for $1 \leqslant X_2 \leqslant 20$), the two functions differ in value strongly, run identically for very low reorientations ($X_2 \leqslant 1$), approach one another for large reorientations ($X_2 \leqslant 20$), and tend to a common maximal value of 1 at complete saturation of electric reorientation ($X_2 \to \infty$). These differences are due to the different methods of measuring $\Delta\varepsilon$ variations and the Kerr effect. Thus, with respect to the Kerr effect, the agent used in measuring the birefringence is a weak light beam, the electric field of which is unable to cause any perceptible optical reorientation of the molecules (but optical reorientation can be observed, measuring the optical birefringence caused by intense laser light [214]). So what one measures with the light beam are changes in refractive indices due only to reorientation of particles caused

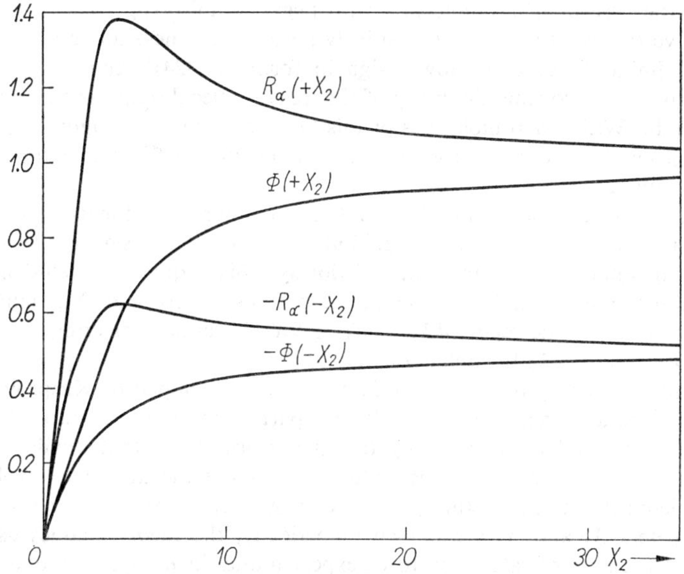

Figure 24 *A comparison of the X_2 dependence of the reorientation function of dielectric saturation $R_\alpha(\pm X_2)$ and of the reorientation function of the Kerr effect $\Phi(\pm X_2)$ for macromolecules with positive and negative anisotropy*

by the strong static electric field. Conversely, when measuring variations in electric permittivity $\Delta\varepsilon$, what we record is reorientation of macromolecules by the strong static electric field and the analysing field concomitantly, provided the frequency of the latter is sufficiently low for the macromolecules to align with it. It is precisely this superposition of orientation effects in the two electric fields applied in the same direction that causes the steeper rise in measured $\Delta\varepsilon$ values.

Figure 21 shows that in solutions of strongly dipolar macromolecules, where reorientation of permanent dipoles predominates strongly over reorientation of induced dipoles ($X_1^2 \gg X_2$), the changes in permittivity are negative in the entire range of variability of the field, up to complete electric saturation. If, however, the situation in the solution is such that the two kinds of reorientation are comparable, the $\Delta\varepsilon$ variation increases or diminishes (and can even exhibit changes in sign) according to the shape of the macromolecules and the strength of the field applied.

Thus, for macromolecules with positive anisotropy, as seen from Figure 22, for low values of the dipole moment ($0 \leqslant X_1^2 \leqslant 4X_2$) the $\Delta\varepsilon$ variations are positive, decrease to zero as the dipole moment increases, and become negative for large dipole moments ($X_1^2 \geqslant 4X_2$). With growing reorientation of permanent dipoles, the value of the negative saturation effect increases steeply, and complete saturation is achieved in the region of values $X_2 \geqslant 5$.

In the case of macromolecules with negative anisotropy, $\Delta\varepsilon$ is at first positive (as was the case for positively anisotropic ones) though twice as small [mind to take the lower sign in formula (324)], and changes to negative with growing dipolarity of the macromolecules and growing field strength. With macromolecules of this kind (as seen from Figure 23) less field strength is necessary for achieving the maximum of $\Delta\varepsilon$ variation and its complete saturation.

It is especially noteworthy that solutions of either kind of macromolecule (prolate as well as oblate) can exhibit inversion of the sign of $\Delta\varepsilon$ with growing applied field strength. Obviously, this requires a well-defined relationship between the reorientation parameters X_1 and X_2; namely, X_1 has to lie in the region $4X_2 \leqslant X_1^2 \leqslant 9X_2$ for prolate macromolecules, and $0 < X_1^2 \leqslant 4X_2$ for oblate ones.

Thus, the theory of non-linear $\Delta\varepsilon$ variations developed by Kasprowicz, Kielich, and Przeniczny [435] yields, in particular, the quadratic field-dependence, and quite generally the divergences from that quadraticity (of a parabolic or hyperbolic nature), including the ability to explain complete saturation, whether positive or negative. Once the reorientation parameters X_1 and X_2 are known numerically, the theory enables us to predict the kind of saturation to be expected and, inversely, it permits the determination of X_1 and X_2 from measurements of $\Delta\varepsilon$, thus enabling us to determine the geometry and electric structure of the macromolecule in solution. Obviously, the theory is applicable as long as the macromolecules do not interact with one another, or if interactions are negligible.

Up to now, experiments have revealed two types of complete dielectric saturation in macromolecular solutions. Parry-Jones, Gregson, and Davies [432] were the first to observe complete negative saturation in a solution of poly-γ-benzyl-L-glutamate in dioxan (see Figure 25). The reorientation parameters for this case are of the order of $X_1 = 10^{-5}\,E$ and $X_2 = 10^{-12}\,E^2$, yielding a ratio of $X_2/X_1^2 \approx 10^{-2}$, and pointing to a decisive role of reorientation of permanent dipoles, whereas saturation is described by values of the function $S_\alpha(X_1, \pm X_2)$ between the curves $9X_2 < X_1^2 \leqslant 100X_2$ in Figure 22. It is obvious that the shape of the experimental curve in Figure 25 also agrees with the curve of $R_p(X_1)$ in Figure 21. In the measurements of Davies *et al.*[432] the application of high electric fields to the solutions gave, in addition to a saturation effect, a large increment in electric permittivity on removal of the field. This anomalous behaviour is thought to be due to aggregation of the PBLG macromolecules.[432b] Recently, Gregson *et al.*[434] performed $\Delta\varepsilon$ measurements on the same solution applying, however, an analysing field of frequency 6 MHz, at which value the contribution of the dipolar component to the $\Delta\varepsilon$ variation of equation (318) is negligibly small. Whereas in the low-frequency measurements (at 1592 Hz) the electric permittivity ε decreased with increasing field strength,[432a] the permittivity ε_∞ (beyond the dispersion range) increased for all values of field strength in conformity with equation (322) and the reorientation function $S_\alpha(+X_2)$ of Figure 22 for $X_1 = 0$. These are

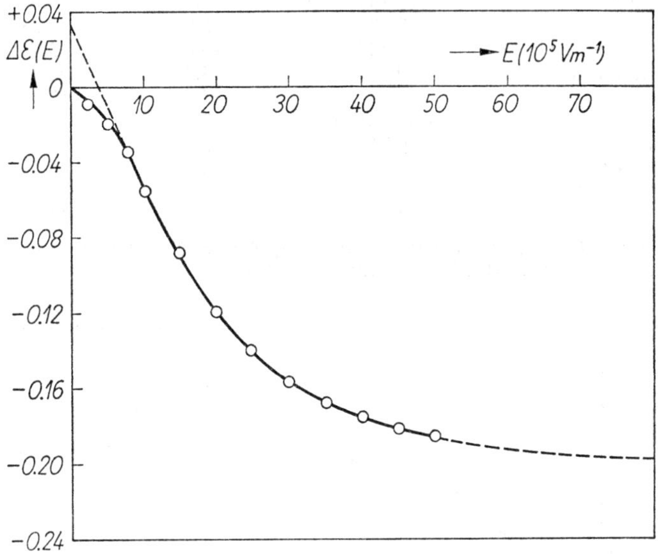

Figure 25 *Measurements by Davies et al.*[432] *of variations* $\Delta\varepsilon$ *as a function of field strength E for a solution of poly-γ-benzyl-L-glutamate (PBLG) in dioxan* (0.09905 g/100 cm^3) *at* 1592 Hz *and* 15.5 °C

highly important measurements, since they have permitted the determination of the electric anisotropy of the PBLG molecule, which amounts to $|\alpha_3 - \alpha_1| = 10^{-36}\ C^2\,m^2\,J^{-1}$, almost six times the value derived from measurements of the Kerr effect.[407] Also, the measurements quoted yield the values of dipole moments and permit their comparison with the values obtained by other methods (see Table 17).

Przeniczny[431] has observed positive $\Delta\varepsilon$ variations in solutions of non-dipolar and weakly dipolar macromolecules (see Figure 26). His measurements are qualitatively explained by equations (322) and (324). Parry-Jones *et al.*[465] have very recently communicated preliminary dielectric studies in the system of egg-yolk lecithin in hexane solution, resulting in a large positive effect $\Delta\varepsilon \approx 10^{-3}$ at a pulsed field strength of the order of $10^6\ V\,m^{-1}$. This work is of much importance to our knowledge of micellar aggregates in biological systems. Recently, Davies[436] has given a discussion

[465] G. Parry-Jones, M. Gregson, and T. Krupkowski, *Chem. Phys. Letters*, 1972, **13**, 266.

Table 17 *Permanent electric dipole moments and electric polarizability anisotropies for some macromolecules and particles*

Macromolecule or particle	*Mean length or diameter*/10^{-6} m	*Electric dipole moment* $p \times 10^{25}$ C m	*Electric anisotropy* $(\alpha_3 - \alpha_1)$ $\times 10^{30}$ C^2 m^2 J^{-1}
Fibrinogen[a]	0.07	0.4	0.0013
Poly-γ-benzyl-L-	0.134	0.9	0.00009
glutamate[b]	0.155	1.3	0.00013
Collagen[c]	0.28	5.0	0.33
Tobacco mosaic virus[d]	0.32	13.6	3.7
Deoxyribonucleic acid[e]	0.4	3.3	22
	0.75	2.4	12
Bentonite[f]	0.15—0.45	33	34.3
Palygorskite[g]	0.24	3.3	20
Attapulgite[h]	1.1	100	26.6
Bacterium coli[g]	1.0	100	311
Laponite clay[i]	0.98	5.7	25

(*a*) Ref. 80; (*b*) Refs. 407*a* and 468; (*c*) Ref. 407*c*; (*d*) Ref. 225*b*; (*e*) Refs. 469 and 470; (*f*) Ref. 471*a*; (*g*) Ref. 471*b*; (*h*) Ref. 472; (*i*) Ref. 473.

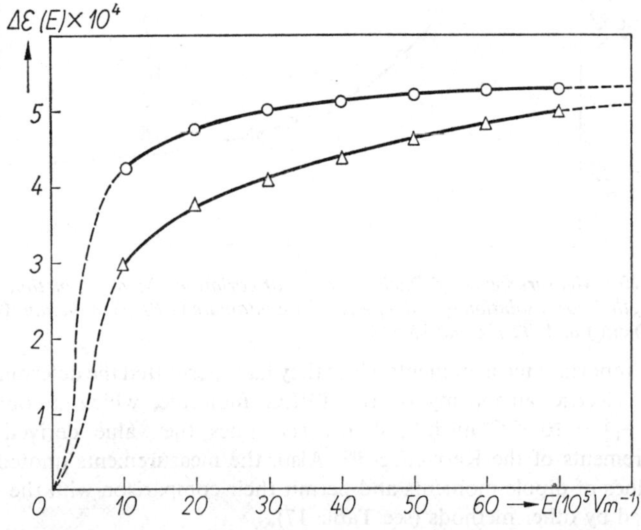

Figure 26 *Variations* $\Delta\varepsilon$ *in electric permittivity versus E for solutions of methyl polymethacrylate in benzene* (○) (0.0973 g/100 cm³) *and* (*isotactic*) *polystyrene in benzene* (△) (0.1104 g/100 cm³) *measured by Przeniczny*[431] *at a frequency of* 6.5 MHz *at* 20 °C. [*Measurements by Gregson have not confirmed the magnitude of these* $\Delta\varepsilon(E)$ *values* (*Senior Reporter*)]

of prospects in studies of electric field effects in biomolecular systems, especially with regard to helical polypeptide structures. Studies have also been made on non-linear electrical effects in lipid bilayer membranes,[466]

[466] D. Waltz, E. Bamberg, and P. Länger, *Biophys. J.*, 1969, **9**, 1150; B. Neumcke and P. Länger, *ibid.*, p. 1160; B. Neumcke, D. Waltz, and P. Länger, *ibid.*, 1970, **10**, 172.

as well as on birefringence of artificial lipidic membranes.[467] Supplementary information is provided by studies of light scattering in electric fields.[468–473]

Finally, we have to ask whether it is feasible to observe deviations from quadratic $\Delta\varepsilon$ variations in molecular systems as well. Such observations would require very strong electric fields on the one hand and a very high measuring sensitivity on the other. The theory proposed above is applicable in approximation to molecular substances on the condition that the reorientation parameters (312) and (313) shall not refer to the isolated molecules but rather to the effective dipole moment and effective electric anisotropy of some region of partial ordering of the molecules, or quasi-domain.[76] Admitting some well-defined structure of these regions, the reorientation parameters X_1 and X_2 could assume values sufficiently large to make $\Delta\varepsilon$ deviate perceptibly from the E^2 law. It is to be doubted, however, whether important deviations from quadraticity of $\Delta\varepsilon$ can be observed owing to molecular reorientation processes alone, without the participation of other mechanisms and effects.

Independently, a highly interesting problem recently revived is that of the alignment of molecules in liquid crystals due to externally applied static and alternating electric fields. The problem was first approached by Jeżewski[474] and Kast,[475] and was developed by Carr,[476] Helfrich,[477] Wysocki *et al.*,[478] and many others,[479] studying various aspects, such as dielectric loss in electric fields and anomalous alignment in the smectic phase and domains,[476] the effect of an electric field on the temperature of mesomorphic–isomorphic phase transitions of liquid crystals,[477] electric

[467] C. Lussan, A. Hou, and P. Bothorel, *Compt. rend.*, 1969, **269**, *B*, 639.
[468] B. R. Jennings and H. G. Jerrard, *J. Phys. Chem.*, 1965, **69**, 2817.
[469] A. Scheludko and S. Stoylov, *Biopolymers*, 1967, **5**, 723; G. Weill, C. Hornik, and S. Stoylov, *J. Chim. phys.*, 1968, **65**, 182.
[470] B. R. Jennings and H. Plummer, *Biopolymers*, 1970, **9**, 1361.
[471] (*a*) S. Stoylov, S. Sokerov, I. Penthanchin, and N. Ibroshev, *Doklady Akad. Nauk. S.S.S.R.*, 1968, **180**, 1165; (*b*) S. Stoylov, Proceedings of the 4th International Congress on Surface Activity, Brussels, 1964, p. 171.
[472] J. Plummer and B. Jennings, *Brit. J. Appl. Phys.*, 1968, **1**, 1753.
[473] J. Schweitzer and B. R. Jennings, *J. Phys.* (*D*), 1972, **5**, 297.
[474] M. Jeżewski, *J. Phys. Radium*, 1924, **5**, 59; *Z. Physik*, 1926, **40**, 153; 1928, **51**, 159; 1929, **52**, 268, 878.
[475] W. Kast, *Ann. Physik*, 1924, **73**, 145; *Z. Physik*, 1927, **42**, 81; 1931, **71**, 39; 1932, **76**, 19.
[476] E. F. Carr, *J. Chem. Phys.*, 1963, **39**, 1979; 1965, **42**, 738; 1965, **43**, 3905; R. P. Twitchell and E. F. Carr, *ibid.*, 1967, **46**, 2765; J. H. Parker and E. F. Carr, *ibid.*, 1971, **55**, 1846; E. F. Carr, *Phys. Rev. Letters*, 1970, **24**, 807; *Mol. Crystals and Liquid Crystals*, 1969, **7**, 253; 1971, **13**, 27.
[477] W. Helfrich, *J. Chem. Phys.*, 1969, **50**, 100; 1969, **51**, 4092; *Phys. Rev. Letters*, 1970, **24**, 201; W. Helfrich and M. Schadt, *ibid.*, 1971, **27**, 561.
[478] J. J. Wysocki, J. E. Adams, and W. Haas, *Phys. Rev. Letters*, 1968, **20**, 1024; W. Haas, J. E. Adams, and J. Wysocki, *Appl. Optics*, 1969, *Suppl.* 3, 196; J. J. Wysocki, J. E. Adams, and D. J. Olechna, *Proc. Amer. Chem. Soc.*, Symposium on Ordered Fluids and Liquid Crystals, New York, 1969, p. 419.
[479] (*a*) F. J. Kahn, *Phys. Rev. Letters*, 1970, **24**, 209; (*b*) L. Melamed and D. Rubin, *Appl. Phys. Letters*, 1970, **16**, 149; (*c*) C. Tani, *ibid.*, 1971, **19**, 241; (*d*) G. R. Luckhurst, *Chem. Phys. Letters*, 1971, **9**, 289; (*e*) W. H. De Jen, C. J. Gerritsma, and A. M. Van Boxtel, *Physics Letters*, 1971, **34A**, 203.

field-induced phase change in cholesteric liquid crystals,[478] electric field hysteresis effects,[479b] and electric field-induced changes in colour.[479a]

Electrostriction and the Electrocaloric Effect.—As mentioned previously, in measurements of the electric properties of dielectrics there intervene effects such as changes in volume V and temperature T of the body under the influence of the applied electric field. Their presence cannot be avoided altogether, though in many cases ingenuity and care in the choice of method makes possible a great reduction in their influence on the value obtained for $\Delta\varepsilon$. In all cases, however, when analysing the results of $\Delta\varepsilon$ measurements, one should strive to have a complete picture of the order of magnitude of the contributions from electrostriction and the electrocaloric effect. We shall now give a discussion of these two effects in a thermodynamical approach.[480, 481] By the way, electrostriction can be dealt with molecularly[482] as well.

Electrostriction. By the thermodynamics of dielectrics,[63] the thermodynamic potential (Helmholtz function F) of a dielectric in a uniform external field E_0 at constant pressure and temperature is:

$$F(E_0) = F_0 - \tfrac{1}{2}E_0 \cdot M(E) \tag{326}$$

where the total dipole moment of the dielectric of volume V immersed in a medium of permittivity ε_0 is:

$$M(E) = V\varepsilon_0 \chi_e \cdot E = V(\varepsilon_i - \varepsilon_0 U) \cdot E \tag{327}$$

ε_i and χ_e denoting respectively the tensors of electric permittivity and susceptibility of the dielectric under consideration, within which there exists a mean macroscopic field E.

By (327), the Helmholtz function (326) becomes:

$$F(E_0) = F_0 - \tfrac{1}{2}\varepsilon_0 VE_0 \cdot \chi_e \cdot E \tag{326a}$$

The volume of the dielectric in the presence of the field is defined as the derivative of Helmholtz's function with respect to the pressure P, at constant T and E:

$$V(E) = \left\{\frac{\partial F(E_0)}{\partial P}\right\}_{T,E} = V(0) - \tfrac{1}{2}\varepsilon_0\left\{\frac{\partial}{\partial P}(VE_0 \cdot \chi_e \cdot E)\right\}_{T,E} \tag{328}$$

We begin by assuming, for simplicity, that the isotropic dielectric changes its volume isotropically under the influence of the field, without undergoing any change in shape, *i.e.* that its deformation can be dealt with as being a uniform, hydrostatic expansion or compression. On this assumption,

[480] M. Abraham and R. Becker, 'The Classical Theory of Electricity and Magnetism', Blackie, London, 1932; E. A. Guggenheim, 'Thermodynamics', North-Holland, Amsterdam, 1949.

[481] (*a*) M. Pauthenier, *Ann. Physique*, 1920, **14**, 239; *J. Phys. Radium*, 1924, **5**, 312; 1925, **6**, 1; *Compt. rend.*, 1924, **179**, 1257; 1926, **182**, 121; (*b*) M. Bruhat and P. Pauthenier, *J. Phys. Radium*, 1925, **6**, 1; (*c*) J. Minard, *J. Rech. CNRS*, 1962, **13**, 60, 253.

[482] (*a*) B. K. P. Scaife, *Proc. Phys. Soc.* (*B*), 1956, **69**, 153; (*b*) J. F. Ely and D. A. McQuarie, *J. Phys. Chem.*, 1971, **75**, 771.

by the definition (328) and the formula:

$$E = \frac{3\varepsilon_0}{\varepsilon + 2\varepsilon_0} E_0 = \frac{3}{\varepsilon_r + 2} E_0$$

we have

$$\Delta V(E) = -\frac{3}{2}\varepsilon_0 \left\{\frac{\partial}{\partial P}\left(\frac{\chi_e V}{\varepsilon_r + 2}\right)\right\}_{T,E} E_0^2 \tag{328a}$$

for the field-induced change in volume of the isotropic dielectric, which we can re-write more extensively as follows:

$$\Delta V(E) = -\frac{\varepsilon_0 V}{2}\left\{\left(\frac{\partial \varepsilon_r}{\partial P}\right)_T - (\varepsilon_r - 1)\left(\frac{\varepsilon_r + 2}{3}\right)\kappa_T\right\} E^2 \tag{329}$$

where $\kappa_T = -V^{-1}(\partial V/\partial P)_T$, the isothermal compressibility.

In approximation, one has:

$$\left(\frac{\partial \varepsilon_r}{\partial P}\right)_T = \kappa_T\left(\rho\frac{\partial \varepsilon_r}{\partial \rho}\right)_T \tag{330}$$

and the change in volume (329) takes the form:

$$\Delta V(E) = -\frac{\varepsilon_0 V}{2}\kappa_T\left\{\left(\rho\frac{\partial \varepsilon_r}{\partial \rho}\right)_T - (\varepsilon_r - 1)\left(\frac{\varepsilon_r + 2}{3}\right)\right\}E^2 \tag{329a}$$

Equation (329) is applicable when the change in permittivity as a function of pressure is known from experiment. Equation (329a) is of more general use, since we more often know the change in permittivity as a function of density. Methods for the determination of (330) have been worked out by Fabelinskii.[27]

On resorting to the Clausius–Mosotti or the Langevin–Debye formula, we obtain:

$$\left(\rho\frac{\partial \varepsilon_r}{\partial \rho}\right)_T = (\varepsilon_r - 1)\left(\frac{\varepsilon_r + 2}{3}\right) \tag{330a}$$

implying the vanishing of the change in volume (329a). The circumstance was noted by Scaife,[482] who at the same time showed that the electrostriction is not zero if the change in permittivity is calculated from the Yvon–Kirkwood or the Onsager theory.[483] The foregoing conclusions have been confirmed by experimental studies performed by various techniques,[484–487] as well as by numerical calculations.[488]

[483] H. Falkenhagen and H. Jacob, *Ann. Physik* (*Leipzig*), 1951, **8**, 105.
[484] (*a*) O. E. Frivold, *Physik. Z.*, 1921, **22**, 603; (*b*) W. E. Danford, *Phys. Rev.*, 1931, **38**, 1224.
[485] (*a*) H. Goetz and W. Zahn, *Z. Physik*, 1958, **151**, 202; (*b*) W. Zahn, *ibid.*, 1962, **166**, 275.
[486] V. A. Petukhov, *Akust. Zhur. S.S.S.R.*, 1958, **4**, 286.
[487] L. Bobrowicz, *Fizyka Dielektryków i Radiospektroskopia* (*Poznań*), 1972, **6/1**, 93.
[488] (*a*) S. Kielich, *Physics Letters*, 1967, **24A**, 435; (*b*) B. Kasprowicz and S. Kielich, *Acta Phys. Polon.*, 1967, **31**, 787.

In the general case, electrostriction is not isotropic even in isotropic dielectrics: a field-induced anisotropy of electrostriction appears [489, 490] and is perceptible also in measurements of the static [481] and optical [491] Kerr effects. This anisotropy of electrostriction is particularly apparent if the field acting on the dielectric is strong, such as the intense electromagnetic field of a light wave,[213] in which case the time-dynamics of the electrostrictive effect have to be studied separately.[492, 493]

The majority of authors [27, 494] do not consider changes in volume in the field but rather the so-called strictive pressure arising in a medium immersed in a strong electric field; they write, by approximation:

$$\Delta P(E) = \frac{\varepsilon_0}{2}\left(\rho\frac{\partial\varepsilon_r}{\partial\rho}\right)_T E^2 \tag{331}$$

The changes in volume $\Delta V(E)$ and pressure $\Delta P(E)$ induced by the field E cause the following changes in permittivity:

$$\Delta\varepsilon_r^V(E) = \left(\frac{\partial\varepsilon_r}{\partial V}\right)_{T,E}\Delta V(E) \tag{332}$$

$$\Delta\varepsilon_r^P(E) = \left(\frac{\partial\varepsilon_r}{\partial P}\right)_{T,E}\Delta P(E) \tag{333}$$

where

$$V\frac{\partial\varepsilon_r}{\partial V} = -\rho\frac{\partial\varepsilon_r}{\partial\rho}$$

The variation (332) assumes the form:

$$\Delta\varepsilon_r^V(E) = -\left(\rho\frac{\partial\varepsilon_r}{\partial\rho}\right)_{T,E}\frac{\Delta V(E)}{V} \tag{332a}$$

or, on inserting the change in volume (329a),

$$\Delta\varepsilon_r^V(E) = \tfrac{1}{2}\varepsilon_0\,\kappa_T\left\{\left(\rho\frac{\partial\varepsilon_r}{\partial\rho}\right)_T^2 - (\varepsilon_r-1)\left(\frac{\varepsilon_r+2}{3}\right)\left(\rho\frac{\partial\varepsilon_r}{\partial\rho}\right)_T\right\}E^2 \tag{334}$$

Likewise, by resorting to the relation (330), one obtains by (333) the following change in permittivity due to electrostrictive pressure:

$$\Delta\varepsilon_r^P(E) = \tfrac{1}{2}\varepsilon_0\,\kappa_T\left(\rho\frac{\partial\varepsilon_r}{\partial\rho}\right)_T^2 E^2 \tag{335}$$

On the assumption of the approximation (330a) the variation (334) vanishes, whereas the variation (335) takes the form:

$$\Delta\varepsilon_r^P(E) = \tfrac{1}{2}\varepsilon_0\,\kappa_T(\varepsilon_r-1)^2\left(\frac{\varepsilon_r+2}{3}\right)^2 E^2 \tag{335a}$$

489 V. A. Zamkov, *Optics and Spectroscopy*, 1963, **15**, 355.
490 S. Kielich, *Acta Phys. Polon.*, 1968, **34**, 1093.
491 F. B. Martin, Doctor's Thesis, University of Bordeaux, 1971.
492 Y. R. Shen, *Physics Letters*, 1966, **20**, 378; Y. R. Shen and Y. J. Shaham, *Phys. Rev.*, 1967, **163**, 224.
493 E. Kerr, *IEEE J. Quantum Electronics*, 1970, **6**, 616.
494 F. Gires, *Ann. Radioélectricité*, 1968, **23**, 281.

The differences between formulae (334) and (335) require some discussion. Strictly, formulae (331) and (332) are valid on the assumption that the volume of the dielectric does not change. If this assumption is not made, the change in pressure due to electrostriction is not given by formula (331) but by the more general formula:[482]

$$\Delta P(E) = -\tfrac{3}{2}\varepsilon_0\left\{\frac{\partial}{\partial V}\left(\frac{\chi_e V}{\varepsilon_r + 2}\right)\right\}_{T,E} E_0^2 \qquad (331a)$$

For small variations in pressure, we obtain by (328a) and (331a) the following, obvious relation:

$$\Delta P(E) = -\frac{1}{\kappa_T}\frac{\Delta V(E)}{V} \qquad (331b)$$

which states that the two processes of change, if calculated strictly, lead in the end to the same result for $\Delta\varepsilon(E)$.

In the case of electric permittivity at optical frequencies, the electrostrictive variations can be determined from isotropic light scattering on fluctuations of density.[488, 490] 'Optico-striction' is one of the more important factors leading to the self-focusing and self-trapping of laser beams.[151, 213]

Adiabatic Electrocaloric Effect. If the system is insulated thermally, the switching on of an external electric field causes its temperature to rise by ΔT. This is known as the adiabatic electrocaloric effect which, with regard to (326a), is defined by the change in temperature of the dielectric:

$$\Delta T(E) = -\frac{T\varepsilon_0}{2C_P}\left\{\frac{\partial}{\partial T}(VE_0 \cdot \chi_e \cdot E)\right\}_{P,E} \qquad (336)$$

C_P being the specific heat capacity at constant pressure.

For an isotropic spherical dielectric sample, formula (336) yields:

$$\Delta T(E) = -\frac{\varepsilon_0 TV}{2C_P}\left\{\left(\frac{\partial\varepsilon_r}{\partial T}\right)_P + (\varepsilon_r - 1)\left(\frac{\varepsilon_r + 2}{3}\right)\alpha_P\right\}E^2 \qquad (336a)$$

where $\alpha_P = 1/V(\partial V/\partial T)_P$ is the cubic expansivity at constant pressure.

The electric-field-induced change in temperature (336) causes a change in permittivity of the dielectric which we define as follows:

$$\Delta\varepsilon_r^T(E) = \left(\frac{\partial\varepsilon_r}{\partial T}\right)_{P,E}\Delta T(E) \qquad (337)$$

or, with regard to (336a), in the following way better adapted to numerical calculations:

$$\Delta\varepsilon_r^T(E) = -\frac{\varepsilon_0 TV}{2C_P}\left\{\left(\frac{\partial\varepsilon_r}{\partial T}\right)_P^2 + (\varepsilon_r - 1)\left(\frac{\varepsilon_r + 2}{3}\right)\left(\frac{\partial\varepsilon_r}{\partial T}\right)_P\alpha_P\right\}E^2 \qquad (337a)$$

In a satisfactory approximation, we have the relation:[27]

$$\frac{1}{\alpha_P}\left(\frac{\partial\varepsilon_r}{\partial T}\right)_P \simeq -\left(\rho\frac{\partial\varepsilon_r}{\partial\rho}\right)_T$$

permitting the reduction of (337a) to the form:

$$\Delta\varepsilon_r^T(E) = -\frac{\varepsilon_0 TV}{2C_P}\alpha_P^2\left\{\left(\rho\frac{\partial\varepsilon_r}{\partial\rho}\right)_T^2 - (\varepsilon_r - 1)\left(\frac{\varepsilon_r + 2}{3}\right)\left(\rho\frac{\partial\varepsilon_r}{\partial\rho}\right)_T\right\}E^2 \quad (338)$$

On using the approximation (330a) this variation vanishes. Many authors use the following, approximate formula for the electrocaloric effect:[27, 494]

$$\Delta\varepsilon_r^T(E) = -\frac{\varepsilon_0 TV}{2C_P}\left(\frac{\partial\varepsilon_r}{\partial T}\right)_T^2 E^2 \quad (338a)$$

or

$$\Delta\varepsilon_r^T(E) = -\frac{\varepsilon_0 TV}{2C_P}\alpha_P^2\left(\rho\frac{\partial\varepsilon_r}{\partial\rho}\right)_T^2 E^2$$

$$= -\frac{\varepsilon_0 TV}{2C_P}\alpha_P^2(\varepsilon_r - 1)^2\left(\frac{\varepsilon_r + 2}{3}\right)^2 E^2 \quad (338b)$$

Dielectric Ellipsoid. The general formulae (328) and (336) permit the calculation of electrostrictive and electrocaloric changes in electric permittivity of a dielectric of arbitrary geometrical shape. To this aim, one has only to resort to the relation between the macroscopic field E existing in the anisotropic dielectric and the electric field in vacuum:

$$\boldsymbol{E}_0 - \{\boldsymbol{U} + (\boldsymbol{\varepsilon}_r - \boldsymbol{U})\cdot\boldsymbol{L}\}\cdot\boldsymbol{E} \quad (339)$$

Assuming the linear approximation, we have:

$$\boldsymbol{M}(\boldsymbol{E}) = \boldsymbol{A}\cdot\boldsymbol{E}_0 \quad (340)$$

whence by (326) we obtain the following change in Helmholtz function:

$$F(E) = -\tfrac{1}{2}\boldsymbol{E}_0\cdot\boldsymbol{A}\cdot\boldsymbol{E}_0 \quad (341)$$

due to a strong external field $\boldsymbol{E}_0$. The analytic form of the electric polarizability tensor $\boldsymbol{A}$ of the dielectric ellipsoid in vacuum is to be had from a comparison of equations (327), (339), and (340):

$$\boldsymbol{A}\cdot\{\boldsymbol{U} + (\boldsymbol{\varepsilon}_r - \boldsymbol{U})\cdot\boldsymbol{L}\} = \varepsilon_0 V(\boldsymbol{\varepsilon}_r - \boldsymbol{U}) \quad (342)$$

From equation (342), the principal tensor elements of the polarizability tensor $\boldsymbol{A}$ are:[139]

$$A_s = \frac{\varepsilon_0 V(\varepsilon_{rs} - 1)}{1 + (\varepsilon_{rs} - 1)L_s} \quad (342a)$$

where ε_{rs} and L_s are, respectively, the principal elements ($s = x, y, z$) of the electric permittivity tensor $\boldsymbol{\varepsilon}_r$ and the shape tensor $\boldsymbol{L}$ of the ellipsoid.

By (342), the change in Helmholtz function (341) is:

$$F(E_0) = -\tfrac{1}{2}\varepsilon_0\left\{\frac{V(\varepsilon_{rs} - 1)}{1 + (\varepsilon_{rs} - 1)L_s}\right\}E_{0s}^2 \quad (342b)$$

leading to the following expressions for the change in volume and temperature due to the field:

$$\Delta V(E) = -\tfrac{1}{2}\varepsilon_0 V\left\{\left(\frac{\partial \varepsilon_{rs}}{\partial P}\right)_T - (\varepsilon_{rs} - 1)[1 + (\varepsilon_{rs} - 1)L_s]\kappa_T\right\} E_s^2 \tag{343}$$

$$\Delta T(E) = -\frac{\varepsilon_0 TV}{2C_P}\left\{\left(\frac{\partial \varepsilon_{rs}}{\partial T}\right)_P + (\varepsilon_{rs} - 1)[1 + (\varepsilon_{rs} - 1)L_s]\alpha_P\right\} E_s^2 \tag{344}$$

For an isotropic dielectric ($\varepsilon_{rs} = \varepsilon_r$) in the shape of a sphere ($L_s = \frac{1}{3}$), equations (343) and (344) reduce to formulae (329) and (336a) respectively.

Formulae (343) and (344) are applicable to samples of various shape. Thus, for a dielectric in the shape of a plane-parallel plate in a perpendicularly applied field ($\varepsilon_{rs} = \varepsilon_r$, $L_s = 1$),

$$\Delta\varepsilon_r^V(E) = \frac{\varepsilon_0}{2\kappa_T}\left\{\left(\frac{\partial \varepsilon_r}{\partial P}\right)_T^2 - \kappa_T\,\varepsilon_r(\varepsilon_r - 1)\left(\frac{\partial \varepsilon_r}{\partial P}\right)_T\right\} E^2 \tag{345}$$

$$\Delta\varepsilon_r^T(E) = -\frac{\varepsilon_0 TV}{2C_P}\left\{\left(\frac{\partial \varepsilon_r}{\partial T}\right)_P^2 + \alpha_P\,\varepsilon_r(\varepsilon_r - 1)\left(\frac{\partial \varepsilon_r}{\partial T}\right)_P\right\} E^2 \tag{346}$$

For a dielectric of the shape of a plane-parallel plate or arbitrary cylinder in a longitudinally applied field ($L_s = 0$), equations (343) and (344) yield:

$$\Delta\varepsilon_r^V(E) = \frac{\varepsilon_0}{2\kappa_T}\left\{\left(\frac{\partial \varepsilon_r}{\partial P}\right)_T^2 - \kappa_T(\varepsilon_r - 1)\left(\frac{\partial \varepsilon_r}{\partial P}\right)_T\right\} E^2 \tag{347}$$

$$\Delta\varepsilon_r^T(E) = -\frac{\varepsilon_0 TV}{2C_P}\left\{\left(\frac{\partial \varepsilon_r}{\partial T}\right)_P^2 + \alpha_P(\varepsilon_r - 1)\left(\frac{\partial \varepsilon_r}{\partial T}\right)_P\right\} E^2 \tag{348}$$

Numerical calculations of changes in electric permittivity due to electrostriction and the electrocaloric effect performed with formulae (334), (335), (338), and (345)—(348) are listed in Tables 18 and 19. We note that for some liquids the two effects have opposite signs, thus lowering the total effect on $\Delta\varepsilon$. It is interesting to note that the changes in $\Delta\varepsilon$ calculated from the strict formulae (334), (338), and (345)—(348) are considerably smaller than those calculated with the approximate formulae (335) and (338a). Formulae (334), (338), and (345)—(348) leave the door open for an inversion in sign of the respective permittivity variations in some bodies. This is the situation in dielectrics, in contrast to electrostriction in conductors, which increases in all cases under the effect of an external field.[63] Hitherto, these formulae have not been subjected to a verification by direct experiment. Tables 18 and 19 contain numerical values of the electrostrictive and electrocaloric effects, which perturb the measurement of changes in dielectric permittivity of dielectrics under the influence of a strong field E.

In view of what has been said, the experimentally measured change in electric permittivity is composed of *at least* three effects:[142]

$$\Delta\varepsilon_r^M(E) = \Delta\varepsilon_r^S(E) + \Delta\varepsilon_r^V(E) + \Delta\varepsilon_r^T(E) \tag{349}$$

Table 18 *$\Delta\varepsilon_r^V(E)$ variations due to electrostriction in dielectric samples of various shapes calculated with the data listed in ref.* 488*b*

Liquid	ε_r (20 °C)	$\kappa_T/10^{-10}$ m² N⁻¹	$\left(\rho \frac{\partial \varepsilon_r}{\partial \rho}\right)_T$	$\Delta\varepsilon_r^V E^{-2}/10^{-20}$ m² V⁻²			
				Eqn. (334)	Eqn. (335)	Eqn. (345)	Eqn. (347)
Carbon tetrachloride	2.24	10.6	1.45	− 0.21	+ 0.99	− 0.91	+ 0.14
Benzene	2.28	9.5	1.65	− 0.18	+ 1.15	− 0.88	+ 0.26
Carbon disulphide	2.63	9.4	2.37	− 0.14	+ 2.34	− 1.89	+ 0.73
Cyclohexane	2.03	11.2	1.29	− 0.06	+ 0.82	− 0.51	+ 0.17
Toluene	2.38	9.5	1.65	− 0.28	+ 1.02	− 1.10	+ 0.13
Chloroform	4.80	8.7	4.48	− 7.16	+ 7.74	− 23.75	+ 1.18
Nitrobenzene	35.5	4.9	64.7	− 5147.8	+ 908.7	− 16292.2	+ 424.1
Acetone	21.4	12.5	30.1	− 2151.6	+ 502.4	− 6779.0	+ 162.1

Table 19 $\Delta\varepsilon_r^T(E)$ *variations due to electrocaloric effect in dielectric samples of various shapes calculated with the data listed in ref.* 488*b*

Liquid	$V/C_P/$ 10^{-6} m^3 K J^{-1}	$\alpha_P/10^{-3}$ K^{-1}	$\left(\frac{\partial\varepsilon_r}{\partial T}\right)_P/10^{-3}$ K^{-1}	$\Delta\varepsilon_r^T E^{-2}/10^{-20}$ m^2 V^{-2}			
				Eqn. (337*a*)	Eqn. (338*a*)	Eqn. (346)	Eqn. (348)
Carbon tetrachloride	0.74	1.21	− 1.71	+ 0.09	− 0.28	+ 0.27	− 0.04
Benzene	0.66	1.21	− 1.92	+ 0.05	− 0.32	+ 0.27	− 0.06
Carbon disulphide	0.80	1.19	− 2.67	+ 0.09	− 0.74	+ 0.67	− 0.20
Cyclohexane	0.69	1.21	− 1.53	+ 0.02	− 0.20	+ 0.14	− 0.04
Toluene	0.66	1.08	− 1.68	+ 0.07	− 0.24	+ 0.27	− 0.03
Chloroform	1.04	1.28	− 5.74	+ 4.15	− 4.45	+ 13.65	− 0.68
Nitrobenzene	0.55	0.83	− 53.71	+ 1168.32	− 206.26	+ 3698.2	− 96.28
Acetone	0.59	1.43	− 43.07	+ 609.39	− 142.28	+ 1919.87	− 45.92

namely an effect of dielectric saturation $\Delta\varepsilon_{\Gamma}^{S}(E)$ due to various statistical–molecular mechanisms, the thermodynamic electrostrictive effect $\Delta\varepsilon_{\Gamma}^{T}(E)$, and an electrocaloric effect $\Delta\varepsilon_{\Gamma}^{T}(E)$. When proceeding to a theoretical analysis of the experimental data, it is of the utmost importance to keep this in mind and not to jump to conclusions, before one is absolutely clear as to the numerical value and sign of each of the contributions to equation (349). Especial circumspection is needed in situations in which the measured $\Delta\varepsilon_{\Gamma}^{M}(E)$ value obeys the quadratic E^2 law as is the case with the majority of molecular substances, where the various contributions can be of the same order of magnitude. In macromolecular substances, dielectric saturation $\Delta\varepsilon_{\Gamma}^{S}$ is sure to predominate. Besides, much depends on the choice of the measuring method and experimental conditions.

6 Concluding Remarks and Recapitulation

Related Electric-field-induced Phenomena.—The studies of the two important phenomena of electrically induced optical birefringence and electrical saturation in isotropic dielectrics are of immense significance for the obtaining of information on the non-linear electro-optical and electric properties of atoms, molecules, and macromolecules, as well as on the structure of media consisting of such microsystems. Contemporary studies, however, bear on yet other non-linear effects, induced in matter by strong electric fields. A fuller discussion of these effects would unduly extend this review article, but we shall concisely state their characteristics and the kind of information available from their study.

Electric Field Dependence of Optical Rotation. For some years now, structural chemists, in addition to methods of optical rotation [19] (o.r.) and circular dichroism [22] (c.d.), have employed extensively optical rotatory dispersion [20, 21] (o.r.d.), both normal and anomalous. These studies have permitted the elucidation of otherwise inaccessible fine details of the structure of molecules and macromolecules. The present rapid development of experimental and theoretical work on natural optical activity is certainly being stimulated by the demands of biochemistry and biophysics. Experimental techniques for o.r., c.d., and o.r.d. measurements have undergone considerable perfection recently, as laboratories have at their disposal ready-made, standardized spectropolarimeters and dichroigraphs. Molecular optical activity generally differs along specific axes of a macromolecule; thus, the specific rotation associated with the major axis of a helical macromolecule differs from that of a transverse or minor axis. In dilute solution, orientation being random, the system as a whole exhibits no anisotropy in optical rotation. However, if the solute macromolecules are partially or completely aligned in the direction of an external static electric field E, the molar optical rotatory power α_m becomes a function of E. Quite generally, the field-induced change in optical rotatory power is:[217, 408]

$$\frac{\alpha_m(E) - \alpha_m}{\alpha_m} = \frac{(2g_{33} - g_{11} - g_{22})}{2(g_{11} + g_{22} + g_{33})}\Phi(X_1, \pm X_2) \tag{350}$$

where g_{11}, g_{22}, g_{33} are the principal elements of the natural gyration tensor.

The theory of quadratic variations in optical activity with respect to the electric field strength was first formulated for macromolecules by Tinoco and Hammerle,[495] and then developed by others.[496] The earliest experiments are due to Tinoco [497] in solutions of poly-γ-L-glutamate in ethylene dichloride; of late, this experiment has been extended to transient optical rotation changes by Jennings and Baily.[498] Also, electric field effects on the optical rotatory power of a compensated cholesteric liquid crystal have been studied.[499]

Electrical Dichroism. Owing to different absorption of right and left polarized light waves, optically active substances (beside rotating the polarization plane) convert linearly polarized light into elliptically polarized light in the region of absorption. A solution containing absorbing molecules, if placed in a strong electric field, exhibits anisotropy of absorption in the absorption band region specific to the solute molecules or macromolecules. This phenomenon of electric dichroism was detected and explained by Kuhn *et al.*[500] Subsequent investigation revealed that the effect, which is due to orientation of the molecules by the strong field, is accompanied by a shift in the absorption band of the solution as a result of changes in dipole moment at excitation. The theory of electric dichroism taking into consideration the shift in absorption band has been worked out by Liptay and Czekalla,[501] to whom is also due a method for the determination of electric dipole moments in excited states.

Another original approach to the application of electric dichroism for the characterization of colloidal particles is due to Tolstoi *et al.*[502] and, with regard to macromolecular solutions, to Dvorkin *et al.*[503] and later authors.[504]

[495] (*a*) I. Tinoco, jun., and W. G. Hamerle, *J. Phys. Chem.*, 1956, **60**, 1619; (*b*) R. W. Woody and I. Tinoco, jun., *J. Chem. Phys.*, 1967, **46**, 4927.

[496] (*a*) N. Go, *J. Chem. Phys.*, 1965, **43**, 1275; *J. Phys. Soc. Japan*, 1966, **21**, 1579; 1967, **23**, 88; (*b*) R. L. Disch and D. J. Sverdlik, *J. Chem. Phys.*, 1967, **47**, 2137; (*c*) Y. N. Chiu, *ibid.*, 1969, **50**, 5336; 1970, **52**, 1042.

[497] I. Tinoco, jun., *J. Amer. Chem. Soc.*, 1957, **79**, 4248; 1959, **81**, 1540.

[498] B. R. Jennings and E. D. Baily, *Nature*, 1970, **228**, 1309.

[499] H. Baessler, T. M. Laronge, and M. M. Labes, *J. Chem. Phys.*, 1969, **51**, 3213.

[500] W. Kuhn, H. Duhrkop, and H. Martin, *Z. phys. Chem.*, 1939, **B45**, 121.

[501] (*a*) W. Liptay and J. Czekalla, *Z. Naturforsch.*, 1960, **15a**, 1072; (*b*) J. Czekalla and G. Wick, *Z. Electrochem.*, 1961, **65**, 727; (*c*) W. Liptay, *Z. Naturforsch.*, 1965, **20a**, 272, 1441.

[502] N. A. Tolstoi, A. A. Spartakov, and G. N. Khilko, *Kolloid-Z.*, 1960, **22**, 705; N. A. Tolstoi and A. A. Spartakov, *ibid.*, 1966, **28**, 580; N. A. Tolstoi, A. A. Spartakov, and A. A. Trusov, *ibid.*, p. 735; 1967, **29**, 584.

[503] G. A. Dvorkin, *Doklady Akad. Nauk S.S.S.R.*, 1960, **135**, 739; G. A. Dvorkin and A. S. Spirin, *ibid.*, 1960, **135**, 987; G. A. Dvorkin, *Biofizika*, 1961, **6**, 403; E. I. Golub and G. A. Dvorkin, *ibid.*, 1964, **9**, 545.

[504] (*a*) Y. Shindo and R. S. Stein, *J. Polymer Sci., Part A-2, Polymer Phys.*, 1969, **7**, 2115; (*b*) F. S. Allen and K. E. Van Holde, *Biopolymers*, 1971, **10**, 865.

Polarization of Fluorescence in an Electric Field. The polarization of the fluorescence from molecules in liquid solution under the influence of an electric field was first observed by Czekalla,[505] who proposed a classical phenomenological theory of the effect and methods for measuring the degree of polarization of fluorescence and determining the dipole moments of molecules in excited states. In isotropic solutions of fluorescent dyes, no direction of vibrations of the molecular oscillators is favoured. In the process of absorption, only those molecules are favoured whose vibration directions lie along the electric vector of the exciting light wave. An externally applied electric field causes partial orientation of the dipolar molecules in solution in the non-dipolar solvent, thus modifying the degree of polarization of the fluorescence light, which is found to depend on the square of the electric field strength.[505, 506] The reader interested in other methods of studying electric dichroism and electric polarization of luminescence is referred to the review articles of Labhart.[507]

Electric-field-induced Spectra. A light-emitting or -absorbing molecule in solution is by no means a free molecule, as is the case in vapours, but is ceaselessly subject to interaction with the solvent molecules surrounding it, leading to changes in its optical and electrical properties. The earliest interpretation of the resulting shift in absorption and fluorescence bands was proposed by Jabłoński,[508] according to whom the molecule of a dye has in general initial energy levels for emission different from those immediately attained during absorption. The optical properties of molecules are accessible to study not only by way of interaction with an external electric field, but also through interaction with the surrounding medium. Bilot and Kawski [509] performed a quantum-mechanical calculation of the interaction between a solute molecule and the solvent by perturbation calculus on the assumption of Onsager's model of an electric reaction field of the solvent acting on the dipole molecule, situated at the centre of a spherical cavity.

The effect of an external electric field on the absorption spectra of solute molecules has been studied by many authors,[510–512] showing these methods to be a tool for determinations of ground-state as well as excited-state dipole moments and transition polarizabilities (linear and non-linear) of such systems.

505 (*a*) J. Czekalla, *Z. Elektrochem.*, 1960, **64**, 1221; *Chimia* (*Switz.*), 1961, **15**, 26; (*b*) W. Liptay, *Z. Naturforsch.*, 1963, **18a**, 705.
506 G. Weber, *J. Chem. Phys.*, 1965, **43**, 521.
507 H. Labhart, *Chimia* (*Switz.*), 1961, **15**, 20; *Tetrahedron*, 1963, **19** (suppl. 2), 223; *Experientia*, 1966, **22**, 65.
508 A. Jabłoński, *Z. Physik*, 1931, **73**, 460.
509 L. Bilot and A. Kawski, *Z. Naturforsch.*, 1962, **17a**, 621; 1963, **18a**, 10, 256; *Acta Phys. Polon.*, 1962, **22**, 289; A. Kawski, *ibid.*, 1964, **25**, 285; *Bull. Acad. polon. Sci., Sér. Sci. math. phys. astr.*, 1964, 12, 179; A. Kawski and L. Bilot, *Acta Phys. Polon.*, 1964, **26**, 41; *Naturwiss.*, 1964, **51**, 82.
510 N. J. Bridge, D. A. Haner, and D. Dows, *J. Chem. Phys.*, 1968, **48**, 4196; 1966, **44**, 3128; K. T. Huang and J. R. Lombardi, *ibid.*, 1971, **55**, 40, 4072.
511 H. N. Spector, *Physica*, 1966, **32**, 1551; 1964, **30**, 1917.
512 (*a*) A. Schweig, *Mol. Phys.*, 1968, **15**, 1; (*b*) C. A. G. O. Varma and L. J. Oosterhoff, *Chem. Phys. Letters*, 1971, **9**, 406.

Electric Field Effect on Light Scattering. As yet, it has not been possible to detect changes in scattered light intensity due to an electric field in molecular substances.[223] However, an electric field effect on the critical opalescence has been reported.[513] In recent years, light-scattering studies in the presence of static or alternating electric fields in macromolecular and colloidal solutions have been pursued increasingly,[225, 468–473] revealing E^2 variations as well as complete electric saturation of scattered light,[225c] permitting the numerical determination of the polarizability anisotropy of macromolecules and its sign.[514] The experimental methods of studying electric light scattering and transient scattering variations induced by static electric fields are discussed in a review article by Stoylov[135] and in other recent papers.[470–473]

Non-linear Polarizabilities of Atoms and Molecules.—The linear electrical and optical properties of atoms and molecules are described by the second-rank tensor $\boldsymbol{a}(\omega)$ occurring in the equation of the first-order dipole moment $\boldsymbol{p}^{(1)}(\omega)$ induced by an electric field $\boldsymbol{E}(\omega)$:

$$\boldsymbol{p}^{(1)}(\omega) = \boldsymbol{a}(-\omega, \omega) \cdot \boldsymbol{E}(\omega) \tag{351}$$

The elements of $\boldsymbol{a}(\omega)$ are determined from studies of molecular refraction, anisotropic light scattering,[515] the Kerr effect,[132] and optical birefringence.[214] Frequency-dependent polarizabilities can also be calculated directly to various orders of approximation by methods of quantum mechanics.[516]

Research has of late been developing on the spectra of depolarized light in simple gases and liquids,[517–522] leading to the determination of pair polarizability of atoms or molecules[523, 524] as well as of the polarizability anisotropies induced by multiple correlations.[523, 525]

In a molecule acted on simultaneously by two electric fields $\boldsymbol{E}(\omega_1)$ and $\boldsymbol{E}(\omega_2)$, oscillating at frequencies ω_1 and ω_2, a second-order dipole moment

513 (*a*) P. Debye and K. Kleboth, *J. Chem. Phys.*, 1965, **42**, 3155; (*b*) P. Debye, C. C. Gravatt, and M. Ieda, *ibid.*, 1967, **46**, 2352; (*c*) M. J. Cooper and R. D. Mountain, *ibid.*, 1968, **48**, 1064.

514 S. Kielich, *Optics Comm.*, 1970, **1**, 345; *Acta Phys. Polon.* (*A*), 1970, **37a**, 447, 719; *J. Colloid Interface Sci.*, 1970, **34**, 228.

515 A. Pacault, The Jerusalem Symposia on Quantum Chemistry and Biochemistry, III, Jerusalem 1971, p. 39.

516 (*a*) H. P. Kelly, *Phys. Rev.*, 1969, **182**, 84; *ibid.*, 1970, **1A**, 274; (*b*) T. Ando and A. Saika, *J. Chem. Phys.*, 1971, **55**, 4203.

517 (*a*) H. B. Levine and G. Birnbaum, *Phys. Rev. Letters*, 1968, **20**, 439; (*b*) J. P. McTague and G. Birnbaum, *ibid.*, 1968, **21**, 661.

518 J. P. McTague, P. A. Fleury, and D. B. Du Pré, *Phys. Rev.*, 1968, **188**, 303.

519 W. S. Gornal, H. E. Howard-Lock, and B. P. Stoicheff, *Phys. Rev.*, 1970, **1A**, 1288.

520 B. Simiz-Glavaski, D. A. Jackson, and J. G. Powles, *Phys. Letters*, 1971, **34A**, 255.

521 M. Thiyeau, G. C. Tabisz, B. Oksengorn, and B. Vodar, *J. Quant. Spectroscopy Radiative Transfer*, 1970, **10**, 839.

522 J. P. McTague and G. Birnbaum, *Phys. Rev.*, 1971, **3A**, 1376.

523 S. Kielich, *Chem. Phys. Letters*, 1968, **2**, 112; *Optics Comm.*, 1971, **4**, 135; S. Kielich and S. Woźniak, *Acta Phys. Polon.* (*A*), 1971, **39**, 233.

524 (*a*) D. B. Du Pré and J. P. McTague, *J. Chem. Phys.*, 1969, **50**, 2024; (*b*) H. B. Levine and G. Birnbaum, *ibid.*, 1971, **55**, 2914.

525 H. L. Frisch and J. McKenna, *Phys. Rev.*, 1965, **139**, A68.

$p^{(2)}(\omega_3)$ at frequency $\omega_3 = \omega_1 + \omega_2$ is induced:

$$p^{(2)}(\omega_3) = \beta(\omega_3)\, b(-\omega_3; \omega_1, \omega_2) : E(\omega_1)\, E(\omega_2) \qquad (352)$$

where the ordering factor $\beta(\omega_3)$ is equal to $\frac{1}{4}$ for $\omega_1 = \omega_2$ and to $\frac{1}{2}$ if the two frequencies are different. In the static case, ($\omega_1 = \omega_2 = 0$), the factor $\beta(0)$ amounts to $\frac{1}{2}$ for the case of a single field and to 1 for that of two distinct fields.

In the absence of electron dispersion and absorption, the tensor of second-order non-linear polarizability $b(-\omega_3; \omega_1, \omega_2)$ can be dealt with as totally symmetric, and the numbers of its non-zero and independent elements are to be found in Table 11. Static values of the non-linear polarizability $b^0 = b(0; 0, 0)$ have been calculated theoretically for some molecules.[290, 293, 526] Tensor elements $b^{\omega} = b(-\omega; \omega, 0)$ are determined from measurements of the Kerr effect[178, 383] or molecular light scattering in liquids.[348]

Values of $b^{2\omega} = b(-2\omega, \omega, \omega)$ are to be determined from non-linear light scattering at doubled frequency by gases and liquids[226] or from second-harmonic generation by a gas in the presence of a static electric field.[220] We obviously have in mind substances composed of dipolar molecules undergoing reorientation by the static field.[222, 527]

The dipole moment of order 3, induced in a molecule by three electric fields oscillating at frequencies $\omega_1, \omega_2, \omega_3$ and analysed at the frequency $\omega_4 = \omega_1 + \omega_2 + \omega_3$, amounts to:

$$p^{(3)}(\omega_4) = \gamma(\omega_4)\, c(-\omega_4; \omega_1, \omega_2, \omega_3) \vdots E(\omega_1)\, E(\omega_2)\, E(\omega_3) \qquad (353)$$

where the ordering factor $\gamma(\omega_4)$ is $\frac{1}{4}$ if all the frequencies are different, $\frac{1}{8}$ if two differ, $\frac{1}{24}$ for $\omega_1 = \omega_2 = \omega_3$, and $\frac{1}{6}$ in the single-field static case.

On the assumption of total symmetry of the tensor of third-order non-linear polarizability $c(-\omega_4; \omega_1, \omega_2, \omega_3)$, its non-zero and independent elements are the same as those of Table 12. Direct theoretical calculations of $c^0 = c(0; 0, 0, 0)$ have been performed for the atoms of inert gases and some simple molecules.[294, 295] Values of the tensor elements $c^{\omega} = c(-\omega; \omega, 0, 0)$ have been determined for numerous molecules from static Kerr effect studies[178, 383, 384] and values of $c^{\omega,\omega_L} = c(-\omega; \omega, \omega_L, -\omega_L)$ from measurements of optical birefringence induced by laser light.[528] Measurements of second-harmonic generation by gases in the presence of a static electric field[220, 221] yield the tensor elements $c^{2\omega} = c(-2\omega; \omega, \omega, 0)$, which can also be obtained from second-harmonic scattering in centro-symmetric liquids.[227, 529] The elements of the tensor $c^{3\omega} = c(-3\omega; \omega, \omega, \omega)$

[526] R. G. Brewer and A. D. McLean, *Phys. Rev. Letters*, 1968, **21**, 271.
[527] H. F. Hameka, *Canad. J. Chem.*, 1971, **49**, 1823.
[528] (*a*) S. Kielich, *Proc. Phys. Soc.*, 1967, **90**, 847; (*b*) S. Kielich, J. R. Lalanne, and F. B. Martin, *Compt. rend.*, 1971, **272**, *B*, 731.
[529] S. Kielich, J. R. Lalanne, and F. B. Martin, *Acta Phys. Polon.* (*A*), 1972, **41a**, 479; S. Kielich and M. Kozierowski, *Optics Comm.*, 1972, **4**, 395.

are determined from studies of third-harmonic generation in gases and liquids,[530, 531] and have recently been calculated for perfect gases.[532]

Although general theories of non-linear optical polarizabilities are available,[533–535] only a few simple cases have been solved numerically.[536] Numerical values of the tensor elements of the non-linear polarizabilities $b(-\omega_3, \omega_1, \omega_2)$ and $c(-\omega_4, \omega_1, \omega_2, \omega_3)$, determined by the above stated methods and calculated theoretically, are listed in Table 9.

Radial and Angular Molecular Correlations.—The non-linear phenomena exhibited by dense bodies and discussed above reveal the finer details of the various molecular interactions which, in the molecular–statistical approach, are described by parameters of the radial and angular kinds.

Radial Correlations. In the approximation of binary correlations, we have to deal with the following parameters of radial correlations:

$$\langle r_{ij}^{-n} \rangle = 4\pi \int r_{ij}^{-n+2} g(r_{ij})\, \mathrm{d}r_{ij} \tag{354}$$

For non-excessively condensed gases, the binary radial correlation functions are approximated by equation (156), *i.e.* in explicit form we have, by equation (119),

$$g(r_{ij}) = \exp\left\{-\ Y_{st}^2\left[\left(\frac{\sigma}{r_{ij}}\right)^s - \left(\frac{\sigma}{r_{ij}}\right)^t\right]\right\} \tag{355}$$

where we use the dimensionless correlation parameter:

$$Y_{st}^2 = \frac{\varepsilon}{kT}\left(\frac{s}{s-t}\right)\left(\frac{s}{t}\right)^{\frac{t}{s-t}} \tag{356}$$

for a potential of the $s - t$ type.

For a potential of the type (355), the mean values (354) reduce to:[458]

$$\langle r_{ij}^{-n} \rangle = \frac{4\pi\rho\sigma^{3-n}}{sY_{st}^4} H_n(Y_{st}) \tag{357}$$

[530] P. D. Maker, R. W. Terhune, and C. M. Savage, 'Third Conference on Quantum Electronics, Paris, 1963', Columbia University Press, New York, 1964, p. 1559.

[531] J. F. Ward and G. H. C. New, *Phys. Rev.*, 1969, **185**, 57.

[532] (*a*) E. L. Daves, *Phys. Rev.*, 1968, **169**, 47; (*b*) R. Klingbeil, V. G. Kaveeshwar, and R. P. Hurst, *ibid.*, 1971, **4A**, 1760; (*c*) B. P. Tripathi, R. K. Laroraya, and S. L. Srivastava, *ibid.*, p. 2076; (*d*) R. E. Sitter, jun., and R. P. Hurst, *ibid.*, 1972, **5A**, 5.

[533] S. Kielich, *Acta Phys. Polon.*, 1964, **26**, 135; *Physica*, 1964, **30**, 1717; *Proc. Phys. Soc.*, 1965, **86**, 709; *J. Physique*, 1967, **28**, 519.

[534] (*a*) S. J. Cyvin, J. E. Rauch, and J. C. Decius, *J. Chem. Phys.*, 1965, **43**, 4083; (*b*) A. D. McLean and M. Yoshimine, *ibid.*, 1967, **47**, 1927.

[535] (*a*) B. J. Orr and J. F. Ward, *Mol. Phys.*, 1971, **20**, 513; (*b*) Tai Yup Chang, *J. Chem. Phys.*, 1972, **56**, 1745, 1752.

[536] (*a*) J. R. Gott, M. J. P. Musgrave, and J. N. Murrell, *Mol. Phys.*, 1967, **12**, 295; (*b*) L. D. Ievleva and T. Ya. Karagodova, *Optika i Spektroskopiya*, 1967, **32**, 991; L. D. Ievleva and M. A. Kovner, *ibid.*, 1969, **36**, 601; (*c*) A. V. Luzanov, Yu. B. Malykhanov, and M. M. Mestechin, *ibid.*, 1970, **38**, 836; (*d*) R. R. Alfano and S. L. Shapiro, *Phys. Rev. Letters*, 1970, **24**, 1217; (*e*) R. W. Hellwarth, A. Ovyoung, and N. George, *Phys. Rev.*, 1971, **4A**, 2342.

where

$$H_n(Y_{st}) = Y_{st}^{2(2s+3-n)/s} \sum_{m=0}^{\infty} \frac{1}{m!} Y_{st}^{2m(s-t)/s} \Gamma\left(\frac{mt+n-3}{s}\right) \quad (358)$$

are functions well adapted for numerical tabulation.

For $s = 12$ and $t = 6$, *i.e.* for a 12–6 Lennard-Jones potential, the functions (358) reduce to the functions (157a) introduced by Pople.[321] For $s = 18$ and $t = 6$ as well as for $s = 28$ and $t = 7$ we have the functions discussed and tabulated by Saxena and Joshi.[322]

In a similar way, ternary, quaternary, and higher radial correlations can be calculated on making assumptions as to the model.[317, 361]

In the liquid case, sufficiently good numerical evaluations are obtained using Kirkwood's model of hard spheres, leading to radial parameters in the form (154). In the general case no analytic forms of the radial correlation functions are known for liquids, and calculations of $\langle r_{ij}^{-n} \rangle$ *etc.* present considerable difficulties. These can be avoided if measurements of molar refraction and molar polarization in liquids are available since, in conformity with formulae (149) and (150), the latter permit the numerical determination of $\langle r_{ij}^{-n} \rangle$. The radial parameters thus obtained can be used for the numerical evaluation of non-linear effects in simple liquids.

Angular Correlations. In real gases and liquids consisting of anisotropic molecules, non-linear phenomena are strongly dependent on the angular correlations described by the parameters (179)—(181) and (271)—(273), with statistical mean values defined in general as:

$$\langle G \rangle = \frac{\int G(\tau^N) \exp\left\{-\frac{U(\tau^N)}{kT}\right\} \mathrm{d}\tau^N}{\int \exp\left\{-\frac{U(\tau^N)}{kT}\right\} \mathrm{d}\tau^N} \quad (359)$$

the integrals extending over all N molecules of the region having configurations $\tau^N = \tau_1, \tau_2, \ldots, \tau_N$ (these are in general positional variables $\boldsymbol{r}^N = \boldsymbol{r}_1, \boldsymbol{r}_2, \ldots, \boldsymbol{r}_N$ and orientational variables $\Omega^N = \Omega_1, \Omega_2, \ldots, \Omega_N$).

The total potential energy $U(\tau^N)$ of interaction between the N molecules is, in some cases, separable into a central (radial) part $U_0(\boldsymbol{r}^N)$ dependent only on position, and a non-central (angular) part $V(\boldsymbol{r}^N, \Omega^N)$ dependent on molecular orientation Ω^N. As stated in Sections 2 and 3, the orientational interaction energy V is generally tensorial in nature if it includes multipole interactions. Although V can be expressed in terms of the angular variables for some simple cases of molecular symmetry, it is nevertheless of an analytic form too highly involved to permit strict calculation of the mean values (359). We thus are left to choose one of the following two procedures for solving the problem numerically: (*i*) either to apply the strict form of V but to calculate the mean values (359) by classical perturbation calculus or (*ii*) to simplify maximally the form of V as a function of the angular variables so as to be able to calculate (359) strictly, without resorting to approximations.

Weak angular correlations. The first procedure applies to weak angular interactions between the molecules, *i.e.* to cases when one has simultaneously $V \ll U_0$ and $V < kT$. The Boltzmann factors in (359) can now be expanded in power series in V/kT and then integrated term by term with appropriate approximation. In particular, restricting the problem to binary correlations, one has to calculate only the two correlation parameters (179b) and (180b), in which the binary angular correlation function can be written as follows:[377]

$$g^{(2)}(\tau_i, \tau_j) = g(r_{ij})\left\{1 - \frac{V_{ij}}{kT} + \frac{V_{ij}^2}{2k^2T^2} - \frac{V_{ij}^3}{6k^3T^3} + \ldots\right\} \quad (359a)$$

It is assumed here that the binary radial correlation function $g(r_{ij})$ is known in explicit form for the case under consideration.

Obviously, in the zeroth approximation of (359a) when no orientational interaction exists, the angular correlation parameters (179b) and (180b) vanish. Hence, by applying potential energies of binary orientational interaction V_{ij} in strict form (discussed in detail in Section 2), one can calculate successive approximations of the correlation parameters in accordance with the expansion (359a). We adduce some calculations thus performed for particular cases:

(*a*) Anisotropic dipolar molecules. The energy of angular interaction between two dipolar molecules being given by formulae (60a) and (87a) we have, on applying the approximation (359a) to the angular correlation parameters (179b) and (180b),

$$\begin{aligned} J_{\mathrm{A}}^{(1)} = {} & \tfrac{2}{75}p_i^3 p_j^3 u^3(\langle r_{ij}^{-9}\rangle + \tfrac{8}{49}u^2 p_i^2 p_j^2\langle r_{ij}^{-15}\rangle + \ldots) \\ & + \tfrac{4}{15}u^2 p_i p_j(\alpha_{vi} p_j^2 + p_i^2 \alpha_{vj})(\langle r_{ij}^{-9}\rangle + \tfrac{12}{35}u^2 p_i^2 p_j^2\langle r_{ij}^{-15}\rangle + \ldots) \end{aligned} \quad (360)$$

$$\begin{aligned} J_{\mathrm{A}}^{(2)} = {} & \frac{u^2}{75} p_i^2 p_j^2(\langle r_{ij}^{-6}\rangle + \tfrac{12}{49}u^2 p_i^2 p_j^2\langle r_{ij}^{-12}\rangle \ldots) \\ & + \frac{u}{25}\{(\alpha_{vi}\kappa_i p_j^2 + p_i^2\alpha_{vj}\kappa_j)\langle r_{ij}^{-6}\rangle \\ & + \tfrac{4}{49}u^2 p_i^2 p_j^2[\alpha_{vi}(7 + 11\kappa_i)p_j^2 + p_i^2\alpha_{vj}(7 + 11\kappa_j)]\langle r_{ij}^{-12}\rangle\} \end{aligned} \quad (361)$$

with the notation $u = (4\pi\varepsilon_0 kT)^{-1}$.

(*b*) Anisotropic quadrupolar molecules. In dealing with non-dipolar molecules, we have to consider only the parameter $J_{\mathrm{A}}^{(2)}$ which, on taking into account anisotropic London dispersional interactions [equation (90a)], is of the form:

$$\begin{aligned} J_{\mathrm{A}}^{(2)} = {} & \frac{3Y_{ij}}{50}\alpha_{vi}\kappa_i\alpha_{vj}\kappa_j \\ & \times\left\{\langle r_{ij}^{-6}\rangle + \frac{3Y_{ij}}{98}[49 - 14(\kappa_i + \kappa_j) + 38\kappa_i\kappa_j]\langle r_{ij}^{-12}\rangle + \ldots\right\} \end{aligned} \quad (362)$$

with the notation:

$$Y_{ij} = \frac{1}{kT}\left(\frac{h\nu_i\, h\nu_j}{h\nu_i + h\nu_j}\right)$$

If the molecules have quadrupole electric moments Θ, the energies are those of formulae (63a) and (89a), yielding additionally the contribution:

$$J_{\mathrm{A}}^{(2)} = \frac{u}{35}\{3(\alpha_{\mathrm{V}i}\,\kappa_i\,\Theta_j^2 + \Theta_i^2\,\alpha_{\mathrm{V}j}\,\kappa_j)\langle r_{ij}^{-8}\rangle$$

$$+\ 4u\Theta_i^2\,\Theta_j^2\langle r_{ij}^{-10}\rangle - \tfrac{6}{7}u\Theta_i\,\Theta_j[\alpha_{\mathrm{V}i}(7 + 2\kappa_i)\,\Theta_j^2$$

$$+\ \Theta_i^2\,\alpha_{\mathrm{V}j}(7 + 2\kappa_j)]\langle r_{ij}^{-13}\rangle + \ldots\} \qquad (363)$$

Since only the radial parameters (354) remain to be calculated in the expressions (360)—(363), we see that numerical evaluations are feasible with appropriately chosen models of the radial correlation functions $g(r_{ij})$.

Strong angular correlations. Numerous liquids exhibit very strong orientational correlations, when $V_{ij} \geqslant kT$, and perturbation calculus can no longer serve for the calculation of mean values (359) (since the exponential factor cannot be expanded in powers of V/kT). As already mentioned, in cases like this one has to express the energy of molecular interaction as a function of one angle θ_{ij}, subtended by the symmetry axes of two molecules i and j, *i.e.* in the form (275). With regard to rigid electric dipoles, the problem has been discussed earlier (see p. 339). On assuming the dipoles to be anisotropically polarizable we have, in place of (274a),

$$\langle\cos^n\theta_{ij}\rangle = (\pm\ 1)^n L_n(Y_1, \pm\ Y_2, \pm\ \ldots) \qquad (364)$$

Here, the Langevin functions of order n are:

$$L_n(Y_1, \pm\ Y_2, \pm\ \ldots) = \frac{\displaystyle\int_1^{-1} t^n \exp(Y_1 t \pm Y_2 t^2 \pm \ldots)\,\mathrm{d}t}{\displaystyle\int_1^{-1} \exp(Y_1 t \pm Y_2 t^2 \pm \ldots)\,\mathrm{d}t} \qquad (365)$$

the parameters $Y_1, Y_2, \ldots$ defining correlations of the respective kinds, and being functions of the temperature and density.

For dipolar molecules with the tendency to array themselves into nearly antiparallel pairs (*e.g.* nitrobenzene [352]), the energy of mutual interaction can be put in the form:[269]

$$-\frac{V_{ij}}{kT} = Y_0 + Y_1\cos\theta_{ij} - Y_2\cos^2\theta_{ij} + \ldots \qquad (366)$$

where

$$Y_1 = \frac{1}{4\pi\varepsilon_0}\left\{\frac{p^2}{kq}\frac{\rho}{T} - \frac{3p\Omega}{kq^{5/3}}\frac{\rho^{5/3}}{T} + \frac{3\Omega^2}{2kq^{7/3}}\frac{\rho^{7/3}}{T}\right\} \qquad (366a)$$

$$Y_2 = \frac{1}{4\pi\varepsilon_0}\left\{\frac{15\Theta^2}{4kq^{5/3}}\frac{\rho^{5/3}}{T} - \frac{3\alpha_v\,\kappa}{kq^2}(p^2 + \tfrac{3}{8}\alpha\kappa h\nu)\frac{\rho^2}{T} - \frac{15\Theta\Phi}{2kq^{7/3}}\frac{\rho^{7/3}}{T}\right\} \qquad (366b)$$

with p, Θ, Ω, Φ denoting respectively the electric dipole, quadrupole, octupole, and hexadecapole moments of an axially symmetric molecule having the electric anisotropy $3\alpha\kappa = \alpha_3 - \alpha_1$. Numerically, the parameter $q \approx \rho d^3$ is to be had from experiment, and is found [90] to lie between 1.2 and 1.5.

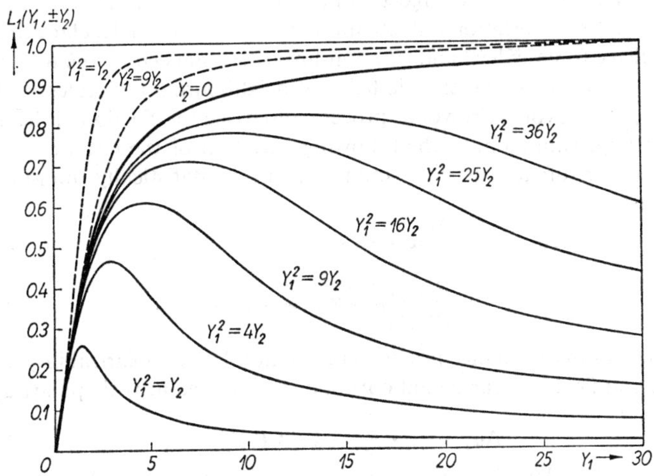

Figure 27 *Generalized Langevin functions* $L_1(Y_1, \pm Y_2)$ versus *the correlation parameter* Y_1. *Dashed curves show* $L_1(Y_1, + Y_2)$*; continuous bold-face line shows* $L_1(Y_1, 0)$*; below the continuous line lie the functions* $L_1(Y_1, - Y_2)$

With regard to (364), the binary correlation parameters (179a) and (180a) for the model (366) become:

$$J_A^{(1)} = \pm zL_1(Y_1, \pm Y_2) \tag{367}$$

$$J_A^{(2)} = \tfrac{1}{2}z\{3L_2(Y_1, \pm Y_2) - 1\} \tag{368}$$

The explicit form of the Langevin functions of degrees 1 and 2 occurring above is given by the formulae (314a) and (234a) on replacing X_1 by Y_1 and X_2 by Y_2. A graph of the Langevin function $L_1(Y_1, \pm Y_2)$ is to be found in Figure 27. The graph of

$$\tfrac{1}{2}\{3L_2(Y_1, \pm Y_2) - 1\} = J_A^{(2)}/z \tag{368a}$$

is identical with that of Figure 13 if X_1 is replaced by Y_1 and X_2 by Y_2.

Formulae (366)—(368) show that higher-order correlations modify considerably the shape of the correlation functions, calculated on the basis of dipole interaction alone. This enables us to achieve better agreement with the experimental data for nitrobenzene and benzene.[269]

Fluctuations of Molecular Fields.—A complete explanation of the non-linearities electrically induced in liquid dielectrics still requires the discussion of fluctuational processes due to the molecular fields within the dielectric.

In atomic and simple molecular liquids, the Kerr effect, electric saturation, *etc.*, are caused not only by non-linearities directly induced in the atoms and molecules by the external electric field (the Voigt effect), but are primarily due to Yvon–Kirkwood fluctuational–statistical processes. One should keep in mind, however, that the intrinsic electric polarizability of non-dipolar molecules is subject to modification by at least three factors: (*i*) the effect of translational fluctuations,[344, 345] (*ii*) intermolecular attractive or repulsive forces,[354, 355] and (*iii*) non-linearities induced by the temporally and spatially fluctuating electric fields of neighbouring molecules.[326, 348]

Thus, the Yvon–Kirkwood process discussed in Section 4 leads, in dipolar approximation, to the following variation of the electric polarizability tensor of a molecule immersed in a statistically non-uniform medium:

$$\Delta \boldsymbol{\alpha}_i = - \sum_{j=1}^{N_A} \boldsymbol{\alpha}_i \cdot \boldsymbol{T}_{ij} \cdot \boldsymbol{\alpha}_j + \sum_{j=1}^{N_A} \sum_{k=1}^{N_A} \boldsymbol{\alpha}_i \cdot \boldsymbol{T}_{ij} \cdot \boldsymbol{\alpha}_j \cdot \boldsymbol{T}_{jk} \cdot \boldsymbol{\alpha}_k - \ldots \tag{369}$$

If the atoms or molecules undergo a non-linear polarizability in the fluctuating field of near neighbourhood, the change in their polarizability tensor is:

$$\Delta \boldsymbol{\alpha}_i = \boldsymbol{\beta}_i \cdot \boldsymbol{F}_i + \tfrac{1}{2}\boldsymbol{\gamma}_i : \boldsymbol{F}_i \boldsymbol{F}_i + \ldots \tag{370}$$

with $\boldsymbol{\beta}_i$ and $\boldsymbol{\gamma}_i$ denoting tensors of the non-linear polarizability due to the molecular field $\boldsymbol{F}_i$ in the absence of external fields. For atoms, $\boldsymbol{\beta}_i = 0$ and electric non-linearity can be caused by the square of the field of London dispersional forces.[326, 377]

Likewise, with regard to dipolar molecules, one should remember that the dipole moment $\boldsymbol{p}$ of a molecule in the ground state undergoes a modification by the fields of its molecular neighbourhood. Hence the change in dipole moment of the molecule due to fluctuational fields is:

$$\Delta \boldsymbol{p}_i = \boldsymbol{\alpha}_i \cdot \boldsymbol{F}_i + \tfrac{1}{2}\boldsymbol{\beta}_i : \boldsymbol{F}_i \boldsymbol{F}_i + \tfrac{1}{6}\boldsymbol{\gamma}_i \vdots \boldsymbol{F}_i \boldsymbol{F}_i \boldsymbol{F}_i + \ldots \tag{371}$$

The electric fields of electric multipoles are extensively discussed in Sections 2 and 3. Here, it should only be noted that in many a case, with the aim of performing rapid numerical evaluations, one can replace the molecular field $\boldsymbol{F}_i$ by Onsager's reaction field which, for dipolar molecules, amounts to:

$$\boldsymbol{R}_i = \frac{2(\varepsilon_r - 1)}{\varepsilon_0(2\varepsilon_r + 1)} \frac{\boldsymbol{p}_i}{a^3} \tag{372}$$

All in all, the molecules of a condensed dielectric polarize non-linearly in an externally applied electric field E as well as in the molecular field F. Since the latter has a field strength of the order of $F \approx 10^8$—10^{10} V m^{-1}, which is two or three orders of magnitude greater than the usually applied external fields $E \approx 10^7$ V m^{-1}, the molecular field is apt to induce large non-linearities in the molecules. The fact should not be neglected if one wishes to have good agreement between experiment and theory.[391, 523]

Prospects in Electric Non-linearity Studies.—*The Kerr Effect.* The study of the Kerr effect over almost 100 years has accumulated a vast body of information regarding the electro-optical properties of isolated atoms, molecules, and macromolecules, as well as the structure of gases, liquids, and solutions. Particularly well founded is the work carried out by the school of Le Fèvre on the Kerr effect in infinitely dilute solutions, which has led to the determination of the linear polarizabilities and anisotropies of molecules.[132] Related studies of the Kerr effect have been developed with regard to solutions of macromolecules by the school of O'Konski,[133, 134] in particular extending to the case of electric saturation of the optical birefringence of conducting as well as non-conducting particles.[407] The centre headed by Buckingham has developed the study of the Kerr effect in simple gases for the direct determination of non-linear polarizabilities of atoms and molecules.[178] Hitherto, however, the Kerr effect has not been sufficiently resorted to for studying angular molecular correlations, though adequate theories are to hand.[141, 142] Recent years have witnessed an extension of Kerr-effect studies from static-electric-field- to laser-light-induced phenomena. This optical Kerr effect, discovered by Mayer,[214] yields information regarding the anisotropy of linear polarizability and non-linear polarizability of atoms and molecules.[528]

In the experimental field, a number of novel, ingenious methods for the measurement of optical birefringence resorting to classical [381] and laser [538] analysing light beams have been evolved. Very much is to be expected from pulsed methods, down to picosecond pulse methods inclusively; they shall surely permit a separation of the effects of molecular reorientation and non-linear molecular polarizability. Indeed, Kerr-effect studies can now be said to provide a fundamental and unfailing method of investigation of the electromagnetic properties of atoms and molecules if extended to embrace non-linear dispersion and absorption.

Dielectric Saturation. Though reaching back to 1920, the investigation of static electric field effects on electric permittivity was not pursued with sufficient intensity until recently, chiefly owing to the circumstance that the accuracy and sensitivity of electric detection are lower than those of optical detection. Accordingly, most existing $\Delta\varepsilon$ measurements relate to liquids and solutions exhibiting high non-linearities in an intense electric field. The wider applicability of the method is inhibited by electric conductivity of the liquids and the impossibility of extending it to very strong fields for fear of dielectric breakdown. Under these restrictions the non-linearities induced in molecular dielectrics are generally small and not easily measurable, but are apparent as complete electric saturation in macromolecular substances.[432]

[537] S. Kielich, *Physica*, 1967, **34**, 365; *J. Physique*, 1968, **29**, 619.

[538] (*a*) G. Delfino, *Optics Comm.*, 1971, **4**, 60; (*b*) M. Paillette and D. Vautier, *ibid.*, 1971, **4**, 292; (*c*) A. Owyoung, R. W. Hellwarth, and N. George, *Phys. Rev.*, 1972, **5B**, 628; (*d*) B. R. Jennings and B. L. Brown, *European Polymer J.*, 1971, **7**, 805.

The principle of measurements of $\Delta\varepsilon$ is a very simple one; nevertheless, the construction of reliable apparatus to this aim is no simple matter, but requires considerable effort and experience in the field of electronics. It would seem that, here, there remains much to be done by the experimenter: existing measuring methods will have to be perfected and, more essentially still, new ingenious methods will have to be devised aimed at the simplest possible solutions, *e.g.* by resorting to integrated circuits, and the like. In order to glean more abundant measurements in shorter periods of time, it will be necessary to introduce automatized techniques like those employed in the various complex devices of modern laboratories. In spite of the progress accomplished in recent years, the existing methods are by no means unanimous in their rendering of the field effect, which is contributed to by a number of composite mechanisms, both of a statistical–molecular and thermodynamical nature. Work in this field of investigation, largely developed by the group of Piekara in Poland, has in more recent years been extended by that of Davies, in Great Britain, to substances consisting of macromolecules.

Of all the hitherto investigated non-linear phenomena in dielectrics, electric saturation is the most highly complex, both with regard to measuring technique and theoretical description, since it involves the full range of processes apt to take place in electric fields but as yet not accessible to a separate treatment one by one. In this respect, it would be desirable to pursue the studies of dispersion of saturation begun by Gregson *et al.*;[432] their extension over a wide range of frequencies of both the analysing and polarizing fields, up to optical frequencies,[539] would be of decisive importance. This indeed should permit the clarification of various outstanding issues. Also, with regard to the method of strong pulsed fields, it is necessary to measure $\Delta\varepsilon$ using sine, rectangular, and reversing pulse shapes, as was done in measurements of the Kerr effect and electric field-induced light scattering,[135] and to reduce the pulse durations from 1 ms, maybe even to the picosecond range.[540]

This complexity of the saturation phenomenon, though a drawback, is at the same time an asset from the theoretical point of view. The reader will have noted from the theory discussed above that electric saturation makes directly apparent the participation of reorientation of the electric dipoles of molecules and macromolecules. In this respect, saturation is unique, owing to the part played in it, in addition to the various molecular processes specific to other effects (*e.g.* the Kerr effect), by entirely new processes providing valuable information on the interactions between molecules in condensed phases and their internal structure. We believe it

[539] S. Kielich, *Acta Phys. Polon.*, 1967, **32**, 405; *J. Colloid and Interface Sci.*, 1970, **33**, 142; **34**, 228.

[540] (*a*) R. G. Brewer and C. H. Lee, *Phys. Rev. Letters*, 1968, **21**, 267; (*b*) R. Cubeddù, R. Poloni, C. A. Sacchi, and O. Svelto, *Phys. Rev.*, 1970, **2A**, 1955; (*c*) I. L. Fabielinski, *Uspekhi Fiz. Nauk S.S.S.R.*, 1971, **104**, 77; (*d*) S. A. Akhmanov and A. S. Chirkin, 'Statistical Effects in Nonlinear Optics', Izd. Moskovskogo Universiteta, 1971.

$$\chi(\omega) = \int_0^\infty c(t)\exp(-\,\mathrm{i}\omega t)\,\mathrm{d}t \qquad (39)$$

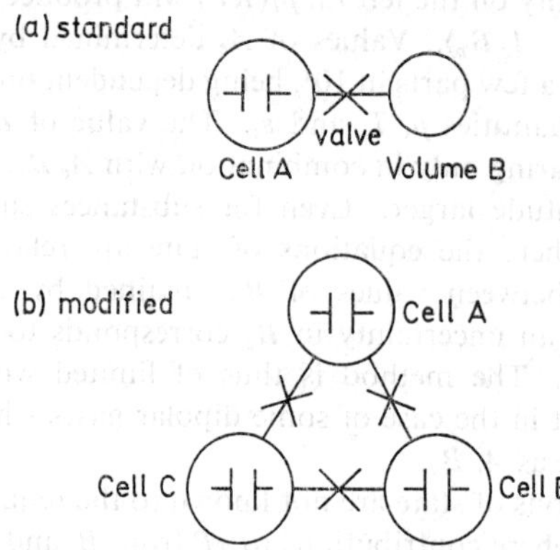

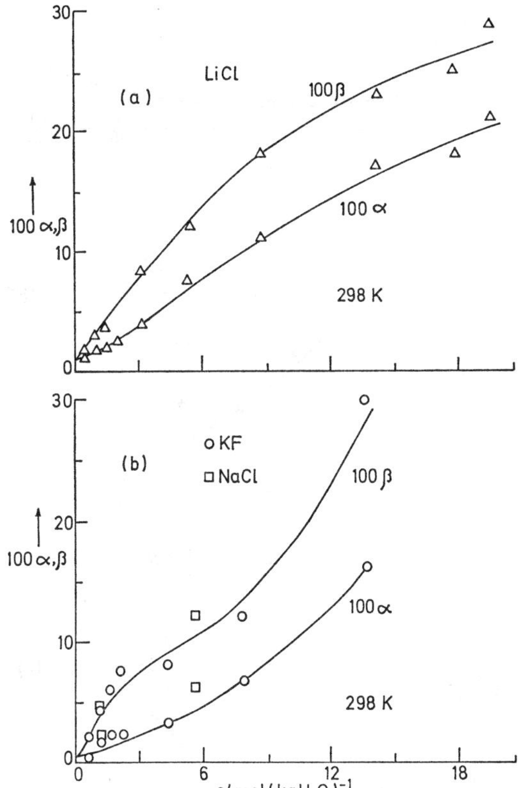

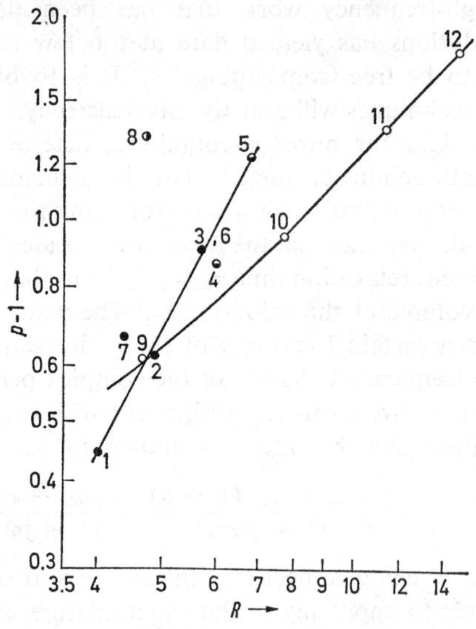

$$p = \alpha \boldsymbol{E}' \quad (11)$$

$$\delta n_i = \nu_i \, \delta \xi \qquad (58)$$

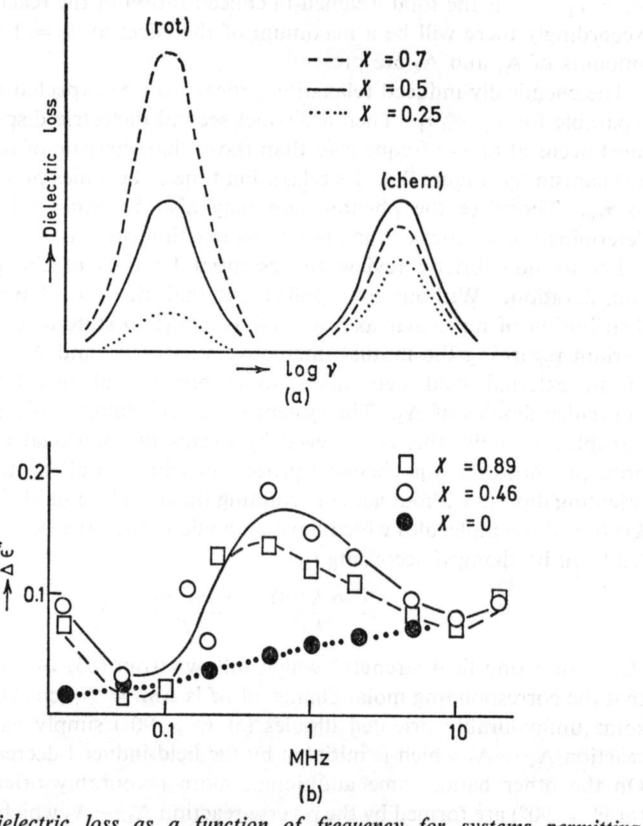

	Symmetry elements								
	Plane	*Symmetry axes*						*Centre*	
Schoenflies symbol	C_s	C_6	C_5	C_4	C_3	C_2	S_n	C_i	*Examples of molecules*
D_{3h}	4				1	3	$S_3(C_3)$		BCl_3, C_3H_6, $C_6H_3Cl_3$, C_2H_6 (eclipsed form)
D_{6h}	7	1				6	$S_3(C_3)$ $S_6(C_6)$	1	C_6H_6, C_6Cl_6
T					4	3			$C(CH_3)_4$ (partly rotated)
T_h	3				4	3		1	
O				3	4	6			
T_d	6				4	3	$3S_4(C_2)$		CH_4, CF_4, CCl_4, $SiCl_4$, $C(CH_3)_4$
O_h	9			3	4	6	$4S_6(C_3)$ $3S_4(C_4)$	1	SF_6, UF_6, $Fe(CN)_6$, $Cr(CO)_6$
Y			6		10	15			
Y_h	15		6		10	15	$6S_5(C_5)$ $10S_3(C_3)$	1	$B_{12}H_{12}^{2-}$
K									
K_h								1	Free atoms
$C_{\infty v}$		1	1	1	1	1			HCl, HCN, HC≡CCl, CN, COS,
$D_{\infty h}$		1	1	1	1		$S_n(C_n)$	1	H_2, CO_2, CS_2, C_2H_2, Cl_2, C_3O_2
D_{5h}	6		1			5	$S_5(C_5)$		Plane symmetrical C_5H_{10}
D_{5d}	5		1			5	$S_{10}(C_5)$	1	Ferrocene
D_{6d}	6	1				6	$S_{21}(C_6)$		
D_{4d}	4			1		4	$S_8(C_4)$		S_8
C_∞		1	1	1	1	1			

	Electric moment			
Molecule	*Dipole* $p \times 10^{30}/\text{Cm}$	*Quadrupole* $\Theta \times 10^{40}/\text{Cm}^2$	*Octupole* $\Omega \times 10^{50}/\text{Cm}^3$	*Hexadecapole* $\Phi \times 10^{60}/\text{Cm}^4$
N_2O	0.60[e]	9.29[e] 10.12[m] − 11.69[h]		
LiH	19.64[f] − 20.04[p]	− 16.70[f] − 14.83[p]	− 15.56[p]	
CH_4	0	0	15.03[f] 19.71[q] 16.70[r] 12.36[s] 8.35[t]	− 20.04[f] − 25.72[q]
CF_4	0	0	14.70[f] 16.03[u] 19.04[v] 45.76[s]	
CCl_4	0	0	50.10[w] 18.37[x]	
SiH_4	0	0	22.04[f]	− 55.44[f]
NH_4^+	0	0	21.04[f]	− 21.04[f]
SF_6	0	0	0	19.37[y] 167.0[z]
$CClF_3$	1.70[z]		3.34[z]	

$$\boldsymbol{p}(\boldsymbol{E}) = 4\pi\varepsilon_0\,\alpha_V\,\boldsymbol{E} = \alpha\boldsymbol{E} \tag{67}$$

$$\begin{aligned} \Delta\alpha_{\|}(E) &= \tfrac{1}{2}\gamma_{\|} E^2 \\ \Delta\alpha_{\perp}(E) &= \tfrac{1}{2}\gamma_{\perp} E^2 \end{aligned} \tag{74}$$

$$\langle r_{ij}^{-6} \rangle = \frac{2\pi^2}{9} \frac{\rho}{v} = \frac{2\pi^2}{9u} \rho^2 \qquad (151a)$$

$$K_{\mathrm{m}\infty} = \frac{4\pi N_{\mathrm{A}}}{45kT}\left\{\delta_{\mathrm{v}2}^{\omega}\left(\delta_2^{e} + \frac{p_2^2}{kT}\right) - \delta_{\mathrm{v}1}^{\omega}\,\delta_1^{e}(1 + 2J_{11}^{(2)})\right\} + \frac{4\pi N_{\mathrm{A}}}{45kT}\left\{\delta_{\mathrm{v}1}^{\omega}\,\delta_2^{e} + \delta_{\mathrm{v}2}^{\omega}\,\delta_1^{e} + \frac{\delta_{\mathrm{v}1}^{\omega}\,p_2^2}{kT}\right\}J_{12}^{(2)} \tag{228}$$

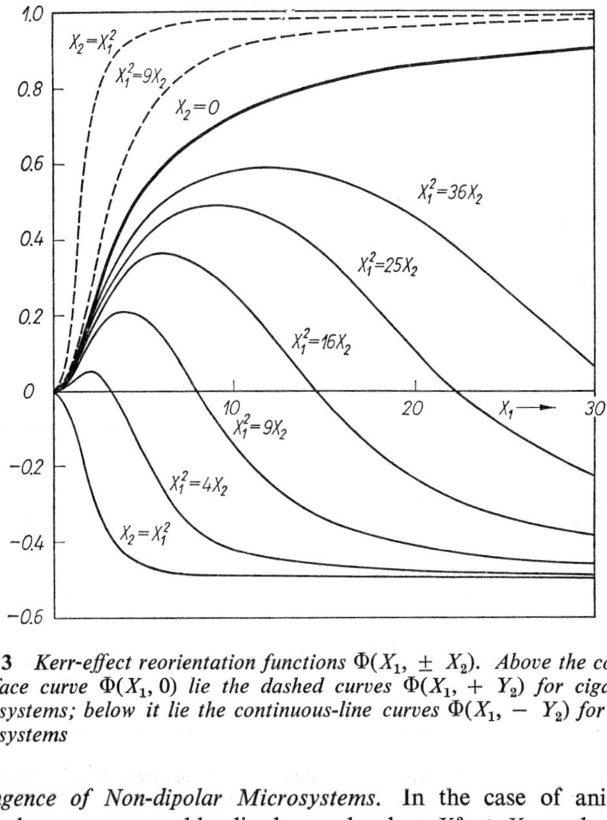

Liquids	C_6H_6	CS_2	$CHCl_3$	C_5H_5N	C_6H_5Cl	$C_6H_5NO_2$
$p \times 10^{30}$/C m	0	0	3.83	7.32	5.17	14.06
α_1 × 10^{40}/C² m² J⁻¹	7.05	6.15	10.00	6.42	8.25	8.22
α_2 × 10^{40}/C² m² J⁻¹	13.66	6.15	10.00	13.19	15.38	15.75
α_3 × 10^{40}/C² m² J⁻¹	13.66	16.81	7.41	12.03	17.26	19.05
$S_m^{(1)}$ × 10^{25}/C² m⁵ J⁻² mol⁻¹	3.26	8.49	0.50	2.94	5.06	2.93
$S_m^{(2)}$ × 10^{25}/C² m⁵ J⁻² mol⁻¹	0	0	− 14.08	11.52	53.95	517.55
$S_m^{(3)}$ × 10^{25}/C² m⁵ J⁻² mol⁻¹	0	0	− 49.78	− 657.35	− 163.57	− 8947.33
S_m^{total} × 10^{25}/C² m⁵ J⁻² mol⁻¹	3.26	8.49	− 63.36	− 642.89	− 104.56	− 8426.85
$V_m \times 10^{-6}$/m³ mol⁻¹	89.4	60.2	80.2	80.45	101.68	102.3
ε_r (20 °C)	2.28	2.64	4.81	13.23	5.7	34.82
$\Delta\varepsilon^{theor}/E_p^2 \times 10^{18}$/m² V⁻²	+ 0.04	+ 0.24	− 6.28	− 1592.4	− 133.89	− 560733.2
$\Delta\varepsilon^{exp}/E_p^2 \times 10^{18}$/m² V⁻²	+ 0.3 [430]	+ 0.1 [420*b*]	− 4.0 [420*b*]		− 1.1 [420*b*]	− 577 [421]
		+ 0.37 [430]	− 3.5 [426]		− 73.9 [421*b*]	+ 361 [424]

would be useful to extend saturation studies to the wider class of as yet untouched pure liquids, as well as to liquid solutions in the low concentration range, applying non-dipolar as well as polar solvents. Most important, too, should prove the application of electric saturation as the sole method of determining non-linear electric polarizabilities of molecules and macromolecules.

The author wishes to express his warmest thanks to Professor Mansel Davies for the invitation and encouragement to write this review article when visiting our Institute in October 1970. Most of the text was written at Bordeaux University, where the author profited largely by the kind hospitality of Professors A. Pacault and P. Bothorel during the academic year 1970/71.

Also, I wish to thank all authors who had the kindness to make available to me their preprints and reprints, thus helping me to write this review.

I am sincerely indebted to K. Flatau for his English translation of my review and his discussions, which have led to a clearer presentation of certain points. I wish to thank R. Zawodny for preparing the Tables of Section 2, Mrs. W. Woźniak for her unfailing care in copying the English version, as well as to Mrs. I. Stróżyk for drawing the Figures. The author is particularly grateful to Mrs. B. Kasprowicz for her help in assembling the Tables of Sections 4 and 5, for completing the full list of References, and for her solicitous reading and correcting of the article as a whole.

Author Index